PHYSICS OF ICE

PHYSICS OF ICE

Proceedings of the International Symposium on
Physics of Ice, Munich, Germany, September 9-14, 1968

Edited by

NIKOLAUS RIEHL
BERNHARD BULLEMER
HERMANN ENGELHARDT

Physik-Department der Technischen Hochschule München

Symposium sponsored by
German Federal Ministry of Scientific Research
Bavarian State Ministry of Education and Culture

Springer Science+Business Media, LLC 1969

Library of Congress Catalog Card Number 72-81281
ISBN 978-1-4899-5575-3 ISBN 978-1-4899-5573-9 (eBook)
DOI 10.1007/978-1-4899-5573-9

Originally published by Plenum Press in 1969.

Softcover reprint of the hardcover 1st edition 1969

LIST OF PARTICIPANTS

A.N. Aufdermaur, Eidg. Institut für Schnee- und Lawinen-forschung, CH-7260 Weissfluhjoch / Davos / Switzerland

A. Bär, Physik-Department, Technische Hochschule München, D-8 München 2 / Germany

Dipl.-Phys. L. Becker-Knoblauch, Physik-Department, Technische Hochschule München, D-8 München 2 / Germany now at: Institut für Physikalische Chemie, Technische Hochschule München, D-8 München 2 / Germany

J. Bilgram, I. Physikalisches Institut, Universität Stuttgart, D-7 Stuttgart N / Germany

P.G. Bishop, Department of Physics, University of Birmingham, Birmingham 15 / England

Dr. J. Brickmann, Institut für Physikalische Chemie, Universität Freiburg, D-78 Freiburg / Germany

Dr. J. Brownscombe, C.S. I.R.O. Division of Radio-physics, Epping, N.S.W. 2121 / Australia

R. Büll, Physik-Department, Technische Hochschule München, D-8 München 2 / Germany

Dr. B. Bullemer, Physik-Department, Technische Hochschule München, D-8 München 2 / Germany

Dipl.-Phys. O. Buser, Eidg. Institut für Schnee- und Lawinenforschung, CH-7260 Weissfluhjoch/Davos / Switzerland

Professor P.R. Camp, Department of Physics, University of Maine, Orono, Maine 04473 / USA

Professor G. Careri, Istituto di Fisica, Università di Roma, I-00100 Roma / Italy

Dr. W. Carnuth, Physikalisch-Bioklimatische Forschungs-stelle Garmisch-Partenkirchen der Fraunhofer Gesellschaft zur Förderung der angewandten Forschung eV., D-81 Garmisch-Partenkirchen / Germany

Dr. J. Cross, Institute of Science and Technology, University of Manchester, Manchester 1 / England

Dr. G. Curtis, Atomic Energy Research Establishment, Harwell, Didcot, Berks. / England

Dr. G. Dantl, I. Physikalisches Institut, Universität Stuttgart, D-7 Stuttgart N / Germany

Dr. J.E. Dye, Institute of Meteorology, University of Stockholm, S-10432 Stockholm 19 / Sweden

J.E. Dykins, Research Civil Engineer, U.S. Naval Civil Engineering Laboratory, Port Hueneme, Calif. 93041 / USA

Dipl.-Ing. U. Eckener, Physik-Department, Technische Hochschule München, D-8 München 2 / Germany

W. Egle, Physik-Department, Technische Hochschule München, D-8 München 2 / Germany

Dr. K. Eiben, Institute of Radiation Chemistry, Nuclear Research Center, D-75 Karlsruhe / Germany

Dipl.-Phys. I. Eisele, Physik-Department, Technische Hochschule München, D-8 München 2 / Germany
now at: Department of Chemistry, University of Kansas, Lawrence, Kansas 66044 / USA

Priv.-Doz. Dr. J. Engel, Max-Planck-Institut für Eiweiss- und Lederforschung, Universität München, D-8 München 15 / Germany

Dr. H. Engelhardt, Physik-Department, Technische Hochschule München, D-8 München 2 / Germany

J.K. Fang, Department of Chemistry, Northwestern University, Evanston, Illinois 60201 / USA

P. Faure, Docteur du 3e Cycle, Faculté des Sciences de Grenoble, F-38 St. Martin-d'Hères / France

Dr. B. Federer, Eidg. Institut für Schnee- und Lawinenforschung, CH-7260 Weissfluhjoch/Davos / Switzerland

Professor E. Forslind, Division of Physical Chemistry, Royal Institute of Technology, Stockholm 70 / Sweden

Dipl.-Phys. A. Gattinger, Physik-Department, Technische Hochschule München, D-8 München 2 / Germany

Dipl.-Phys. M. Gautschi , Laboratorium für Féstkörperphysik, Eidg. Technische Hochschule, CH-8006 Zürich / Switzerland

Dr. J.W. Glen, Department of Physics, University of Birmingham, Birmingham 15 / England

H.P. Glockmann, M.Sc., Physik-Department, Technische Hochschule München, D-8 München 2 / Germany

Dr. W. Good, Eidg. Institut für Schnee- und Lawinenforschung, CH-7260 Weissfluhjoch/Davos / Switzerland

Professor P. Gosar, Institut "Jozef Stefan" and University of Ljubljana / Yugoslavia

Professor H. Gränicher, Laboratorium für Festkörperphysik, Eidg. Technische Hochschule, CH-8006 Zürich / Switzerland

Dipl.-Phys. K. Guckelsberger, CENG / PCBT, F-38 Grenoble / France

Dr. E. Hahne, Institut für Technische Thermodynamik, Technische Hochschule München, D-8 München 2 / Germany

Dipl.-Ing. H. Haltenorth, Physik-Department, Technische Hochschule München, D-8 München 2 / Germany

Dr. D. Helmreich, Physik-Department, Technische Hochschule München, D-8 München 2 / Germany

Professor A. Higashi, Department of Applied Physics, Faculty of Engineering, Hokkaido University, Sapporo / Japan

Professor P.V. Hobbs, Cloud Physics Laboratory, Department of Atmospheric Sciences, University of Washington, Seattle, Washington 98105 / USA

Professor G.L. Hofacker, Department of Chemistry, Northwestern University, Evanston, Illinois 60201 / USA
present address: Institut für Theoretische Chemie, Technische Hochschule München, D-8 München 2 / Germany

Dr. P. Hoffmann, Lehrstuhl für Kernchemie, Eduard-Zintl-Institut, Technische Hochschule Darmstadt, D-61 Darmstadt / Germany

H. Huber, Physik-Department, Technische Hochschule München, D-8 München 2 / Germany

Professor C. Jaccard, Institut de Physique, Université de Neuchâtel, CH-2000 Neuchâtel / Switzerland

Dr. A.C. Jason, Torry Research Station, Ministry of Technology, Aberdeen AB 9 8 DG / Great Britain

Dr. D.A. Johnson, Meteorological Office, Bracknell, Berkshire / England

Dr. S.J. Jones, Department of Physics, University of Birmingham, Birmingham / England
now at: Inland Waters Branch, Department of Energy, Mines and Resources, Ottawa / Canada

Professor A. Kahane, Feculté des Sciences de Grenoble, F-38 St. Martin-d'Hères / France

Professor B. Kamb, California Institute of Technology, Geological Sciences, Pasadena, Calif. 91109 / USA

V. Kern, Physik-Department, Technische Hochschule München, D-8 München 2 / Germany

Dipl.-Ing. J. Klinger, Faculté des Sciences de Grenoble, F-38 St. Martin-d'Hères / France

Dr. W. Klöpffer, Battelle-Institut e.V., D-6 Frankfurt/M / Germany

Professor H.O. Kneser, I. Physikalisches Institut, Universität Stuttgart, D-7 Stuttgart N / Germany

Dipl.-Phys.L. Knoblauch, see: Becker-Knoblauch

Dr. G. Kvajić, Ice Research Project, Physics-Department, McGill University Montreal, Montreal / Canada

Dipl.-Ing. I. Lackinger, Institut für Grundbau und Bodenmechanik, Technische Hochschule München, D-8 München 2 / Germany

U. Mitzdorf, Physik-Department, Technische Hochschule München, D-8 München 2 / Germany

Dr. O.E. Mogensen, Laboratory of Applied Physics II, Technical University of Denmark, Lyngby / Denmark

Professor H. Moser, Sektion Physik, Universität München, D-8 München / Germany

H. Müller-Krumbhaar, Physik-Department, Technische Hochschule München, D-8 München 2 / Germany

Professor J. Muguruma, Snow and Ice Section, Division of Building Research, National Research Council of Canada, Ottawa 7, Ontario / Canada
on leave from: Department of Applied Physics, Hokkaido University, Sapporo / Japan
now at: Cavendish Laboratory, Cambridge / England

Professor S.O. Nielsen, Atomic Energy Commission, Research Establishment Risö, Roskilde / Denmark

Dipl.-Phys. G. Noll, I. Physikalisches Institut, Universität Stuttgart, D-7 Stuttgart N / Germany

Professor L. Onsager, Department of Chemistry, Sterling Chemistry Laboratory, Yale University, New Haven, Connecticut 06520 / USA

J. Paren, Scott Polar Research Institute, Cambridge / England

Dr. K. Philberth, Destouchesstrasse 14,
D-8 München 23 / Germany

M. Philippe, Faculté des Sciences de Grenoble,
F-38 St. Martin-d'Hères / France

M. Pick, Physik-Department, Technische Hochschule
München, D-8 München 2 / Germany

Dr. S.W. Rabideau, Los Alamos Scientific Laboratory,
University of California, Los Alamos, New Mexico 87544 /
USA

M. Ratner, Department of Chemistry, Northwestern
University, Evanston, Illinois 60201 / USA

Dipl.-Met. O. Reinwarth, Bayerische Akademie der
Wissenschaften, Kommission für Glaziologie, München,
D-8 München 22 / Germany

Dr. R. Reiter, Physikalisch-Bioklimatische Forschungs-
stelle Garmisch-Partenkirchen der Fraunhofer Gesellschaft
zur Förderung der angewandten Forschung eV.,
D-81 Garmisch-Partenkirchen / Germany

Dipl.-Phys. K.-B. Renker, Institut für Angewandte
Kernphysik, Kernforschungszentrum Karlsruhe,
D-75 Karlsruhe / Germany

Professor N. Riehl, Physik-Department, Technische
Hochschule München, D-8 München 2 / Germany

Dr. M. Roulleau, State University of New York at Albany
A.S.R.C., Scotia, New York 12302 / USA
on leave from: Laboratoire de Météorologie et Physique
de l'Atmosphère, F-75 Paris / France

Dipl.-Phys. R. Ruepp, I. Physikalisches Institut,
Universität Stuttgart, D-7 Stuttgart N / Germany

Professor L.K. Runnels, Coates Chemical Laboratories,
Louisiana State University, Baton Rouge, Louisiana
70803 / USA

Professor E.W. Rusche, Department of Physics, University
of Missouri-Rolla, Rolla, Missouri 65401 / USA

C. Schröder-Etzdorf, Physik-Department, Technische
Hochschule München, D-8 München 2 / Germany

Priv.-Doz. Dr. E. Schwarzmann, Anorganisch-Chemisches
Institut, Universität Göttingen, Göttingen / Germany

Dr. R.G. Seidensticker, Westinghouse Research
Laboratories, Pittsburgh, Pa. 15235 / USA

Dr. G. Siegle, I. Physikalisches Institut, Universität
Stuttgart, D-7 Stuttgart N / Germany

Dr. R. Siksna, Institutet för Högspänningsforskning, Uppsala Universitet, Uppsala / Sweden

Dr. M.P. Sixou, Laboratoire de Physique Electronique, Faculté des Sciences d'Orsay, F-91 Orsay / France

P.A. Speare, Institute of Science and Technology, University of Manchester, Manchester 1 / England

Dr. G. Statz, Iwan N. Stranski-Institute of Physical Chemistry, Technische Universität Berlin, D-1 Berlin 12 / Germany

Dr. H. Stiller, Institut für Festkörper- und Neutronenphysik, Kernforschungsanlage Jülich, D-517 Jülich / Germany

Dr. J.A. Sussmann, Israel Atomic Energy Commission, Soreq Nuclear Research Center, Yavne / Israel

Dipl.-Phys. R. Taubenberger, Laboratorium für Festkörperphysik, Eidg. Technische Hochschule, CH-8006 Zürich / Switzerland

Dipl.-Ing. H. Vetsch, Laboratorium für Festkörperphysik, Eidg. Technische Hochschule, CH-8006 Zürich / Switzerland

Professor J.J. Weiss, School of Chemistry, University of Newcastle upon Tyne, Newcastle upon Tyne Ne1 7 RU / England

Dipl.-Phys. M. Weithase, I. Physikalisches Institut, Universität Stuttgart, D-7 Stuttgart N / Germany

Dr. E. Whalley, Division of Applied Chemistry, National Research Council of Canada, Ottawa 7 / Canada

Dr. R.W. Witworth, Department of Physics, University of Birmingham, Birmingham 15 / England

Dr. O. Wörz, Department of Chemistry, Brown University, Providence, Rhode Island 02912 / USA

Professor H. Wolff, Physikalisch-Chemisches Institut, Universität Heidelberg, D-69 Heidelberg / Germany

Professor E.J. Workman, Cloud Physics Observatory, University of Hawaii, Hilo, Hawaii 96720 / USA

Dipl.-Ing. H.R. Zelsmann, Faculté des Sciences de Grenoble, F-38 St. Martin-d'Hères / France

LIST OF CONTRIBUTORS

Arnold, D.
Aufdermaur, A.N.

Bishop, P.G.
Blanckenhagen v., P.
Brajovic, V.
Bullemer, B.

Camp, P.R.
Carnuth, W.
Cole, R.H.
Cross, J.D.

Dantl, G.
Dykins, J.E.

Eiben, K.
Engel, J.
Engelhardt, H.

Finch, E.D.
Fukuda, A.

Glen, J.W.
Glockmann, H.P.
Gosar, P.
Gränicher, H.
Grigull, U.

Hahne, E.
Haltenorth, H.
Hamilton, W.C.
Helmreich, D.
Higashi, A.
Hobbs, P.V.
Hofacker, G.L.

Jaccard, C.
Johnson, D.A.
Jones, S.J.

Käss, M.
Kahane, A.
Kamb, B.
Kiszenick, W.
Klinger, J.
Kvajic, G.

Laplaca, S.J.
Levi, L.
Lippert, E.
Longini, R.L.

Mascarenhas, S.
Mogensen, O.E.
Mounier, S.
Müller-Krumbhaar, H.
Muguruma, J.

Noll, G.

Onsager, L.

Pick, M.
Pounder, E.R.
Prakash, A.

Rabideau, S.W.
Ratner, M.A.
Reiter, R.
Renker, B.
Riehl, N.
Roulleau, M.

Ruepp, R.
Runnels, L.K.

Seidensticker, R.G.
Seki, S.
Siegle, G.
Siksna, R.
Sixou, P.
Statz, G.
Suga, H.

Sugisaki, M.
Sussmann, J.A.

Weiss, J.J.
Weithase, M.
Whalley, E.
Wörz, O.
Wolff, H.
Workman, E.J.

PREFACE

Interest in the physical properties of ice has grown considerably during the past decades. Much of its increasing importance as an object of research is due to the fact that ice, apart from its obvious significance to glaciologists and meteorologists, is well suited for investigating the behaviour of protons in "hydrogen bonds". Ice, therefore, is a good model substance for many physical, physico-chemical and biochemical problems. For this reason, workers in all these different disciplines are interested in structure and physical behaviour of ice.

Thus it seemed reasonable to arrange an international symposium with emphasis on the physical and chemical aspects for all scientists interested in ice physics. The present volume contains the Proceedings of this third International Symposium on "Physics of Ice" which was held from September 9-14, 1968, at Munich, Germany. The first meeting of this kind took place at Erlenbach, Switzerland, in 1962. It had been organized by Professor H.Gränicher. The next conference of this series was arranged by Professor Z.Yosida in 1966 at Sapporo, Japan. The program of the Sapporo meeting was wide in scope, including field glaciology and a parallel conference on cryobiology. Dr. J.W.Glen, University of Birmingham, Great Britain, so the participants of the Munich symposium agreed, shall coordinate the planning of the next international symposium on ice physics.

An essential part of the Munich symposium dealt with electrical and mechanical processes originating in proton transfer and relaxation in ice. Other important parts were devoted to structure problems and lattice dynamics as well as to phase transitions, particularly second order transitions. Treatment of applications of ice physics to various fields, such as glaciology, meteorology or to biological problems, had to be restricted to a few survey lectures given by prominent representatives of these disciplines.

Quite a number of interesting results have already been gained from ice research. The extremely high mobility of protons being transferred by way of hydrogen bonds, the quasi-liquid state of the ice surface, the peculiar phase transformations, the dielectric properties and the indications for ferroelectric behaviour, the appearance of space charges during crystallization as well as some other effects are stimulating for many different fields of science.

Our thanks go to the German Federal Ministry of Scientific Research and to the Bavarian State Ministry of Education and Culture for supporting the Symposium, to all contributors, to all colleagues who helped to arrange it and, particularly, to Dr. D.Helmreich for compiling and preparing this volume.

N.Riehl
B.Bullemer
H.Engelhardt

Munich
January, 1969

CONTENTS

REVIEW ON PROBLEMS OF THE PHYSICS OF ICE

H. Gränicher

Eidg. Technische Hochschule

Zürich / Switzerland

1. Introductory and historical survey

About 12 years ago Professor Urey, the discoverer of deuterium, came to see our laboratory. When he learnt that we were working on the electrical properties of ice crystals, he asked astonished: "What, isn't there everything known about ice?". This widespread opinion perhaps reflects, why the study of physical properties of ice has been neglected for a long time. However, for the sake of justice I have to admit that it was the progress in the understanding of lattice imperfections in general and the advent of new experimental techniques such as neutron diffraction and nuclear magnetic resonance which gave ice physics a successful start in the fifties. Ice research at the ETH Zürich dates back to 1926 when Peter Debye and Paul Scherrer were young professors there. Obviously the study of dielectric properties was in the foreground at that time. The question was raised whether the observed dielectric dispersion were due to the relaxation of water dipoles or whether it should be explained by a Maxwell-Wagner mechanism /1/. This phenomenon occurs whenever layers of dielectrics of varying conductive properties are stacked together. In his thesis Oplatka in 1934 was able to show that highly purified and degassed water produced ice crystals, in which there were no space-charges present and in which the potential drop between the electrodes was linear.

Much later only it was proved that the dielectric relaxation is in fact an inherent bulk property of ice, indirectly by the calculation of the static dielectric constant by Powles in 1952 and by the electric studies of doped crystals. But curiously enough the lattice imperfection mechanisms which were shown around 1957 to be the correct explanation of the electrical kinetic properties of ice, brought the Maxwell-Wagner mechanism back into the picture, but no longer on macroscopic but on the molecular scale.

In his important paper of 1951 Bjerrum /2/ had suggested that the dielectric relaxation rate be due to the migration of orientational lattice defects. These imperfections of the hydrogen bond network which we call Bjerrum defects now, are by themselves only able to move hydrogen atoms within the same water molecule, thus producing changes of the orientation of the electric moment. (Readers not familiar with structure and properties of defects are referred to /3/ and /4/). The puckered layers of water molecules are linked by hydrogen bonds which in the simplest example can be parallel to the applied electric field. Through layers of such hydrogen bonds the "conductivity" is considerably lower. This is due to the scarcity of ion state defects, which alone are able to shift protons along the bonds. When Adolf Steinemann /5/ started his investigations on why and how impurities act so strongly on the electric properties, it became clear to us that we should search for a doping material which would replace water molecules in the lattice substitutionally and at the same time affect the concentration of hydrogen. Hydrofluoric acid HF proved to be a good choice. It led to experimental results which could be interpreted theoretically by the use of the law of mass action for the various equilibrium reactions and of the reaction rate theory. By such an approach we obtain the correct temperature and doping concentration dependences for the quantities of interest. It was possible to conclude that the predominant electrically active reorientation mechanism is the Bjerrum mechanism in pure ice. Quasistatic quantities such as the dc conductivity are controlled by the minority process which is the ion state mechanism in pure ice at not too low temperatures. It is noteworthy that this assignment would not have been possible if experimental data of pure crystals alone were available. This is due to the fact that both types of defects (Bjerrum defects and ion state (ionic) defects) carry charges and that they are complementary with respect to their elementary proton motions.

While electrical investigations on pure and doped ice crystals were going on at the ETH, Professor Kneser and his coworkers in Stuttgart studied the mechanical relaxation dispersion /6/ in the same audio frequency range as the dielectric dispersion. In a theoretical paper his coworker Bass /7/ showed that Bjerrum defects, acting as catalysts for changing the hydrogen configurations, enable the crystal to reach arrangements which are energetically more favorable in the presence of a mechanical stress. Eigen and De Maeyer /8/ in Göttingen taking extreme care to use water of the highest purity measured the dc conductivity in low and high fields. They established the existence of saturation currents and of the dissociation field effect. This work which was later improved in cooperation with Spatz /9/ produced data on the rates of dissociation and recombination in the dissociation equilibrium of water into the ion states. The concentration of these ionic defects turned out to be smaller than the one of Bjerrum defects at least in the temperature range of these original studies by a factor of 10^5 in pure ice. This handicap of the ions is relieved however by the astonishingly high mobility of the positive ion states or excess proton states as they are called sometime. Whereas Bjerrum defects need a substantial energy of activation in order to be able to diffuse in the lattice (0.235 eV), excess protons apparently move by quantummechanical tunneling. Onsager and Dupuis in 1960 /10/ published a theory of the electrical phenomena on the same bases as the work in Zürich, but emphasizing the defect interactions in a Debye-Hückel type approach. Therefore, at the turn of the decennial 1960, not without a certain proud, we thought that the ice problem was in principle solved and that our concepts explained all the phenomena known at that time. If our optimism were justified, probably no new conference on ice physics would have been necessary. But as a matter of fact the community of ice physicists has increased considerably since then. The new endeavour has enlarged our knowledge

a) by stretching the range of experimental parameters such as purity, temperature and
b) by studying new effects, e.g. thermoelectricity, the Hall effect and selfdiffusion.

In view of the great amount of new information it becomes very important that we do not loose the essential aim out of sight. By this I mean we should obtain a consistent molecular picture which contains all the various aspects of the ice problem. I am convinced that we have to make an effort in this respect now. Our model concepts should converge rather than multiply. An example of a situation which we should try to avoid, is the surface problem of ferroelectric barium titanate. There one speaks of a chemical surface, an electrical surface and of a domain nucleating surface. In this case we have three different pictures which cannot be brought to coincidence and these pictures obviously are related to the same and one physical reality. I am glad that we can have this conference. Certainly it will confront us with all facets of ice properties and it will be helpful to our mutual understanding.

So far I have resumed that state of our knowledge as it existed at the time of the Erlenbach Colloquium six years ago. In the following main part of my talk I want to discuss a number of topics, in which important progress has been achieved and/or in which major problems are still unsolved. Of course the choice and the emphasis given to the various topics reflects my own interest and experience.

2. Crystal growth

All experimental work on ice starts with the problems of water purification and crystal growth. I think every laboratory has developed more or less its own scheme. With the availability of ion exchange resins and plastic tubings one can do it in a relatively elegant way. The choice and the cleaning of the materials need great care and it is questionable whether one can avoid organic impurities from getting into the ice. Also, we do not know what their effect in the ice might be. Prolonged multiple distillation is certainly a more straightforward technique, but it requires quite elaborate installations.

The purification process should also eliminate dissolved gases. Permanent gases such as nitrogen produce bubbles in the growing material. Carbon dioxide is a great nuisance as it has strong effects on the electrical properties . Therefore, it is essential to handle the purified water without further contact with air. A recent publication /11/ seems to indicate that a zone refining process brings the ultimate in high purity. The growth itself, if properly controlled, is always one step of purification by segregation of impurities. The crystals tend to reject all impurities. Early work /12/ dealt with the axial distribution (along the direction of growth) of impurites. Later it became apparent in studies done in Munich /13/ that one also has to worry about the radial distribution. The latter depends strongly on the shape of the growing interphase /14/. The amount of inclusion of imperfections is a function of growth direction, growth speed and of the environmental electrical conditions. In addition there is a charge separation during the solidification process which is called the Costa-Ribeiro effect /15/. In particular with growth from ionic solutions, high growth potentials develop (Workman-Reynolds effect /16/) which are interrelated with the segregation of impurities. As a result of these effects a freshly grown crystal in general will have an imhomogeneous distribution of imperfections (space charges in the case of ionic impurities). Ice nucleation and the charging phenomena mentioned above have been studied extensively mainly because of their meteorological relevance for thunderstorm electricity and hailstone formation /17/.

3. Electrical electrodes

Once good high purity single crystals have been produced there is the problem of bringing them to the correct geometrical shape required and to apply electrodes for electrical investigations. Of course all this has to be done without contaminating the crystals. At the ETH we have worked out the technique of gold plating the crystals in high vacuum at liquid nitrogen temperatures. The surface on which gold is to be condensed, is cleaned by sublimation immediately before the evaporation process.

The deeper concern is the impossibility to produce truly ohmic contacts due to the dissociation potential of water. From experience one knows that never all charge carriers are discharged at the electrodes even if high potentials are applied to the electrodes. More or less pronounced space charges are built up close to the electrodes which disturb all measurements at low and zero frequencies /18/. One way out consists of the application of completely blocking electrodes which are thought to lead at least to a clear physical situation /19/. Ion exchange membranes have been used several times /8, 20/ which act as a strong source or sink for electrically active defects. Therefore, a diffusion potential must be expected to build up between the bulk of the ice and the outer membrane surface. An alternative possibility are palladium electrodes charged with hydrogen which have injecting properties as the Munich group has shown /21/. In high fields the question arises whether the injected particles are really excess protons (H_3O^+ states) or whether protons are shot into interstitial positions and travel along interstitial channels which are relatively wide in ice.

4. Ageing effects, possibility of stacking faults

Ageing effects have been found long ago e.g. by Tammann in studies of the thermal expansion of ice. There is still no explanation for them. Recent work of the Stuttgart group established their role in dielectric and elastic investigations. Dantl /22,23/ concludes that the discrepancy between his own determinations of the elastic moduli and the values of other authors is due to differences in the ageing state. New measurements showed that the density of freshly grown ice may be up to 0.3 % higher than the density of aged ice.

If we can assume the ice to be chemically of high purity, the anomalous high density of fresh ice must correspond to an unstable high degree of lattice disorder, specifically to a concentration of interstitial water molecules exceeding notably the concentration of molecular vacancies. The density of interstitial molecules decreases during the ageing process to the value of the thermal equilibrium concentration. This hypothetical explanation accounts for the density anomaly, i.e. for the ageing effects at high temperatures, but not for the age dependence of dielectric behaviour and of the spin-lattice relaxation time found in the region of -50 °C by R.Ruepp and G.Siegle respectively /this conference/.

It is well known that the oxygen positions in the structure of hexagonal and cubic ice correspond to the zinc and sulphur positions in ZnS which occurs in the hexagonal wurtzite and the cubic zincblende structure. In the family of the isomorphous sulphides and selenides stacking faults are found quite frequently. In view of the low value of the difference in lattice energy between cubic and hexagonal ice stacking faults seem to me very probable for ice too. Admittedly they would be much harder to detect than in the ZnS type crystals. As no density anomaly associated with stacking faults is to be expected, they will influence only a restricted class of properties, e.g. the mobility of excess protons.

5. Surface effects

The possibility that the crystal surface properties might influence bulk properties has been disregarded till very recently. It is the merit of the Munich group to have pioneered in this area. Details of these investigations will be given later in this conference. It was found that the electrical surface conductivity may lead to incorrect bulk conductivity values obtained with simple electrode configurations in the temperature range above -30 °C. Reliable data can be got by guard ring measurements or - even better - by guarded four-probe measurements. It is necessary in any case to eliminate the as-grown surface of the crystals which is usually done mechanically. If possible one should further increase the purity of the surface by sublimation prior to the measurements and it is mandatory to contain the ice sample in a controlled inert gas atmosphere. On the other hand a forced sublimation during the experiment is to be avoided, because as shown recently by Jaccard /24/ such a sublimation affects the conductivity presumably by releasing defects on the surface. As in the case of injecting electrodes, in this situation we loose our knowledge of the defect concentration. Customarily one derives these concentrations from the law of mass action under the assumption of thermal equilibrium and of intrinsic conditions.

As an influence of the surface properties on bulk properties can never be fully eliminated, it is a task for future research to study the ice surface in all its aspects such as the real structure, surface defect density and conductivity. Ice surface problems have attracted the interest of meteorologists long ago already, e.g. due to their relation to friction on ice and to electrification by friction or contact.

6. Piezoelectric effect

The piezoelectric effect is a very controversial issue and it was already so at the last meeting in 1962. There are many experiments pro and contra /23/. I think it is essential to distinguish between the following two things: the true piezoelectric effect and the pseudo-piezoelectric effect.

There is no doubt that the pseudo-piezoelectric effect exists. By this I mean an effect due to inhomogenities in the distribution of impurities or sessile charged defects. The experiment described by Deubner et al. /25/ was done with ice which had been grown very rapidly. The piezoelectric effect, which they observed, disappeared after a few days, i.e. in a time adequate for diffusion of impurities. As indicated in section 2 such a crystal is to be expected to have an inhomogeneous impurity and hence charge distribution. A change of polarization is to be expected, if such a crystal is compressed and elastically deformed.

In my opinion the true piezoelectric effect has never been observed. Recent tests with well aged crystals were negative /23/. What can we expect? The oxygen sublattice is centrosymmetric and down to very low temperatures /26/ no phase transition has been observed. Therefore, no true piezoelectric effect is to be expected from the oxygen network alone. There is of course the possibility that at very low temperatures an ordered polar hydrogen arrangement occurs which needs not be ferroelectric, but which should be accompanied by an appropriate specific heat anomaly at the ordering temperature. In pure ice the rate of transformation is much to slow at temperatures where an ordered phase would be stable. There are slight indications of anomalies in the 120 ^{o}K region /M.Pick, this conference/ and in doped ice crystals, but it cannot be claimed that full ordering has been observed.

7. Lattice vibrations

There is extensive literature on intra- and intermolecular vibrations as derived from infrared absorption and Raman spectra and from inelastic neutron scattering. These data have been related to the specific heat and have been used in the interpretation of the second moment of the proton magnetic resonance line /27/. I have to leave it to the specialists to judge the level of knowledge reached in this field. Personally I hope that there will be more of an impact of these data on our understanding of the defect mechanisms.

8. Elastic properties

The elastic moduli of ice /23/ are the quantities which are known with the highest accuracy among all ice data. This is the fruit of the great efforts of the Stuttgart group over many years. The mechanical relaxation at audio frequencies has been mentioned already. It is observed for deformation vibrations other than those of rotational symmetry. High frequency losses have been found and interpreted as scattering of phonons on microstructure boundaries.

Brill and Camp /28/ had shown that plastic deformation does not affect the dielectric behaviour of the ice. This result had been checked in our laboratory by Jaccard, because it is of vital importance to the interpretation of the dielectric dispersion in terms of Bjerrum defects. However, the statement cannot be reversed. Jones /29/ recently found that the creep rate of fluoride doped ice is increased relative to pure ice. Glen /30/ has interpreted this result on the grounds that dislocations can only move if the hydrogen atoms that would give rise to the formation of defects are reoriented ahead of the moving dislocation by defects of the majority mechanism.

9. Selfdiffusion

This topic really has become the home domain of Professor Riehl and his group. Prior to 1960 there was only one isolated observation by Kuhn and Thürkauf /31/ which indicated that deuterium and O^{18} have the same diffusion constant. The work of the Munich group /32,33/ led to the following results:

- Deuterium, tritium and oxygen^{18} have within the limits of accuracy the same diffusion constant and an activation energy of 0.63 ± 0.03 eV. These results are confirmed by tritium measurements of Ramseier /34/ at CRREL, USA with the exception that his diffusion constants are about 30 % lower than the ones obtained in Munich.
- In both laboratories an anisotropy of about 10 % of the diffusion constant was found, the diffusion perpendicular to the c-axis being larger.
- The diffusion constant turned out to be independent of doping with either HF or NH_4F in a concentration range covering more than two orders of magnitude /33/.

These results rule out that Bjerrum and ionic defects contribute appreciably to the selfdiffusion process. One has to accept that in selfdiffusion entire water molecules are migrating. This might be due to one or a combination of the following mechanisms: molecular interchange of neighbouring molecules, molecular Schottky vacancies and interstitial molecules. Based on an estimate of the energy of formation /35/ the Schottky vacancy process seems to be predominant. No real proof of it has been given. It is suggested /36/ that the study of the pressure dependence of the selfdiffusion might enable to establish the nature of the relevant process.

10. Quasistatic equilibrium properties

Nothing overwhelming can be reported on quantities such as lattice energy, specific heat, entropy, and static dielectric constant, but there has been a major process in theoretical calculations. Lattice energy computations are fairly good on an absolute scale. However, they are still inadequate to answer satisfactorily the question of the energy differences among the various hydrogen configurations. A very elaborate study has been published by Campbell et al. /37/ who has attacked the problem of a computer calculation of the static dielectric constant by the use of the Fröhlich theorem. In a similar attempt Coulson and Eisenberg /38/ have derived the dipole moment of water molecules in ice. They found an average value of 2.6 D. The increase of the dipole moment over the vapor phase value of 1.85 D is due to hydrogen bond formation and is to be expected on theoretical grounds.

In his classical paper Pauling had given an approximate calculation of the number of hydrogen configurations of an ice crystal of N molecules. He deduced a value for the disorder entropy of $S_o = k \ln(3/2)^N$ which is in good agreement with the experimental residual entropy value. In recent years the number of configurations has been determined with high accuracy. For the twodimensional square ice lattice approximate computations using different methods by Nagle /39/ and Suzuki /41/ led to values which differed only by one digit in the fifth significant figure from the exact result of Lieb /40/. A rigorous solution for the hexagonal ice lattice has not yet been achieved but the value obtained by Nagle is accurate at least to the fifth decimal. His result is slightly larger than Paulings value. As a consequence the theoretical disorder entropy value has become higher by 1.1 % and is in an even better agreement with the experimental value. But this is merely an incidence as the experimental value is only known with an accuracy of about ± 5 %. Certainly Pauling's interpretation of the residual entropy as disorder entropy is well founded now.

11. Properties involving hydrogen rearrangements

The dielectric relaxation data which are quoted so often /42,43/ were measured prior to 1953. In selfdiffusion and proton magnetic resonance work we have seen the activation energy values grow by about 15 % as the crystal quality improved over the years. I began to question our comparing these new data with old dielectric data. I am glad to see that new dielectric dispersion studies have been made and will be presented in this conference. Ruepp has determined a value for the activation energy of the relaxation time of 0.61 ± 0.01 eV compared to 0.575 eV of /42,43/.

Therefore, once more the activation energies of the dielectric relaxation and of the selfdiffusion are the same within the limits of accuracy. But this is an incidence. The completely different dependence of these properties on doping rules out the possibility of a common mechanism as envisaged by Haas /44/ at the 1962 colloquium.

In section 5 I had mentioned already that the dc conductivity measurements had to be revised because the effect of the surface conductivity had been ignored in earlier work. The activation energy comes to an average of 0.37 eV (8.5 kcal/mole), considerably lower than the previously accepted value (0.48 eV or 11 kcal/mole) /9/. Details of the new experiments will be given in the papers of Bullemer and Kahane later in the conference.

In all measurements of conductivities and relaxation times a change to a lower activation energy is observed at low temperatures. The bending over in the usual graphs occurs at temperatures which depend on the sample purity. Hence, the activation energy (typically 0.26 eV or 6 kcal/mole) of the low temperature region is considered as extrinsic behaviour. However, with the high purity crystals available today a completely different variation of the activation energies has been found. From a comparison of the summaries of the papers to be presented to this conference by Bullemer, by Kahane, and by Ruepp the situation can be summarized as follows:
The activation energy values for the majority mechanism (dielectric relaxation time) are 0.61 eV (14 kcal/mole) at temperatures above -40 to -60 °C and change to 0.43 eV (10 kcal/mole) at lower temperatures. On the other hand the dc conductivity which is due to the minority process increases from the high temperature value stated above (0.37 eV) to a value of 0.57 eV (13 kcal/mole) at temperatures below -70 °C. These changes suggest that Bjerrum and ionic defects interchange their role at temperatures between -40 and -70 °C depending on sample. Or more specifically Bjerrum defects which provide the majority at high temperatures become the minority mechanism at low temperatures where ionic defects act as majority process. This suggested working hypothesis may serve as a guide for new experimental investigations and requires a confirmation.

CRYSTAL PLASTICITY	DIFFUSION	ELECTRIC PROPERTIES
creep	selfdiffusion of (H), D, T, O^{18} diffusion of HF	dielectric relaxation conductivity saturation current Hall effect thermoelectric effect MECHANICAL RELAXATION
	PROTON SPIN RELAXATION	
DISLOCATIONS	POINT IMPERFECTIONS	HYDROGEN BOND NETWORK IMPERFECTIONS
	Schottky vacancies (Frenkel mechanism)	Bjerrum defects Ion states

Table 1 Physical properties of ice crystals and their interpretation in terms of lattice imperfections

Table 1 summarizes the physical properties which are due to imperfection mechanisms, together with the suggested processes. It has been pointed out that the separation into the three categories is not absolute, as the rate of dislocation motion is governed by the majority process of the hydrogen network defects. The spin-lattice relaxation time T_1 of proton magnetic resonance requires some comment:

The proportionality of T_1 to the dielectric relaxation

time had led to the conclusion that the rate determining proton relaxation process is the one of the dielectric relaxation and theoretical calculations showed that the dielectric process appears to be not efficient enough by a factor of about 3 /11,45/. It is clear that the self-diffusion process may be effective for spin relaxation, but it had been argued on theoretical grounds /35/ that selfdiffusion is not effective for the dielectric relaxation. The strong dependence of the dielectric properties on doping in contrast to the doping independence of the selfdiffusion /33/ strongly supports this view.

As indicated at the beginning of this section the modern values of activation energies are almost identical and can no longer serve as a decisive argument. If absolute values of the spin-lattice and the dielectric relaxation time are to be compared, one has to keep in mind that according to the now customary definition the dielectric relaxation time is the decay time of the macroscopic specimen and is not identical with the relaxation time of individual dipoles /46/. The only straightforward argument is the dependence of the relaxation times on doping. As the spin-lattice relaxation time is strongly affected e.g. by the presence of HF /47/, T_1 is determined by the electrically active process. Admittedly this argument is weak in the limit of vanishing doping concentration. There it cannot be excluded that the selfdiffusion process slightly overtakes.

Table 2

UNSOLVED PROBLEMS

Actual structure of defects: Bjerrum defects
Ion states

Kinetics of defect formation and diffusion (mobilities, selective and collective excitations)

Defect interactions and aggregates

Traps for defects, particularly traps for charge carriers

Diffusion in ice: predominant mechanism
pressure dependence

Mechanism of ageing effects

12. Unsolved problems

Some open questions have been discussed already. Table 2 lists them together with new ones which cannot be explained here. However, I would like to give some thoughts to one problem, the actual structure of Bjerrum defects. The simplest representation of Bjerrum defects which shows all relevant features is the rigid lattice model. It served often as illustration in papers but it was not meant to be realistic /35, see p.456/.

At the time of the Erlenbach Colloquium theoretical chemists began to worry in particular about the D-defect. The repulsion energy of the close lying protons in the rigid model would be excessive.

Dunitz /49/ and Cohan et al. /50/ suggested models which bring the hydrogen to stable positions of lower repulsion energy holding the oxygen framework fixed. Alternatively Eisenberg and Coulson /51/ showed that a stable colinear structure is energetically possible if the lattice is allowed to relax elastically in the surrounding of the defect. This discussion came to a stillstand: The three models b), c) and d) of table 3 gave energies of formation in agreement with the empirical value. The quantum-chemical calculations are not accurate enough to rule out any one of the models. Hence, one has to get experimental evidence on the actual structure. At the ETH investigations are in progress along two lines: the study of the pressure dependence of electrical properties /36/ and EPR studies of electrons trapped by D-defects or holes trapped by L-defects.

The defect aggregates of Bjerrum defects coupled to either interstitial molecules or to vacancies have been proposed. They couple the selfdiffusion to the dielectric relaxation mechanism. The investigation of these aggregates is not very advanced and in the light of the discussion given in section 11 there seems to be no experimental evidence to postulate the presence of such aggregates.

It may be added that the problem of the L-defect is not at all trivial. Neighbouring molecules will tend to shift toward the centre of the vacancy but this will be counteracted by the repulsion between the electron lone pairs. Once we know more about the structure of the defects there will be hope to get a deeper understanding of the details of their kinetics.

BJERRUM (ORIENTATIONAL) DEFECTS

D = positive defect (doubly occupied bond)

L = negative defect (vacant bond)

a) rigid lattice model

b) rotated D-defect (Cohan et al. 1962)

c) X-defect (Dunitz 1963)

d) relaxed lattice model (Eisenberg and Coulson 1963)

e) DI and LI nterstitials (Haas 1962)

f) DV and LV acancies (Runnels 1963 / Kopp 1965/7)

Table 3 List of proposed models of Bjerrum defects

13. Final remarks

All what I had to say refers to the normal hexagonal ice. Studies of D_2O ice have aided our understanding at various points, but our knowledge of D_2O ice lags much behind. The cubic form is well established by now, but its stability range at low temperatures makes it not very attractive for investigations of particular interest in the case of hexagonal ice.

A door to a completely new and exciting world of ice physics has been opened by Dr. Whalley and his colleagues at the National Research Council of Canada, the whole family of high pressure forms of ice, and Professor Kamb from Caltec and Dr. Rabideau from Los Alamos have worked on the structural aspects of these polymorphs. I am convinced that their papers in this symposium and future work on the polymorphic forms will contribute a host of valuable information. Hopefully by comparison of the properties of these various structures much more can be learnt of the nature of the hydrogen bond than from studies of hexagonal ice alone.

This introductory talk as any sightseeing tour had to be incomplete and superficial. But if I have succeeded to stimulate your interest for the rest of the symposium, then I have reached more than I could expect.

Acknowledgment: The research on ice physics at the ETH Zürich has been supported since 1962 by the Schweizerische Nationalfonds zur Förderung der Forschung.

References

/1/ Gränicher, H., F. Jona, Helv.Phys.Acta Suppl.V 50-60 (1960), where a complete list of references to the early Zürich work may be found

/2/ Bjerrum, N., Kgl.Danske Videnskab.Selskab. Mt.-fys.Medd. 27, No.1, 56p (1951)

/3/ Onsager, L., M. Dupuis, in "Electrolytes", p.27-46, Pergamon Press (1962); Runnels, L.K., Scientific American, 118-126, Dec. (1966)

/4/ Gränicher, H., Physik kondens.Materie 1, 1 (1963); Z.Kristallogr. 110, 432-471 (1958)

/5/ Steinemann, A., Helv.Phys.Acta 30, 553-610 (1957)

/6/ Kneser, H.O., S.Magun, G. Ziegler, Naturwiss. 42, 437 (1955); Schiller, P., Z.Physik 153, 1 (1958)

/7/ Bass, R., Z.Physik 153, 16 (1958)

/8/ Eigen, M., L. De Maeyer, Proc.Roy.Soc. (London) A 247, 505 (1958)

/9/ Eigen, M., L. De Maeyer, H.C. Spatz, Ber.Bunsenges. Phys.Chemie 68, 19 (1964)

/10/ Onsager, L., M. Dupuis, Rendiconti S.I.F. X Corso, p.294 (1960); Dougherty, T.J., J.Chem.Physics 43, 3247-3252 (1965)

/11/ Barnaal, D.E., I.J. Lowe, J.Chem.Phys. 48, 4614 (1968)

/12/ Jaccard, C., L. Levi, ZAMP 12, 70-6 (1961)

/13/ Blicks, H., H. Egger, N. Riehl, Phys.kondens. Materie 2, 419-422 (1964)

/14/ Jaccard, C., Phys.kondens.Materie 4, 349-354 (1966)

/15/ Pinatti, D., S. Mascarenhas, J.Appl.Phys. 38, 2648 (1967)

/16/ Workman, E.J., S.E. Reynolds, Phys.Rev. 78, 254-259 (1950)

/17/ Workman, E.J., Physics of Ice, Proc.Intern.Symp. Munich, Plenum Press, New York (1969)

/18/ Gränicher, H., Physics of Ice, Proc.Intern.Symp. Munich, Plenum Press, New York (1969)

/19/ Mounier, S., P. Sixou , Physics of Ice, Proc.Intern. Symp. Munich, Plenum Press, New York (1969)

/20/ Kahane, A., Physics of Ice, Proc.Intern.Symp. Munich, Plenum Press, New York (1969)

/21/ Engelhardt, H., N. Riehl, Phys.kondens.Materie 5, 73-82 (1966)

/22/ Dantl, G., I. Gregora, Naturwiss. 55, 176 (1968)

/23/ Dantl, G., Phys.kondens.Materie 7, 390-397 (1968)

/24/ Jaccard, C., in "Physics of Snow and Ice", part 1, 173-179 (1967), Inst.Low Temp.Sci., Sapporo

/25/ Deubner, A., R. Heise, K. Wenzel, Naturwiss. 47, 600 (1960)

/26/ Brill, R., A. Tippe, Acta Cryst. 23, 343 (1967)

/27/ Barnaal, D.E., I.J. Lowe, J.Chem.Phys. 46, 4800-4809 (1967)

/28/ Brill, R., P.Camp, Nature (London) 179, 623 (1957)

/29/ Jones, S.J., Phys.Letters 25A, 366 (1967)

/30/ Glen, J.W., Phys.kondens.Materie 7, 43-51 (1968)

/31/ Kuhn, W., M. Thürkauf, Helv.Chim.Acta 41, 938 (1958)

/32/ Delibaltas, P., O. Dengel, D. Helmreich, N. Riehl, H. Simon, Phys.kondens.Materie 5, 166-170 (1966)

/33/ Blicks, H., O. Dengel, N. Riehl, Phys.kondens. Materie 4, 375-381 (1966); Dengel, O., E. Jacobs, N. Riehl, Phys.kondens. Materie 5, 58-59 (1966)

/34/ Ramseier, R.O., J.Appl.Phys. 38, 2553-2556 (1967)

/35/ Gränicher, H., Z.Kristallogr. 110, 432-471 (1958)

/36/ Gränicher, H., Physics of Ice, Proc.Intern.Symp. Munich, Plenum Press, New York (1969)

/37/ Campbell, E.S., G. Gelernter, H. Heinen, V.R.G. Moorti, J.Chem.Phys. 46, 2690-2707 (1967)

/38/ Coulson, C.A., D. Eisenberg, Proc.Roy.Soc. A291, 445-453; and 454-459 (1966)

/39/ Nagle, J.F., J.Math.Phys. 7, 1484-1491 (1966)

/40/ Lieb, E.H., Phys.Rev. 162, 162-172 (1967)

/41/ Suzuki, Y., in "Physics of Snow and Ice", Inst. Low Temp.Science Sapporo, p.21-41 (1967)

/42/ Auty, R.P., R.H. Cole, J.Chem.Phys. 20, 1309 (1952)

/43/ Humbel, F., F. Jona, P. Scherrer, Helv.Phys.Acta 26, 17-32 (1953)

/44/ Haas, C., Phys.Letters 3, 126-128 (1962)

/45/ Onsager, L., L.K. Runnels, Proc.Nat.Acad.Sci. 50, 208-210 (1963); and J.Phys.Chem. (in press)

/46/ Fatuzzo, E., P.R. Mason, Proc.Phys.Soc. 90, 729-740; and 741-750 (1967)

/47/ Kopp, M., D.E. Barnaal, I.J. Lowe, J.Chem.Phys. 43, 2965 (1965)

/48/ Dunitz, J.D., Nature (London) 197, 860-861 (1963)

/49/ Cohan, N.V., M. Cotti, J.V. Iribarne, M. Weissmann, Trans.Faraday Soc. 58, 490-498 (1962); and Nature (London) 201, 490 (1964)

/50/ Eisenberg, D., C.A. Coulson, Nature (London) 199, 368 (1963)

STRUCTURE PROBLEMS OF ICE

E. Whalley

Division of Applied Chemistry, National

Research Council of Canada, Ottawa / Canada

1. Introduction

The ice that we are familiar with is only one of many different kinds of solid water. Water forms more solid phases than any substance we know. The phase diagram is shown in fig. 1. There are several kinds of lines on the diagram which indicate different kinds of phase boundaries, as explained in the caption.

Most of this conference is concerned with the common form, ice Ih. There is another phase similar to ice Ih, but with the corresponding cubic structure and commonly known as Ic, which occurs in the region of ice Ih but appears to be always metastable relative to it. It was discovered by König /1/ in 1943.

All the measurements described by Gränicher in the preceding article were made along the zero-pressure line. Very few persons have ventured, if I can use mountaineering terminology, out on to the face. The earliest of these who made significant contributions was Tammann /2/ at Göttingen in 1900 when he investigated the phase diagram to about 3 kbar, and discovered the phases known as II and III. Bridgman /3/ continued the investigation to about 18 kbar at Harvard in 1912 and discovered ice V and VI. He had a few measurements that were probably of a metastable phase in the region of V, but it was not fully characterized until D_2O was studied in 1935 /4/. In 1937 /5/ he extended his measurements to about 45 kbar and discovered ice VII. For about 30 years no other phases were added, until ice VIII /6/ and ice IX /7/

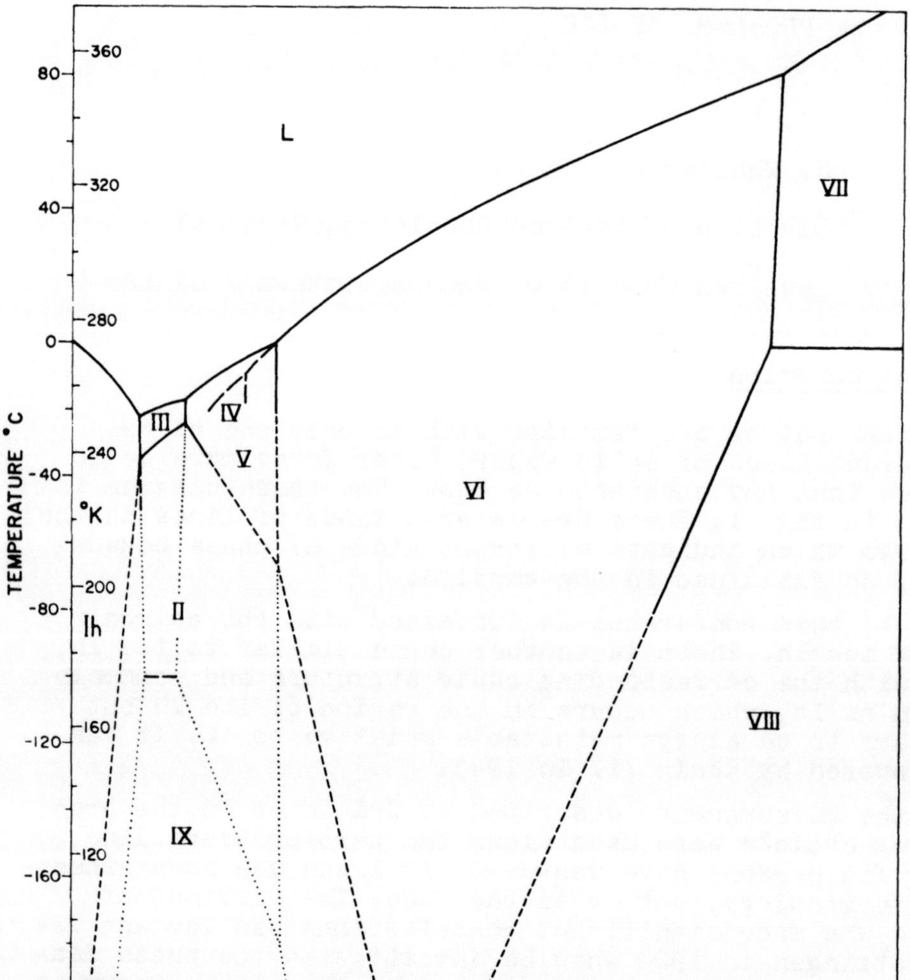

Fig.1 Phase diagram of ice. The solid and long-dashed lines are directly measured stable and metastable lines respectively, and the short-dashed and dotted lines are extrapolated or estimated stable and metastable lines respectively

were added in 1966 and 1968 as a result of work that we did in Ottawa.

There are a number of other phases that we might best describe as incompletely characterized that have never been thoroughly investigated and reported. We ourselves have evidence /8/ for some so I feel reasonably confident that some exist, and there are probably other phases as well. In addition there is a vitreous or amorphous phase /9/ that is formed by condensing the vapor slowly at liquid -nitrogen temperature. There is also a large number of phases known as the clathrate hydrates /10/, which are essentially phases of ice which have big holes in them, and these holes are filled with molecules like ethylene oxide, xenon, argon, etc. Apparently, these additional molecules have no function other than to prevent the big holes from collapsing. If we could follow the phase diagram of ice into the region of negative pressures some of these phases that are less dense than ice I might be stable; but unfortunately this does not seem to be possible.

Little has been done on the high-pressure phases until the past decade or so, except for the phase boundary measurements of Tammann /2/ and Bridgman /3-5/. Recently, however, a number of people have become interested, and several investigations have been reported. But our knowledge, by comparison with what we know about the hexagonal form, is still meagre. Since we can talk for a week almost entirely on ice Ih at zero pressure, how many weeks could we talk once we get out onto the pressure-temperature face and begin to understand it?

2. General considerations about the structures

The structural problems of these phases are obviously many and varied, and cannot be reviewed in detail in a small space. Only a few interesting or important features can be included. There are two important general considerations that are true of all the phases of ice. The first is that all of them are made of water molecules, and the water molecules are not destroyed when any of the phases are formed.

The evidence for this comes from a number of sources, but perhaps the most direct is the infrared spectrum. As an example, the infrared spectrum of ice II from 4000 to 350 cm^{-1} is shown in fig. 2. A water molecule in the vapor phase has three internal vibrations which are schematically represented in fig. 3. There is a symmetric and an antisymmetric stretching vibration, ν_1 and ν_3 ,

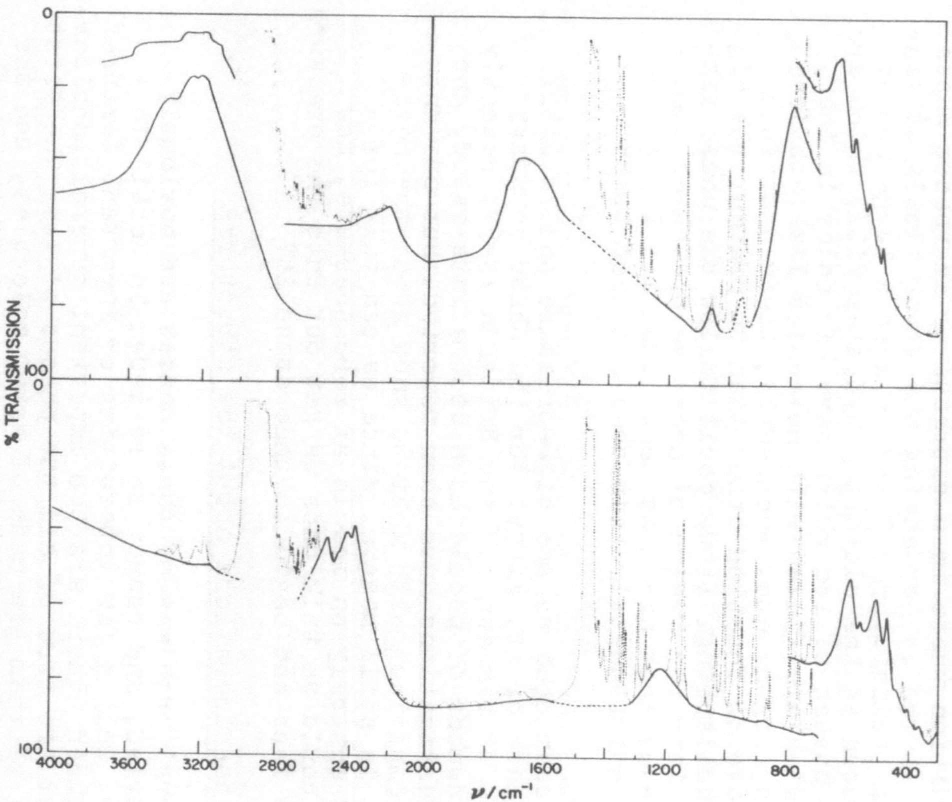

Fig.2 Infrared spectrum of ice II in the range 4000 to 350 cm^{-1} at 100 ^{o}K, zero pressure. The upper frame is of H_2O and the lower frame is of D_2O. The dotted lines represent absorption due to the mulling agent

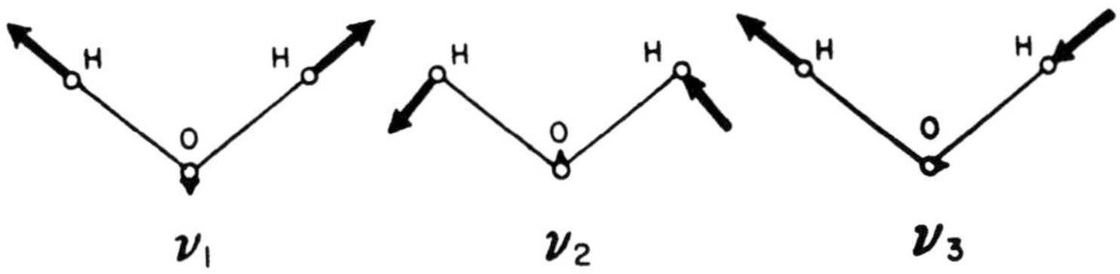

Fig.3 Schematic representation of the symmetric ν_1 and antisymmetric ν_2 stretching vibrations and the bending vibration ν_3 of an isolated water molecule. (After Herzberg /12/)

which are about 3657 cm^{-1} (121.9 THz) and 3756 cm^{-1} (125.2 THz) /12/. In ice II_1 there is a strong absorption band centred about 3300 cm^{-1} which is quite certainly due to these vibrations. They are decreased in frequency because the molecules are hydrogen bonded, but they are not decreased enough for us to consider that the water molecule is destroyed. The band is a broad one because the vibrations of different molecules are coupled by molecular interaction. The bending vibration ν_2, the form of which is shown schematically in fig. 3, is at 1595 cm^{-1} in the vapor /12/ and is about 1690 cm^{-1} in the solid although its exact value is a little uncertain because it overlaps overtones of lower-frequency vibrations. Clearly again, the water molecules have not been destroyed, although they have of course been perturbed.

The second general consideration is the following. In all of the ten phases that have so far been described, and in all the clathrate hydrates except one or two, the structures are dominated by two themes: four coordination and the possibility of orientational disorder. Both these themes are caused by the unique property of the water molecule that it can form four hydrogen bonds, in two of which it donates its hydrogen atoms and in the other two it accepts hydrogen atoms from other molecules. The first theme is that the hydrogen bonds have a tetrahedral arrangement as shown in fig. 4. It is generally accepted that the hydrogen bonding is caused by the two polarizable lone electron pairs of a water molecule. Unfortunately we know little at the present time about the extension of these lone pairs in space, although we frequently draw them as rather prominent, at least partly to account for the directional properties of hydrogen bonds. Experiment cannot tell where the electrons are, as there is no way

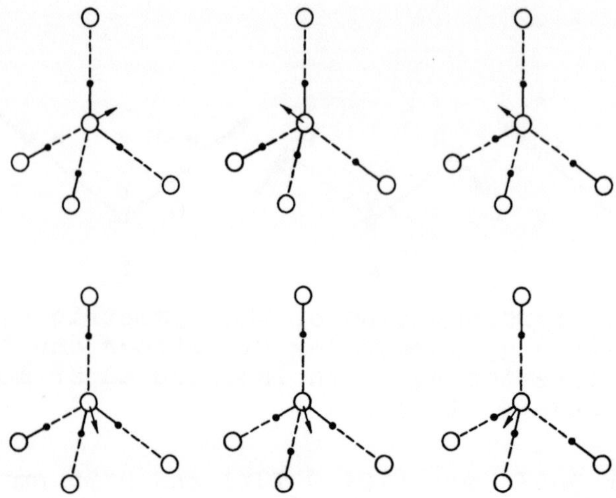

Fig.4 The four co-ordination and the six possible orientations of a water molecule in ice (after Owston/13/)

at present of measuring electron densities to the required accuracy. It would be very useful if those who calculate wave functions of water were to examine the properties of the lone pairs: it would probably help greatly in understanding hydrogen bonding.

The second theme is also illustrated in fig. 4. It is clear that the requirement of tetrahedral hydrogen bonding, with two hydrogen atoms close to the central oxygen and two close to two of the neighboring oxygen atoms, allows the molecular axis of the central water molecule to point in six directions. Consequently, there can be orientational disorder.

3. Techniques for investigating the structures

Many techniques have been used for investigating the structure of ice, and no technique used so far has told us all we want to know. X-ray, electron, and neutron diffration have all been used, but apart from papers in this conference, neutron diffraction has been reported so far only on ice Ih /14/. These techniques tell primarily about the oxygen positions, and neutron diffraction of course tells about the hydrogen positions as well.

The entropy is important in determining whether the orientations are ordered or disordered, and was first used by Pauling /15/, who discussed the residual entropy of ice Ih extrapolated to 0 °K from calorimetric measurements. Calorimetric measurements have not been done on the high-pressure phases, but Kamb /16/ was the first to point out in print that the entropies of transformation were important sources of information on the relative configurational entropies. The main idea is that the vibrational entropies of different phases of ice on an equilibrium line are very similar. The reason for this is that the vibrational entropy depends primarily on the gross features of the vibrational spectrum, and in particular on the spectrum of the translational lattice vibrations. The gross features of the spectrum depend in turn primarily on the gross features of the intermolecular potentials involved, and these are primarily the nearest neighbor interactions, the number of them and their strength. There are four nearest neighbors, all hydrogen bonded, in all the phases of ice and these provide the dominant forces, and determine the general features of the density of vibration states. The fine features are determined of course by the details of the crystal structure, but the fine details do not affect the entropy greatly. Hence, if there are entropy differences more than 0.1 or 0.2 cal deg^{-1} $mole^{-1}$ between phases on an equilibrium line, it is rather likely to be largely configurational in origin, and this has been a fruitful source of information. The existence of ice VIII was in fact inferred /17/ from some early measurements /5/ of the ice VI-VII line by means of these considerations.

Infrared and Raman measurements are very important in detecting hydrogen bonding. The decrease in O-H stretching frequency caused by hydrogen bonding is well known. It is exemplified by the infrared spectrum of ice II in fig. 2. The centre of the O-H stretching band is about 3300 cm^{-1}, compared with 3700 cm^{-1} in the gas phase. Another important infrared band is that due to the rotational vibrations of water molecules. In ice II (fig. 2) and in other phases of ice it is centred about 800 cm^{-1}. This corresponds to a force constant for a rotational displacement of about 0.75 mdyn $Å^{-1}$ if for rough purposes the rotational vibrations are considered as uncoupled. Such a high force constant implies a strongly directed intermolecular potential, which is characteristic of hydrogen bonding.

Vibrational spectroscopy also gives information about order and disorder. The general theme here is that the excitations of an ordered crystal can be expanded in plane waves, and the only plane waves that can interact with radiation as fundamentals are those whose wave length is the same as the wave length of the radiation. There are perhaps 10^{25} vibrations or more per mole of crystal but only a handful of these can interact with light as fundamentals. In a disordered crystal, no such selection rule holds, the excitations cannot in general be expanded in plane waves, and neither can the dipole moment density. Hence all 10^{25} (or so) vibrations are infrared and Raman active. In general, then, ordered crystals have sharp absorption bands, whereas disordered crystals have broad bands. Ice II for example is orientationally ordered and has a far infrared spectrum /18/ at 77 °K and 1 atmosphere pressure containing a few sharp bands, as shown in fig.5; they are caused by the few long-wave-length vibrations in which water molecules move in translation. Ice I, by contrast, is orientationally disordered, and as the spectra /19/ in fig. 6 show, there is a broad continuous absorption with some features, but not the sharp bands characteristic of ice II. Other infrared bands that are characteristic of ordered crystals are the sharp features to the low-frequency side of the rotational vibrational band of ice II shown in fig. 7. These features are appreciably broadened when a small amount of isotopic impurity (H_2O in D_2O or D_2O in H_2O) is added. The broadening by a small amount of isotopic disorder can only mean that the pure crystal is well ordered. The disordered phases of ice have no such sharp features.

Dielectric measurements are also important. They give a correlation number g /20/ invented by Kirkwood which is defined by fixing attention on a particular molecule, and finding the angle γ_i between the chosen molecule and all other molecules i (including the chosen molecule). g is then given by the sum of the cosines of γ_i averaged over all allowed configurations

$$g = \langle \sum_i \cos \gamma_i \rangle .$$

Its value tells us something about the kind of disorder that might be present.

Besides knowing the gross crystal structure and whether it is ordered or disordered, like Oliver Twist, we always want more. There are many other fine details that we would like to know about the structure; for example, how much is the water molecule distorted in the crystal? If the crystal is disordered orientationally then the molecular positions are also disordered, and we would like to describe the disorder. The only information on this point has been

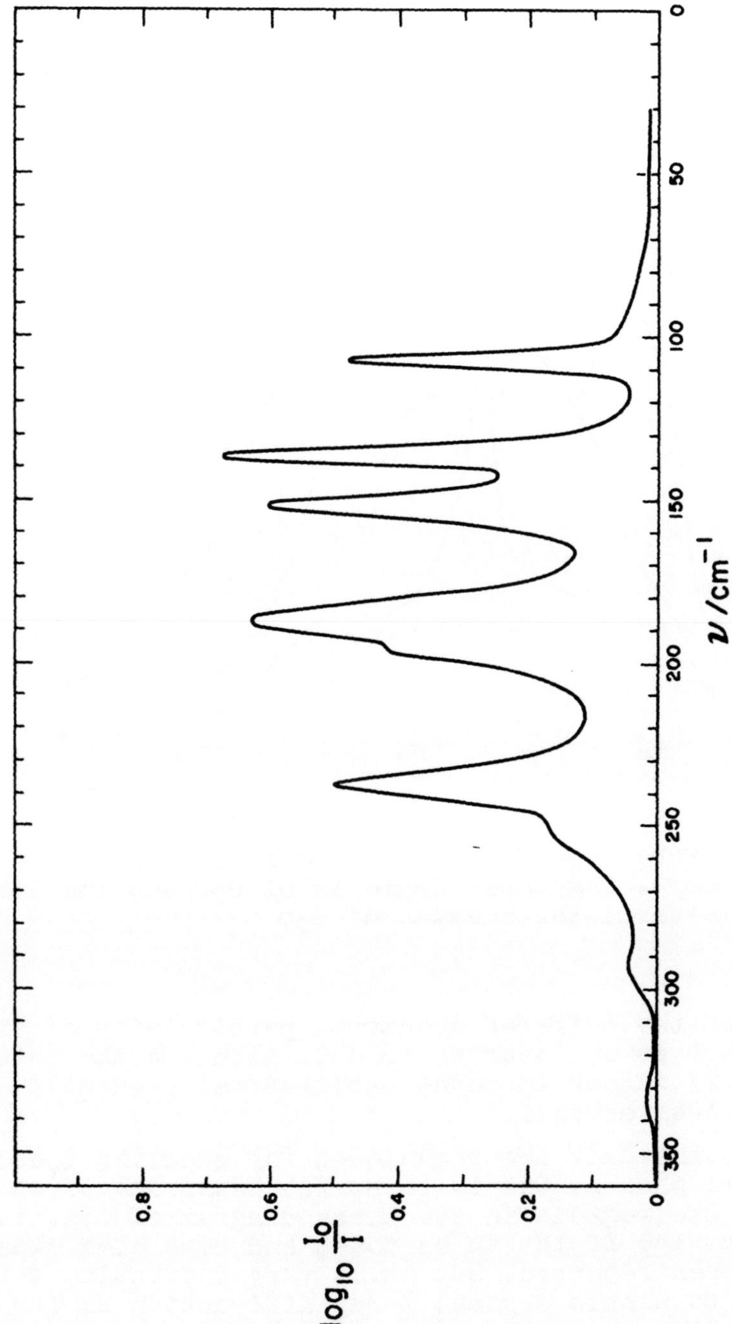

Fig.5 Far infrared spectrum of ice II in the range 350 to 50 cm^{-1}

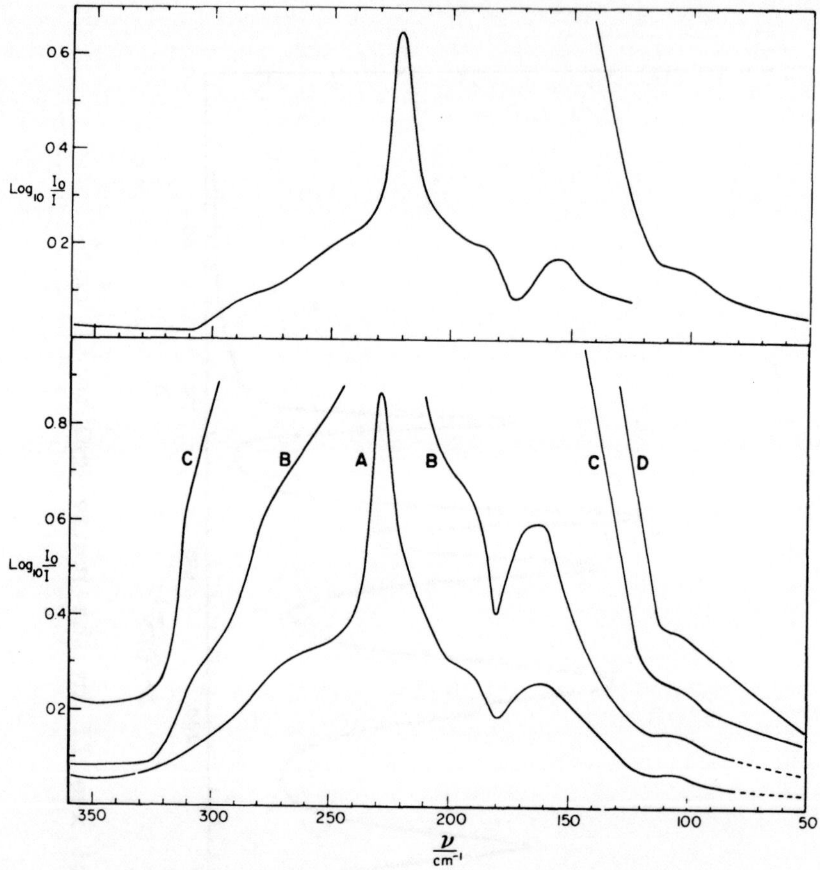

Fig.6 Far infrared spectrum of ice Ih in the range 350 to 50 cm^{-1}. The upper frame is of D_2O and the lower is of several thicknesses of H_2O

obtained from the infrared spectrum, particularly of small amounts of HDO as an impurity in H_2O. Although the interpretation /21/ is not in doubt qualitatively, quantitatively it's less certain.

There are essentially two techniques for handling these high-pressure phases. One is to do the measurements under pressure in the regions in the phase diagram of fig. 1. This is very nice if it can be done, and some nice experiments have been reported. But it is very difficult, for example, to do single crystal X-ray diffraction or far-infrared spectroscopy at the 20 or 30 kbar required to

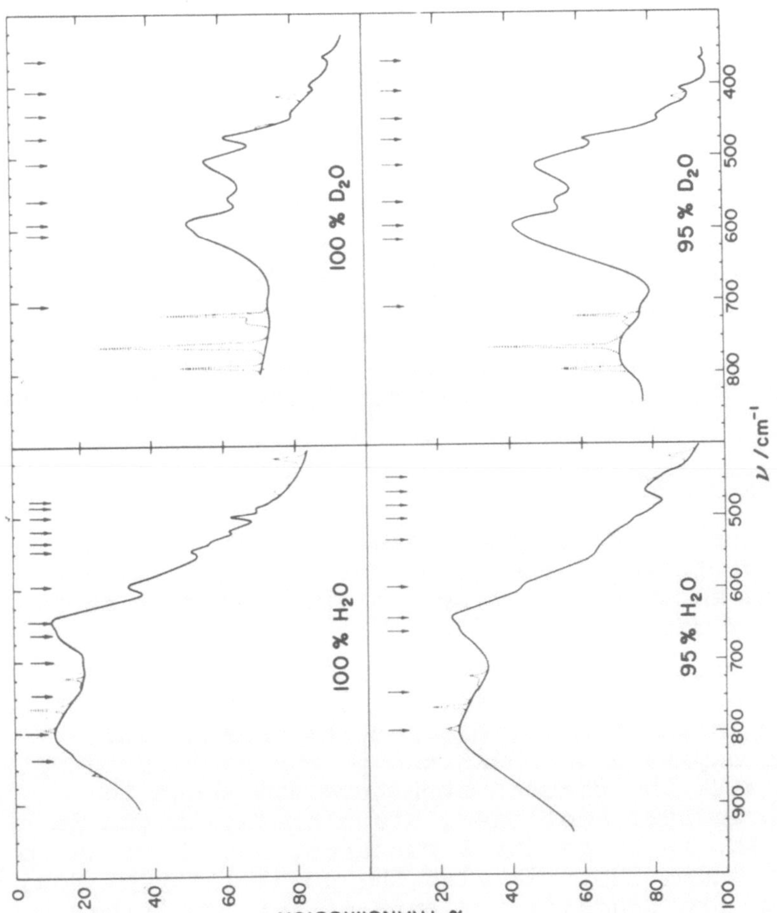

Fig.7 Absorption band due to the rotational vibrations of ice II. The decrease in intensity of the sharp features when 5% of isotopic impurity is added is consistent with ordered orientations

investigate ice VII. An alternative technique is to recover a phase at low temperatures in a metastable state. We are all familiar with many examples of metastable things; we ourselves of course in an oxygen atmosphere are highly metastable. Diamond is a metastable substance that is handled quite readily at room temperature. The phases of ice cannot be handled in this way at room temperature, but they can be handled at the temperature of liquid nitrogen, as first shown by Tammann /22/ 60 years ago, when he recovered ice II and IX from a high-pressure vessel. MacFarlan /23/ in the middle 1930s applied this technique to X-ray diffraction, and in fact managed to recover and examine single crystals of ice VI. Unfortunately he was not able to determine the structure. In the recovery, the molecular motions are of course frozen enough that the transformation is frozen. If the transformation is frozen, then many other kinds of molecular motion like diffusion, reorientation, etc. are also frozen, and these motions cannot generally be examined. Then the measurements must be done at higher temperatures under pressure. However, if the required measurements can be done at low pressures and temperatures, they can often be done easier and better.

4. Crystal structures

It is quite impossible to summarize all that is known about the structures of ice in a short account. I have therefore chosen to pick out some of what seem to me to be interesting aspects.

The crystal structure of hexagonal ice, the oxygen atom positions of which are shown in fig. 8, has been described by Gränicher in the preceding paper, and needs little comment. The oxygen atom positions of the corresponding cubic phase with the diamond structure are shown in fig. 9. As Gränicher mentioned, stacking faults can in principle exist in these two structures, but there is no evidence for them. There is also the possibility of real crystals that are stacking variants of this cubic-hexagonal system. In hexagonal ice, the layers are repeating in a sequence abab..., and in the cubic phase abcabc..., where a, b, and c represent the hexagonal layers in slightly different aspects. Many other stacking sequences are possible and many have been observed /24/ in silicon carbides and other materials. It is a little surprising to me that they have never been observed in the ice system. Cubic ice has been made by heating at least six different solids, namely several of the high-pressure phases /25, 26/ and the vitreous phase /1/, without

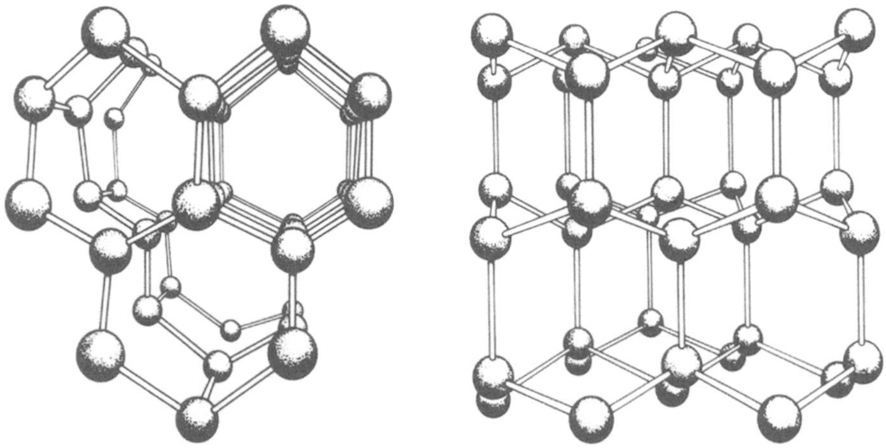

Fig.8 Oxygen atom positions in two views of ice Ih

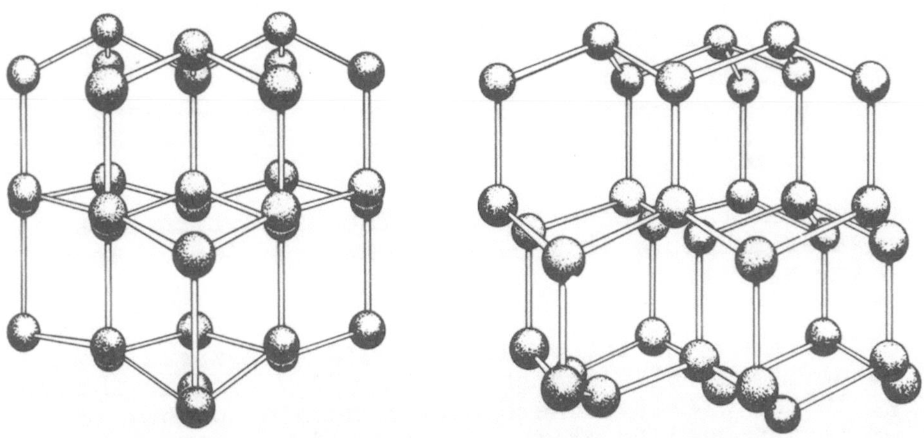

Fig.9 Oxygen atom positions in two views of ice Ic

evidence of stacking variants being obtained. Ammonium fluoride, on the other hand, which has a very similar structure to ice Ih, has two high-pressure phases that can be recovered at low temperatures. When these phases are heated they transform /27/, not to the cubic structure, but to two stacking variants, different ones from the two different structures.

Ice Ih, of course, is well-known to be disordered; ice Ic is also disordered although the evidence is less extensive than for ice Ih. The main diffraction evident so far is the electron diffraction measurements of Honjo and Shimaoka /28/ which they claimed were consistent with disorder but perhaps could not rule out some ordered structures. The infrared spectrum is very strong proof of disorder. The infrared spectra of ice Ih and Ic do not differ within experimental error in the range 4000 to 30 cm^{-1} /21, 29/. Both spectra are typically rather broad and amorphous, which undoubtedly proves that the two phases are disordered.

The question of how much the water molecule is distorted in the crystal is an interesting one. Peterson and Levy /14/ reported some 10 or 12 years ago based on single-crystal neutron diffraction that the oxygen-hydrogen bond was lengthened about 0.05 Å from the vapor value. In terms of a simple model, this is very high as the following considerations show. We consider the O_1-H_1 bond represented in fig. 10 as an elastic spring.

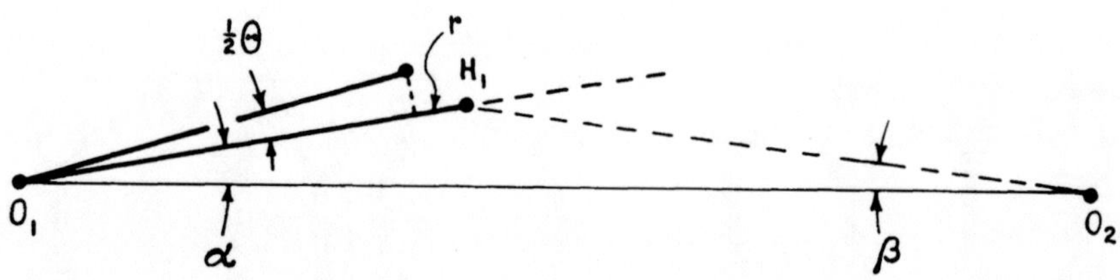

Fig.10 Schematic representation of the distortion of a water molecule in ice. Only one O-H bond is shown for clarity

The configuration of the water molecule is assumed symmetric about its axis, although only one O-H bond is shown in fig. 10 for clarity. The hydrogen bonding of oxygen atom O_2 is supposed to exert a central force f on H_1 which lengthens the O-H distance by r and changes the H-O-H angle by θ. The model is no doubt crude, but it serves for our purpose. The angle of departure of the O_1-H_1 line from the O_1-O_2 line is α. The elastic energy stored in the O_1-H_1 bond is, assuming a harmonic potential $1/2 \times k_r r^2$ where k_r is the harmonic force constant for the O-H bond. If the force constant is taken as that of the vapor molecule, 8 mdyn $Å^{-1}$, the stored energy per mole of water corresponding to a lengthening of the O-H bond of 0.05 Å is 2.8 kcal, which is about $\frac{1}{4}$the heat of sublimation. It is surprising that such a large elastic energy should be stored.

Experimentally it is not known whether the H-O-H angle in ice differs from the vapor value of 104.5°. If it does, then the orientational disorder causes each hydrogen atom to be in one of three positions around the O-O line and 120° apart. This disorder would add to the amplitude of the proton, and would be difficult to disentangle from it.

A partial answer can perhaps be obtained from theory. By resolving forces on H_1 (fig. 10) parallel to the O-H line we have

$$k_r r = f \cos (\alpha + \beta) ,$$

where k_r is the force constant for the O-H bond, and by resolving perpendicular to the O-H line we have

$$\frac{1}{2} a k_\theta \theta = f \sin (\alpha + \beta) ,$$

where a is the O-H distance and k_θ the force constant for the H-O-H angle. Hence by division

$$\theta = 2 \frac{k_r}{k_\theta} \frac{r}{a} \tan (\alpha + \beta) .$$

If r is taken as 0.05 Å and k_r and k_θ the values for the solid, 6 and 0.75 mdyn $Å^{-1}$, then approximately

$$\frac{1}{2} \theta = 0.6 \alpha .$$

This result suggests that if the hydrogen-bonding forces are enough to lengthen the O-H bond by 0.05 Å, they are enough to distort the H-O-H angle if O_2 is off the original O_1-H_1 line.

Ice Ih is disordered and so if it is cooled it should become metastable relative to an orientationally ordered phase based on the same molecular positions. This has never yet been observed and so it must occur quite slowly, and the difficulty is then to detect the transition. One way of doing this is to measure the limiting high-frequency permittivity, which is expected to change at an order-disorder transformation because a large part of it is caused /19/ by the infrared-active translational lattice vibrations. These would change greatly at an order-disorder transformation because in the ordered phase only a few sharp bands are active whereas in the disordered phase the whole of the band is active. It is possible now with commercial bridges to measure the permittivity to 1 or 2 parts in a million. Consequently, very small changes in the amount of order or disorder in a crystal might be detectable. An attempt is being made /30/ to find an ordering transformation in hexagonal ice by this method.

At 123.3 ^{o}K it should have been possible to detect the initial stages of an ordering transformation if it took of the magnitude of 50 years for completion, but there was no evidence for it. At this temperature the relaxation time of the orientational polarization, extrapolated from measurements above -50 ^{o}C /31/, is about 6 years.

There are however phases of ice with ordered orientations. The first of these to be recognized was ice II, almost simultaneously by infrared /11/ and X-ray diffraction /16/ evidence. I have however chosen to describe ice IX, which is obtained /7/ by cooling ice III, as an example of an ordered phase. X-ray diffraction of single crystals of the phase obtained by cooling ice III to liquid-nitrogen temperature was reported by Kamb and Datta /32/ in what is probably the first paper in the recent revival of interest in these phases, and deduced the oxygen positions shown in fig. 11. The structure consists essentially of four-fold spirals of the O_I oxygen atoms. The spirals are joined by the O_{II} atoms in such a way that each O_{II} is bonded to four spirals. The infrared spectrum /11/ showed that ice III when cooled to low temperatures was ordered, while the dielectric properties /33/ and the entropy of transition from ice I /16,34/ showed that it was orientationally disordered at high temperatures. The ordering transformation was detected by dielectric methods as indicated in fig. 12, in which the limiting low-frequency permittivity ε_o and the correlation parameter g derived from it are plotted against temperature /7/. There is clearly a transformation to an ordered phase with g = o in the region -70 to -108 ^{o}C. The fact that g becomes zero and there is no sign of ferroelectricity leads us to conclude that the low-temperature phase, ice IX, is antiferroelectric.

If the hydrogen atoms fit into the same unit cell as the oxygen atoms, and the infrared spectrum /11/ is consistent with this interpretation, then it is possible to guess where the hydrogen atoms might be in ice IX. We expect that the water molecule will prefer to be in O-O-O bonds nearest to 104 $\frac{1}{2}^{o}$, which is the H-O-H angle in the vapor. The nearest one (see fig. 11) is 103°, and if two are put there then the others fall by symmetry into the places shown in fig. 11.

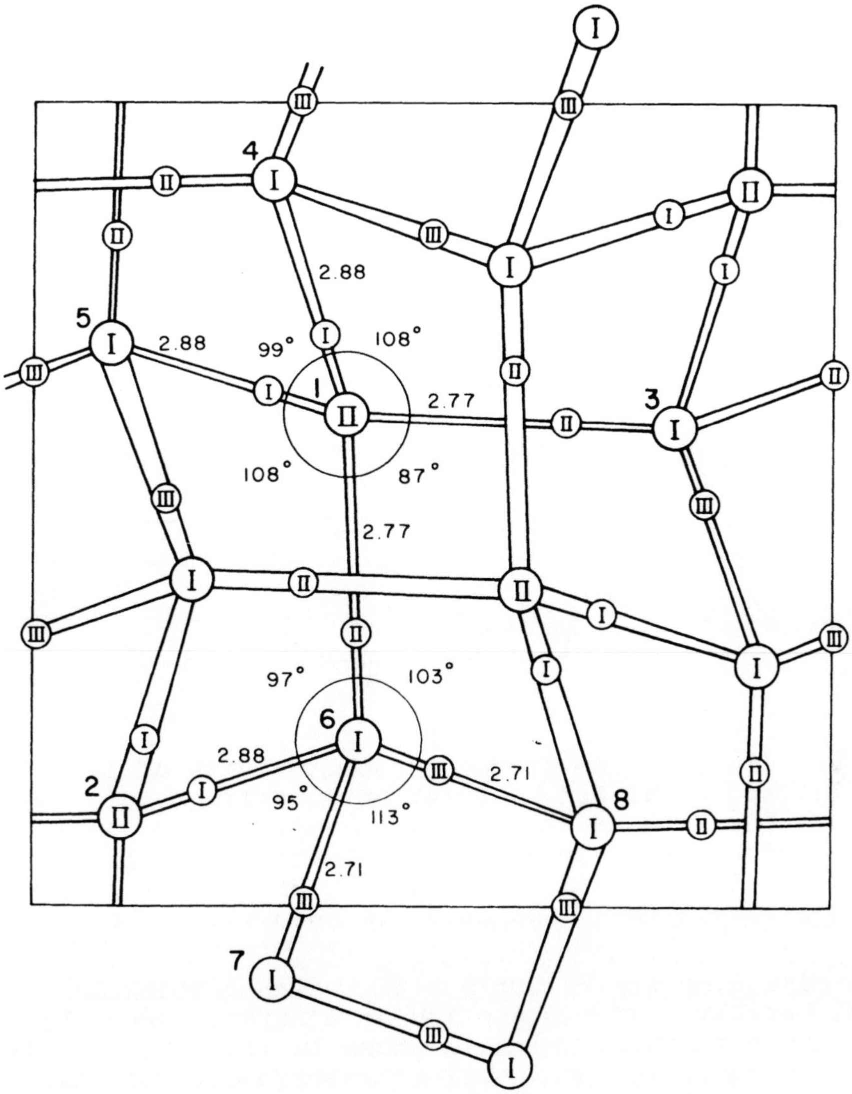

Fig.11 View of the structure of ice IX along the c-axis. The angles not shown are 315 and 416, 129°; 167, 144°; 268, 92°

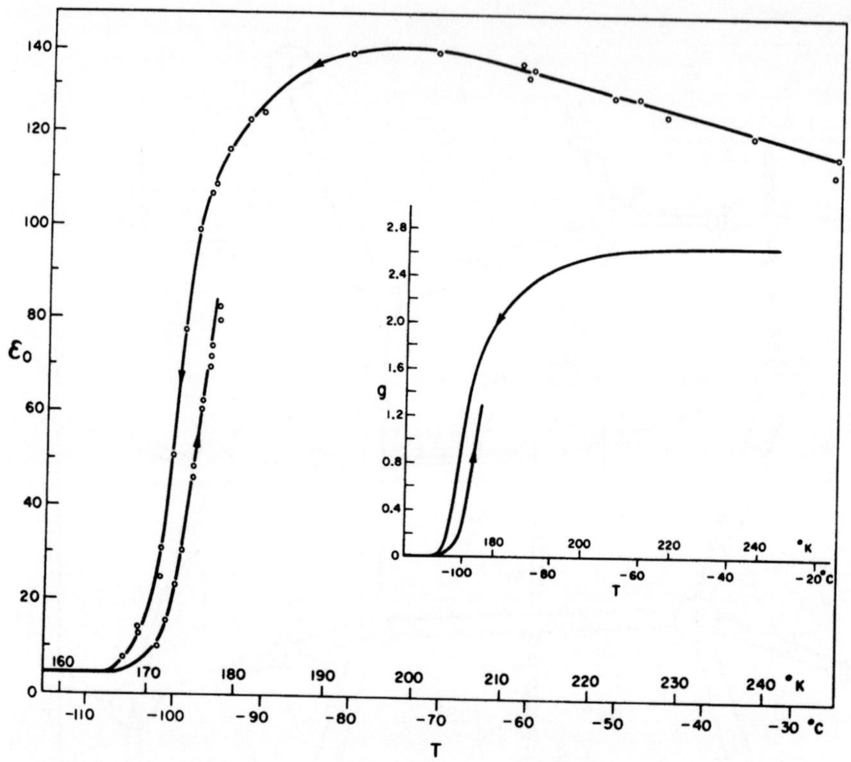

Fig.12 The limiting low-frequency permittivity of ice III and IX at 2.3 kbar for both cooling and heating. The orientational correlation parameter g calculated from smoothed permittivities is shown in the insert figure. The arrowheads show the direction of change of temperature

I shall pass over ice V, but ice VI, which I think is the most beautiful of all the ice structures, cannot be missed. The structure, which is shown in fig. 13, is also due to Kamb /35/. The unit cell is tetragonal, and fig. 13 is a view down the four-fold axis of a model. The corners of the tetragonal cell are shown by the white dots on the balls representing the oxygen atoms. The structure is based on a rather simple unit which consists of an oxygen atom I, hydrogen bonded to four oxygen atoms II, two of which II' are above I and the two others II" are below I in a tetrahedral arrangement that is obtained from a regular tetrahedron by elongation parallel to the vertical (in fig. 11) two-fold (four-fold inversion) axis. The oxygen atoms I are at the corners of the tetragonal cell, and the unit I' with two II'" atoms above and

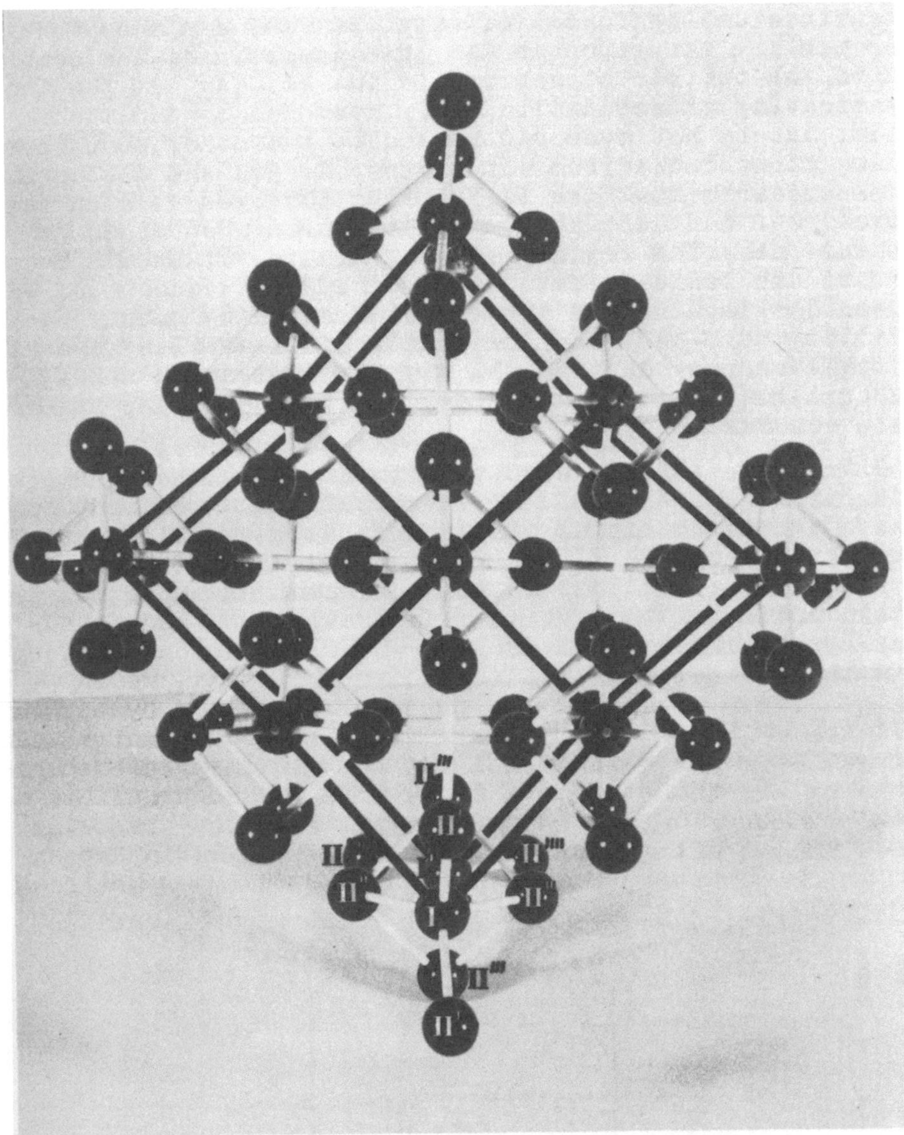

Fig.13 Views of the structure of ice VI along the c-axis. The hydrogen atoms are not shown

two II"" atoms below is on the adjacent corner in the direction <001>. These tetrahedral units are bonded in chains parallel to the <001> direction, by hydrogen bonds between, for example, II" and II'". Each chain is hydrogen bonded in the <100> and <010> directions to neighboring chains.

The structure so formed is very open and can accommodate another similar structure in its interstices. The combination of the two structures is ice VI. Each of the interpenetrating structures is fully hydrogen bonded within itself, but is not hydrogen bonded to the other structure. There is no doubt from both infrared /36/ and dielectric /33/ measurements that ice VI is disordered at high temperatures, and also at low temperatures when it is cooled at the rate of a few degrees per minute. It should become ordered at low temperatures if cooled slowly enough. We have been looking for the transformation by measuring the permittivity, and find /8/ that there is a slow reversible transformation near 123 °K. The transformation takes several months for completion, and the new phase has not yet been examined.

Ice VII and VIII are also very interesting. Weir, Block, and Piermarini /37/ did a very beautiful experiment some years ago in obtaining single crystal X-ray diffraction photographs of ice VII at 25 kbar and 25 °C in a diamond cell. They showed that it was body-centred cubic, a structure which in fact Kamb and Davis /38/ had found earlier by cooling ice VII to -50 °C at 25 kbar to form a phase that we now call VIII. Earlier still, we in Ottawa had shown /26/ that the phase obtained by quenching ice VII to 100 °K is only approximately body-centred cubic. The oxygen atom positions in ice VII and the presumed hydrogen bonding are shown in fig. 14. It can be described as two interpenetrating ice Ic structures, that are fully hydrogen bonded within themselves but are not hydrogen bonded to one another, except perhaps at imperfections.

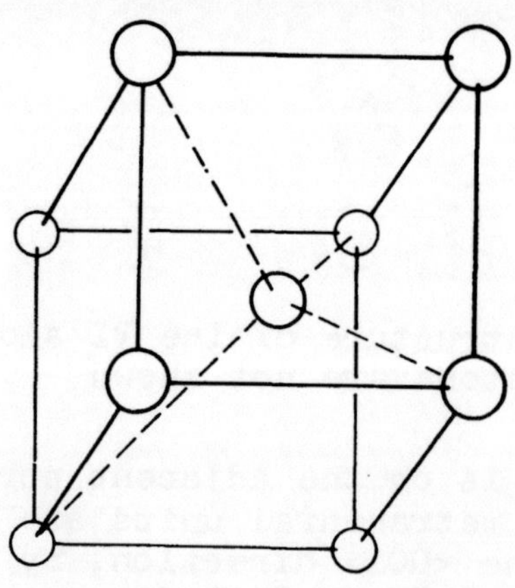

Fig.14 Structure of ice VII. The hydrogen atom positions are not shown

One of the phases of ammonium chloride has a structure /39/ related to that of ice VII, the cube corners being occupied by chloride ions and the body centres by ammonium ions (or of course the reverse).

When ice VII is cooled below 0 °C there is a sharp change in its dielectric properties /6/. Both the real and the imaginary parts of the complex capacitance are shown as a function of temperature in fig. 15. Clearly there is a transformation to a phase that has no orientation polarization, and which has been designated ice VIII. Ice VIII is orientationally ordered, rather likely in an antiferroelectric arrangement, though the details of the hydrogen atom positions are not yet known.

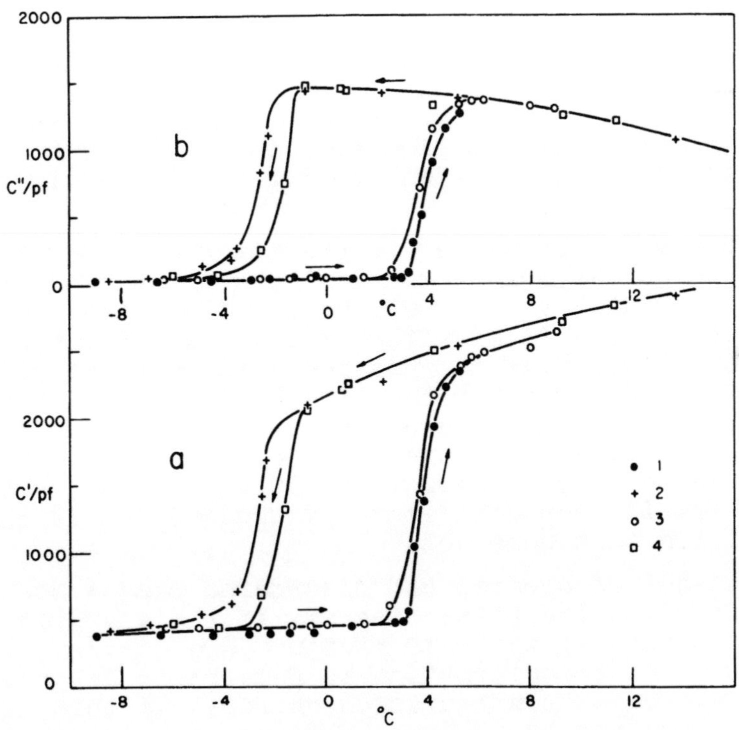

Fig.15 Real and imaginary parts of the complex capacitance through the transformation ice VII-VIII, at 30.5 kbar and several frequencies indicated in kHz by the numbers attached to the symbols

5. Some general comments about the phases

The structure of these phases of ice are rarely simple, and it is probably almost impossible to predict new phases or to predict structures. Who would have had the imagination and the courage to predict that the ice VI structure could exist? However, Kamb and Davis /38/ have predicted in a nice tour de force that in fact a phase should not exist. This is the cubic phase close-packed or face-centred cubic phase. One might expect on naive grounds that if ice VII were squeezed hard enough, it should go to a cubic close-packed phase. Kamb and Davis have predicted that this will not occur, and their prediction has been strikingly vindicated very recently by Holzapfel and Drickamer /40/ who reported X-ray diffraction measurements up to what they claim was 230 kbar, and found no change in the crystal structure.

The relation between crystal structures and intermolecular forces can of course be discussed from the reverse direction, and known structures used to learn about the forces. And again Kamb /41/ has pioneered in this by comparing ice Ic and the two interpenetrating ice Ic lattices of ice VII.

It might perhaps be easier to predict the properties of an order-disorder transformation than those of a major change of crystal structures. Here again, Kamb /16/ has predicted the ice II structure should have a disordering energy of 3 kcal $mole^{-1}$. This seems to me to be high, because it is much greater than the heat of melting of ice II as obtained from the heats of transition /3/ II-III and III-liquid, and because all the other phases have a disordering energy of less than 200 cal $mole^{-1}$; but nevertheless ice II does not become disordered and is the only phase of ice that does not.

The existence of ordered and disordered phases has a profound effect on the phase diagram. If the vibrational entropies of phases at the same pressure are similar, then the entropy of transformation is determined largely by the difference of configuration entropy. If this difference is zero and the volume of transformation ΔV is not zero then from the Clausius-Clapeyron equation for the pressure p and temperature T on an equilibrium line,

$$(\partial p/\partial T)_{equil.} = \Delta S/\Delta V ,$$

we conclude that the equilibrium line is nearly vertical on the scale of fig. 1. If the configurational entropy differs because one phase is ordered and one orientationally disordered, then the equilibrium line slopes appreciably. This explains the unusual appearance of the field of ice II in fig. 1, and why the fields of the ordered phases II, VIII, and IX spread out at low temperatures and dominate the phase diagram.

An interesting speculation is the following. Bridgman pointed out 60 years ago that the I-II line extrapolates to about 0 °K, 0 bar, and suggested that at 0 °K ice II and ice Ih have about the same stability. Suppose now that ice I and II had the same configurational entropy, but the energetics remained essentially as they are so that the I to II line would be nearly parallel to the temperature axis and at a low (or even at a negative) pressure. This is of course not a scientific supposition because if you say the world is not as it is you have to change many things in a consistent manner; nevertheless, it is perhaps worth a brief consideration. At the bottom of the Arctic Ocean, ice II would be more stable than ice I and so the Ocean would be frozen from the bottom up. Great undersea glaciers of ice II would flow from the arctic basin, and so all the oceans would be frozen from the bottom for the greater part of their depth. Salt is almost insoluble in ice, and so the sea would be very salty. Since ice II is about 20% denser than ice I, the average depth of the oceans would be about 20% less than they are, and the shorelines would be down off the continental shelves. The effect of all this on our world would be very profound. It does not happen only because of the 0.8 cal deg^{-1} $mole^{-1}$ orientational entropy of ice I. Now you know the story of the horseshoe nail!

N.R.C. No. 10540

References

/1/ König, H., Z.Krist. 105, 279 (1943)

/2/ Tammann, G., Ann.Phys. 2, 1 (1900)

/3/ Bridgman, P.W., Proc.Am.Acad.Arts Sci. 47, 441 (1912)

/4/ Bridgman, P.W., J.Chem.Phys. 3, 597 (1935)

/5/ Bridgman, P.W., J.Chem.Phys. 5, 964 (1937)

/6/ Whalley, E., D.W. Davidson, J.B.R. Heath, J.Chem. Phys. 45, 3976 (1966)

/7/ Whalley, E., J.B.R. Heath, D.W. Davidson, J.Chem. Phys. 48, 2362 (1968)

/8/ Whalley, E., J.B.R. Heath, unpublished work

/9/ Burton, E.F., W.F. Oliver, Proc.Roy.Soc. A 153, 166 (1935)

/10/ For a review of their structures see G.A. Jeffrey and R.K.McMullan, Prog.Inorg.Chem. 8, 43 (1967)

/11/ Bertie, J.E., E. Whalley, J.Chem.Phys. 40, 1646 (1964)

/12/ Herzberg, G., Infrared and Raman Spectra. D.van Nostrand Co.Inc. 1945, p.171

/13/ Owston, P.G., Quart. Rev. 5, 344 (1951)

/14/ Peterson, S.W., H.A. Levy, Acta Cryst. 10, 70, 344 (1957)

/15/ Pauling, L., J.Am.Chem.Soc. 57, 2680 (1935)

/16/ Kamb, B., Acta Cryst. 17, 1437 (1964)

/17/ Whalley, E., D.W. Davidson, J.Chem.Phys. 43, 2148 (1965)

/18/ Bertie, J.E., H.J. Labbé, E. Whalley, J.Chem.Phys. 49, 775 (1968)

/19/ Bertie, J.E., E. Whalley, J.Chem.Phys. 46, 1271 (1967)

/20/ Kirkwood, J.G., J.Chem.Phys. 7, 911 (1939)

/21/ Bertie, J.E., E. Whalley, J.Chem.Phys. 40, 1637 (1964)

/22/ Tammann, G., Z.Anorg.Allgem.Chem. 63, 285 (1909)

/23/ McFarlan, R.L., Rev.Sci.Instr. 7, 82 (1936) J.Chem.Phys. 4, 60, 253 (1936)

/24/ Verma, A.R., P. Krishna, Polymorphism and Polytypism in Crystals. John Wiley and Sons. Inc. New York, 1966

/25/ Bertie, J.E., L.D. Calvert, E. Whalley, J.Chem.Phys. 38, 840 (1963)

/26/ Bertie, J.E., L.D. Calvert, E. Whalley, Can.J.Chem. 42, 1373 (1964)

/27/ Nabar, N.A., L.D. Calvert, E. Whalley, unpublished work

/28/ Honjo, G., K. Shimaoka, Acta Cryst. 10, 710 (1957) K. Shimaoka, J.Phys.Soc. Japan 15, 106 (1960)

/29/ Bertie, J.E., H.J. Labbé, E. Whalley, J.Chem.Phys. submitted for publication

/30/ Whalley, E., J.B.R. Heath, unpublished work

/31/ Auty, R.P., R.H. Cole, J.Chem.Phys. 20, 1309 (1952)

/32/ Kamb, W.B., S.K. Datta, Nature 187, 140 (1960)

/33/ Wilson, G.J., R.K. Chan, D.W. Davidson, E.Whalley, J.Chem.Phys. 43, 2384 (1965)

/34/ Whalley, E., D.W. Davidson, J.Chem.Phys. 43, 2148 (1965)

/35/ Kamb, B., Science 150, 205 (1965)

/36/ Bertie, J.E., H.J. Labbé, E. Whalley, J.Chem.Phys. 49, 2141 (1968)

/37/ Weir, C., S. Block, G. Piermarini, J.Res.Natl.Bur. St. C, 69, 275 (1965)

/38/ Kamb, B., B.L. Davis, Proc.Nat.Acad.Sci. 52, 1433 (1964)

/39/ Goldschmidt, G.H., D.G. Hurst, Phys.Rev. 83, 88 (1951)

/40/ Holzapfel, W., H.G. Drickamer, J.Chem.Phys. 48, 4798 (1968)

/41/ Kamb, B, J.Chem.Phys. 43, 3917 (1965)

DEUTERON ARRANGEMENTS IN HIGH-PRESSURE FORMS OF ICE

W. C. Hamilton, B. Kamb, S. J. LaPlaca, A. Prakash

Brookhaven National Laboratory, Upton, New York 11973

and

California Institute of Technology, Pasadena, Calif. 91109

Abstract

Single-crystal neutron diffraction on quenched D_2O samples at 110°K gives the following results. The dense ice phases are tetrahedrally hydrogen bonded, but with considerable distortion from the ideal tetrahedral coordination seen in ice I. Both ice II and ice IX (low-temperature form of ice III) are fully proton ordered. Direct evidence is found in ice II for a twinned structure that is closely related to the presence of proton order; no corresponding effect occurs in ice IX. The choice of water molecule orientations in ice II and ice IX is such as to match the molecular D—O—D angle as nearly as possible with the available O···O···O coordination angles. The deuterons tend to lie near the O···O centerlines, but other factors (such as acceptor relationships) also influence the molecular orientation in detail. Measured D—O—D angles do not differ significantly from the free-molecule value, 104.5°. Individual O—D distances are not accurate enough to reveal the small effects of different local H-bonding environments, but the mean O—D distance for each of the phases is significantly (though only slightly) increased over the free-molecule distance of 0.957 Å. The mean O—D distances correlate well with spectroscopic data. In terms of this correlation, the O—D distance determined for ice I in earlier work appears anomalously long. Quenched ice V has a partially proton-disordered structure. Some deuterons appear almost "full", some approximately "half", and some almost missing. A partially disordered structure of this general type should also occur in ice III over the temperature range of its transition to ice IX.

INTRODUCTION

The purpose of the work summarized below is to enlarge our understanding of the nature of hydrogen bonding between water molecules, by studying the effects of pressure and temperature in altering the system of bonding from one ice phase to another. The interrelationships of structure and properties among the various ice phases shed significant light on the particular phase ice I, which is the main subject of the present symposium. The dense forms of ice clearly show phenomena of proton order and disorder, which may serve as models against which to compare the much more subtle order-disorder effects that have been proposed for ice I. The work described below helps to provide a structural picture of the order-disorder phenomena in the ice phases.

A brief summary of structural information for the various ice phases, as determined by x-ray diffraction methods, is given by Kamb /1/. This summary may serve in conjunction with the preceding paper by Dr. Whalley as the point of departure for the present discussion.

To obtain reliable experimental information on the positions of the protons in ice structures, and to discuss in detail the extent and nature of long range order in the proton arrangement, it is necessary to study the structures by neutron diffraction. The present study uses for this purpose single-crystal neutron diffraction, which leads to the most certain and accurate results that are possible by diffraction methods. We note here that within the greater uncertainties of the powder diffraction method, our results appear to be in agreement with those reported by Rabideau and Finch in the following paper.

The results given below for ices II, V, and IX are part of a broader study in which we are investigating by single-crystal neutron diffraction all of the high-pressure forms of ice that are accessible to study by the quenching technique. None of the work has yet been published in its final, completed form, and the report below is to be considered a preliminary summary.

The work was supported by a grant from the U.S. National Science Foundation to the California Institute of Technology, and by the U.S. Atomic Energy Commission under contract to Brookhaven National Laboratory. Opportunity to present the results at this symposium was made possible by a travel grant from the U.S. National Science Foundation.

EXPERIMENTAL METHODS

The ice samples were prepared from D_2O, in a high-pressure bomb containing a straight bore fitted with two movable pistons.

After preparation of the desired form of ice, and after quenching by immersion in liquid nitrogen, the samples were expelled from the bomb and examined at atmospheric pressure. Sample preparation, selection of single crystals, and alignment by x-ray methods were carried out in Pasadena. The samples were then transported by air to Brookhaven, for the neutron diffraction measurements and subsequent crystallographic calculations.

The neutron diffraction measurements were made in the High Flux Beam Reactor, Brookhaven National Laboratory. Once the crystallographic orientation was established, measurement of diffracted intensities was carried out automatically, under computer control. During neutron irradiation, samples were maintained at a temperature of 110°K by a stream of cold nitrogen gas. Details of experimental procedures and of calculations for reducing and interpreting the diffraction measurements will be given in the full publication of the final results.

We first describe separately the experimental results for ice II, IX, and V, and then summarize some general conclusions from the work.

ICE II

Although the structure of ice II is related to that of ice I, the protons in the ice II structure are completely ordered, in contrast to the proton disorder in ice I. The ice II structure is shown in fig. 1. Puckered hexagonal rings of water molecules can be seen, similar to those in ice I. These rings stack one above another to form columns similar to those in ice I, but linked together differently.

X-ray study /2/ revealed indications that the structure is proton ordered. Certain features of the structure, as determined by x-rays, allowed a prediction of the actual arrangement of protons. Neutron diffraction confirms the existence of proton order and also the particular proton arrangement that was predicted It is shown in fig. 1, and the refined crystallographic parameters are listed in table 1. We discuss here the structural features that led to these successful conclusions, and their relations to the neutron diffraction results.

1. Space group and twinning. To a first approximation, the ice II structure can be described in the rhombohedral space group $R\bar{3}c$. However, the *c* glide-plane extinction condition is violated, showing that the symmetry is lower. An indirect argument /2/ leads to the conclusion that the space group is $R\bar{3}$. This then implies that the crystals must be twinned, to explain the fact that they show Laue symmetry $\bar{3}m$, instead of merely $\bar{3}$. Any departure

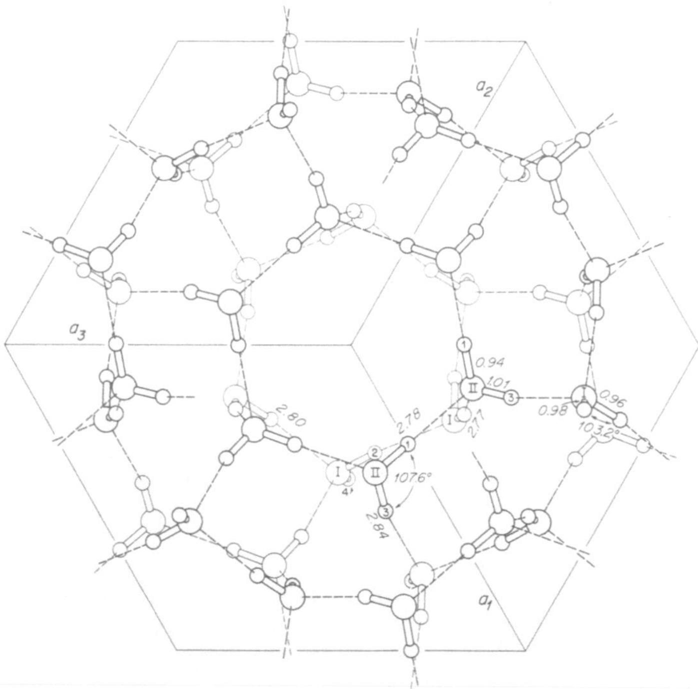

Fig. 1 Structure of ice II, viewed along the hexagonal c-axis. Hydrogen bonds between the water molecules are shown as dashed lines. Each puckered hexagonal ring consists of water molecules of one of the two non-equivalent types, as indicated. Rings of the two types, stacked alternately one above the other, are linked together into columns by H-bonds in the same way as in ice I. Adjacent columns are linked together in a more tightly fitting way than in ice I. Measured bond distances (in Å) and angles are indicated.

from volume equality of the twinned counterparts in a sample should result in violation of the mirror plane m in the apparent Laue symmetry, and the violation should be stronger in neutron than in x-ray diffraction. These ideas are confirmed by neutron diffraction as follows. Although the neutron diffraction intensities of pairs of reflections $hk\ell$, $kh\ell$ are approximately equal, as expected from the m of the apparent Laue symmetry, there are in fact small but systematic differences between them, which correlate with the differences expected from the structure in fig. 1. They can be accounted for by assuming that each reflection contains the combined intensities from a single crystal of volume fraction α and its twinned counterpart of volume fraction $(1-\alpha)$. For a complete set of neutron diffraction data from one crystal, we find $\alpha = 0.565 \pm 0.005$, the volume fraction thus deviating by a small but significant amount from the value 1/2 that would result strictly in an apparent Laue symmetry $\bar{3}m$. For a second crystal, which even with x-rays showed some violations of the *m* in the apparent Laue symmetry, neutron intensities of pairs $hk\ell$ and $kh\ell$ differed by as much as a factor of 5, thus demonstrating in an impressive way the absence of the *m*. For this crystal, observed intensities correspond to $\alpha = 0.792 \pm 0.015$. The variation of α from sample to sample accords with the existence of twinning on a scale comparable to the sample size (~1 mm).

2. *Symmetry degradation and proton order.* The approximate structure in space group $R\bar{3}c$ involves 12 molecules, all equivalent by symmetry, in the rhombohedral unit cell. In the actual $R\bar{3}$ structure there is a slight distortion of the $R\bar{3}c$ bond framework, and the 12 molecules separate into two non-equivalent groups of 6 each. If the hydrogen bonding were accomplished by a disordered proton arrangement, introduced into the undistorted bond framework, the environments of all oxygen atoms would on the average be the same, and there would be no reason for the water molecules to separate into two non-equivalent groups. Within the bond framework, ordering of the water molecule orientations can occur only at symmetry lower than $R\bar{3}c$. Thus the observed symmetry degradation implies proton ordering. Neutron diffraction confirms this argument in two ways. (1) The intensities of reflections $hk\ell$ with ℓ odd, in relation to general intensities $hk\ell$, are much stronger for neutron diffraction than for x-ray diffraction, indicating that the protons do not even approximately satisfy the *c*-glide relation, while the oxygen atoms do. (2) Detailed refinement of a disordered (half-deuteron) model of the ice II structure gives only poor agreement (residual 32%) with the neutron intensities, and implies unrealistically large thermal motions for some of the deuterons, whereas the ordered arrangement of fig. 1 refines to reasonable agreement (residual 17%), with reasonable thermal motions for all atoms.

3. *Nature of structural distortion and choice of molecular orientations.* The water molecule orientations in the ice II structure were originally predicted on the basis of two closely related assumptions: (1) The water molecules assume orientations that match the H—O—H angle of 104.5° of the free water molecule as closely as possible against available O···O···O angles presented by the tetrahedral environments in the $R\bar{3}c$ structure. (2) The distortion from the $R\bar{3}c$ to the $R\bar{3}$ structure occurs in such a way as to improve the match further. Out of 13 possible proton arrangements in space groups $R\bar{3}$ and $R3$, the above assumptions single out the arrangement in fig. 1, which the neutron data show to be the correct one. The other arrangements either refine with large shifts of some deuteron coordinates, leading back ultimately to the arrangement of fig. 1,

or else they do not refine satisfactorily. Correctness of the explanation given for the way in which the structure distorts from $\underline{R}\bar{3}c$ to $\underline{R}\bar{3}$ is confirmed by its success in predicting the proton arrangement.

A complete refinement of the ice II structure in fig. 1, using a full set of neutron diffraction data, gives an agreement residual of 8% for the intensities of 167 reflections strong enough to be definitely observed. The refined coordinates can be used to discuss details of the molecular configuration, which will be summarized in the concluding discussion.

ICE III AND ICE IX

Ice III when cooled through the temperature range from -65° to -108°C shows a gradual change in dielectric properties, indicative of a progressive development of proton order /3/. The high-temperature phase is non-quenchable, and the low-temperature phase, which Dr. Whalley has designated ice IX, is the one obtained for study by quenching in our experiments. The ice IX structure has the unique property, among known ice phases, that it admits both ordered and disordered proton arrangements without any change in the size or symmetry of the tetragonal unit cell (space group $\underline{P}4_12_12$). Single-crystal x-ray diffraction data at 110°K /4/ revealed the protons, but equally good agreement (residual 11% on $\underline{F}^2$) could be achieved for a disordered (half-hydrogen) arrangement and for one particular ordered arrangement of the protons, of the four ordered arrangements that fit the oxygen positions.

Neutron diffraction shows unambigously that the deuterons are fully ordered, in the particular ordered arrangement compatible with the x-ray results. It is shown in fig. 2, and is the arrangement predicted by Whalley, Heath, and Davidson /3/.

The diffraction data for ice IX are the best that we have obtained for any phase of ice, and the results are the most reliable in detail. The structure can be refined to an agreement residual of 6.7% (on $\underline{F}^2$). A few of the details of molecular configuration are shown in fig. 2. The agreement between oxygen coordinates as determined by x-ray and by neutron diffraction (table 2) indicates a reliability of about ±0.0005 for the coordinate values determined by either method.

A change from proton disorder to order within the bond framework of ice IX can account for the transition from ice III to ice IX, observed dielectrically. Direct evidence that the oxygen arrangement is not altered in the transition should be obtainable by x-ray diffraction under pressure, but unfortunately in our study of this

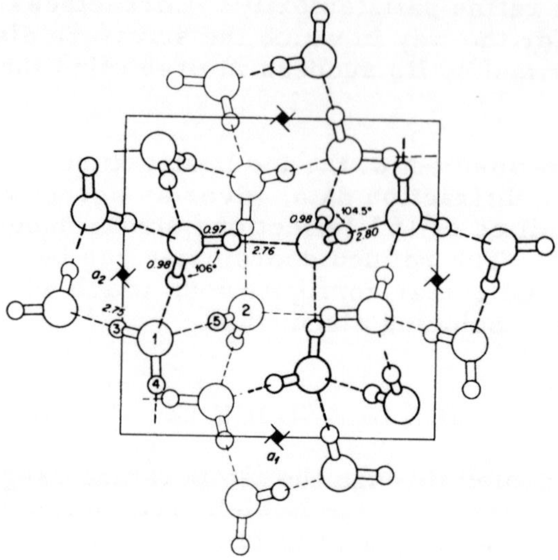

Fig. 2 Structure of ice IX, viewed along the tetragonal c axis, with same conventions as in fig. 1. Water molecules of type 1 form helical chains around the 4_1 axes parallel to the c axis. These helices are linked together laterally by molecules of type 2.

type /5/, we did not try to observe the phase ice III. A major reorganization of the oxygen arrangement and bond network would require a first order transition, like the transition from ice II to III, and it seems certain that the continuous, progressive change in dielectric constant observed over a range of about 40°C represents progressive ordering of the protons without change in the basic bond framework or oxygen positions.

The ice III structure (as so inferred) has a feature that should be strongly conducive to proton order: the presence of an abnormally large O···O···O bond angle of 144°, at the water molecule of type 1. This angle is avoided in the ordered structure of ice IX. There is no direct information as to whether it is in fact occupied by water molecule orientations in ice III. Avoidance of this angle does not require the complete proton order that is achieved in ice IX, and other factors must therefore contribute to the ordering process in the transition from III to IX.

The O···O···O angles actually occupied in the ice IX structure (99°, 101°) are the best that are available for use in an arrangement with the cell size and symmetry indicated by x-ray data. A perhaps slightly better angle (106.5°) could be used in structures of lower symmetry, but no violations of the extinction conditions for space group $\underline{P}4_12_12$ were found, nor any indications of a supercell.

ICE V

Electrical properties indicate that ice V is proton disordered within its stability field /6/, and the infrared absorption spectrum suggests that the structure remains proton disordered at 77°K /7/. Complete proton ordering cannot occur within the monoclinic structure (space group $\underline{A}2/\underline{a}$) found by x-ray diffraction at 110°K /8/, because the water molecules must form asymmetric hydrogen bonds across 2-fold axes and across symmetry centers. Complete proton ordering could occur only with an increase in cell size or a lowering of symmetry. Neutron diffraction is more sensitive than x-rays to any resulting effects of proton order. In making the neutron diffraction measurements, we searched for violations of the extinction conditions for space group $\underline{A}2/\underline{a}$, but the results were negative.

Nevertheless, the neutron data indicate significant proton ordering in ice V at 110°K, but of a type not observed previously in other ice phases. Fourier synthesis gives peaks of highly variable height at the expected deuteron positions. This indicates a need for variable population factors at the different sites, expressing a variable probability of occupancy of each site by a deuteron. Refinement of the structure with adjustable population factors for all deuteron sites yields the results shown in fig. 3. Population factors from 0.07 to 0.94 are found at the various sites. The Bernal-Fowler rules are obeyed to an accuracy of generally better than ±0.05 in the sums of the population factors, hence the variations in population factors must be significant at least to this level of accuracy, and do not result simply from random experimental errors.

To a first approximation, the deuteron sites separate into those with population factor ~0, those with ~1/2, and those with ~1. The population factors of 1/2 might be only apparent values, if the structure were in reality fully ordered in space group $\underline{A}\,\underline{a}$, which cannot be distinguished from $\underline{A}2/\underline{a}$ by extinction conditions. There are three ordered arrangements in $\underline{A}\,\underline{a}$ that might correspond to the approximately observed population factors of 0, 1/2, and 1, when analyzed on the assumption that the space group is $\underline{A}2/\underline{a}$. A test of these three possibilities by refinement in $\underline{A}\,\underline{a}$ shows that none of them is correct: none of them gives good agreement with the data, the agreement residuals for all remaining at about 60% (on $\underline{F}^2$). These results indicate that the proper space group is $\underline{A}2/\underline{a}$, for which a much better agreement (residual 12%) is obtained. Atomic coordinates for ice V are given in table 3.

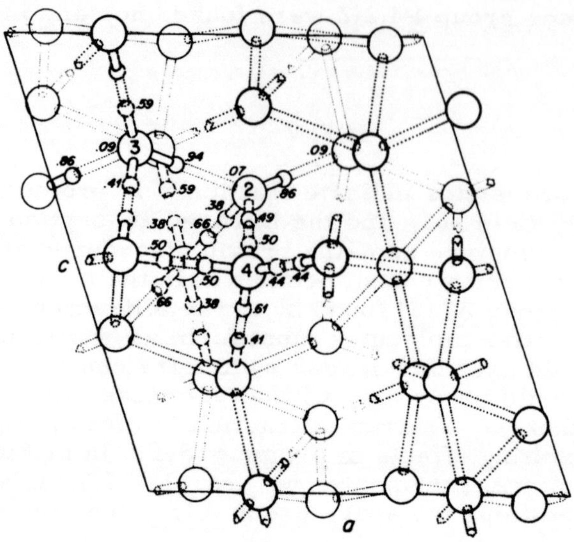

Fig. 3 Structure of ice V at 110°K, viewed along the monoclinic b axis. Oxygen atoms are shown as large balls. Some of the deuteron positions are indicated by small balls, accompanied by the corresponding population factor. The nearly "full" deuterons are shown solid, the approximately "half" deuterons are dashed, and the nearly absent deuterons are not drawn. The dotted rods are the hydrogen bonds. To avoid excessive clutter in the drawing, the details of bond lengths and angles are omitted.

The type of structure implied by the foregoing considerations is one in which there is long-range proton order, indicated by population factors differing from 1/2, but in which the long-range order is not complete, since no set of translation symmetry elements can be found for which the population factors are 0 and 1 only. The structure is thus statistically disordered, but the six different molecular orientations at each water molecule site do not occur with equal probability. If the population factors were strictly 0, 1/2, and 1, we could conclude that at water molecule sites 2 and 3 only two of the six possible orientations occur, and these two with equal probability, whereas at sites 1 and 4, all six orientations occur with equal probability. But in a statistical, partially ordered structure of this type, it seems unlikely that the restrictions on molecular orientation will operate so rigorously as to exclude certain orientations completely, while allowing the remaining orientations to occur with equal probability. From this point of view, no special merit attaches to population factors of 0, 1/2, and 1, and in fact the measured population factors deviate significantly from these values.

We have no evidence to show positively that within its stability field, ice V is fully disordered (all deuteron population factors strictly 1/2), and that the partial order observed by neutron diffraction develops only at low temperatures. However, a state of partial order would probably be temperature dependent, and would therefore cause anomalies in the static dielectric constant as a function of temperature, such as are observed in the transition region between ice III and ice IX /3/. Dr. Whalley has mentioned in the previous paper that some dielectric anomalies are found in ice V near -150°C, far below the actual stability field of this phase. These anomalies have not yet been clarified in detail, but they may correlate with the onset of the partial deuteron order observed by neutron diffraction.

CONCLUSIONS

Neutron diffraction shows that the four short O···O distances (2.75-2.87 Å) about each oxygen atom in the ice structures are hydrogen bonded: each such distance contains either a single proton near the O···O centerline, asymmetrically located near one end of the bond or the other, or else it contains two asymmetrically located proton positions occupied statistically with probabilities summing to 1. This conclusion had been reached earlier from x-ray study /2, 4, 8/, and the presence of hydrogen bonding was also indicated spectroscopically /7/. The neutron diffraction results give a detailed picture of the systems of hydrogen bonding.

The prevalent concept of water as a tetrahedral molecule is reinforced in the sense that 4-fold, H-bonded coordination is retained in the dense ice phases. However, the distortions from ideal tetrahedral geometry are large. In ice IX, the O···O···O donor angles (coordination angles into which the water molecules are oriented to donate deuterons O···D—O—D···O) are little distorted (99°, 101°), but the other coordination angles are greatly distorted from tetrahedral (89° to 144°), showing that the tetrahedral constraint on the orientations of H-bonds accepted by the water molecules is weak. In ice II the donor angles are distorted somewhat more than in ice IX (89°, 99°); the range of the other O···O···O angles is again large (80° to 129°). In ice V, where individual donor angles cannot be clearly singled out, the range of O···O···O angles is 83° to 135°.

At the level of accuracy obtained in our results, the water molecules show no measurable response, in size and shape, to the local bonding environments within the individual ice phases. The D—O—D angles are to within experimental error the same as in the free water molecule (104.5°). The scatter of individual O—D distances away from the mean O—D distance for each phase is not experimentally significant. This is confirmed by the fact that the measured O–D distances do not correlate properly with the O···O bond lengths in which they lie /9/. Infrared spectra /7/ show that the crystallographically distinct H-bonds correspond to measurably different O–D stretching frequencies, so that at the level of sensitivity of the spectroscopic method there are detectable differences between the various O–D bonds. Random experimental errors in the individual O–D distances determined by neutron diffraction tend to be compensated when all of the crystallographically distinct distances in each structure are averaged. The mean O–D bond lengths for the different phases show a systematic variation that correlates with mean bond strength as measured spectroscopically (table 4). The lengthening of the mean O–D bond length over that in water vapor is small, but probably significant. In comparing the various ice phases with water vapor (table 4), it appears that the O–D bond length measured earlier in ice I by Peterson and Levy /10/ is anomalously long in relation to the bond lengths in the other ice phases, when judged against the O–D stretching frequencies.

The individual water molecules have orientations that tend to place the deuterons close to the O···O centerlines, or that give only a small bending of the H-bonds as measured by the departure of the O–D···O angles from 180°. These orientations do not in general permit the unshared electron pairs (located tetrahedrally with respect to the deuterons of each molecule) to be directed closely along the bond centerlines, which is another indication that there is only limited validity to the standard tetrahedral concept of the water molecule. However, in detail most of the molecules depart significantly from a perfectly symmetrical orientation in the O···O···O donor angle, an orientation that would strictly minimize the bending of the O–D···O angles away from 180°. This shows that the orientations are slightly but definitely affected by geometrical conditions at the acceptor ends of the bonds, or by long-range electrostatic forces between non-nearest neighbors. A result of these departures from symmetrical orientation is that the extent of bond bending in ice II (bend angles 2.1°, 12.1°, 13, 6°, and 13.8° as seen at the deuterons) is about the same as in ice IX (bend angles 5.1°, 12.9°, 14.9°), in spite of the fact that the O···O···O donor angles in ice II (89°, 99°) are definitely worse than in ice IX (99°, 101°).

In distinguishing between proton-ordered and proton-disordered ice structures, the neutron diffraction results confirm the conclusions reached by Bertie and Whalley /7/ from infrared spectra. The spectroscopic method does not, however, seem able to distinguish between a state of complete proton disorder and a state of only partial disorder, such as that found by neutron diffraction in ice V.

The partial proton disorder in ice V is a state of affairs that can be expected in ice structures where the tetrahedral coordination is distorted. The six different water molecule orientations at each site will involve different amounts of bond bending, required by the different O···O···O angles. They will therefore have different energies, and will occur with different probabilities. A similar type of partial proton disorder must occur in ice III over the range of temperatures where dielectric measurements /3/ indicate progressive proton ordering, as the phase IX is approached. The deuteron population factors must be a strong function of temperature over this range, and must approach 0 and 1 at the low-temperature limit where the completely ordered structure of ice IX is reached. It is not certain that the population factors approach 1/2 at the high-temperature limit of the transition range, but above this limit, they must become essentially constant, so that the dielectric capacitivity shows a normal dependence on temperature /3/.

It would be important to understand why the increasing proton order in ice III at lower temperatures leads ultimately to the complete order in ice IX, whereas in ice V the ordering achieved is only partial. This difference in behavior is probably related to the fact that the temperature-dependent partial order in the ice III - ice IX transition region is readily detected dielectrically, whereas correspondingly clear effects in ice V have not yet been found. For reasons that are not yet obvious structurally, the proton ordering energy achievable in the ice V framework seems to be so low that significant ordering begins only at temperatures where the dielectric relaxation time is too long to allow reliable measurement of the static capacitivity. Presumably the further increase in relaxation time at lower temperatures then "quenches" the water molecule orientations before a fully ordered arrangement becomes stable. Extrapolation of existing dielectric measurements /6, 3/ suggests that at about -150°C, where some dielectric anomalies have been detected (as noted earlier), the relaxation time is already several hours.

A structural difference between ice III and ice V that probably contributes to their different dielectric and proton-ordering behavior is the fact, noted earlier, that in ice III both disordered and fully ordered arrangements can occur within the unit cell provided by the basic hydrogen-bond framework, whereas in ice V the basic symmetry requires some proton disorder. In the ice III unit cell, the water-molecule orientation probabilities can vary continuously from 1/6, in the high-temperature limit where all orientations occur with equal probability, to values of 0 or 1 that represent complete order. In ice V, certain of the proton probabilities can begin to deviate from 1/2 only when some of the high-temperature symmetry elements are dropped. This discontinuous change in symmetry represents the development of a perturbation in the bond framework, which must propagate through the entire crystal so as to achieve long-range order. It seems likely that the need to develop such a perturbation, for which the bond framework itself provides no motivation, is an inhibiting factor for long-range proton ordering, by comparison with a case like ice III where long-range order can develop without it. A similar argument has been used to rationalize the widely held view that a proton-ordered structure does not develop in ice I at low temperatures. However, the distinction between proton ordering with and without symmetry change is not in itself sufficient to account for the observed differences in behavior among the various ice phases. This is shown by the situation in ice II. Here, as noted previously, a symmetry change is again necessary for ordering, and yet it occurs so forcefully that this structure shows no transition to a proton-disordered arrangement even at the highest accessible temperatures /6/.

Many aspects of the thermodynamics, phase stability, and order-disorder phenomena in the ice phases need to be given a more detailed structural explanation before these substances can be considered well understood. The present study is a contribution in this direction.

References

/1/ Kamb, B., Ice polymorphism and the structure of water in: Structural Chemistry and Molecular Biology (A. Rich and N. Davidson, eds.): Freeman and Co., San Francisco (1968)

/2/ Kamb, B., Acta Cryst. 17, 1437 (1964)

/3/ Whalley, E., J.R.B. Heath, D.W. Davidson, J. Chem. Phys. 48, 2362 (1968)

/4/ Kamb, B., A. Prakash, Acta Cryst. B24, 1317 (1968)

/5/ Kamb, B., B. L. Davis, Proc. Natl. Acad. Sci. U. S. 52, 1433 (1964)

/6/ Wilson, G.J., R.K. Chan, D.W. Davidson, E. Whalley, J. Chem. Phys. 45, 3976 (1965)

/7/ Bertie, J.E., E. Whalley, J. Chem. Phys. 40, 1646 (1964)

/8/ Kamb, B., A. Prakash, C. Knobler, Acta Cryst. 22, 706 (1967)

/9/ Hamilton, W.A., J.A. Ibers, Hydrogen Bonding in Solids: Benjamin, New York, p. 52 (1968)

/10/ Peterson, S.W., H. Levy, Acta Cryst. 10, 70 (1957)

Table 1

Atomic Coordinates in Ice II

Space group $R\bar{3}$. $\underline{a}$ = 7.78 Å, α = 113.1°
All atoms in general positions.

Atom	x	y	z
O(I)	.272	.026	-.147
O(II)	.480	.757	.339
D(1)	.728	.404	.403
D(2)	.149	.041	-.202
D(3)	.742	.198	.371
D(4)	.423	.195	-.016

Table 2

Atomic Coordinates in Ice IX
Comparison of Results of Neutron (n) and X-ray (x) Diffraction

Space group $P4_12_12$. $\underline{a}$ = 6.73, $\underline{c}$ = 6.83 Å

Atom	Position	Method	x	y	z
O(1)	8b	n	.1094	.3012	.2859
		x	.1096	.3016	.2873
O(2)	4a	n	.3927	--	--
		x	.3932	--	--
D(3)	8b	n	-.012	.332	.214
D(4)	"	n	.114	.158	.297
D(5)	"	n	.300	.360	.106

Table 3

Atomic Coordinates in Ice V

Space group $A2/a$. $\underline{a}$ = 9.22, $\underline{b}$ = 7.54, $\underline{c}$ = 10.35 Å, β = 109.2°.
All atoms in position 8f, except O(1) in 4e.

Atom	x	y	z	Population Factor
O(1)	.250	-.189	.000	1.00
O(2)	.463	.057	.155	"
O(3)	.277	-.347	.248	"
O(4)	.399	.359	-.014	"
D(5)	.336	-.109	.045	0.66
D(6)	.249	-.253	.080	0.38

Table 3 (continued)

Atom	x	y	z	Population Factor
D(7)	.398	-.034	.092	0.39
D(8)	.333	.114	.234	0.07
D(9)	.469	.161	.105	0.49
D(10)	.559	.005	.199	0.86
D(11)	.255	-.297	.155	0.59
D(12)	.424	-.491	.310	0.09
D(13)	.312	-.264	.319	0.41
D(14)	.196	-.385	.257	0.94
D(15)	.431	.265	.050	0.50
D(16)	.310	.386	-.001	0.50
D(17)	.370	.304	-.103	0.61
D(18)	.464	.459	-.006	0.44

Note: The atomic coordinates listed here were obtained in an isotropic least-squares refinement in which the deuteron population factors were constrained to be 1/2, and the population factors listed resulted from a further refinement in which this constraint was removed.

Table 4

Comparison of O–D Bond Length and Stretching Frequency

Phase	Mean O–D (Å) Measured	Mean O–D (Å) Corrected†	Mean ν_{OD}§ (cm^{-1})
Ice I*	0.996	1.008	2421
Ice IX	0.975	0.987	2454
Ice II	0.970	0.982	2472
Ice V	0.964	0.976	(2461)
Vapor	--	0.957	2727

§Mean of individual ν_{OD}(HDO) frequencies measured by Bertie & Whalley /7/. For ice V, the value listed is the maximum of a broad absorption band and is correspondingly inaccurate.

*Neutron diffraction results for ice I are from Peterson & Levy /10/.

†Bond lengths corrected for thermal motion by adding a constant correction equal to that used for ice I by Peterson & Levy /10/. These corrected values are to be compared with the equilibrium O–D distance determined spectroscopically in water vapor, which is listed in the same column.

STRUCTURAL STUDIES OF ICE POLYMORPHS BY NEUTRON DIFFRACTION, PROTON AND DEUTERON NUCLEAR MAGNETIC RESONANCE*, **

S.W. Rabideau and E.D. Finch***

University of California, Los Alamos Scientific Laboratory, Los Alamos, New Mexico / USA

Abstract

Neutron diffraction studies of flat samples in transmission were made on the powdered polycrystalline ices Ih, Ic, II, V, and IX at 80 °K. Lattice parameters obtained from these spectra by computer methods have been determined for: (a) ice Ic, peak centroid for which (between 15° and 62°) give a weighted value of a_o=6.353 ± 0.001 Å; (b) ice II, when indexed on the rhombohedral unit cell give a_R=7.743±0.002 Å, α=113.09±0.03°; and (c) ice IX, the tetragonal, pseudocubic lattice parameters for which were found to be $a_o \approx c_o$=6.745±0.004 Å. The ice V neutron diffraction spectrum consists of a large number of low intensity lines; the analysis of this pattern has not been completed. The existence of proton disorder in ice Ic has been shown. The oxygen and proton structural arrangements in the proton-ordered ices II and IX have been established. Proton and deuteron nuclear magnetic

---

* This work was done under the auspices of the U.S. Atomic Energy Commission.

** This review is based in part on papers to be published in J.Chem.Phys.: (a) Ice IX., by S.W. Rabideau, E.D. Finch, G.P.Arnold, and A.L. Bowman. (b) Ice II., by E.D. Finch, S.W. Rabideau, R.G. Wenzel, and N.G. Nereson. (c) Ice Ic., by G.P. Arnold, E.D. Finch, S.W. Rabideau, and R.G. Wenzel. (d) Proton and Deuteron NMR of Ice Polymorphs by S.W. Rabideau and E.D. Finch.

*** Present address: Massachusetts General Hospital, Boston, Mass. 02114.

resonances of several ice polymorphs prepared from H_2O and D_2O have been recorded. Proton nuclear resonances have been recorded under pressure for ices Ih, II, III, and V. Rigid lattice proton second moments have been obtained at 75 °K at atmospheric pressure for ices Ih, Ic, II, V, and IX. Deuteron quadrupole coupling constants for ices Ic , II, V, and IX have been estimated from derivative of dispersion mode traces by comparison with the ice Ih quadrupole split resonance under similar experimental conditions. The relatively constant values of e^2qQ/h suggest that the O-D bond distance is essentially unchanged in the different ices in agreement with neutron diffraction results.

1. Introduction

Studies of ice in our Laboratory over the last five years have been concerned principally with the recording and the interpretation of the proton, deuteron, and oxygen-17 nuclear magnetic resonances (NMR) in polycrystalline and single crystal specimens of hexagonal (Ih) H_2O and D_2O ices. In the present review of the extension of our work to cubic ice and the high pressure polymorphs, emphasis will be placed upon the crystal structures, the proton and oxygen arrangements, as revealed by neutron diffraction, proton and deuteron NMR.

2. Experimental

A heat-treated piston and cylinder as illustrated in fig. 1 were fabricated from BeCu (Berylco No. 25) for the preparation of the high pressure ices. As shown, this cylinder was fitted with an electrical connection leading to an rf coil so that proton NMR measurements could be made of the ices under pressure. One end of the coil was attached to a brass ring which in turn was grounded to the cylinder wall. The other end was joined to a BeCu rod, the upper portion of which was insulated from the cylinder with a Kel-F cone. This cone and rod design was patterned after that of Dickson and Meyer /1/. A manganin coil was substituted for the rf coil for pressure- calibration purposes. A low-level Robinson /2/ oscillator was built and used in the recording of absorption derivative proton resonances. Most measurements were made at 15.5 MHz. The audio signal was fed to a PAR Model HR-8 phase-sensitive detector. The magnetic field was modulated at 280 Hz with a peak-to-peak amplitude of ca. one gauss. A second cylinder of the same dimensions but without the electrical lead provision was used for the pre-

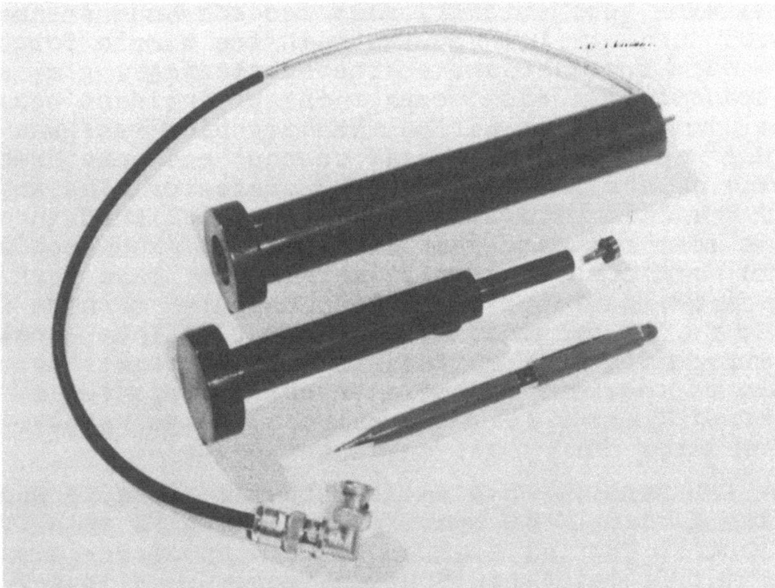

Fig.1 View of pressure piston and cylinder made of Berylco No. 25 BeCu alloy fitted with rf lead and cable

paration of quantities of the ice polymorphs for NMR and neutron diffraction studies.

The various polymorphs were conveniently identified by the determination of the buoyant force exerted on powdered ice samples as they were immersed in liquid nitrogen. These identifications were confirmed with neutron diffraction observations. The ices were ground in a steel mortar under liquid nitrogen in a nitrogen atmosphere dry box to less than 250 microns. It was not necessary to grind cubic ice, Ic, prepared by the transformation of the high pressure forms since the entire sample passed through the sieve without grinding. The powdered ices were kept under liquid nitrogen during the loading operations of the neutron sample holder.

A flat sample holder with threaded cover 8 cm in diameter was used in transmission to minimize multiple scattering and absorption corrections. The holder was fabricated from Ti-Zr "null-matrix" alloy /3/. This composition was selected so that no coherent scattering was contributed by the sample holder. A neutron wavelength of 1.142 Å was provided by scattering from the (220) planes of a lead single crystal monochromator used in transmission.

The patterns were automatically recorded and scanned in steps of 0.05° or 0.1° in 2θ, with the sample rotating so that its normal bisected the scattering angle. A thin ^{235}U foil transmission counter in the incident beam controlled the counting period. The scanning rate was usually 1.5° per hour. Automatic readout equipment was used to record the counts from a $^{10}BF_3$ detector. The angular accuracy of the diffractometer was established through the use of a standard sample of niobium oxide powder. The zero and absolute errors were shown to be less than 0.01° in 2θ between 22° and 75°. Computer methods were applied to the original neutron count data to locate the peak centroids, half widths, and line intensities. Through the use of the known dispersion characteristics of the neutron diffractometer, it was possible to resolve overlapping lines into their component parts.

Proton NMR measurements at atmospheric pressure were made with the Varian DA-60 spectrometer at 75 °K at a frequency of 60 MHz. Rf fields of 20 to 35 micro-gauss were required to obtain unsaturated signals. The long T_1 for protons in ice at the temperature of liquid nitrogen was apparent for under saturation conditions the signals disappeared and a period of approximately one hour was required for recovery. The Varian Model V-4200 wideline spectrometer was used to observe the deuteron resonances in the D_2O ices. Because the deuteron resonances could not be observed in any of the ices at 75 °K, presumably because of saturation difficulties, it was necessary to bring the samples to a point several degrees below the respective phase-transition temperatures to record the dispersion mode signals.

3. Results and Discussion

Shown in fig. 2 is a phase diagram of H_2O which will be useful for reference in this discussion. The boundaries have been delineated principally through the work of Bridgman /4/, but recent contributions have also been made to our knowledge of the phase relationships in the H_2O system /5,6/. We have accepted the nomenclature suggested by Whalley of designating as ice IX the proton ordered form prepared by cooling ice III under pressure. Thus, there are ten known crystalline modifications of ice in addition to a vitreous phase-a remarkably complicated "simple system". In this review, the neutron diffraction results will be given first followed by the proton and deuteron NMR results obtained at atmospheric pressure, and finally, the NMR results obtained for ice under pressure will be discussed.

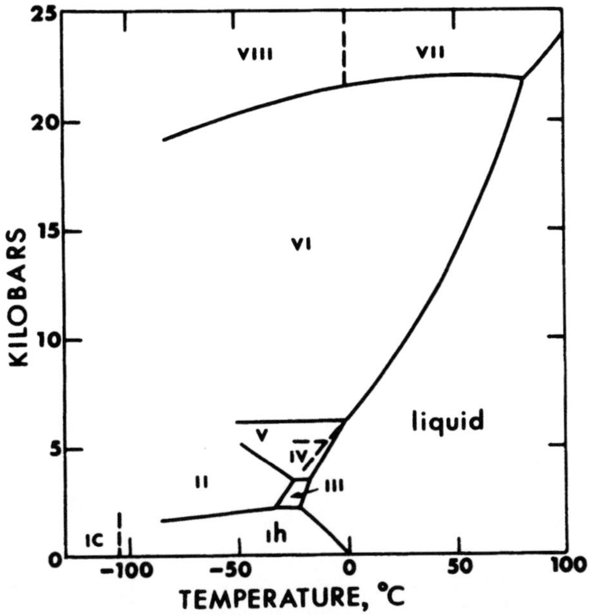

Fig.2 Phase diagram of H_2O

3.1 Neutron diffraction

Ice Ic*- The neutron diffraction spectrum of D_2O ice Ic at 80 °K obtained by the transformation of ice II is given in fig. 3. The cubic cell lattice parameter, a_o, was found to be 6.353±0.001 Å for D_2O cubic ice at this temperature. The calculated positions which correspond to this value of a_o are indicated by vertical lines placed above the (hkl) symbols. The asymmetry of the (111) peak in the ice Ic spectrum shown in fig. 3 is apparent both on the low and high angle sides of the peak centroid. All the ice Ic lines are significantly broadened as has been determined from the known diffractometer dispersion characteristics. In fig. 4 is shown the (111) peak of cubic ice prepared from the high pressure ices II, V, and IX. The transformations of these ices to ice Ic were made in a temperature-controlled n-pentane bath. In no instance did the ices come in contact with the coolant. The temperatures at which the transitions were made were 166 °K for ices V and IX and 177 °K for ice II.

* Group Fd3m
O_h^7

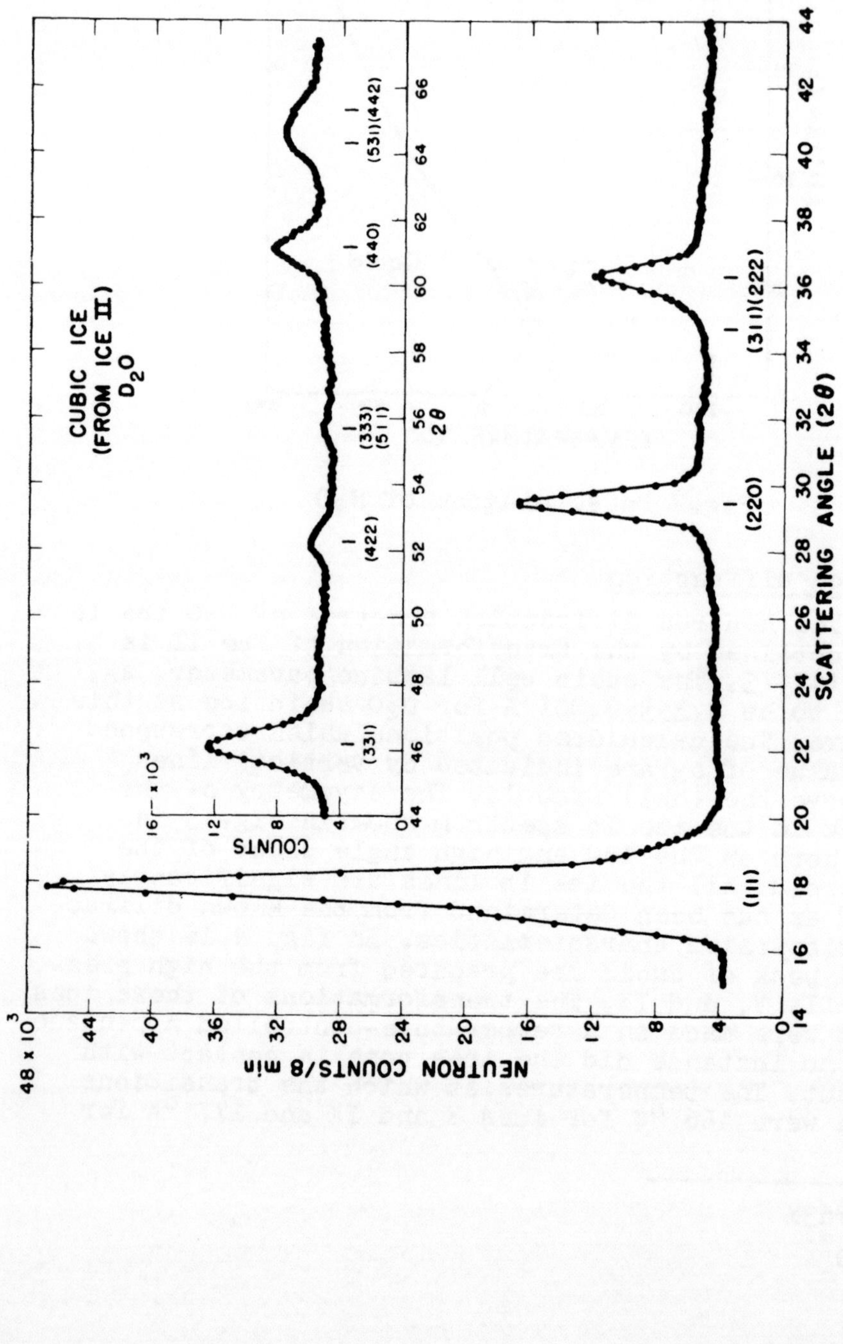

Fig.3 Neutron diffraction powder pattern of cubic ice (Ic) at 80°K prepared by transformation of D_2O ice II

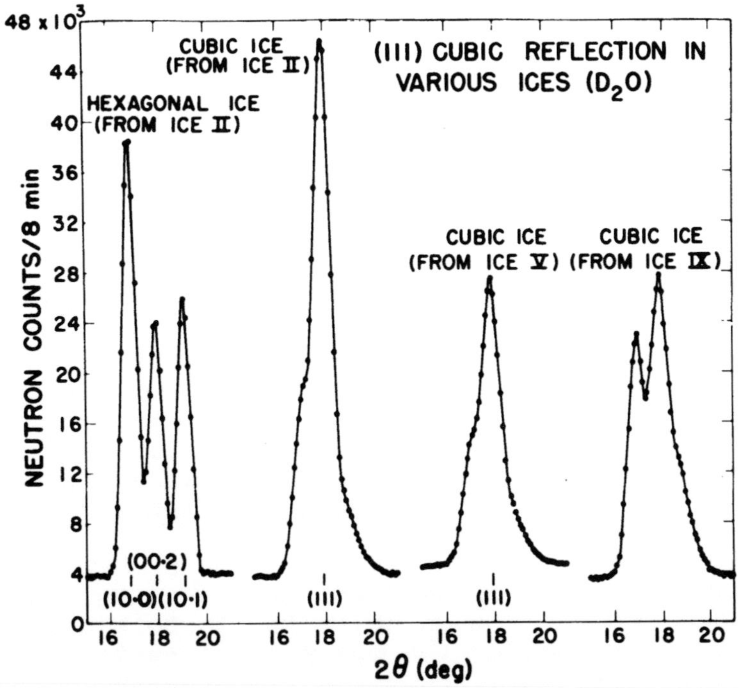

Fig.4 The (111) cubic reflection from ice Ic prepared by the transformation of ices II, V, and IX. Hexagonal ice prepared by the transformation of cubic ice

It appears from the observed asymmetry of the (111) peak that an impurity of hexagonal ice is present. Fig. 4 shows that the triplet in the D_2O hexagonal ice spectrum is clearly resolved while the cubic ice lines are not resolved even though approximately the same angular range of 2θ is covered. This lack of resolution has been attributed to small crystal size. An analysis made with the application of the Scherrer formula indicates a mean particle size of about 130 Å for ice Ic made from ice II. In the transformation of ice Ic to ice Ih in several stages, 15 min at -65 °C, 15 min at -55 °C and finally 30 min at -52 °C, the particle size of the ice Ih as indicated by the (10·0) linewidth gradually increased from approximately 310 Å to 610 Å. If crystal growth occurred at 75° to 80 °K, it was very slow since the ice Ic and Ih spectra were found to be reproducible even after the elapse of several days at this temperature.

In an attempt to demonstrate the existence of proton disorder in ice Ic, a calculation of the expected neutron diffration spectrum was made for a proton-ordered tetragonal arrangement, space group $I4_1md$. If the Bernal Fowler requirements are met, the number of proton ordered possibilities is severely restricted. For this proton ordered model (no other models were conceived), the very much poorer quality of fit of the calculated and experimental intensities strongly suggests that ice Ic has a proton disordered arrangement.

It has been reported /7/ that aqueous solutions of $FeCl_2$ when quenched in liquid nitrogen form cubic ice rather than the expected hexagonal phase. This conclusion was based on differences observed in the two-line Mössbauer spectrum characteristic of a ^{57}Fe nuclear quadrupole hyperfine interaction. Neutron diffraction results which have been obtained for D_2O solutions of $FeCl_2$ quenched in liquid nitrogen have shown a typical ice Ih spectrum. A quantitative intensity ratio measurement for the (10·0), (00·2), and (10·1) peaks in the $FeCl_2$-doped ice proved to be essentially identical with those observed for pure hexagonal ice. An examination of samples of ice prepared from H_2O solutions of $FeCl_2$ was also made by neutron diffraction methods to be certain that there was no difference between D_2O and H_2O as solvents. In this case also there was no evidence for the presence of cubic ice. Clearly the observed difference in the quadrupole splitting signifies a change in the electric field gradient, but this does not appear to be attributable to a phase change. Conceivably, some of the altered electric field gradient could have its origin in the crystal strains introduced by the quenching prodedure.

Ice II - The neutron diffraction spectrum of polycrystalline D_2O ice II obtained at 80 ^oK is shown in fig. 5. The ice II was prepared by the transformation of ice Ih at 3 kbar at a temperature of 195 ^oK; to be certain that the transformation had taken place, the temperature was raised to 233 ^oK at this same pressure. As judged by the rate of piston displacement, the transformation did not occur rapidly at the lower temperature. A small amount of hexagonal ice was present as an impurity as indicated by the weak peaks at 2θ scattering angles of 16.8^o and 29.4^o. These are lines of major intensity in the hexagonal ice spectrum and have been identified as (10·0) and (11·0) peaks. It was possible to subtract the hexagonal ice background to obtain the corrected ice II diffraction pattern.

Kamb /8/ has been able to infer a proton ordered arrangement for ice II from structural considerations reached through the examination of single crystals of ice II by X-ray methods. A space group $R\bar{3}$ was assigned to ice II

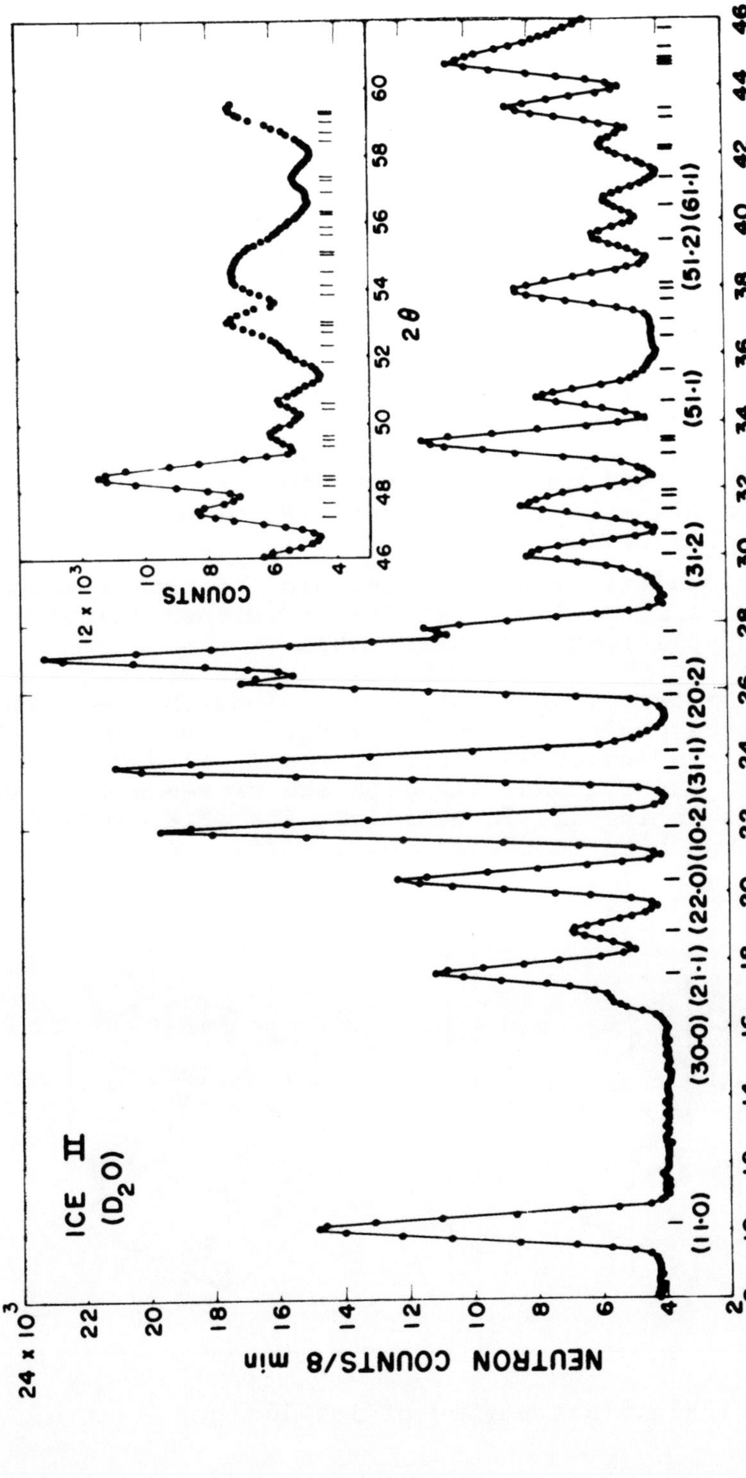

Fig.5 Neutron diffraction spectrum of polycrystalline ice II (D_2O) at 80 °K

with 12 molecules in the rhombohedral unit cell. To simplify calculations in this work, hexagonal coordinates have been used throughout in the treatment of the ice II data. The lattice parameters for D_2O ice II at 80°K have been found to be a_H=12.920±0.003 Å, and c_H=6.234±0.002 Å. These values are in good agreement with the results reported by Bertie et al./9/ which were obtained from an analysis of ice II X-ray powder patterns. An ordered proton arrangement has also been suggested for ice II on the basis of dielectric relaxation measurements, residual entropy considerations, and infrared spectra /10-13/.

There are two non-equivalent oxygen atoms and four non-equivalent deuterium atoms in the D_2O ice II lattice. In table 1 are given the computer evaluated final coordinate positions, bond distances and bond angles for the oxygen and deuterium atoms. The high degree of precision with which Kamb has located the oxygen /8/ in the ice II lattice suggested that it was useful to adopt these values. With the parameters of table 1, a residual, $\underline{R}$, of 0.041 was obtained where $\underline{R}$ is defined by the ratio $\sum_i w_i|I_o-I_c|/\sum w_i I_o$. The excellent quality of the fit between the experimental and calculated spectra indicated that a further adjustment of the oxygen positions was unwarranted.

It is to be noted that the O-O-O bond angles are 88° and 99°, but the deuterons were not placed along the O-O bonds. Instead, initial angles of 105° were chosen for the D-O-D angles, and as indicated in table 1, the final angles are 106°±3°. Thus, the evidence suggests the presence of bent hydrogen bonds in ice II. In addition, the existence of proton order has been demonstrated. In fig. 6 a stereoview

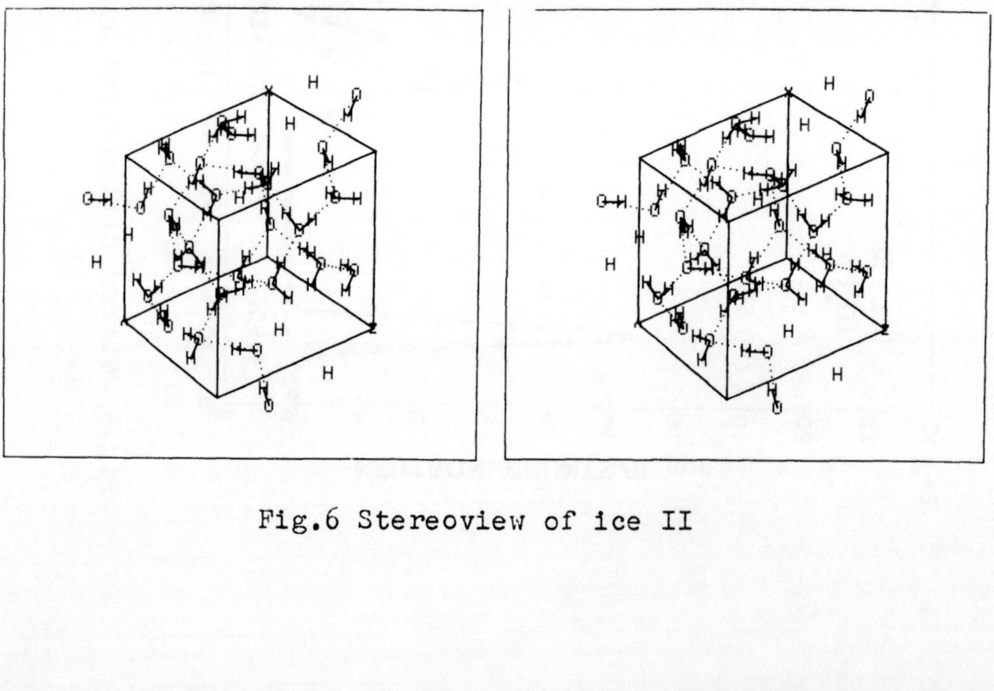

Fig.6 Stereoview of ice II

Table 1 Position parameters for oxygen and deuterium atoms, bond distances and angles in ice II at 80 °K*

Atom	x	y	z
H(I 2)	0.151 ± 0.003	0.200 ± 0.002	0.983 ± 0.006
H(I 4)	.223 ± .002	.214 ± .003	.202 ± .005
H(II 1)	.219 ± .002	.112 ± .002	.517 ± .004
H(II 3)	.302 ± .002	.065 ± .002	.442 ± .006
O_I	$.222_7$	$.196_3$	$.050_3$
O_{II}	$.188_0$	$.230_0$	$.480_0$

Proton-oxygen pair		bond distance	bond angle
H(I 2)-O_I	O_I	1.04 ± 0.04 Å	
H(I 4)-O_I		.98 ± .03 Å	
H(I 2)-O_I-H(I 4)			106 ± 3°
H(II 1)-O_{II}	O_{II}	.98 ± 0.03 Å	
H(II 3)-O_{II}		.96 ± .04 Å	
H(II 1)-O_{II}-H(II 3)			106 ± 3°

* Oxygen coordinates are those reported in X-ray work of Kamb /8/. The isotropic thermal parameter, B, was found to be 0.70 $Å^2$; all atoms were required to have the same value. Atom and angle designations correspond to those listed in Kamb's paper /8/.

of the proton (deuteron) and oxygen atom positions as calculated by computer methods is given.

<u>Ice IX</u> - The deuterated ices have been examined in this work by neutron diffraction rather than the protonated forms (except in special circumstances, see ice Ic) because of the large spin-incoherent scattering cross section of protons compared with deuterons. Ice IX was prepared by first transforming ice Ih to ice III at -30°C at 2.1 kbar, raising the pressure to 3 kbar, and then quenching the sample in liquid nitrogen /9/. The observed neutron diffraction pattern of ice IX is shown in fig. 7.

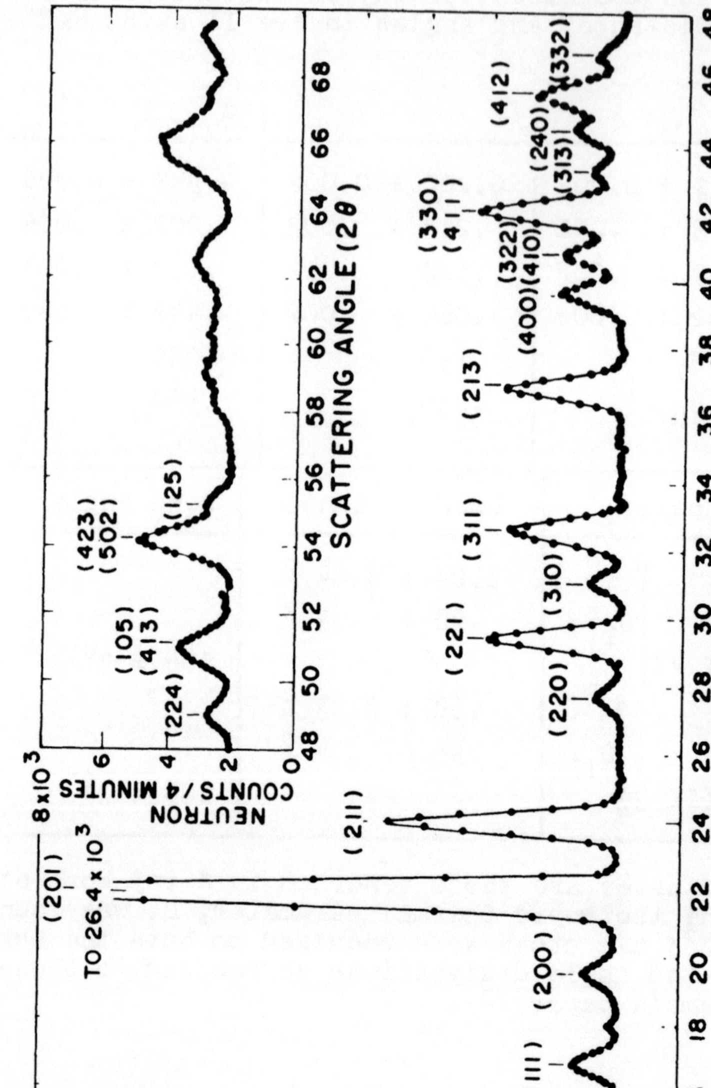

Fig.7 Neutron diffraction pattern of ice IX (D_2O) at 80 °K

Through single crystal X-ray diffraction data, Kamb and Datta /14/ found that ice IX is dimensionally cubic but symmetrically tetragonal with a_o=c_o = 6.80 Å. In the present work, these parameters for D_2O ice IX at 80°K have been found to be 6.745±0.004 Å. Below each (hkl) symbol is a vertical mark which indicates the calculated centroid position with the use of these lattice parameters. The space group $P4_12_12$ has been proposed /14/ for ice IX with four oxygens in the special position (x,x,0) and eight oxygens in the general position (x,y,z). There are twentyfour deuterium atoms in three eight-fold general positions (x,y,z); accordingly there are 13 position parameters to be determined.

With the ixe IX neutron diffraction peak intensities as input parameters, the final deuterium and oxygen positions were obtained with the use of least squares computer procedures. The O-O-O bond angles were calculated from the final oxygen positions; these are given in table 2 together with the bond distances and angles for the O_I and O_{II} classes of oxygen atoms. In the determination of the peak centroids, line halfwidths and integrated intensities, Gaussian lineshapes were fitted to each peak in the ice IX spectrum. The program was capable of obtaining line intensities for incompletely resolved peaks such as those at high angles by fixing peak positions and halfwidths.

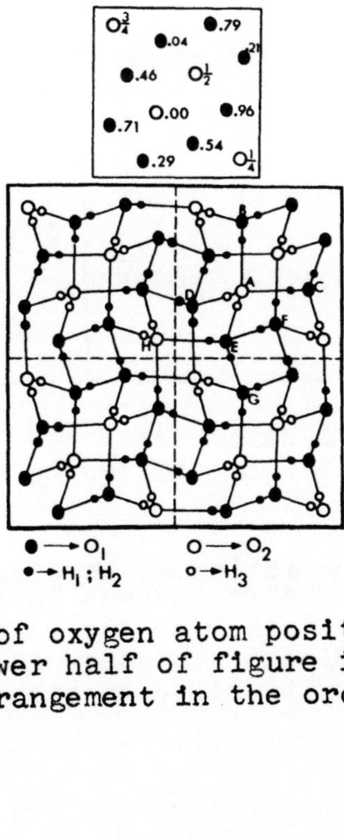

Fig.8 Projection of oxygen atom positions in the ice IX lattice. Lower half of figure indicates preferred deuteron arrangement in the ordered structure

Table 2 Final atom positions, bond distances and angles for ice IX *

Atom	x	y	z
H_I	0.980 ± 0.007	0.326 ± 0.003	0.221 ± 0.007
H_{II}	.118 ± .005	.168 ± .003	.286 ± .004
H_{III}	.304 ± .004	.362 ± .004	.096 ± .003
O_I	.108 ± .005	.305 ± .006	.290 ± .009
O_{II}	.407 ± .006	.407 ± .006	.000 ±

Molecule		O-O-O Bond Angle
D-A-E	H_{III}-O_{II}-H_{III}	97.3 °
B-A-C		94.7 °
H-E-G	H_I-O_I-H_{II}	97.2 °
H-E-F		144.9 °
A-E-G		93.0 °
A-E-F		95.3 °

Proton-oxygen Pair		Bond Distance	H-O-H Bond Angle
O_I-H_I	O_I	0.99 ± 0.06 Å	
O_I-H_{II}		.93 ± .04 Å	
H_I-O_I-H_{II}			101 ± 4 °
O_{II}-H_{III}	O_{II}	1.00 ± .04 Å	
H_{III}-O_{II}-H_{III}			90 ± 4 °

* The isotropic thermal parameter B was found to be 1.0 $Å^2$; all atoms were constrained to have the same value.

In fig.8 is shown a projection of the ice IX unit cell along the c-axis with twelve oxygens in general and special positions. The number to the right of each atom represents the z-coordinate expressed as the fraction or percentage of the unit cell dimension. In the lower portion of fig. 8, four of these projections are drawn adjacent to one another to indicate the bonding arrangement. There are four possible ordering arrangements of the deuterons in ice IX if compliance with the Bernal-Fowler rules is a prerequisite. Theoretical powder patterns were calculated for each of the four deuteron ordering arrangements. An O-D bond length of 1.01 Å and a D-O-D bond angle of 103° were used. All calculated and observed intensities were adjusted relative to the intensity of the (201) peak which was arbitrarily assigned an intensity of 100 units. The atom arrangement of fig. 8 was superior to the other possibilities as indicated by a residual $\underline{R}$ value of 0.065. A computer-generated stereoview of ice IX is shown in fig. 9 from the final atom positions of table 2.

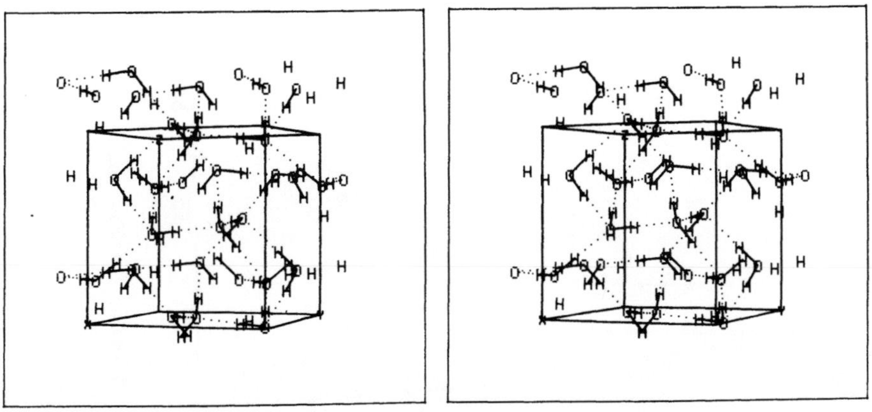

Fig.9 Stereoview of ice IX

The final bond distances in ice IX calculated from the atom positions appear to be reasonable with the comment that the O_I-H_{II} distance could approach 0.97 Å as an upper limit. As indicated, the O-O-O bond angles are approximately 97° for the three oxygen atoms along which the deuterons are positioned. The H_I-O_I-H_{II} angle of 101±4° is in accord with expectations. The H_{III}-O_{II}-H_{III} angle of 90°±4° appears to be too small since it is several degrees smaller than the O-O-O angle. This result may reflect the assumption of the same isotropic thermal parameter for both deuterium and oxygen atoms.

Ice V - Kamb et al. /15/ have made structure studies of ice V by X-ray diffraction methods and have found spacings in general agreement with those reported by Bertie et al. /9/. In the present work, ice V was prepared by the transformation of ice III at 243°K followed by quenching in liquid nitrogen. The neutron diffraction pattern of this ice polymorph shown in fig. 10 is characterized by many over-

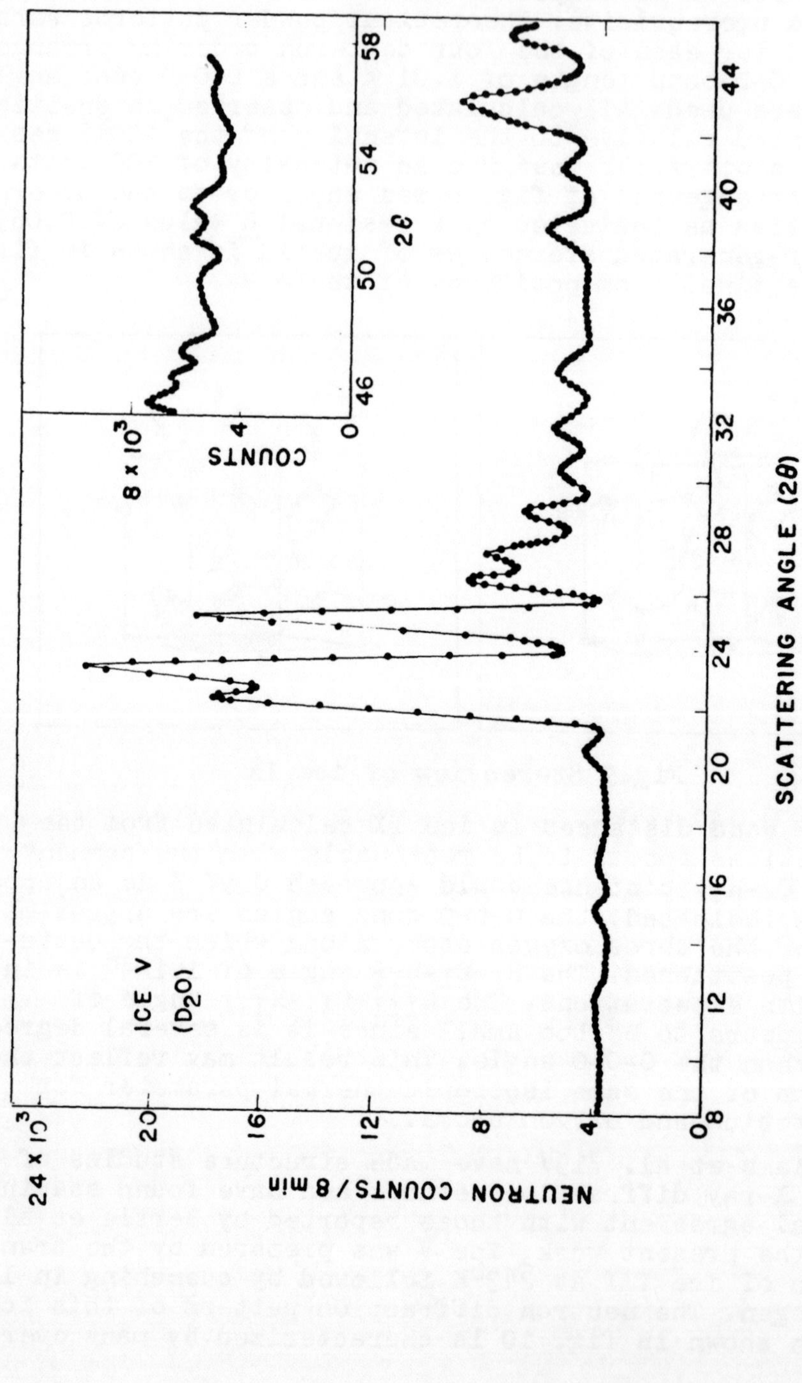

Fig.10 Neutron diffraction spectrum of ice V (D_2O) at 80 ^{o}K

lapping lines of low intensity which make the analysis difficult. Counts were obtained at every 0.05° of scattering angle (2θ) for this pattern. The structure analysis of this ice is still in progress.

3.2 Nuclear magnetic resonance

Proton resonances - A typical derivative of absorption proton magnetic resonance signal obtained for polycrystalline ice II at 75 °K is shown in fig. 11. The measured linewidths and modulation-corrected second moments are given in table 3. To make judgements of the proton arrangements in the various ice forms, it was necessary to obtain the interproton distances to calculate the theoretical second moments for various bond distances and angles. These distances for the proton-ordered ices II and IX were obtained with existing computer programs /16/. A complication is introduced in the consideration of the theoretical second moments for ices Ih and Ic in that it is necessary to consider proton occupation probabilities for all possible proton positions to evaluate the intermolecular contributions. The rigid lattice second moment values have been corrected for the vibrational and librational motion after the manner of Pederson /17/. In table 3 are also listed the intra- and the intermolecular contributions to the second moments for the various ices for several bond angle and distance combinations.

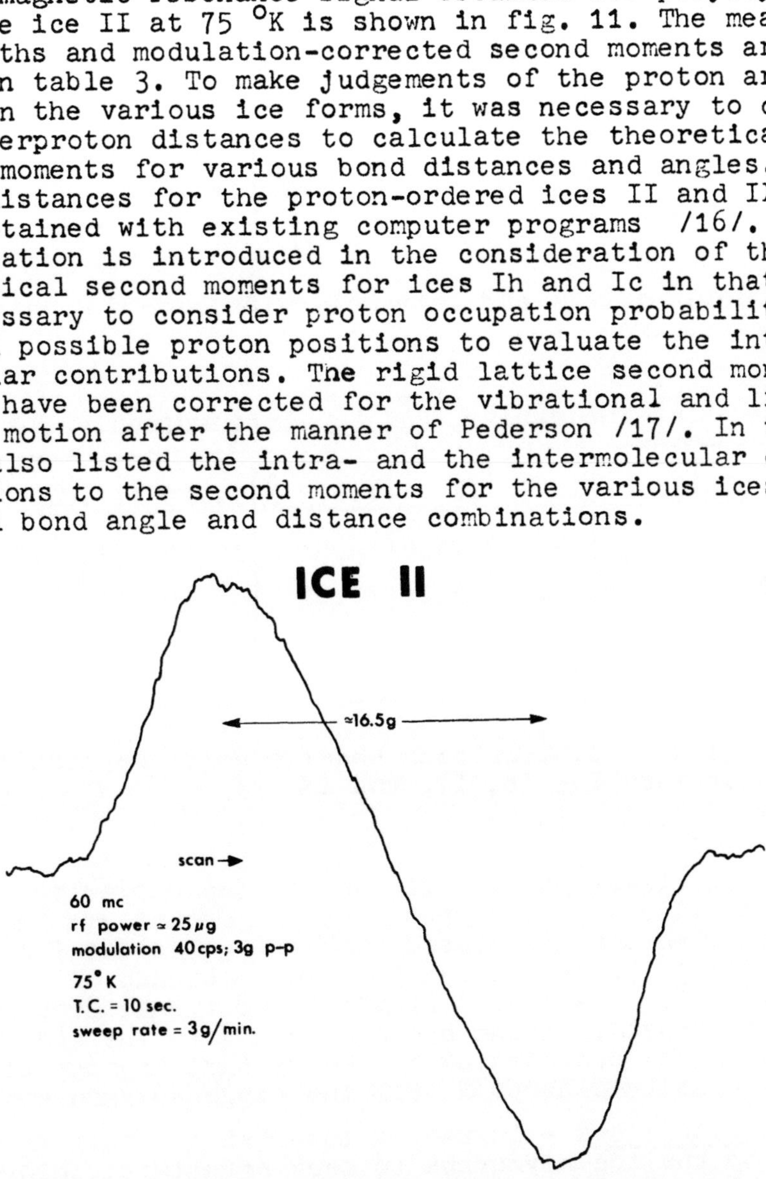

Fig.11 Proton magnetic resonance derivative of absorption spectrum for polycrystalline ice II at 75 °K

Table 3 A. Experimental proton linewidths and second moments for ice polymorphs at 75 °K

Ice	Linewidth, G	Second Moment, G^2
Ih	16.0 ± 0.2	32.4 ± 1.1
Ic	16.0 ± .2	32.2 ± 1,2
II	16.5 ± .3	34.1 ± 1.3
IX	16.8 ± .3	35.6 ± 1.3
V	16.9 ± .4	36.2 ± 1.5
VI	17.4 ± .4	37.9 ± 1.5

B. Calculated inter- and intramolecular second moment contributions

	Intermolecular, G^2			Intramolecular, G^2 *		
H-O-H Angle	Ih	Ic	II	1.005 Å	1.010 Å	1.015 Å
109 °	12.94	13.17	15.24	16.31	15.84	15.37
107 °	12.86	13.09	15.16	17.61	17.09	16.59
105 °	12.79	13.01	15.07	19.05	18.49	17.91
103 °	12.73	12.94	14.99	20.68	20.06	19.48

* Intramolecular contribution considered to be equally valid for ices Ih, Ic, II, and IX.

For all four ices, Ih, Ic, II, and IX the corrected experimental values of the proton second moment are in best agreement with the calculated theoretical values for an H-O-H angle of 104±1° and an O-H bond distance of 1.010±0.005 Å. This result provides additional support for the proposal that the hydrogen bonds in the water molecule in the ice lattice are bent; that is, the protons are displaced about 2° off the oxygen-oxygen bonds.

Deuteron resonances - It was of interest to apply NMR methods to the ice polymorphs to seek structural information which might be revealed by details of the deuteron nuclear resonances. No deuterium signals could be recorded for any of the ices at 75 °K presumably because of saturation problems. Bertie et al. /18/ have reported transition temperatures for the various ices; these are ∿ 150 °K for ices V and IX, ∿ 170 °K for ice II, and ∿ 200 °K for ice Ic. The D_2O ices were warmed to ca. 10° below their respective transition temperatures. Deuterium dispersion mode resonances were successfully recorded for ices Ih, Ic, II, V, and IX. Previously, the value of e^2qQ/h for deuterons in ice Ih has been determined from single crystal data /19/, and it has been found to be temperature independent within 1 kHz between 143 °K and 263 °K. A typical dispersion trace for the deuterium resonance in ice II at 143 °K is shown in fig. 12.

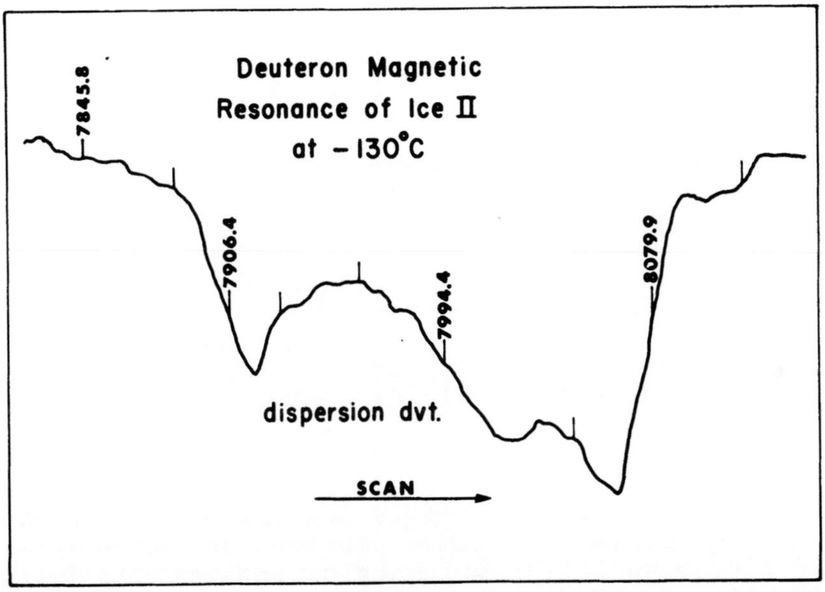

Fig.12 Deuteron magnetic resonance derivative of dispersion signal for ice II at -130 °C

It has been observed that the position of the first peak in the dispersion derivative signal did not change for ice Ih between 143 °K and 263 °K, and also, it was not changed in position by the use of different (50 to 400 milligauss) rf field settings. The quadrupole split resonances were measured from the frequency of the first peak to the Larmor frequency for the liquid; this separation was independent of the direction of scan. Estimates of the deuteron quadrupole coupling constants are given in table 4 for the various ices. It appears that the field gradient at the deuteron site is not appreciably different for any of the ices. An alteration of the O-D bond length by a maximum of 0.005 Å would suffice to bring the values of e^2qQ/h in accord.

Table 4 Deuteron quadrupole coupling constants for ice polymorphs *

Ice	(e^2qQ/h), kHz
Ih	192 ± 2.7
Ic	185 ± 3.6
II	179 ± 5.4
V	183 ± 7.7
IX	178 ± 9.3

* Polycrystalline powder pattern values; single crystal value for ice Ih=215±1 kHz /19/.

3.3 Proton NMR of ices under pressure

Saturated proton resonances were obtained for ice Ih in the pressure vessel shown in fig. 1 at pressures of 0.5, 1.0, and 1.5 kbar with rf fields near 5 milligauss. This rf field was about the lower stable limit of the oscillator. Signals which were obtained for ice Ih doped with $1.2 \cdot 10^{-4}$ mole/l HF were somewhat less saturated. As the ice Ih sample was transformed to ice III at -26 °C and 2.1 kbar, the proton resonance narrowed and as shown in fig. 13 displayed little evidence of saturation. Neither pure ice III nor HF-doped ice III showed any change of linewidth with changing pressure.

Proton linewidths were recorded for $1.2 \cdot 10^{-4}$ mole/l HF-doped ice III at a pressure of 3.0 kbar as a function of temperature. In the temperature region -22 °C to -42 °C, linewidth measurements were made at 5 °C intervals, with 30 to 40 min allowed for equilibration at each temperature. The linewidth increased from a mean (peak-to-peak) value of 7.5 gauss to about 9.7 gauss over this range. At -45°C, the proton linewidth suddenly became very broad and saturated. As the sample was warmed, saturation persisted at -40 °C and at -35 °C, but at -25 °C the relatively narrow absorption derivative signal was again observed.

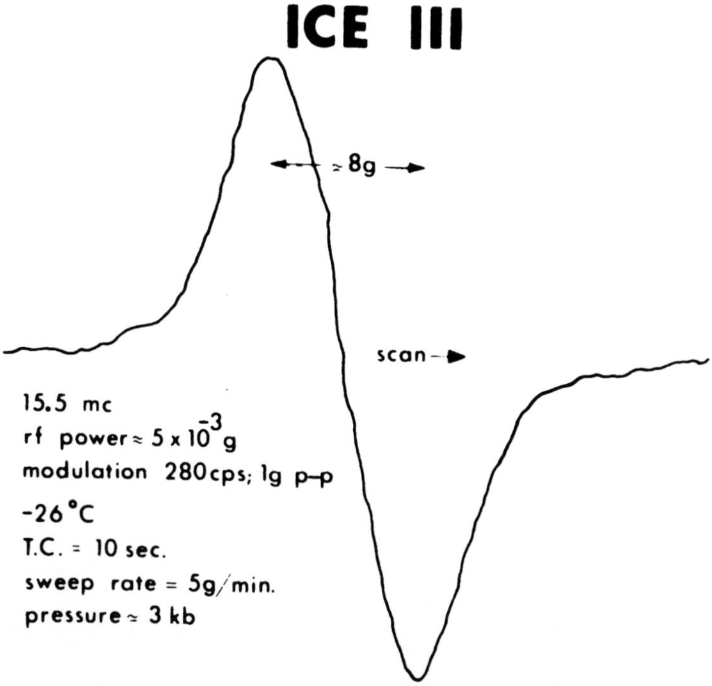

Fig.13 Derivative of absorption proton magnetic resonance of ice III at 3 kbar

It is conceivable that the sudden line broadening could be related to a proton ordered phase transformation (either II or IX).

As HF-doped ice V was cooled at 5 kbar, the proton line-width gradually increased from ∿ 8.8 gauss at -22 °C to about 13.5 gauss at -70 °C. (A value of 16.9 gauss was obtained at 75 °K at atmospheric pressure, table 3). Thus, in contrast to ice III, there was no abrupt line broadening for ice V as a function of temperature.

References

/1/ Dickson, S.A., H. Meyer, Phys.Rev. 138, A1293 (1965)

/2/ Robinson, F.N.H., J.Sci.Instr. 36, 481 (1959)

/3/ Sidhu, S.S., L.Heaton, M.H. Mueller, F.P. Campos, J.Appl.Phys. 27, 1040 (1956)

/4/ Bridgman, P.W., Proc.Am.Acad.Arts Sci. 47, 441 (1912); J.Chem.Phys. 5, 964 (1937)

/5/ Kell, G.S., E. Whalley, J.Chem.Phys. 48, 2359 (1968)

/6/ Pistorius, C., M.C. Pistorius, J.P. Blakey, L.J. Admiraal, J.Chem.Phys. 38, 600 (1963)

/7/ Nozik, A.J., M. Kaplan, J.Chem.Phys. 47,2960 (1967)

/8/ Kamb, B., Acta Cryst. 17, 1437 (1964)

/9/ Bertie, J.E., L.D. Calvert, E. Whalley, J.Chem.Phys. 38, 840 (1963)

/10/ Chan, R.K., D.W. Davidson, E. Whalley, J.Chem.Phys. 43, 2376 (1965)

/11/ Wilson, G.J., R.K. Chan, D.W. Davidson, E. Whalley, J.Chem.Phys. 43, 2384 (1965)

/12/ Whalley, E., D.W. Davidson, J.Chem.Phys. 43, 2148 (1965)

/13/ Bertie, J.E., E. Whalley, J.Chem.Phys. 40, 1646 (1964)

/14/ Kamb, W.B., S.K. Datta, Nature 187, 140 (1960)

/15/ Kamb, B., A. Prakash, C. Knobler, Acta Cryst. 22, 706 (1967)

/16/ Larson, A.C., R.B. Roof, Jr., D.T. Cromer, LASL Report 3309 (1965):
An Integrated Series of Crystallographic Computer Programs. VII.Interatomic Distance and Angle Calculations.

/17/ Pederson, B.,J.Chem.Phys. 41, 122 (1964)

/18/ Bertie, J.E., L.D. Calvert, E. Whalley, Can.J.Chem. 42, 1373 (1964)

/19/ Waldstein, P., S.W. Rabideau, J. Jackson, J.Chem. Phys. 43, 119 (1965)

STUDY OF THE SURFACE OF ICE WITH A SCANNING ELECTRON MICROSCOPE

J. D. Cross

Institute of Science and Technology

University of Manchester, England

Abstract

The "Stereoscan" electron microscope operates over the magnification range 20× to 50000× and has a depth of field several hundred times greater than conventional optical microscopes. Solid specimens can be studied directly without the use of thin films or replicas and therefore such a microscope is very suitable for the study of ice crystals. Electron micrographs of single crystal and polycrystalline ice, with magnifications of up to 10000× are presented. The micrographs show that evaporating polycrystalline ice has a complex fibrous surface while single crystal samples exhibit the expected hexagonal symmetry. The implications of these results for charge transfer during evaporation are outlined.

1. Introduction

This investigation of the surface of evaporating ice was carried out primarily to provide the knowledge of the ice surface necessary to understand the charge transfer that has been shown to accompany the evaporation of ice. It was pointed out by Latham /1/ that the removal of single charged molecules from the ice surface would require so much energy to overcome electrostatic forces that no charging should be observed. This discrepancy between observed and expected behaviour prompted a closer examination of the surface of evaporating ice. The use of optical microscopy is severely limited by the depth of focus of optical microscopes as well as by the limit of resolution. The scanning electron microscope has a wide range of magnifications, 20 to 50000 and has a depth of focus several hundred times greater than optical microscopes, it therefore is an ideal instrument for examining surface of solid samples. The scope of the instrument is however limited by the need for the sample surface to be electrically conducting and by the need to maintain a vacuum better than 10^{-4} torr within the specimen chamber. It was found that ice is a sufficiently good electrical conductor to prevent charging by the electron beam and that if the ice sample is mounted on a thermally insulating support, in this case a fine glass tube, the latent heat of vaporization is sufficient to hold the ice sample at a low enough temperature to bring its vapour pressure below 10^{-4} torr ($< -80^{\circ}C$). By this technique it is possible to observe the surface of evaporating ice directly. The pumping system of the Cambridge Instruments "Stereoscan" is able to handle the gas load from ice specimens up to approximately 10 mg. Samples of ice produced in various ways were mounted on glass capillaries attached to the normal microscope sample holders and the combination was cooled to $-40^{\circ}C$ so that the sample did not reach melting point in the time taken to fit it into the specimen chamber and start pumping. The evaporation rate was such that these samples could be studied for periods up to one hour before the sample completely sublimed. Samples of single and polycrystalline ice formed from triple-distilled water have been studied. Single crystal ice specimen were grown by allowing water to freeze in an undisturbed state with the natural temperature gradient acting. Also crystals of various shapes were grown in a diffusion chamber. Polycrystalline specimens were produced by freezing small water drops on the end of the glass capillaries used to insulate the specimen and also by cutting sections from ice formed by rapidly freezing water in a shallow dish. The ice specimens were attached to the capillary by wetting the end of the capillary before placing the sample upon it. The sample and capillary were then frozen together.

2. Results

A selection of typical micrographs of the surface of single crystal ice are shown in figure 1 to 8. Figures 1 and 2 show the surface of a specimen after three minutes evaporation. Figure 2 is a close-up of figure 1. It can be seen that the surface has a smooth appearance, all features being rounded. After fifteen minutes evaporation the sample has developed marked steps and facets;

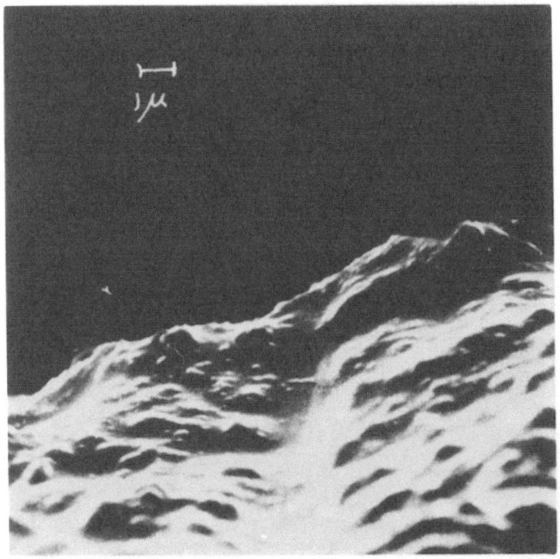

Fig.1 The surface of single crystal ice after 3 minutes evaporation. Instrument magnification ×1350

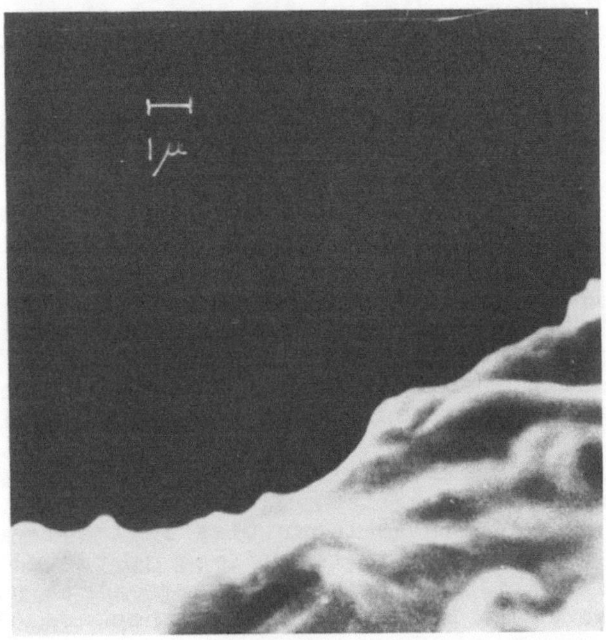

Fig.2 Close-up of the surface in figure 1. Instrument magnification ×6750

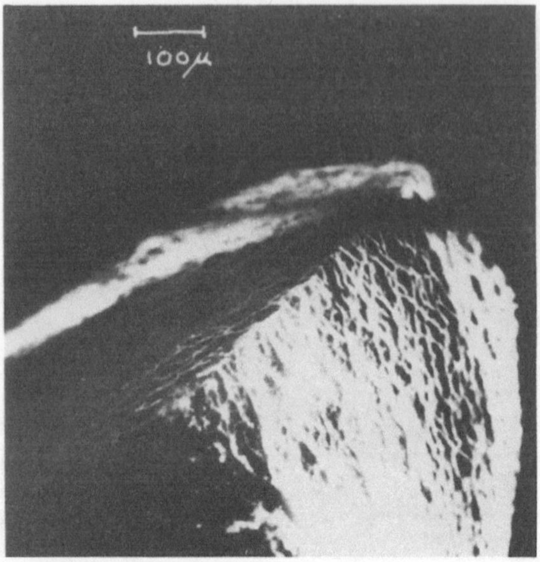

Fig.3 Overall view of single crystal ice with facetting visible. Instrument magnification ×120

figure 3 is an overall view of the sample at this stage. This facetting continues as evaporation proceeds until after approximately thirty minutes the stepped nature of the surface is very clear as shown in figures 4, 5 and 6. These figures show different magnifications of the same region. This stepping results from the difference in the evaporation rate of the basal and prismatic planes in the ice crystal. The prismatic planes are more rapidly removed exposing the basal plane perpendicular to the c-axis as a series of steps. Figures 4 and 5 show clearly that ice is almost amorphous in the basal plane, the prismatic faces being only slightly visible in some sections of figure 4. The basal plane is smooth. Figures 5 and 6 show a collection of small particles less then 0.5µ across at the intersection of the basal plane and prismatic faces. The particles, often spheroidal in appearance, are frequently seen in high magnification electron micrographs of ice. It is not known at present whether these are due to impurities in the ice or whether they are produced by the effect of the elctron beam. As these particles always appear in greatest numbers at discontinuities it is probable that they are due to impurities. Figures 7 and 8 show a feature observed infrequently in the study of single crystals; this is a line of small hexagonal etch pits, the pits change size

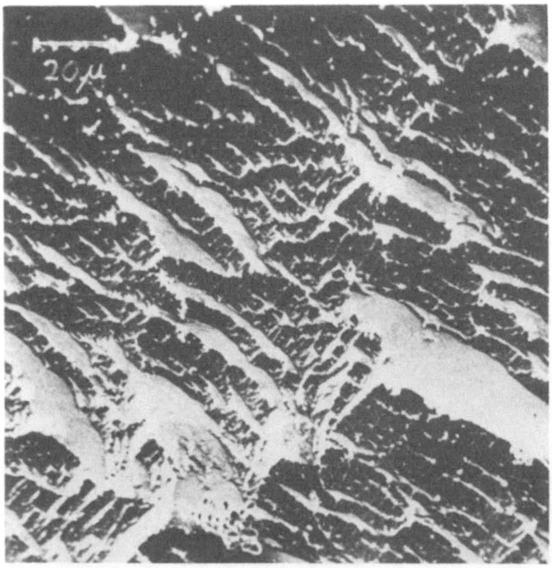

Fig.4 The surface of single crystal ice after 30 minutes evaporation. Instrument magnification ×500

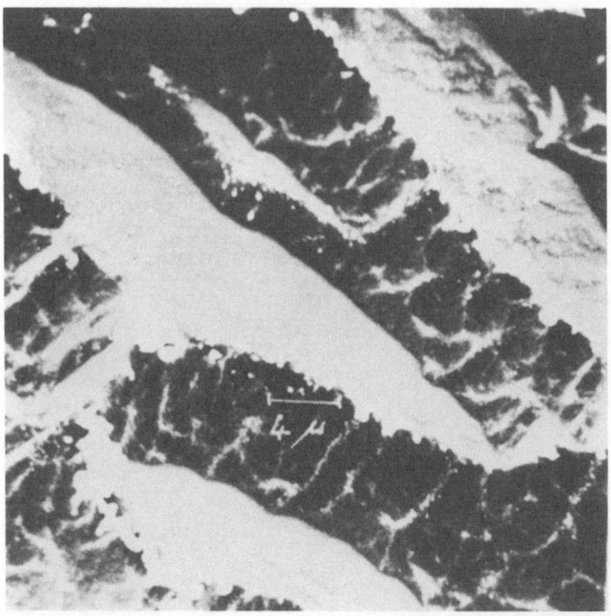

Fig.5 The surface of single crystal ice (close-up of part of figure 4). Instrument magnification ×2500

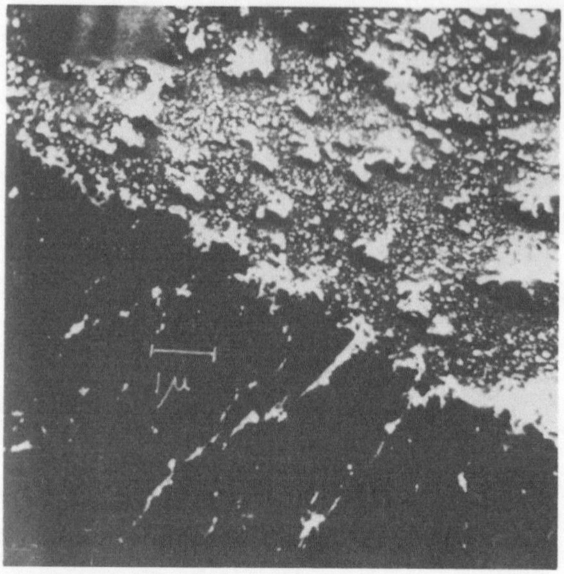

Fig. 6 Close-up part of figure 5. Instrument magnification ×10000

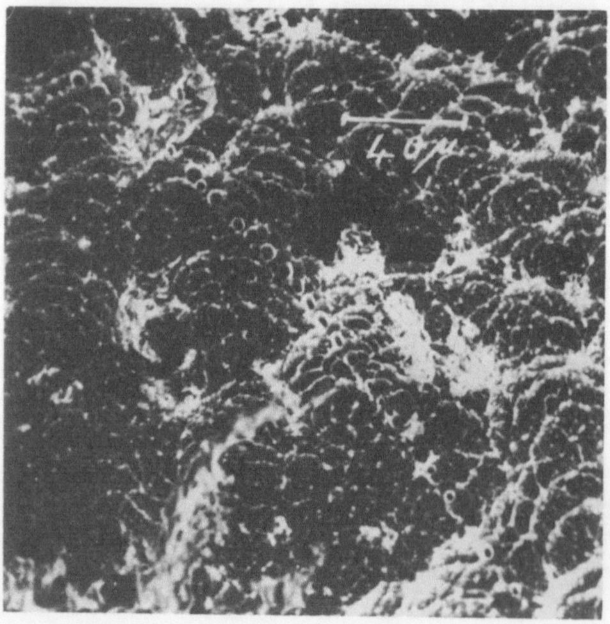

Fig.7 Hexagonal etch pits on single crystal ice. Instrument magnification ×400

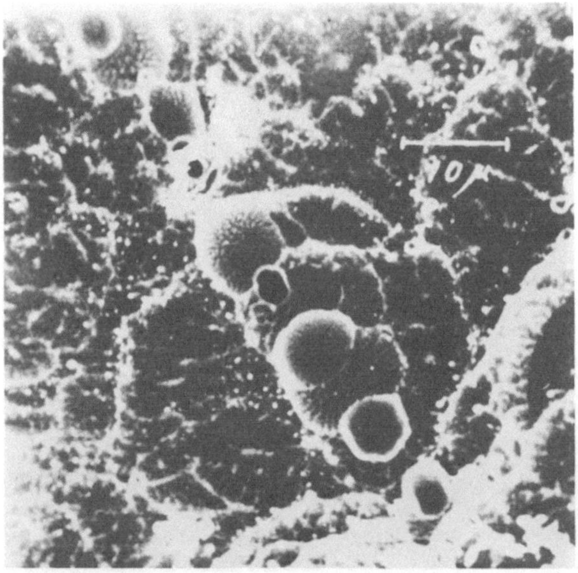

Fig. 8 Hexagonal etch pits on single crystal ice. Instrument magnification ×1600

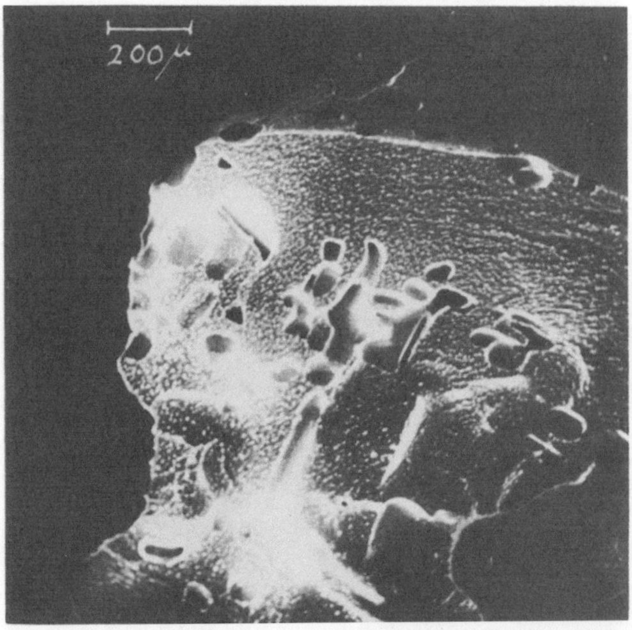

Fig.9 Overall view of planar ice crystal. Instrument magnification ×60

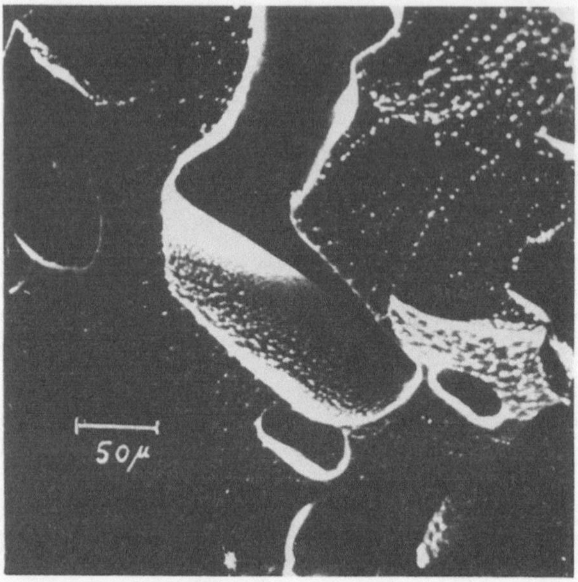

Fig.10 Etch pits on planar crystal. Instrument magnification ×240

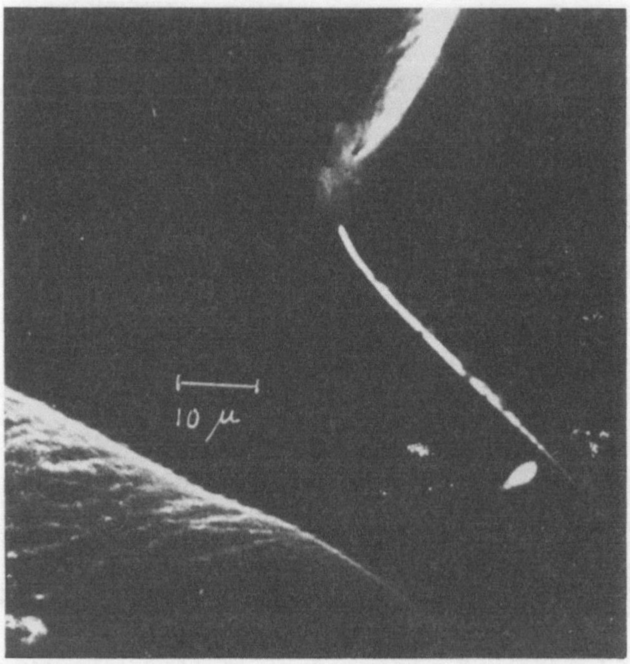

Fig. 11 Detail of etch pits ×1200

rapidly and attempts to observe them at high magnifications are unsuccessful due to the etching effect of the electron beam. These pits have been observed only on surfaces perpendicular to the c-axis and are thought to be due to a dislocation or grain boundary in the crystal.

Figures 9, 10, 11 show the appearance of a plate crystal grown in a diffusion chamber. Figure 9 shows a collection of etch pits, distributed in a roughly hexagonal pattern and figures 10 and 11 are higher magnification views of one pit. This type of behaviour is typical of the evaporating platelets studied. Columns and sections of dendrites had a similar appearance to the cleaved single crystal specimens.

The micrographs of polycrystalline ice are shown in figures 12-19. It might be expected that a polycrystalline ice sample would simply appear as a collection of small single crystals, each crystallite being etched in a similar manner to the larger cleaved specimens. Figures 12 and 13 show a small crystal of this type in a frozen drop on a glass capillary. This was, however, seldom seen and in cases where such small crystals did occur they formed only a small part of the sample. All polycrystalline samples studied were characterised by the development of fine fibrous surface structure overlaying a surface with no obvious crystalline features. Figures 14, 15, 16 are typical micrographs of this type of structure. It can be seen that the fine fibrous strands are considerably less than one micron thick and would not be visible to an optical microscope. This fibrous structure developed as evaporation continued, in some cases large sections became detached from the main sample, indicating that the structure has little mechanical strength. In some cases the fine structure took the form of whiskers. Figures 17 to 19 show the development of these whiskers. Figure 17 is a view of a frozen drop after five minutes evaporation. The surface is comparatively smooth. After a further ten minutes evaporation the micrograph in figure 18 was taken; it can be seen that the surface is now covered with whiskers. Figure 19 is a high magnification view of one whisker. The main column of the whisker is approximately 8μ across the fine structure at the upper end is less than 1μ in thickness.

3. Discussions

The most striking feature of these results is the difference between polycrystalline and single crystal ice. Experimental studies of ice have often been carried out

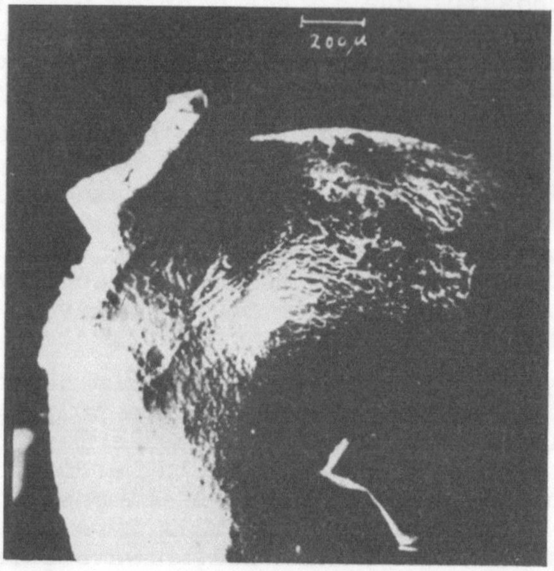

Fig. 12 Overall view of polycrystalline ice containing a region of single crystal. Instrument magnification ×50

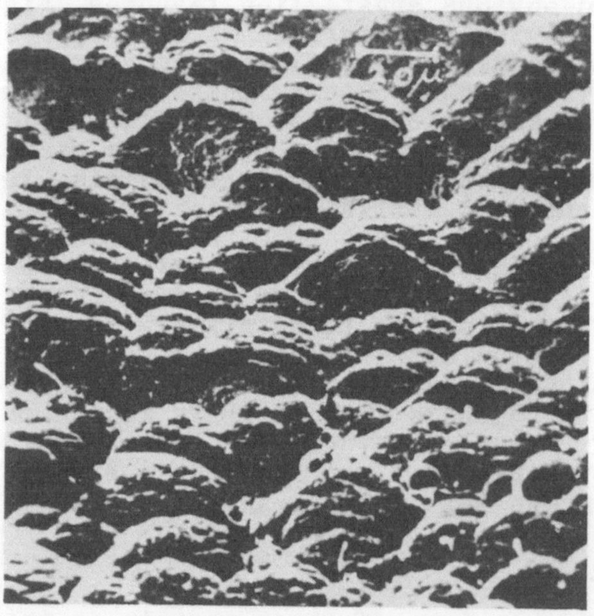

Fig. 13 Facets of single crystal region of figure 12. Instrument magnification ×500

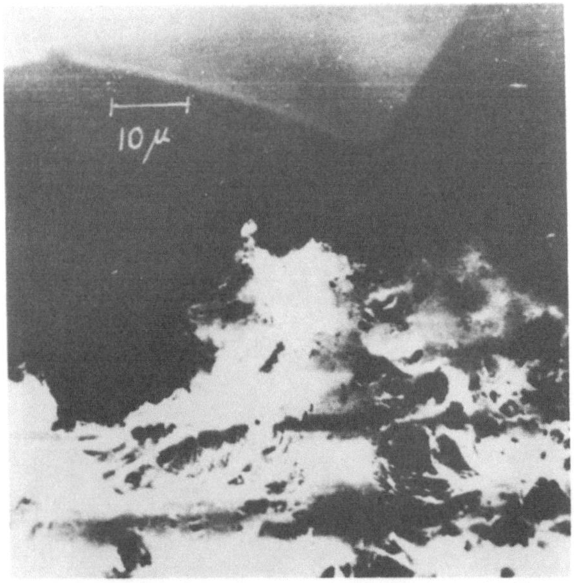

Fig. 14 Surface structure on polycrystalline ice. Instrument magnification ×1200

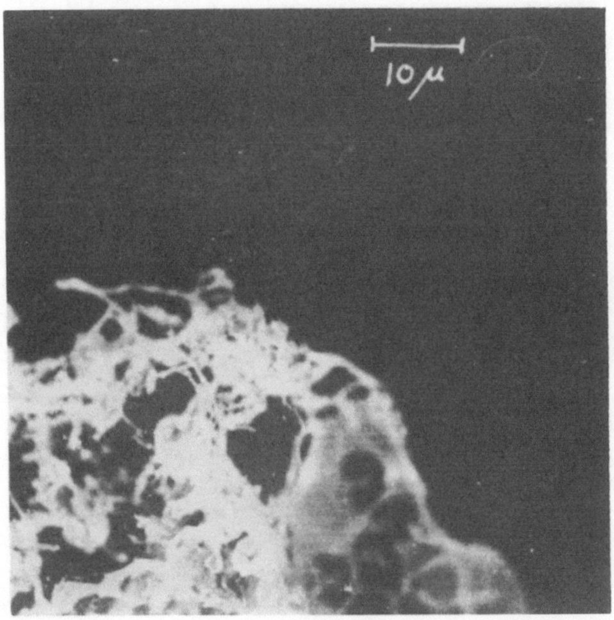

Fig. 15 Surface structure on polycrystalline ice. Instrument magnification ×1200

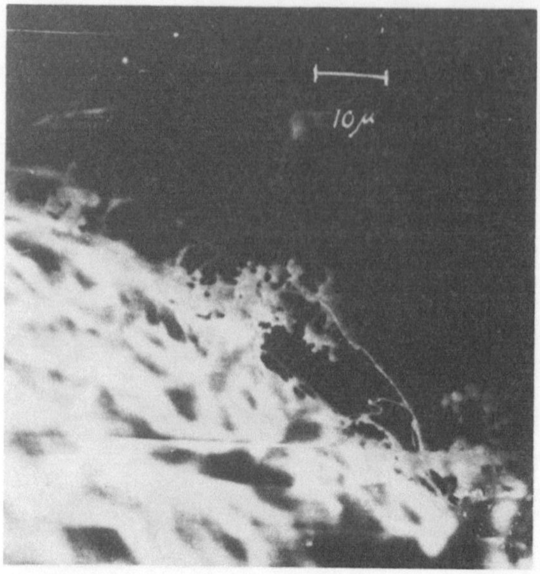

Fig. 16 Surface structure on polycrystalline ice. Instrument magnification ×1200

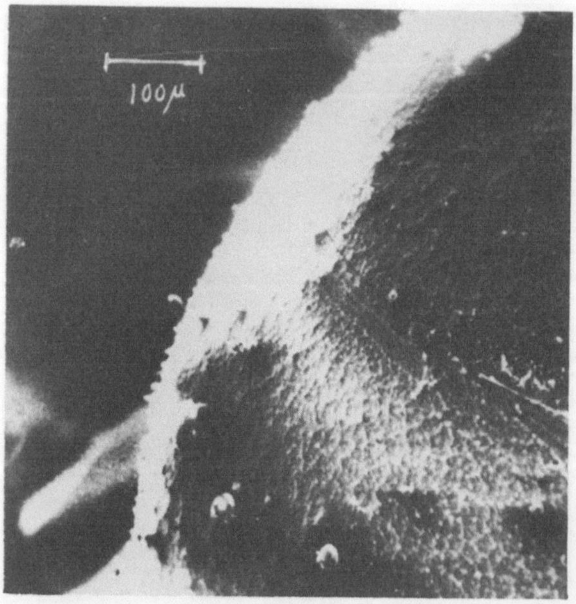

Fig. 17 Overall view of polycrystalline ice after five minutes evaporation. Instrument magnification ×150

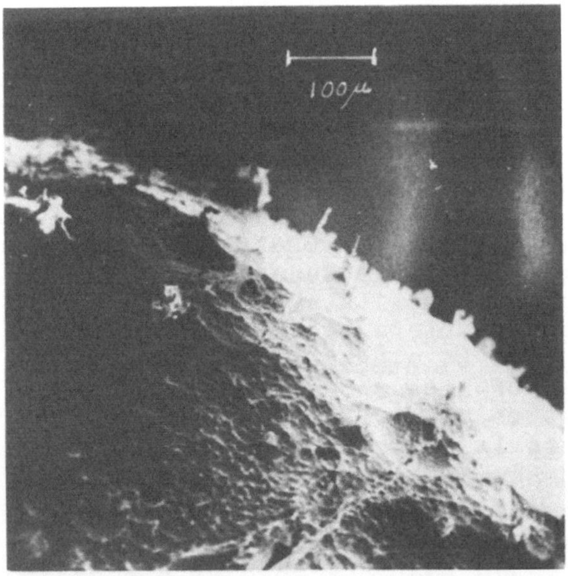

Fig. 18 Sample in figure 17 after 15 minutes evaporation. Instrument magnification ×150

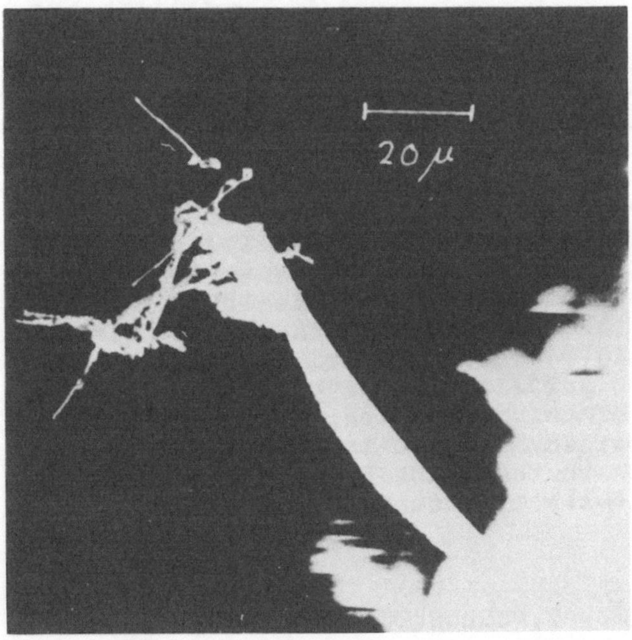

Fig. 19 Close-up of one whisker on figure 18. Instrument magnification ×750

on polycrystalline samples with the assumption that the polycrystalline samples do not differ significantly from the single crystals. In the case of evaporation studies such as those by Latham and Stow /2/ the use of polycrystalline ice must greatly affect the result. In the case of evaporation in an air or nitrogen stream the fine surface structure would be carried away. The temperature gradient in the ice surface would result in fibres being charged by the thermo-electric effect and the evaporation process would result in a charge separation exactly equivalent to the charge separation produced by disintegration of frost deposits by an air stream. As the charge is carried by splinters broken off, there are no energetic problems due to electrostatic forces. Recent work in the laboratory has shown that no measurable charge separation takes place if polycrystalline ice evaporates in vacuo. This is to be expected as the surface is not disturbed in the same way as in air evaporation. The occasional falling off of some splinters in vacuo would produce a charge separation but it would be very small. It is to be expected that the regular and solid surface of single crystal ice would not produce a charge separation on evaporation in air owing to the electrostatic forces opposing the removal of ions from a plane surface and experiments are at present being carried out to check this.

Obviously the actual appearance of ice evaporating at low pressure will differ greatly from the surface in air but the actual structure revealed will exist within the ice when exposed to a saturated environment. All the micrographs of single crystal ice were taken of samples cut from a single crystal and otherwise untreated. If a single crystal sample is mechanically polished before evaporation the surface initially shows the characteristic fibrous appearance of polycrystalline ice. After a few minutes this disappears and the normal single crystal structure is revealed. The rate of evaporation is such that the polycrystalline layer on the surface must be of the order of ten microns thick. The practice of polishing single crystals may lead to incorrect results in any experiment where the quantity being observed could depend upon the crystalline state of the surface.

References

/1/ Latham, J., C.D.Stow, J.Atmos.Sci. 23, 245-247 (1966)

/2/ Latham, J.,C.D.Stow, J.Atmos.Sci. 22, 320-24,(1965)

THE PLANAR GROWTH OF ICE FROM THE PURE MELT

P.V. Hobbs and W.M. Ketcham

Cloud Physics Laboratory, Department of

Atmospheric Sciences, University of Washington

Seattle, U.S.A.

Abstract

A review is given of recent work which has been carried out in our laboratory on the growth of ice into slightly supercooled pure water. The ice apparently grew by the propagation of steps across the ice-water interface. Some of the steps originated at grain boundaries while others grew in a spiral fashion.

The shapes of the grain boundary profiles between adjacent ice crystals which are in contact with (a) the liquid phase and (b) the vapor phase have been measured as a function of the degree of mismatch between the crystallographic orientations of the crystals. These results are used to estimate the magnitudes of the interfacial energies between ice and water. Finally, the development of preferred crystallographic orientations as ice grows slowly from the liquid phase is explained in terms of the shapes of the grain boundary profiles between adjacent grains.

1. Introduction

In recent years ice physicists have paid increasing attention to the surface properties of ice about which much less is known than the bulk properties. In this paper we summarize the results of a series of experiments which have provided new information on several subjects related to the surface properties of ice.

In all of the experiments to be described, the water was frozen by extracting heat through the ice that had already formed. In this case the ice-water interface remaines essentially planar. Due to the slow growth rates that can be achieved in planar growth, the surface structure of the ice may be observed much more readily than in the case of dendritic growth in strongly supercooled water.

The results are described in three parts. In the following section the mechanism by which ice grows from the pure melt is discussed and new experimental evidence is presented. In § 3 an experimental determination of the surface free energies of ice is described. Finally, in § 4 the preferred crystallographic orientations in the growth of ice from the melt is described and a theory is presented to explain this phenomenon.

2. Mechanism of growth.

Several workers have obtained indirect experimental evidence which indicates that when ice grows in the direction of the c-axis into supercooled water it does so by the propagation of steps across the basal plane /1,2/. Hillig, Turnbull /3/ found that the growth velocity of ice normal to the c-axis was proportional to the 1.7th power of the supercooling of the water; they concluded from this result that growth in this direction also was probably controlled by step growth. We now describe experiments in which 'steps' have been directly observed to move across the freezing interface of ice growing by planar growth from pure melt.

The experimental apparatus is shown in fig.1. It consisted of a stainless-steel base with a glass coverplate.

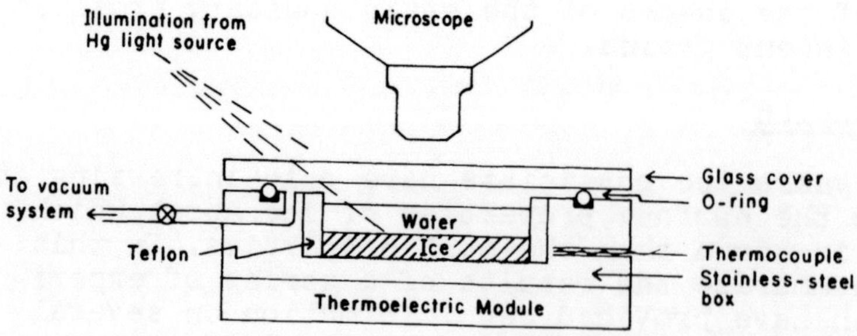

Fig.1 Experimental apparatus

The base was cooled by means of a thermoelectric module attached to the base of the base, and the entire assembly was mounted on the micrometer stage of a microscope so that the ice-water interface could be conveniently viewed.

Prior to freezing, a vacuum system was used to remove all air from the box. The water was singly distilled and was passed through an ion exchange column. The conductivity of the water was 8.3×10^{-7}(ohm cm)$^{-1}$. After the water has been slightly supercooled it was nucleated either by giving the box a sharp rap or by passing a rod cooled in liquid nitrogen over the surface of the water. A short time after the freezing was initiated the ice-water interface became parallel to the bottom of the box, and the polycrystalline ice consisted of grains about 0.3 mm^2 in area. The freezing rates were in the range of 0.1 to 1 micron/sec.

When the water above the ice was essentially isothermal, steps were observed to move over the entire freezing interface. Some of these steps originated at grain boundaries while others grew in a spiral fashion. The concentration of the spiral forms was typically one per mm^2 which was two to four times greater than the concentration of grain boundary sources. The steps were not restricted to the grain from which they originated, although frequently there was a change in the direction of propagation of a step after it had traversed a grain boundary (see fig.2).

The heights of the steps were determined by dividing the vertical distance that a freezing interface advanced in a given time by the number of steps that

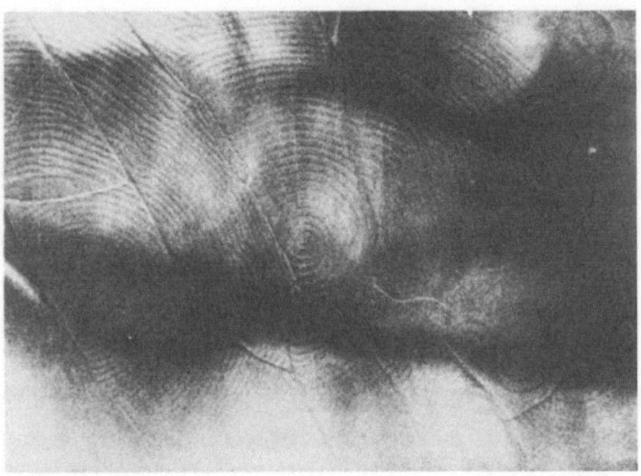

Fig.2 Steps on ice growing from the melt
(Magnification: x133)

crossed the interface in the same time. The smallest step height that was observed was 0.1 microns and the largest was 4 microns. The spacing between the steps ranged from 5 to 20 microns, and the velocity of the steps from about 1 to 10 microns/sec. It should be noted, however, that the height, spacing, and horizontal velocity of a step were very dependent on the temperature gradient in the water. If the glass coverplate was warmed slightly, the step height and the spacing between steps could be reduced to the point where the steps were invisible, even though the vertical growth velocity of the freezing interface remained unchanged. For any particular steady-state freezing conditions the step height and spacing between the steps were approximately the same over the entire freezing interface.

These experimental observations appear to demonstrate conclusively that the planar growth of ice from the pure melt occurs by a step mechanism. However, two of the experimental observations are somewhat disturbing. Firstly if we assume that the spiral steps originate from screw dislocations then the heights of the steps are extraordinarily large. Secondly, it is quite surprising that the steps traverse grain boundaries (it should be noted however, that the depth of a grain boundary was generally less than the height of a step). It is clear that further work is needed on this subject in order to establish the exact nature of the steps observed in these experiments.

3. An experimental determination of the surface free energies of ice

For a pure material four surface free energies may be defined: liquid-vapor (γ_{LV}), a solid-liquid (γ_{SL}), solid vapor (γ_{SV}) and the grainboundary energy (γ_{gb}). The surface free energy between the liquid and the vapor phases of a material is relatively easy to measure by methods which involve deforming the liquid; for water at 0°C it is 75.7 erg/cm^2. The other three surface free energies are more difficult to measure. We now describe the first direct experimental measurements of γ_{SL}, γ_{SV}, and γ_{gb} for the water-substance.

For any polycrystalline material grain boundary grooves form at the solid-vapor interface. If γ_{SV} is isotropic a grain boundary groove is in thermodynamic equilibrium when the following relationship is satisfied:

$$\frac{\gamma_{SV}}{\gamma_{gb}} = \frac{1}{2\cos(\theta_{SV}/2)} \tag{1}$$

where θ_{SV} is the solid-vapor grain boundary groove angle. An equation similar to (1) may be written for the solid-liquid interface provided that the solid-liquid grain boundary groove angle, θ_{SL}, is greater than 0°. At equilibrium and for isotropic γ_{SL} this relation is:

$$\frac{\gamma_{SL}}{\gamma_{gb}} = \frac{1}{2\cos(\theta_{SL}/2)} \tag{2}$$

From (1) and (2) it can be seen that if θ_{SV} and θ_{SL} are measured experimentally the ratios γ_{SV}/γ_{gb}, and γ_{SV}/γ_{SL} may be deduced.

In order to obtain an absolute measure of γ_{SV}, γ_{SL}, and γ_{gb} use is made of Young's equation which relates γ_{SV}, γ_{SL}, and γ_{LV} to the equilibrium contact angle ϕ made at the intersection of the solid-vapor, solid-liquid, and liquid-vapor surfaces. For the case in which the liquid does not completely wet the solid surface, that is $0^o < \phi < 180^o$, the condition of equilibrium at the intersection of the three interfaces requires that:

$$\gamma_{SV} = \gamma_{SL} + \gamma_{LV}\cos\phi \,. \tag{3}$$

For this case (1) and (2) may be substituted into (3) to obtain the following equation.

$$\gamma_{SV} = \frac{\gamma_{LV}\cos\phi\,\cos(\theta_{SL}/2)}{\cos(\theta_{SL}/2)-\cos(\theta_{SV}/2)} \tag{4}$$

Since γ_{LV} is accurately known, γ_{SV} may be evaluated from (4) upon the substitution of experimentally determined values of θ_{SV}, θ_{SL}, and ϕ. By the proper combination of (1), (2) and (3) equations similar to (4) may be derived for γ_{SL} and γ_{gb}, thereby permitting these surfaces free energies to be evaluated.

For the second case, where the liquid completely wets the solid surface (i.e. $\phi = 0$), the change in total interfacial free energy for a change in solid-vapor area is less than or equal to zero. Consequently, for complete wetting the relationship between γ_{SV}, γ_{SL}, and γ_{LV} is in the form of an inequality, namely:

$$\gamma_{SV} \geq \gamma_{SL} + \gamma_{LV} \tag{5}$$

For this case γ_{SV} must be evaluated from the following equation which is obtained by substituting (1) and (2) in (5).

$$\gamma_{SV} \geq \frac{\gamma_{LV} \cos(\theta_{SL}/2)}{\cos(\theta_{SL}/2) - \cos(\theta_{SV}/2)} \tag{6}$$

In this case therefore only a lower limit can be placed on the value of γ_{SV}. Equations similar to (6) may also be written for γ_{SL} and γ_{gb}, and upon substitution of the values of γ_{LV}, θ_{SV}, and θ_{SL} lower limits may also be calculated for these surface free energies.

In order to evaluate γ_{SV}, γ_{SL}, and γ_{gb} for a particular material by the method described above, θ_{SV}, θ_{SL}, and ϕ must be determined. We now describe the experiments which were performed to obtain these quantities.

The experimental apparatus was the same as shown in fig.1. Prior to use the stainless steel box was cleaned by first heating to a temperature of 1000°C to drive off volatiles, then boiling in a solution of sodium hydroxide and ethyl alcohol, and finally boiling in distilled water. The glass cover and the teflon were cleaned by boiling in sodium hydroxide and ethyl alcohol and then boiling in distilled water. Polycrystalline ice was formed in the box by first supercooling doubly distilled water to a temperature of -1 to -2°C and then nucleating the water by passing a glass rod chilled in liquid nitrogen over the surface of the water. Prior to nucleation a vacuum system was used to remove all air from the box. A short time after nucleation the ice-water interface became parallel to the bottom of the box, and the cooling of the box was then adjusted so that the freezing velocity of the planar interface was about 0.1 microns/sec. After all the water had frozen the thickness of the ice was about 3 mm and the average ice-vapor grain area was about 4 mm^2. The absolute pressure inside the box during the after freezing was equal to the equilibrium vapor pressure of the ice.

After all the water had frozen, the ice was maintained at a constant temperature of -3°C for a period of two days in order to allow the grain boundaries to reach an equilibrium shape. A replicating solution of 5% by weight of formvar dissolved in ethylene dichloride was then applied to the ice-vapor surface. After a period of about 10 minutes the solvent evaporated and the formvar replica was carefully peeled off the ice and placed on a clean glass slice. The replica was lightly silvered and observed under an interference microscope. The fringe contours were aligned either parallel or perpendicular (fig.3) to the grain boundary, and 35 mm pictures were taken of the many grain boundary grooves on each replica. In the case of perpendicular fringe pattern, Θ_{SV} was determined from the following equation /4/:

$$\tan (\Theta_{SV}/2) = \frac{2S}{\lambda M} \tan \frac{(\eta)}{2} \qquad (7)$$

where, S is the fringe spacing, λ the wavelength of the monochromatic light used in the interference microscope, M the linear magnification, and η the fringe angle measured from the photograph. For a parallel fringe pattern Θ_{SV} was determined from the equation:

$$\tan (\Theta_{SV}/2) = \frac{2S}{\lambda M} . \qquad (8)$$

There is an error involved in the use of the interference microscope the source of which has been discussed by Tolman and Wood /5/. In order to determine the error involved in the use of our interference microscope a razor blade was cut at precisely 161° and observed under the interference microscope. It was found that tan ($\Theta_{SV}/2$) determined through the use of the interference microscope overestimated the true value of tan ($\Theta_{SV}/2$) by 6,8%. Consequently, all values of Θ_{SV} determined from the interferograms were corrected by this amount.

Initially attempts were made to measure Θ_{SL} by the same method wich was used to determine Θ_{SV}. However, it was quickly discovered that Θ_{SL} could not be determined in this way because it was less than the smallest angle that could be resolved by the interference microscope. However, an alternative method was found whereby Θ_{SL} could be determined directly. In this method the ice, originally colder than 0°C, was slowly warmed. When the temperature of the ice reached 0°C a small column of water would form at the intersection of three grains (fig.4). Water forms first at the intersection of three grains because the equilibrium melting temperature of the curved surfaces in the vicinity of each of these intersection points is less than 0°C. The ice continued to melt at these intersection points until the radius of curvature became sufficiently large that the equilibrium melting temperature of the curved region was

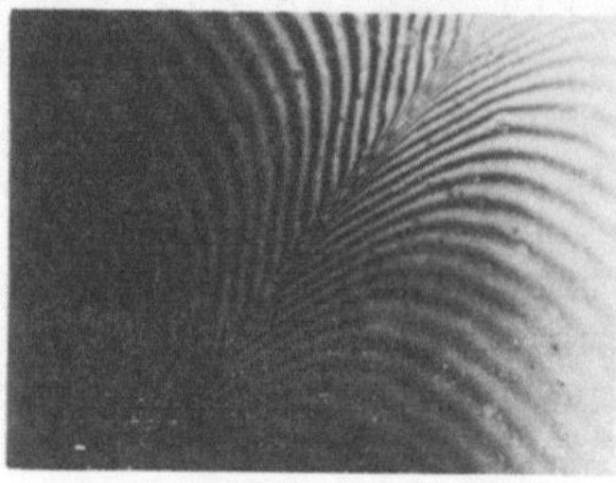

Fig.3 An interferogram of a grain boundary at an ice-water vapor interface when the fringes are normal to the grain boundary (Magnification: x1070)

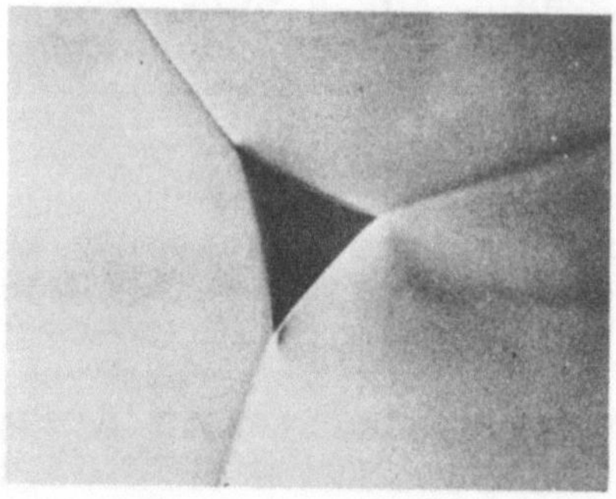

Fig.4 Preferential melting of ice at the intersection of three grains (Magnification: x133)

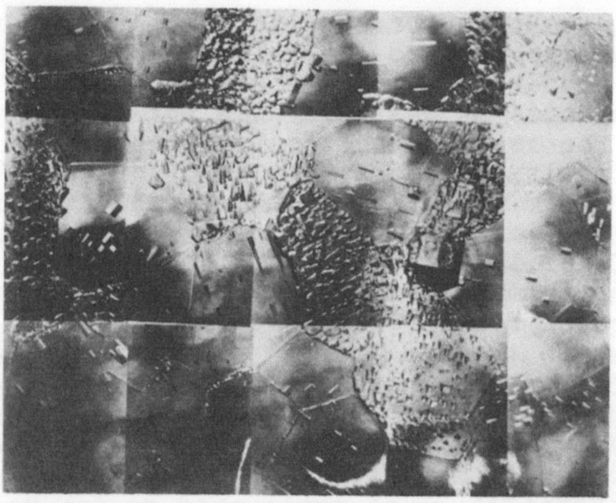

Fig.5 Etch pits on an ice-water vapor interface (Magnification x15)

closely equal to the equilibrium melting temperature of a plane surface. After this preferential melting had occured 35 mm pictures were taken of many of these intersection points and Θ_{SL} was measured directly from these photographs.

The crystal orientation of each ice grain was determined in order to be able to take into account any variation of Θ_{SV} and Θ_{SL} which might result from the anisotropy of γ_{SV}, γ_{SL}, and γ_{gb} with crystal orientation. The method which was used consisted of applying to the ice-vapor surface a solution of 2% formvar dissolved in ethylene dichloride /6/. The solvent was evaporated away leaving a thin slightly porous plastic film over the ice. The atmosphere above the ice was then maintained at sub-saturation with respect to ice by continuous evacuation of the box by the vacuum system. Under these conditions the ice evaporated at random sites causing numerous etch pits to form over the entire surface. The etch pits, which reveal the crystallographic orientation of the crystals, formed in about 10 minutes when the ice was at $-6^{\circ}C$. After satisfactory etch pits had formed 35 mm pictures were taken of the entire etched ice surface. These pictures were then mounted on a board to form a magnified view of the etched surface (fig.5). From this large mosaic the crystal orientation of each ice grain could easily be determined.

In the past the contact angle of water on ice has usually been assumed to be zero degrees. However, recent experiments by Knight /7/ indicate that it may be as large as 12°. Since we questioned whether the angle measured by Knight was in fact the equilibrium angle which exists at the intersection of the ice-vapor and water-vapor interfaces, we have independently determined the magnitude of the contact angle of water on ice. This was done through the use of the following equation (Appendix A) which relates the equilibrium contact angle to the height H of a liquid puddle resting on a plane solid surface:

$$\cos = 1 - \frac{\rho g H^2}{2\gamma_{LV}} \quad (9)$$

where, ρ is the density of the liquid and g the acceleration due to gravity. Since ρ, g, and γ_{LV} are known, determination of H allows the contact angle ϕ to be deduced.

The height of a water puddle resting on ice was measured by first freezing water in the apparatus shown in fig.1 until a puddle of water about 1 cm^2 in area covered a portion of the ice. The temperature of the thermoelectric module was then adjusted so that neither melting nor freezing could be observed. Under these equilibrium conditions the microscope was focused on a grain boundary which was covered with water. The water puddle was then frozen and the microscope refocused on the point at which the grain boundary intersected the ice-vapor surface. From the difference in focusing distances in the two cases, the thickness of the water which had covered the ice was found to be (40 ± 10) microns. From this measured value of H the contact angle of water on ice calculated from (9) is found to about 1°. Since ϕ is greater than 0°, equation (4) may be used to determine γ_{SV} for water.

The experimentally determined values of Θ_{SV} and Θ_{SL} as a function of the acute angle β between the c-axes of the two adjacent grains are shown in fig.6 and 7. From fig.7 it is seen that Θ_{SL} is fairly constant for β between 10° and 80° and is equal to (20 ± 10°). For a similar range of values of β, the magnitude of Θ_{SV} is seen (fig.6) to be equal to (145 ± 2°).

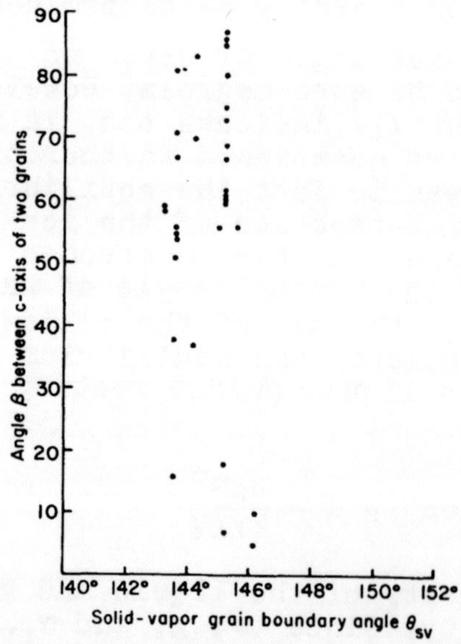

Fig.6 The ice-water vapor grain boundary groove angle Θ_{SV} as a function of the acute angle β between the c-axes of the two grains

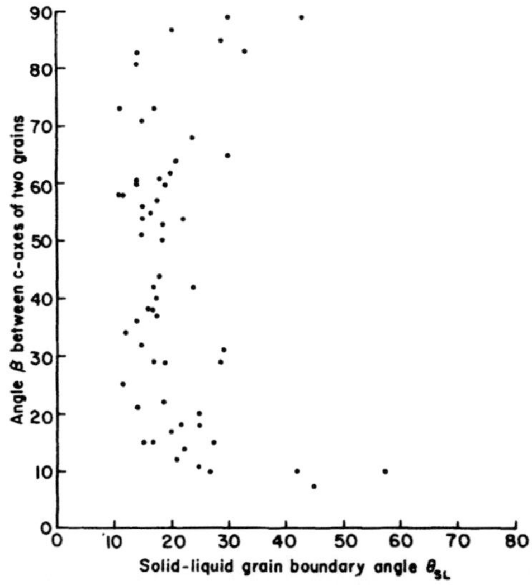

Fig.7 The ice-liquid water grain boundary groove angle Θ_{SL} as a function of the acute angle β between the c-axes of the two grains

With the experimentally determined values of Θ_{SV}, Θ_{SL}, and ϕ , all the necessary information is at hand to determine γ_{SV}, γ_{SL}, and γ_{gb} from equations (1, 2 and 3). At a temperature of $0^{\circ}C$, and for $10^{\circ} < \beta < 80^{\circ}$, the experimentally determined values of

$$\phi = 1^{\circ}, \quad \frac{\Theta_{SL}}{2} = (10 \pm 5)^{\circ}, \quad \frac{\Theta_{SV}}{2} = (72.5 \pm 1)^{\circ},$$

and $\gamma_{LV} = 75.7$ erg /cm^2 yield the following surface free energies:

$$\gamma_{SV} = (109 \pm 3) \text{ erg /cm}^2$$
$$\gamma_{SL} = (33 \pm 3) \text{ erg /cm}^2$$
$$\gamma_{gb} = (65 \pm 3) \text{ erg /cm}^2.$$

In fig.6 and 7 Θ_{SV} and Θ_{SL} are plotted as a function of β so that the anisotropy of γ_{SV}, γ_{SL}, and γ_{gb} can be seen. It should be noted that β is not the true grain boundary mismatch parameter. Figures 5 and 6 show that γ_{gb} varies with β. For small values of β, Θ_{SV} and Θ_{SL} become large. This increase in Θ_{SV} and Θ_{SL} for decreasing β is to be expected because Θ_{SV} and Θ_{SL} must approach 180° as the true mismatch parameter goes to zero. It also appears that γ_{gb} decreases slightly as β approaches 90°. We believe that this decrease is real and is a result of a rather good match of the molecules of one grain with the molecules of the other for values of β close to 90°.

Since the scatter in $\cos(\theta_{SV}/2)$ and $\cos(\theta_{SL}/2)$ is small and since many different ice-water vapor and ice-water surface orientations were included in the results shown in fig.6 and 7, it appears that the anisotropy of γ_{SV} and γ_{SL} is small and is within the experimental error. However, it should be emphasized that only about 3% of the total data points include cases in which the basal plane was exposed at the grain boundary groove. It was observed that when the basal plane was exposed at a grain boundary the groove was generally asymmetrical, while for grain boundaries which did not expose the basal plane the groove was usually symmetrical. This asymmetry indicated that the surface free energy of the basal plane of ice is slightly lower than for the other planes in ice. We will return to this point in § 4.

The value of the ice-water surface free energy deduced above (33 ± 3 erg/cm^2 at 0°C) is somewhat higher than the values which have been obtained previously by applying the theory of homogeneous nucleation to the freezing of water droplets. However, values of γ_{SL} deduced from the theory of homogeneous nucleation are not very accurate. This point can best be illustrated by showing that with a judicious (but by no means unreasonable) choice of parameters, the value of γ_{SL} for water deduced from homogeneous nucleation theory can be made to agree quite closely with the value deduced in this paper. The frequency of formation I of a critical nucleus in a super-cooled liquid due to homogeneous nucleation is given by /8/:

$$I = \frac{nkT}{h} \exp\left(-\frac{\Delta G_a}{kT}\right) \exp\left(\frac{C\gamma_{SL}^3}{\Delta G_v^2 kT}\right) \qquad (10)$$

where, n is the number of molecules per unit volume in the liquid phase, k Boltzmann's constant, h Planck's constant, T temperature, ΔG_a the activation energy for transporting a molecule across the solid-liquid interface, ΔG_v the difference in free energy per unit volume between the liquid and the solid phases, and C a factor related to the shape of the critical nucleus. We choose the following values for water at temperature of -40°C: $I = 10^{12}$ nuclei/cm^3 sec /9/, $\Delta G_a = 2 \times 10^{-13}$ erg /10/, $\Delta G_v = 68 \times \Delta T/T$ cal/g where ΔT is the supercooling of the droplet /11/, and $C = 16\pi/3$ /12/. With these values the magnitude of γ_{SL} determined from equation (10) is 24 erg/cm^2 at -40°C. If we now take

$d\gamma_{SL}/dT$ = 0.2 erg/cm^2/^{o}C /13/, γ_{SL} for water at 0°C is found to be 32 erg/cm^2 which is extremely close to the value deduced in this paper. It should also be noted that with the above values for the quantities in equation (10) $\frac{dI}{dT}$ is equal to -9.3 nuclei/^{o}C/cm^3/sec at -40°C. This result is in good agreement with the experimental observation of Mason /14/ in which the nucleation rate of supercooled droplets was found to increase by about an order of magnitude for every degree decrease in temperature.

We conclude from this discussion that due to the inherent uncertainties in deducing γ_{SL} from the theory of homogeneous nucleation, the value of the surface free energy of the ice-water interface at 0°C which we have deduced by a direct experimental method is probably the most reliable value available at this time.

4. The preferred orientation in the growth of ice from the pure melt

Ice exhibits a preferred crystallographic orientation when it forms under natural conditions in lakes, and also when it is grown in the laboratory by the extraction of heat in one dimension through the ice. In lake ice there is a progressive increase in the number of grains with horizontally oriented c-axes with increasing depths below the ice-air interface. Also, in the growth of large single crystals of ice from the melt by the technique introduced by Jaccard /15/, the c-axis of the single crystal is generally horizontally.

A detailed experimental investigation into the development of this preferred orientation in ice has been carried out by Ketcham and Hobbs /16/.It was concluded that when ice grows from the melt by the extraction of heat through the ice, grain A will encroach (i.e. wedge-out) grain B only if grain B satisfies both the following two conditions:

1. B must have its c-axis tilted towards the line formed by the intersection of the grain boundary between A and B and the ice-water interface (hereafter referred to as 'line L'). This condition is illustrated in fig.8. In order for grain A to wedge off grain B, the c-axis of B must be in the quadrant bounded by the lines Ox and Oy.

2. The projection on the ice-liquid interface of the c-axis of B must be perpendicular to line L.

Provided the c-axis of grain A does not lie in the quadrant bounded by OX and OY, or the projection of the

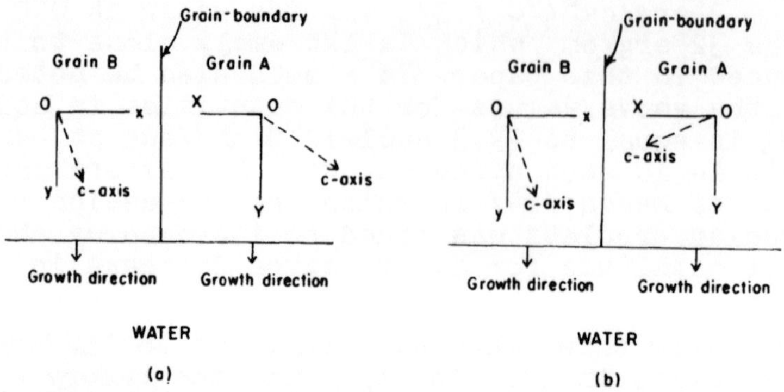

Fig.8 Conditions for grain A to encroach on grain B (actual shape of the grain boundary is not indicated)

c-axis of grain A is not perpendicular to line L, there is no restriction on the orientation of A (fig 8a). However, if the c-axis of grain A does lie in the quadrant bounded by OX and OY and the projection of the c-axis of A is perpendicular to line L (fig. 8b), grain A will encroach onto grain B only if the c-axis of A makes a larger angle with the growth direction than the c-axis of B. In both cases, the smaller the angle between the c-axis of grain B and the direction of growth the faster will grain A encroach onto grain B.

It can be seen from these rules that the larger the angle between the c-axis of a grain and the growth direction the greater is the probability that the grain will be preferred. Therefore as a piece of polycrystalline ice grows from the melt, those grains in which the c-axis makes a large angle with the growth direction will become increasingly dominant. However, it is important to note that in the case shown in fig. 8a grain A will wedge grain B even if the c-axis of grain A makes a smaller angle with the growth direction than does the c-axis of grain B. The above rules are therefore compatible with the experimental results, namely, that the preferred orientation in the growth of ice from the melt is generally one in which the c-axis is perpendicular to the growth direction, but exceptions to this rule can occur.

It should be noted that the total length of the grain boundaries that satisfies the above two conditions for preferred growth might initially be quite small. However, any portion of a grain boundary that does satisfy the two conditions, and which is concave towards an unpreferred grain, will move towards the unpreferred grain as freezing proceeds and in doing so will progressively increase the length of the grain boundary which satisfies the conditions for preferred growth.

Ketcham and Hobbs have explained the above results in the following way. Consider two adjacent ice grains that expose different faces at the ice-water interface, one of which has its c-axis tilted towards the line in which the grain boundary intersects the ice surface (line L) and the other has its c-axis tilted away from line L, as shown in fig.8a. In fig.8a the two grains are shown in an idealized configuration, such as might exist if they had just been brought into contact. However, in order for the surface tension forces to come to equilibrium a groove must form at the grain boundary as shown in fig.9a. Since the basal plane of ice is a low energy surface, it is reasonable to assume that grain B will

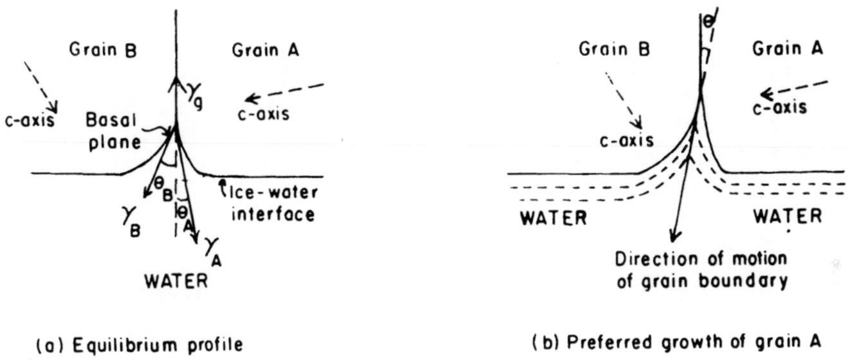

Fig.9 Suggested mechanism for preferred growth in ice (case 1)

adopt a shape which exposes its basal plane at line L thereby minimizing the surface energy. Grain A, on the other hand, cannot expose a low energy basal plane at line L so its shape will be determined by the requirement that the surface tension forces be in equilibrium. This will be the case when $\gamma_A \sin \theta_A = \gamma_B \sin \theta_B$, and since $\gamma_A > \gamma_B$, $\theta_A > \theta_B$. Hence, an asymmetric grain boundary profile forms at line L as observed experimentally (§3). As growth takes place from the melt this asymmetric profile will cause grain A to encroach onto grain B as shown in fig.9b. This theory therefore provides an explanation for the fact that grain A is observed to encroach onto grain B only if grain B has its c-axis tilted towards line L and the projection of the c-axis of B on the ice surface is perpendicular to line L. Moreover, it is easy to show that $d\theta_B = d\theta$, so that as the angle between the c-axis of grain B and the normal to the ice surface decreases, both θ_B and θ increase and the rate at which grain A wedges off grain B also increases, as found experimentally.

Fig. 10a illustrates the situation corresponding to fig. 8b, both grains have their c-axis tilted towards line L, but the c-axis of grain B makes a smaller angle with the normal to the ice surface than does the c-axis of grain A. In this case as the groove begins to form at the grain boundary, grain B will expose the low energy basal plane first and grain A will assume a profile which exposes a higher energy surface and brings the surface tension forces at line L into equilibrium. Moreover, this configuration probably minimizes the total surface free energy. As before, $\theta_A < \theta_B$ and grain A will encroach onto grain B as observed experimentally (fig. 10b).

To summarize the theory described above ascribes the appearance of preferred orientations in the growth of ice from the melt to the asymmetrical profiles that form at certain grain boundaries. Asymmetrical profiles form between two grains when one grain is oriented so that it can expose the low energy basal plane at the grain boundary while the other grain exposes a surface of higher energy.

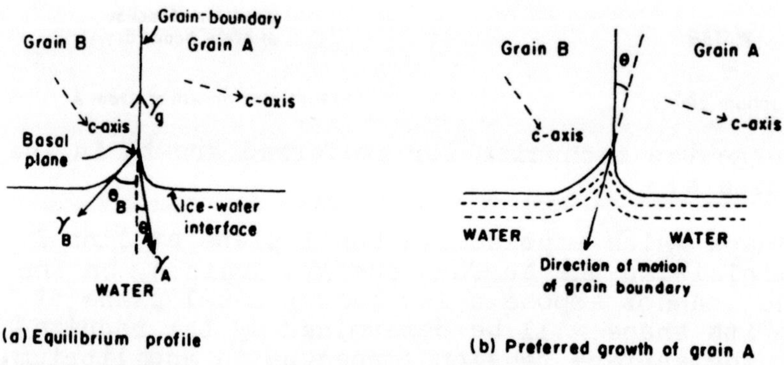

Fig.10 Suggested mechanism for preferred growth in ice (case 2)

APPENDIX A

Equilibrium of a large liquid puddle resting on a plane solid surface

Equation (9) may be derived as follows. Consider a large liquid puddle resting on a plane solid surface (fig.11). If it is assumed that the perimeter between the liquid, solid, and vapor is circular and that the radius r is much greater than the thickness H, the volume V of the puddle is given by

$$V = \pi r^2 H. \tag{A1}$$

At equilibrium the change in energy E of this system for an infinitesimal increase in the area A of the solid surface which is covered by the liquid must be equal to zero. Consequently,

$$\frac{dE}{dA} = 2\pi r\, dr\, (\gamma_{SL} + \gamma_{LV} - \gamma_{SV}) + \frac{\rho g V}{2}\, dH = 0 \tag{A2}$$

where, ρ is the density of the liquid and g is the acceleration due to gravity. Since the volume of the liquid is constant

$$dV = 2\pi r\, H dr + \pi r^2\, dH = 0 \tag{A3}$$

that is,

$$2\pi r dr = -\frac{V}{H^2}\, dH. \tag{A4}$$

Substituting (A4) into (A2) we obtain

$$\gamma_{SV} - \gamma_{SL} - \gamma_{LV} = -\frac{\rho g H^2}{2}. \tag{A5}$$

Also from Young's equation

$$\gamma_{SV} - \gamma_{SL} = \gamma_{LV} \cos\phi. \tag{A6}$$

Substituting (A6) into (A5) and rearranging yields equation (9), namely

$$\cos\phi = 1 - \frac{\rho g H^2}{2\gamma_{LV}}.$$

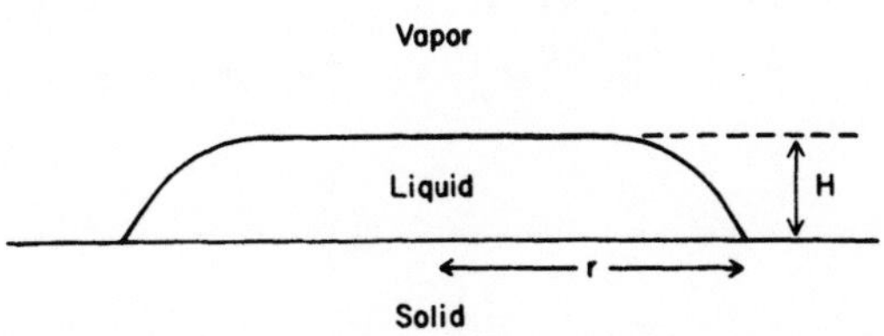

Fig.11 A liquid water puddle on ice

ACKNOWLEDGEMENTS

This research was supported by Grants GP-3809 and GA-780 from the **Atmospheric** Sciences Section of the **National** Science Foundation, U.S.A.

REFERENCES

/1/ Hillig, W.B., 1958, Growth and Perfection of Crystals, edited by R.H.Doremus, B.W. Roberts and D. Turnbull (New York: John Wiley and Sons), p. 350

/2/ Michaels, A.S., P.L.T. Brian, P.R. Sperry, J.Appl. Phys. 37, 4649-4661 (1966)

/3/ Hillig, W.B., D.Turnbull, J.Chem.Phys. 24. 914 (1956)

/4/ Amelinckx, S., N.F. Binnendijk, W. Dekeyser, Physica 19, 1173-1177 (1963)

/5/ Tolman, F.R., J.G. Wood, J. Sci. Instrum. 33, 236-238 (1956)

/6/ Higuchi, K., Acta Met. 6, 636-642 (1958)

/7/ Knight, C.A., J. Colloid and Int. Sci. 25, 280-284 (1966)

/8/ Turnbull D., J.C. Fisher, J. Chem. Phys. 17 71-73 (1949)

/9/ Mossop, S.C., Proc. Phys. Soc. London B68, 193-208 (1955)

/10/ Dufour, L., R. Defay, 1963, Thermodynamics of Clouds. New York: Academic Press

/11/ McDonald, J.E., J. Meteorol. 10, 416-433 (1953)

/12/ Turnbull, D., J. Appl. Phys. 21, 1022-1028 (1950)

/13/ Jacobi, W., Z. Naturforsch. 10a, 322-330 (1955)

/14/ Mason, B.J., Quart. J. Roy. Met. Soc. 78, 22-27 (1952)

/15/ Jaccard, C., Helv. Phys. Acta 32, 89-93 (1959)

/16/ Ketcham, W.M., P.V. Hobbs, J. Cryst. Growth 1, 263-270 (1967)

SEGREGATION OF AMMONIUM FLUORIDE INTO ICE SINGLE CRYSTALS

G. Noll

I. Physikalisches Institut

Universität Stuttgart / Germany

Abstract

The composition of ice single crystals, grown from dilute aqueous solutions of ammonium fluoride (concentrations $c_0 = 4\cdot10^{-4}...5\cdot10^{-2}$ mole/l) was investigated by measuring the electrical conductivity and the pH-value of the melts. The ammonium fluoride is not taken up by the crystals in the stoichiometric ratio of the NH_4F-molecule; the acid components predominate.

1. Introduction

A number of properties of solids are controlled by imperfections. In ice two special types of defects - orientational or Bjerrum defects D and L, and ionic defects H_3O^+ and OH^- - determine the electrical, mechanical, and viscoelastic behaviour /1/. Generating these defects artificially, e.g. by doping the ice crystals with ammonia NH_3, hydrofluoric acid HF, and ammonium fluoride NH_4F, their influence can be studied. NH_3, HF, and NH_4F have great similarity to ice as far as their crystallographic, physical and chemical properties are concerned. That is why it is generally believed that they replace H_2O-molecules in their normal lattice sites, NH_3 producing D-defects, HF L-defects. The configuration of ammonium fluoride in ice has already been discussed by Steinemann and Gränicher /2/. The possible segregation forms - $NH_3\cdot HF$, $NH_4^+\cdot F^-$, $(NH_2)^-\cdot(H_2F)^+$ - represent different types of imperfections. To analyze the compositions of the crystals, we have again investigated the segregation of ammonium fluoride into ice single crystals.

Usually the doped crystals are grown from dilute aqueous solutions of the doping agents. A very common arrangement has a tube of lucite with a single crystalline disk of ice as seed, frozen onto the metal bottom. The vessel is filled with the solution and is dipped into a cooling bath very slowly. The data of our arrangement are: Ø of the tube: 110 mm, length: 350 mm, growth velocity: 0.2 μm/s. The crystal grows upwards. At first only a small part of the doping agent is built into the crystal. The greater part is refused by the growing ice surface. Now the concentration of the solution ahead of the growing crystal increases and gradually the concentration in the crystal itself increases too.

2. Composition of the NH_4F-solution and segregation

In dilute aqueous solutions NH_4F is almost completely dissociated into NH_4^+- and F^--ions. By hydrolysis the base NH_3 and the acid HF are formed, the dissociation equilibrium of water and the activities and concentrations, respectively, of H_3O^+- and OH^--ions are changed. In the solution there is

$$NH_4^+,\ F^-,\ NH_3,\ HF,\ H_3O^+,\ OH^-, \text{ and } H_2O.$$

The activity a_{H_2O} of water is effectively constant = 55.35 mole/l at 25 °C.

The analysis of the composition of the solution shows:

1. There are no detectable NH_4F-molecules in the solution.
2. Neither the concentrations of NH_4^+ and F^-, nor of NH_3 and HF, nor of H_3O^+ and OH^- are equal:

$$c_{NH_4^+} \neq c_{F^-},\ c_{NH_3} \neq c_{HF},\ c_{H_3O^+} \neq c_{OH^-},$$

because of different hydrolysis of NH_4^+ and F^-.

3. The activities of H_3O^+ and OH^- differ from those in pure water markedly.

Thus, for the segregation and the concentrations in the ice crystal the following conclusions can be drawn:

1. It is improbable that NH_4F-molecules exist in ice.
2. We expect that ammonium fluoride is built into ice either in form of ions NH_4^+ and F^-, or in form of NH_3 and HF.
3. It is not sure whether an NH_4^+-ion in a normal H_2O site has an F^--ion in a neighbouring lattice site, or whether it is accordingly NH_3 and HF. An NH_4^+-ion, built into ice, can be neutralized by either

F^- or by OH^-. Therefore, the total concentrations of ammonia ($c_{NH_4^+} + c_{NH_3}$) and of fluoride ($c_{F^-} + c_{HF}$) may differ in ice; that means: ammonium fluoride has not the stoichiometric composition of NH_4F.

3. Experimental results

The composition of the ice crystal influences the pH-value of its melt. We have measured the electrical conductivity and the pH-value of melted ice single crystals, grown from dilute aqueous NH_4F-solutions. The ice crystals were divided into parts of about 15 cm^3 and these parts were melted separately and measured. The values were compared with the data of NH_4F-solutions of stoichiometric composition.

Fig.1 shows the electrical conductivity σ of the melted ice samples as a function of the length coordinate of the crystals grown from NH_4F-solutions with different concentrations c_o. The shape of the conductivity curves results from the growth method and the segregation behaviour of the solute. It agrees with the theory of Jaccard and Levi /3/.

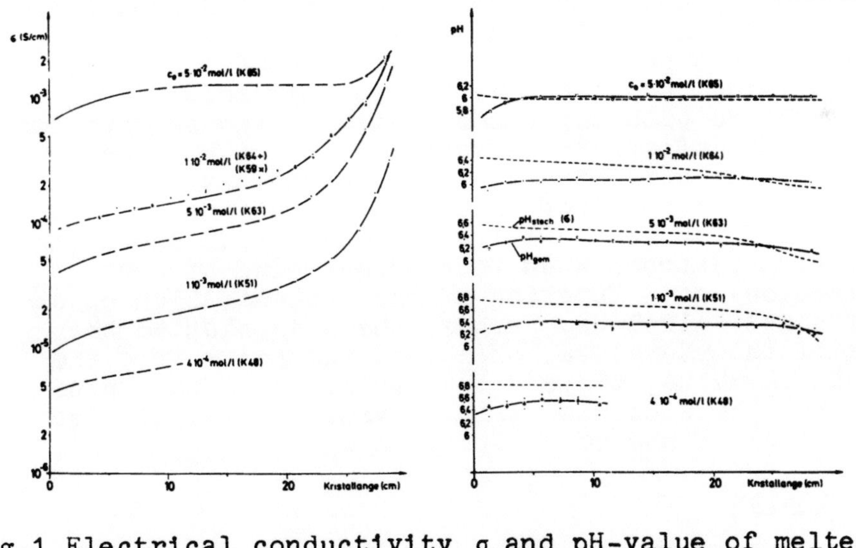

Fig.1 Electrical conductivity σ and pH-value of melted ice - ammonium fluoride - single crystals as a function of the length coordinate of the crystals.
—— pH_{gem}, measured pH-value of the melted ice samples.
---- $pH_{stoich}(\sigma)$, pH-value calculated from the conductivity

The conductivity rises with growing crystal. It does not reach a constant value, because the volumes of the solutions were limited and consequently, during growth of the crystals, their concentrations rose too, particularly at the end of the growth. If we suppose that the melts of the ice single crystals included the ammonium fluoride in stoichiometric composition, we should find pH-values, pH_{stoich}, which are represented by the dashed curves in the right part of the figure. Corresponding to the shape of the conductivity curves, their values should decrease as a function of the length coordinate of the crystals. In fact we measured pH-values, pH_{gem}, which were essentially smaller than the "stoichiometric" pH-values and the shape of the pH-curves differ markedly.

The difference is greatest at the beginning of the crystallization. It decreases during the growth and at the ends of the crystals the two pH-curves cross each other.

The pH-measurements indicate that the ice single crystals do not contain ammonium fluoride in stoichiometric composition. Due to this we can not determine the concentration of ammonium fluoride from the conductivity σ of the melts. But if we extrapolate the conductivity curve to the beginning of crystallization and divide this extrapolated value by the conductivity σ_o of the original solution, we get an expression which we can compare with the concentrations and segregation coefficients of earlier authors /3, 4/. (They have determined their concentrations from conductivity measurements of the melts. From these concentrations we have recalculated the conductivity values). Fig.2 shows the values σ/σ_o (extrapolated to the beginning of crystallization) as a function of the concentration c_o of the original solutions. Further there is plotted the change of the pH-values, pH_{gem}, of the melts from the original pH-value, pH_o, and the deviation of the value pH_{gem} from the stoichiometric pH-value, pH_{stoich}, also as functions of the concentrations c_o.

4. Discussion

We have already mentioned that the ammonium fluoride, built into the ice single crystals, does not have a stoichiometric composition. Obviously at the beginning of the crystallization the acid components are preferred, and the basic components remain in solution. In the middle part of the crystals an equilibrium for

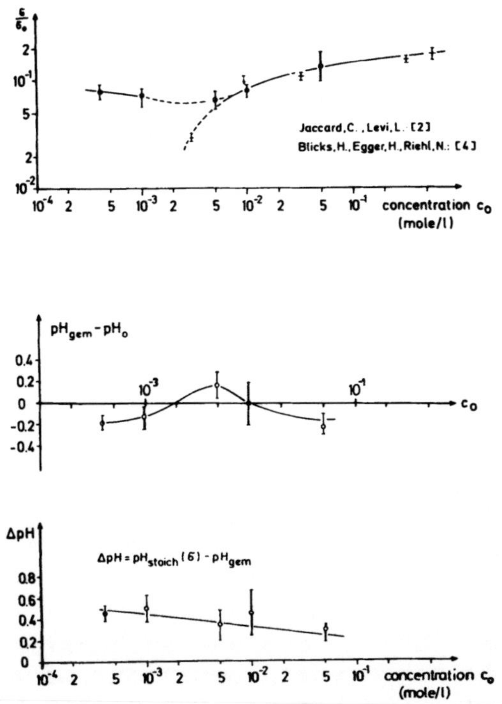

Fig.2 Dependence of the extrapolated values on concentration c_0 of the original NH_4F-solution.
a) Ratio of conductivity σ of the melts to conductivity σ_0 of the original NH_4F-solution.
b) Difference between measured pH-value, pH_{gem}, and pH-value pH_0 of the original NH_4F-solution.
c) Difference between measured pH-value, pH_{gem}, and "stoichiometric" pH-value, pH_{stoich}, $\Delta pH = pH_{stoich}(\sigma) - pH_{gem}$.
All values extrapolated to the beginning of crystallization

the segregation of the different components is reached. During the freezing of the rest of the solution all basic components accumulated in the solution are enclosed into the crystals.

Measuring the electrical conductivity σ and pH-value of the melts is insufficient in order to determine the concentration of each component. Nevertheless it shows that the acid components predominate in ice. There are two possibilities:

1. Components of the hydrofluoric acid predominate. For the activities a we can write

$$a_{H_3O^+} + a_{HF} > a_{OH^-} + a_{NH_3} .$$

2. If we use Bronsted's definition of acid and base, we also have the possibility

$$a_{H_3O^+} + a_{HF} + a_{NH_4^+} > a_{OH^-} + a_{F^-} + a_{NH_3} .$$

From these equations we cannot decide, whether we have an excess of F^-- or NH_4^+-ions in ice. But probably, we have a higher concentration of HF than of NH_3. NH_3 only has a small chance of being built into the ice crystal, because its molecular radius is much larger than those of the other components (NH_3: 1.7 ... 1.8 Å, H_2O: 1.38 Å, NH_4^+: 1.37 Å, F^-: 1.28 Å, coordination number 4 /5, 6/). This agrees with the segregation measurements of Jaccard and Levi for NH_3 /3/.

Till now we do not have a complete explanation for the non-stoichiometric segregation behaviour of ammonium fluoride. Certainly the size of the components has a great influence. But possibly the Workman-Reynolds-potential /7/ or a thermoelectric potential arising during growth /8/ at the interface ice/solution also influences the segregation. To get more detailed information of the segregation behaviour of the various components, we began to study the segregation of hydrofluoric acid HF from dilute aqueous solutions into ice single crystals, also by measurements of the electrical conductivity and pH-value of the melts. From these data, we will determine the concentration of each component of the system H_2O-HF.

Acknowledgements

The author is indebted to Prof. Dr. H.O. Kneser for his interest in this work and to Oberbaurat M. Käss for valuable assistance and discussions.

The work was supported by Deutsche Forschungsgemeinschaft.

References

/1/ Gränicher, H., Phys. kondens. Materie 1, 1-12 (1963)

/2/ Steinemann, A., H. Gränicher, Helv. Phys. Acta 30, 553-580 (1957)

/3/ Jaccard, C., L. Levi, ZAMP 12,70-76 (1961)

/4/ Blicks, H., H. Egger, N. Riehl, Phys. kondens. Materie 2, 419-422 (1964)

/5/ Landolt-Börnstein, Zahlenwerte und Funktionen Bd I/4. Berlin: J. Springer, 1955

/6/ Bernal, J.D., R.H. Fowler, J. Chem. Phys. 1, 515-548 (1933)

/7/ Workman, E.J., S.E. Reynolds, Phys. Rev. 78, 254-259 (1950)

/8/ Latham, J., B.J. Mason, Proc. Roy. Soc. A 260, 523-536 (1961)

REJECTION OF IMPURITIES BY GROWING ICE FROM A MELT

G. Kvajić[1], V. Brajović[2], E.R. Pounder[1]

[1]Ice Research Project, Physics Dept., McGill University Montreal, Canada

[2]Boris Kidrich Institute of Nuclear Sciences, Beograd, Yugoslavia

Abstract

Distribution of radioactive impurities in the liquid and solid phases was measured during the growth of ice under conditions of one-dimensional heat flow. Concentration of impurities along the specimen was proportional to the intensity of gamma rays detected by NaI(Tl) photomultiplier and single-channel analyzer. Large impurity accumulation took place ahead of a crystal growing vertically upwards, while for a downward growth no detectable accumulation of impurities was observed. The partition coefficient for Cs^{134} impurities was calculated in terms of the rate of ice growth. The different behaviour of vertical growth in opposite directions is discussed in terms of convective transfer of heat and solute in the liquid phase.

1. Introduction

It is a well known fact that many physical properties of an ice crystal depend on its content of impurities and their distribution throughout the volume. The amount of impurity and its distribution is, in general, strongly dependent on the freezing parameters, i.e., rate of crystal growth, temperature gradient in the liquid phase close to the crystal, mechanism of solute transfer in the melt, amount of impurity present in the liquid phase, nature of the solute etc. None of these parameters have so far been systematically examined. This work presents some results of preliminary experiments on the problems of impurity segregation in growing ice.

There are several papers on impurity segregation in ice. Jaccard and Levi /1/ calculated the impurity distribution in the direction of growth for an arbitrary initial distribution in the liquid. Their results agree qualitatively with measurements obtained with HF used as an impurity. Stirring the liquid phase allowed a satisfactory determination of the segregation coefficient for HF, NH_3 and NH_4F as a function of concentration of the solution. Riehl et al. /2/ grew single crystals from water doped with HF and NH_4F. In the finished crystals, the concentration showed a marked gradient in radial direction. Weeks and Lofgren /3/ measured the effective distribution coefficient k for ice growing uni-directionally downwards into salt water (from 0.1% to 10% by wt.). They observed an increase in k with growth rate. For small freezing rates k tended to 0.26.

2. Experimental technique and observations

A diagram of the apparatus used for the measurement of the impurity distribution in the solid and liquid phase is given in figure 1. Freezing of water takes place in

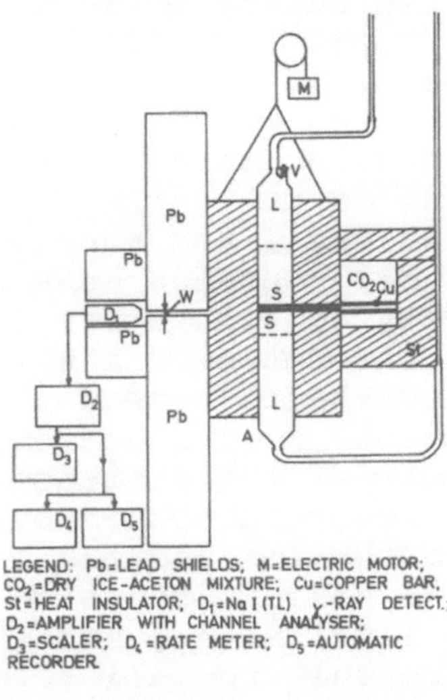

Fig.1 Diagram of the apparatus for measurement of the impurity distribution in solid and liquid phases during growth of ice

a plastic tube (diamenter 1.5 cm, length 50 cm) by means of a cold copper bar (Cu) one end of which is inserted into a cold bath (CO_2) consisting of dry ice and acetone. The other end enters the middle of the plastic tube (L) laterally, causing simultaneous freezing of water upwards and downwards. The actual procedure for starting an experiment is as follows: A desired concentration of impurity (NaCl) is mixed with some amount of radioactive tracer (Cs^{134}). $5 \cdot 10^{-6}\%$ $CsCO_3$ is sufficient to give a reasonable counting rate. This solution is poured into the freezing tube through a valve(V) until the water fills the tube and enters into the capillaries. Now the dry ice-acetone mixture is poured into the container with the copper bar. Several seconds later, the water in the capillaries starts to rise due to the fact that ice has a larger volume than the same amount of water by about 8.3%. In the meantime the electric motor (M) moves the whole freezing apparatus alternatively up and down, so that the solid and liquid phases in the upper and lower parts of the freezing tube pass by the 4.5 mm wide window (W) constructed with 12 cm thick lead blocks. These lead blocks shield the NaI(Tl) detector (D_1) from gamma rays originating in that portion of the freezing tube which is not in front of the window. The intensity of radiation from all the parts of the solid and liquid phases is automatically plotted by the recorder (D_5). Figure 2 illustrates one such plot at the freezing time of about ten hours for the case of distilled water. About the same intensity of radiation is observed in the lower (L↓) and upper (L↑) liquid phases. A large peak is present at the solid-liquid (S-L) interface in the upper part of the tube, while the lower liquid-solid (L-S) interface shows no detectable accumulation of impurity. The amount of impurity trapped in the lower part of the ice (S↓) is small and hardly detectable, while the upper part (S↑) contains much more impurity. We see that the impurity concentration first increases and then decreases as the detector approaches the S-L interface. The actual distance between the lower and upper ice interfaces (LS-SL) as a function of time is obtained from measurements of the distance (LS-SL)on such plots which are being recorded continuously, the known speed of the electric motor (M),and the rate of recording paper advance. Since we measure the time dependence of the levels in the capillaries (h_1) and (h_2) and the lengths (LS-SL), we are able to calibrate the capillaries in terms of ice thickness. In this way we can measure the ice thickness to within ± 0.1 mm as a function of freezing time and also the rate of freezing as a function of either time or ice thickness (fig.3).

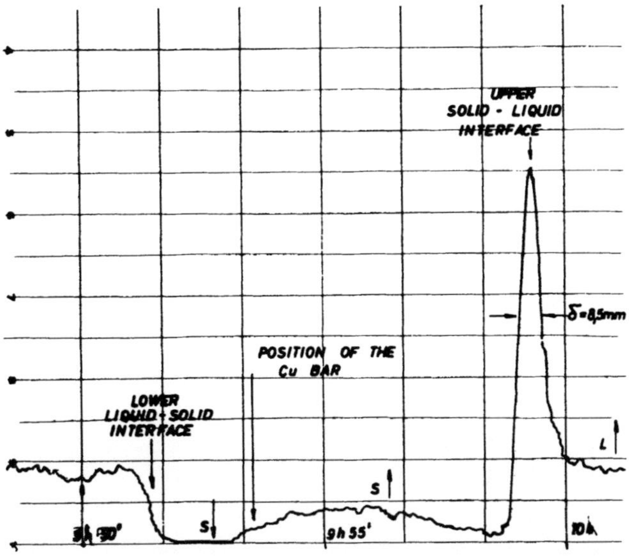

Fig.2 Record of concentration in solid (S) and liquid (L) phases for upper↑ and lower↓ growth (arrows indicate directions of freezing); numbers along the horizontal line represent freezing times (exp.A)

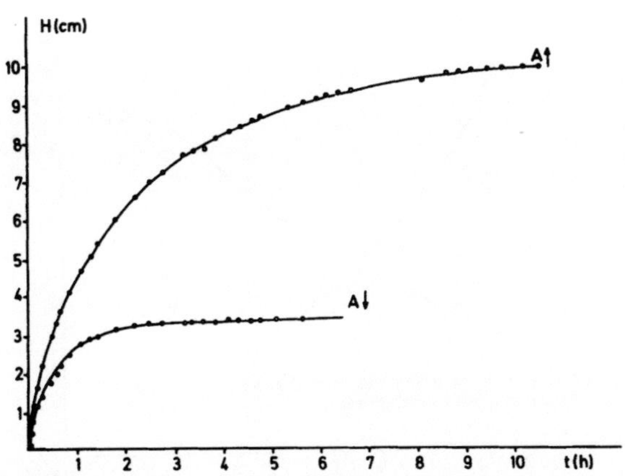

Fig.3 Ice thickness (in cm) vs. freezing time (in hours) (exp.A). The arrows ↑ and ↓ indicate upward and downward growth

Intensity measurement of the gamma radiation from the accumulated concentration C_i ahead of the S-L interface is done by scaler (D_3) at the time when the interface passes by the window (W) of detector (D_1) (see figure 1). The peak values are denoted by dots in fig.4 and the heights of the peaks read from the recorder are denoted by crosses. A continuous curve drawn through these points gives an average value of the peaks as a function of freezing time. Using fig.3 and 4 we can represent the impurity concentration at the interface C_i in a more useful form as a function of ice thickness H (fig.5).

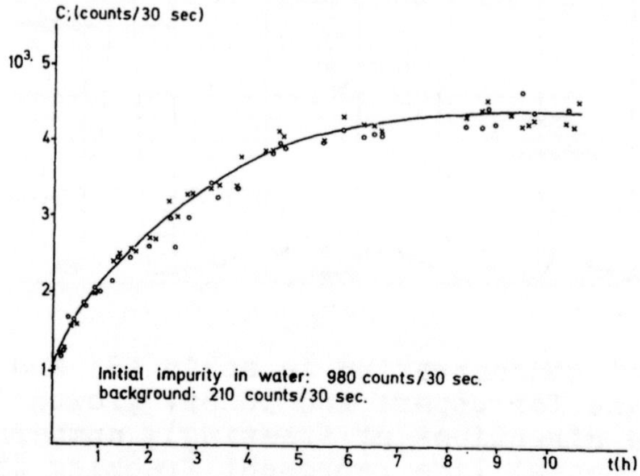

Fig.4 Concentration of impurity C_i at the solid-liquid interface vs. freezing time t (exp.A)

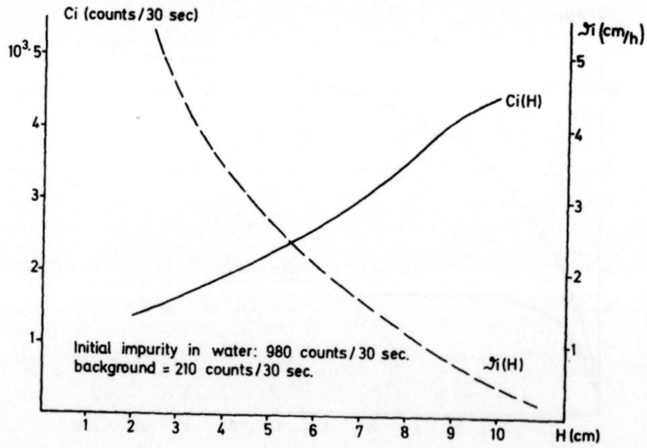

Fig.5 Concentration of impurity at the solid -liquid interface and rate of freezing vs. ice thickness (exp.A)

A satisfactory measurement of the concentration C_s in the upper and lower solid phases is done by stopping the electric motor (M) at fixed positions along the ice sample and counting the gamma radiation for a sufficiently long period of time in order to achieve reasonable accuracy (usually better than 5%). Fig. 6 represents such a measurement for distilled water after freezing. The points indicate the measured intensities, reduced for convenience by a factor of four. The same figure shows the L-S interface and the concentration a few cm further into the lower liquid phase.

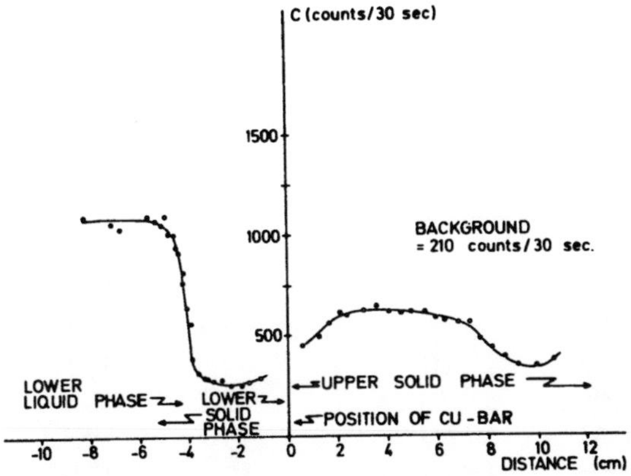

Fig.6 Concentration of impurity along the upper solid, lower solid and lower liquid phases (exp.A)

The interface distribution (or partition) coefficient K is obtained (using figures 5 and 6) by dividing the corrected intensity in the solid $C_s(z)$ at a position z along the axes of ice by the intensity of a peak $C_l(H)$ of the ice thickness H = z. Corrections have to be applied for background and "edge effect". The latter correction is necessary due to the "leaking" of radiation to the detector from impurities in the sample sections adjacent to the window. This correction is negligible when the peak is directly in front of the window, but is very important when C_s is measured and the window is close to the large peak. The last column of table I shows that correction of the values for C_s is especially significant for small freezing rates. Results of such calculations of K are plotted in fig.7 for distilled water, and for two different concentrations of NaCl added to the previous (very dilute solution) of distilled water and $CsCO_3$ ($5 \cdot 10^{-6}$% by wt.). Actually, for the cases of freezing in downward direction, the calculated K values were not interface, but effective, distribution coefficients

t (hours)	N (cm)	V_i (cm/h)	(C_L,C_i-210) (counts/30sec)	C_s-210	$(C_s-210)^+$ corrected	K
FREEZING DOWNWARDS (A ↓)						
11	0.96	4.12	860	77	69	0.080
18	1.3	3.03	860	65	49	0.057
29	1.66	2.60	860	52	43	0.050
35	1.97	2.36	860	52	41	0.047
40	2.14	1.96	860	52	40	0.046
50	2.38	1.33	860	53	40	0.046
65	2.62	0.995	860	60	46	0.053
FREEZING UPWARDS (A ↑)						
18	2.48	5.26	1195	402	399	0.334
35	3.76	3.62	1540	402	397	0.260
40	4.06	3.36	1620	402	396	0.240
50	4.62	3.00	1770	390	383	0.220
65	5.22	2.54	1960	388	379	0.193
77	5.66	2.28	2100	378	344	0.166
85	6.05	2.05	2190	365	352	0.120
109	6.82	1.68	2420	316	299	0.120
150	7.85	1.23	2790	205	181	0.060
165	8.10	1.10	2890	180	154	0.054
191	8.62	0.93	3070	147	115	0.037
202	8.74	0.86	3140	140	104	0.033
215	8.95	0.78	3720	127	89	0.027
230	9.14	0.73	3300	125	83	0.025
245	9.28	0.67	3390	125	80	0.024
261	9.50	0.61	3470	120	71	0.020

Table I: Results for the cases of freezing upwards A↑ and downwards A↓ ; H(t) is taken from fig.3, v_i(H) and C_i(H) from fig.5, and C_s from fig.6; H is the ice thickness, v_i is the rate of freezing, C_i is the concentration at the interface (A↑), C_s is the concentration in the solid H cm away from the beginning of freezing, C_L is the concentration in the lower liquid phase (A↓); K is the distriburtion coefficient, $(C_s-210)^+$/ (C_L-210) for the freezing A↓ and $(C_s-210)^+$/ (C_i-210) for the freezing A↑.

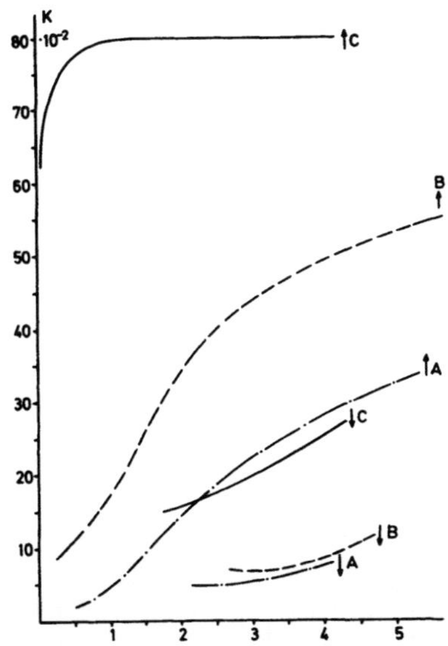

Fig.7 Interface distribution coefficient vs. rate of freezing for three different initial impurity concentrations; exp.A: $5 \cdot 10^{-6}\%$ C_sCO_3 in distilled water saturated with air; exp.B: $5 \cdot 10^{-2}\%$ NaCl by wt. added to the solution A; exp.C: $5 \cdot 10^{-1}\%$ by wt. NaCl added to the solution A; arrows ↑ and ↓ indicate upward and downward freezing

commonly defined as the average concentration in the liquid at the time when the solid portions were formed. We have observed that the impurity distribution is uniform along the whole lower liquid phase and along that portion of the upper liquid which lies above the peak.

3. Solution of the continuity equation for the diffusion of the impurity through the liquid phase

The continuity equation for particles in a fluid in the one-dimensional case can be written

$$\frac{\partial c}{\partial t} = D \frac{\partial^2 c}{\partial z^2} - v \frac{\partial c}{\partial z} ,$$

where v(z, t) is the velocity of an impurity particle (assumed to be equal to the velocity of the fluid surrounding the particle) and D is the coefficient of diffusion of the impurity particles through the liquid phase.

Considering the case of pure diffusion, i.e. neglecting convective transfer of solute and heat, we may use for D a typical value of 10^{-5} cm^2/sec, and v(z,t) = v(t) is taken equal to the negative value of the rate of crystal growth. We have used the observed boundary conditions for the concentration of impurity at the S-L interface $C_1(t)$ from fig.4, and the initial impurity concentration at a distance of 4 cm from the interface. The latter boundary condition is approximately correct since the concentration at a distance of 4 cm from the crystal increased only slightly during freezing. The partial derivatives of z have been replaced by the corresponding finite differences, and v(t) is obtained from fig.3, curve A. This problem has been solved with an analog computer for the time dependence of the concentration at ten different points of the liquid phase. Fig.8 displays ten distribution curves as functions of distance from the S-L interface at ten equal freezing time intervals resulting from these computations. Comparing the distribution curve for t = 10 hours of freezing with the peak in fig.2 we find a large difference in width at half maximum. This is sufficient to rule out diffusion as the sole mechanism of solute transfer through the upper liquid phase.

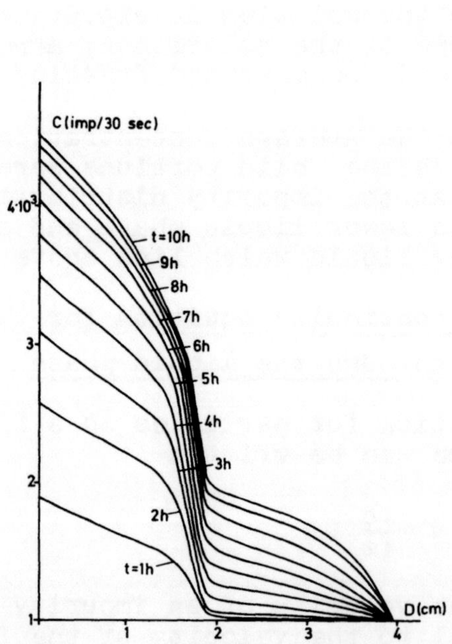

Fig.8 Distribution along ↑ liquid phase obtained by solution of the diffusion equation for ten different freezing times

4. Discussion

Let us infer some information from the fact that the amount of ice grown upwards is three times as large as that grown downwards (fig.3). At a thickness of about 3 cm, the rate of growth is 0.1 and 3 cm/h for downward and upward freezing resp. The heat flux through the lower melt corresponding to a difference in these two growth rates is about 0.1 cal/sec. Since typically the temperature gradient in the melt is $0.5^{\circ}C/cm$ and the corresponding heat conduction is 10^{-3} cal/sec we can reach the conclusion that heat transfer through the lower liquid is 100 times larger than it would be by pure conduction. This is accomplished by vigorous convection, obviously caused by a positive temperature gradient. Since the convection increases the heat transfer by a factor 100, it increases the solute transfer by a factor 1000. The latter value is due to the fact that the heat conductivity of water is ten times larger than solute diffusivity and we assume that under vigorous convection heat and solute transfer are equal. Thus, the effective coefficient D for the lower melt is 10^{-2} cm^2/sec. This strong convection in the lower melt is sufficient to prevent the accumulation of any detectable amount of impurity ahead of the lower L-S interface. The strong convection also causes the observed uniform distribution along the lower melt.

Comparing the computed distribution (fig.8) for the case of pure diffusion with the measured one (fig.2), we can see that the observed peak is much narrower than the theoretical one and that the uniform distribution in the melt above is again a consequence of the presence of convection. This convection is localized in the portion of the upper melt between the peak and that layer of water which has a temperature of about $5^{\circ}C$. Obviously, this flow instability is caused by a positive density gradient present in water in this temperature range. The measured value for the width of the peak is about 8.5 mm and seems independent of freezing rate and impurity concentration. This measured number for the peak should not be taken too seriously, since the window is too wide (4.5 mm) for precise measurements.

Fig.5 shows that the peak $C_1(H)$ rises fastest when the freezing rate is about $v = 1$ cm/h. For rates greater than this, $C_1(H)$ rises more slowly with H, due to increased entrapment in the solid. For smaller rates the peak rise is limited by the solute transfer through the melt.

Fig.6 shows that the concentration $C_s(H\uparrow)$ in the upper solid initially rises with ice thickness ($H\uparrow$), then slowly decreases until $H\uparrow = 7.5$ cm ($v_i = 1.3$ cm/h). In the last stage of freezing C_s decreases at first quickly and then remains constant. The initial rise in concentration could be explained by an initially large temperature gradient in the melt near the interface and a small peak $C_i(t)$. In terms of constitutional supercooling theory, this would mean that a small amount of liquid ahead of the interface is not sufficiently supercooled constitutionally to cause the formation of a well developed dendritic morphology. For an ice thickness between 2 and 7 cm, the distribution slowly decreases in spite of the rapid rise in the peak due to the decrease of the freezing rate. With a decrease in freezing rate v_i, the number of dendritic platelets forming ice crystals decreases and therefore, interdendritic space where most of the impurity is stored also decreases. For $H\uparrow > 7.5$ cm the concentration at first decreases quickly and then remains constant although the rate of freezing still decreases. The initially fast decrease in C_s appears at about the same time as the rate of peak C_i increase is largest ($v_i = 1.2$ cm/h; see fig.5). Probably, at this stage of freezing the S-L interface changes its structure turning from a dendritic to a cellular morphology or, maybe, at this rate "milky" ice turns into transparent ice.

The distribution of impurity in the lower liquid phase first decreases and then remains constant to about 2.5 cm of ice thickness; then it increases. This increase is certainly not due to an actual rise in concentration in the solid itself but to the liquid phase filling the space between ice and wall of the plastic tube. At small freezing rates the heat supply through the tube wall is not negligible (as it is at high rates) producing a markedly conical S-L interface.

Fig.7 shows the distribution coefficient K as function of the freezing rate v_i for three different concentrations. It shows an increase in K with the rate of freezing and initial impurity concentration. For the two parameters held constant, K decreases with an increase in strength of convection, or equivalently, an increase in temperature gradient in the melt at the interface. Since the initial concentration of $CsCO_3$ is very small ($5 \cdot 10^{-6}$% by wt., case A), the main impurities controlling the interface morphology are air molecules. Under normal conditions water is saturated with $4 \cdot 10^{-4}$% of air by wt. For freezing cases B and C, the dominating impurity is

NaCl. Since the physical properties of Na^+ are similar to those of Cs^+ the measured distributions $C_s(H)$ and $C_1(t)$ and the calculated value of $K(v_1)$ are the same for both solutes. Assuming that the mechanisms of entrapment of Cs^+ and air molecules are the same, the dependence of $K(v_1)$ measured for Cs^+ solute is also valid for air molecules. The distributions $C_s(H)$ and $C_1(t)$ observed for Cs^+ are probably not the same as for air molecules. Three main reasons can be stated: 1) Cs^+ is a positive ion and electrostatic interaction takes place at the interface; 2) air molecules are several times as big as Cs^+. Hence, Cs^+ diffuses more easily through the melt than air molecules; 3) the formation of the peak at the interface is continuously interrupted by formation of air bubbles, and so probably strongly affects interface morphology.

5. Conclusion

1) The existence of different degrees of convection in the melt is the main reason for observing accumulation of impurity for the case of freezing in upward direction and the absence of the peak in downward freezing.

2) The distribution coefficient increases with initial concentration and rate of freezing and decreases with strength of convection present in the melt, i.e., decreases with increased temperature gradient at the interface.

3) A relatively large amount of impurity in the solid phase and the large K for low freezing rates imply probably stable cellular morphology. At a freezing rate of about 1 cm/h, for case A, we observed a significant transition in distribution of C_s and C_1; the partition coefficient K tends to a steady state value of about 0.02. This is associated with a transition from dendritic to cellular morphology or more probably from "milky" to "freshwater" ice.

References

/1/ Jaccard, C., L.Levi, Z.Angew.Math.Phys. 12, 70-6, (1961)

/2/ Riehl, N., H.Blicks, H.Egger, Phys.Kondens.Materie 2, 419-422, (1964)

/3/ Weeks W.F., G.Lofgren, private communication (1966)

NEUTRON AND GAMMA ACTIVATED NUCLEATION OF TYNDALL-FLOWERS IN ICE

H. Müller-Krumbhaar

Physik-Department

Technische Hochschule München / Germany

Abstract

Lattice disturbances are created in otherwise "perfect" ice single crystals by neutron and γ-irradiation. These imperfections act as nucleation centers for internal melting when IR energy is absorbed. The shape of these Tyndall-flowers as well as their distribution and annealing properties exhibit different behaviour characteristic of the causative irradiation.

Internal melting of clear ice crystals allows to localize visually lattice disturbances of different nature and origin /1,2,3/. Infrared energy from heat lamps is absorbed in the entire volume of single crystals to produce internal melting.
Interface energy raises the binding energy of lattice disturbances over that of regular lattice points. For this reason the region around a defect begins to melt first during application of IR-energy, while the undisturbed lattice still remains solid.
Different melting figures are observed because of the anisotropy of the ice with respect to thermal and molecular properties and because of the different kinds of disturbances. Point dislocations impurities, e.g., are nucleation centers for the so called Tyndall-flowers, thin structures shaped like snow crystals.

The diameter of these nucleation centers is of the order of hundreds of lattice points as it is shown by an

approximate calculation of the size of nucleation centers for Tyndall-flowers produced by radioactive irradiation /4/ .
By careful control of growth conditions we have succeeded in obtaining ice crystals which exhibit comparatively few Tyndall-flowers on IR-irradiation.
Lattice disturbances in these good, dislocation-free crystals were generated by fast neutron irradiation. They have been made visible by internal melting.

The cylindrical single crystals of ice had a length of 20 cm parallel to the c-axis and a diameter of 8 cm. They were cut into 1 cm thick slices and sealed into polyethylene foil.
Early samples were irradiated at the Munich reactor station. The total dosis was about 10^{14} n/cm^2 plus about 10^{16} gamma-photons/cm^2.

Fig. 1 shows a single Tyndall-flower generated in a non-irradiated crystal, exposing its distinct dendritic structure. The original size of the pictures is about 3 x 3.5 cm^2. The specimen of figure 2 was irradiated by 10^{14} n/cm^2. After the irradiation the crystal was clear but lost its transparency almost completely on IR-irradiation. No discrete Tyndall-flowers are visible anymore.

Other samples were irradiated by an almost pure Pu-Be neutron source at the "Gesellschaft für Strahlenforschung, Neuherberg", in order to minimize effects caused by gamma-rays. The average energy of these neutrons is about 4 MeV, the corresponding mean free path of the neutrons in ice is about 5 cm. The neutron dosis was only between 10 and 30 n/cm^2, giving about 1 to 4 reactions per cm^2. Even this low dosis produced an effect clearly visible by Tyndall-flowers, as shown in figure 3 and 4.

The sample of figure 3 has not been neutron irradiated, the sample of fig.4 received a dosis of about 25 n/cm^2.
In a control experiment, other slices of ice were exposed to gamma rays from a Co^{60} source. The total number of irradiation quanta was about 100 times larger than the neutron dosis in the pure neutron experiment. The effect, as evidenced by Tyndall-flowers, was less than with neutron irradiation.(Fig.5, non-irradiated; fig.6, gamma-irradiated).

So far, only the phenomenology of this effect has been investigated. It might be applied to the study of radiation damage in solids, to structural studies of the disturbed ice lattice /4/ and to detection of fast neutrons and other particles.

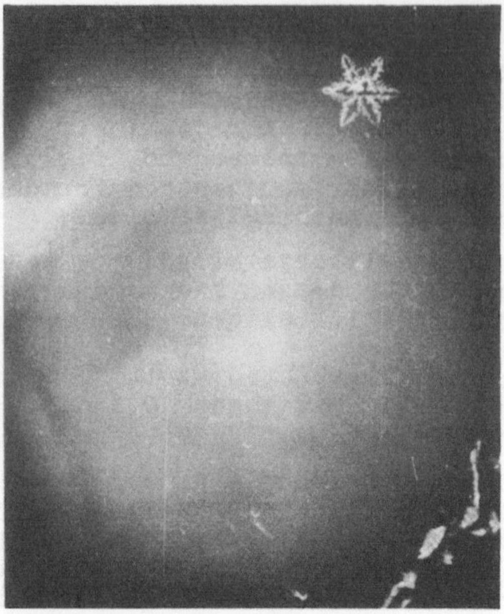

Figure 1

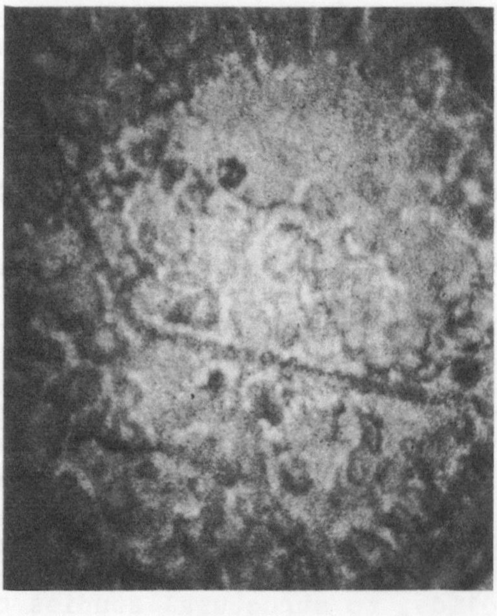

Figure 2

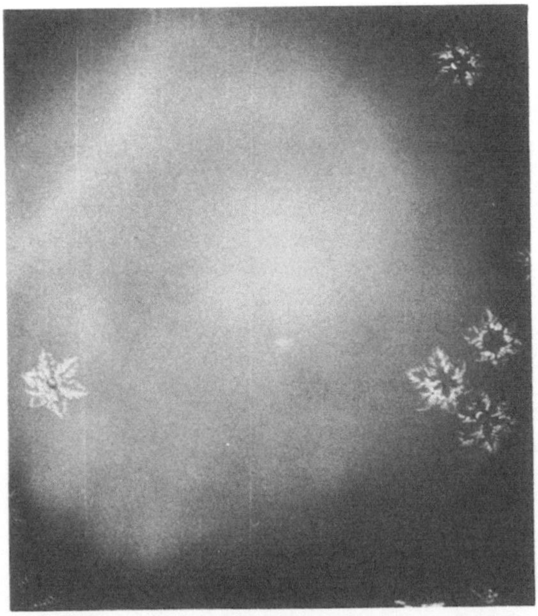

Figure 3

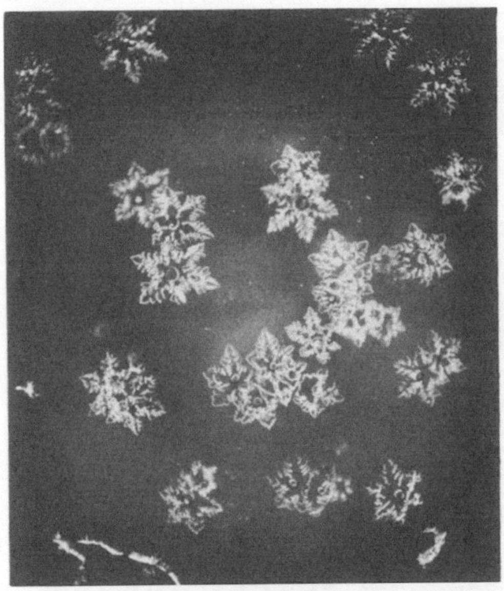

Figure 4

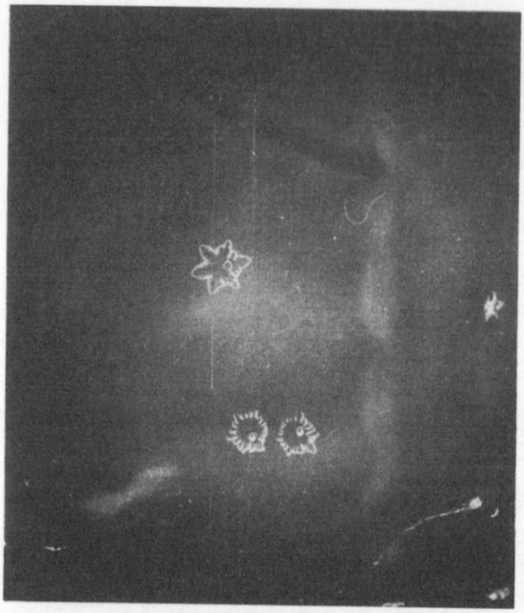

Figure 5

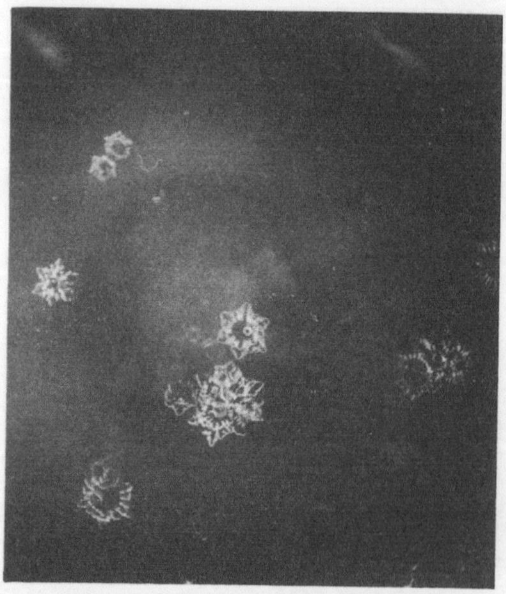

Figure 6

More detailed investigation will require ice crystals which are free from point dislocations which serve as nucleation centers for Tyndall-flowers.
Recombination of nucleation centers has been observed in crystals which have been kept in a cold box for several months at about -16°C.
This might be a possibility to obtain crystals with reasonable expenditure of time which are almost free from dislocations.

References

/1/ Nakaya, U., Snow, Ice and Permafrost Research Paper 13 (SIPRE). Wilmette, Ill., U.S.A. (1956)

/2/ Käss, M., S.Magun, Z.Kristallogr. 116, 354 (1961)

/3/ Niklaus, J.P., Master Thesis, Techn.Hochschule München (1962)

/4/ Müller-Krumbhaar, H., Master Thesis, Techn.Hochschule München (1968)

HYDROGEN BONDS IN BIOLOGICAL SYSTEMS

J. Engel

Max Planck Institut für Eiweiß- und Lederforschung; Universität München / Germany

This paper is a short review of the stabilizing and dynamic action of hydrogen bonds in biological systems. The representation is rather phenomenological. No attempt will be made to review the enormous number of publications which have offered proof of the existence of hydrogen bonds in biological systems by methods which are the same as in organic or inorganic chemistry. The number of substances and molecular organizations which are of biological importance is very large. The discussion will be limited to a few interesting aspects connected with hydrogen bonds in two classes of biological molecules: in proteins and in nucleic acids.

The family of protein molecules contains members which serve simple construction purposes and which make up a large part of plant and animal life. There are also members which serve very sophisticated functions. Enzymes (e.g. lysozyme) catalyse reactions. Others bind and release certain substances very specifically (e.g. myoglobin and hemoglobin). Many of these molecules have the property of being accessible to the regulatory action of other molecules. Feedback and other control mechanisms are possible, and a very effective collaboration is achieved between different molecules by appropriate sterical arrangements. In general, one may view proteins as very important building blocks and functional units of the living world, which, like the units of an electronic circuit (transistors, etc.), may satisfy very complicated requirements if combined in a proper way. Considering the complex task which is carried out by a single protein

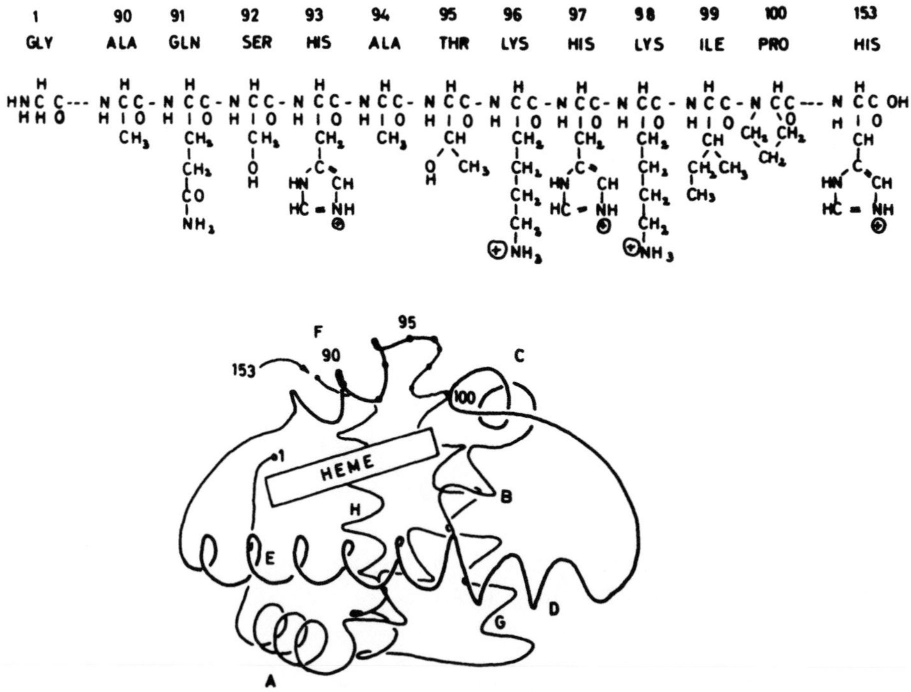

Fig.1 Part of the amino acid sequence of (human) myoglobin. (Hill cited in Braunitzer /1/). The lower part of the figure shows a drawing of the three dimensional folding of the peptide chain of (sperm whale) myoglobin whose structure was solved by Kendrew et al. /14/. Numbers indicate the positions of the amino acid residues; the letters are the names for the portions of α-helical configuration which show up in myoglobin

molecule, we would not expect it to have a simple structure. The primary structure of proteins, however, is very simple. The near 20 different amino acids which are linked by peptide bonds differ only in their side chains. Fig.1 shows, as an example, an arbitrarily selected part of the amino acid sequence of myoglobin (human). The backbone of the peptide chain is the same in all proteins. Each amino acid residue in this backbone includes an NH group as a possible donor and a CO group as a possible acceptor of a hydrogen bond. The hydrogen bonds between these groups are believed to stabilize the α-helix and

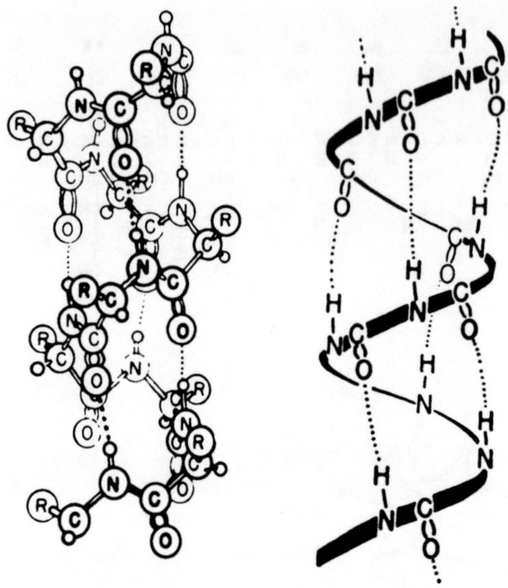

Fig.2 Right-handed α-helix after Corey and Pauling /2/ (Drawing by Kauzmann /13/)

the ß-structures. In our example (fig.1), we see that about 3/4 of the peptide chain of myoglobin has the secondary structure of an α-helix. (For reviews on the sequence and structure of myoglobin and hemoglobin see Perutz /22/, Schroeder and Jones /29/, and Braunitzer /1/.

Before turning to the tertiary structure of myoglobin and to the hydrogen bonds between the side chains of amino acid residues, we may discuss the participation of hydrogen bonds in the stabilization of the α-helix, (fig.2), in detail.

Taking <u>only</u> into consideration that all possible hydrogen bonds must be formed between the NH and the CO groups of the peptide bonds, many regular helical arrangements are possible.

$$\overset{\text{H} - - - - - - - - - - - \text{O}}{-\underset{|}{\text{N}} - (\text{CO} - \text{CHR} - \text{NH})_n - \underset{\|}{\text{C}} -} \qquad N = 3n + 4$$

$$\overset{\text{H} - - - - - - - - - - - - - - \text{O}}{-\underset{|}{\text{N}} - \text{CHR} - (\text{CO} - \text{NH} - \text{CHR})_n - \underset{\|}{\text{C}} -} \qquad N = 3n + 5$$

Helices designated as S_N-helices may be formed where S = number of residues per turn and N = number of atoms in the hydrogen bonded ring. However, only the α-helix 3.6_{13}-helix) and the 3_{10} helix are actually observed. This shows that other factors must also play a role. Obvious limitations which must be noted in constructing a helix (and any other conformation) are: 1) the fixed bond length and bond angles; 2) the trans-conformation of the peptide bond; and 3) the sterical restrictions imposed by the Van der Waals radii of the atoms. If one calculates the permissible conformations which are in agreement with the above three limitations (ignoring the presence of hydrogen bonds completely), one then arrives at very interesting conclusions. We may consider, as an example, the polyamino acids: polyglycine, and poly-L-alanine.

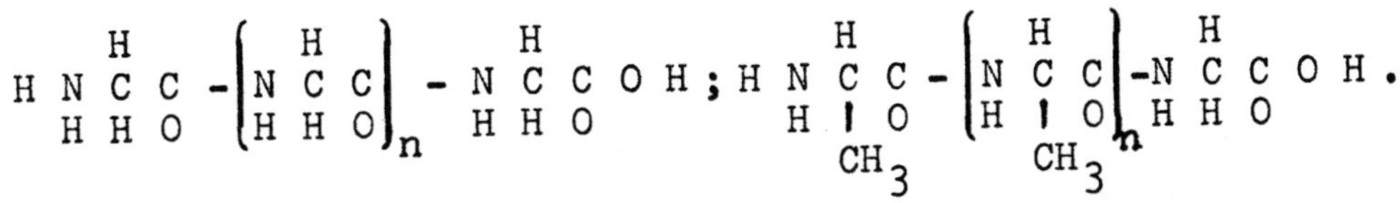

Polyamino acids are good models for our purpose, because they have a simple primary structure (all side chains identical) and exhibit the basic structures of proteins α-helix, β-structure, random coil) in pure forms. One may describe each conformation of the peptide backbone by two angles of rotations e.g., the angle Ψ between the α-C atom and the N atom and the angle Φ between C_α and CO. Scheraga et al. /28/ calculated the permissible combinations of these angles for our two polyamino acids. It is interesting to note how much the simple sterical effects of the side chains (H in polyglycine and CH_3 in polyalanine) influence the possible folding of the chain. Very many configurations are possible for polyglycine, and it was experimentally found that polyglycine does not form an α-helix. The number of possible conformations for poly-L-alanine, however, is already much limited. Very interesting to our discussion is the fact that the coordinates of the α-helix, in which poly-L-alanine was actually found to exist, fall in a small allowed area of the diagram Φ versus Ψ. A more sophisticated calculation of the Van der Waals conformational energy of poly-L-alanine (De Santis et al. /3/) has shown that at the coordinates of the α-helix, there exists a deep minimum in energy (which is deeper for the right-handed α-helix than for the left-handed). These calculations do not consider hydrogen bonds at all and show that the α-helix may be predicted by sterical considerations alone. We can see from the atomic coordinates of the α-helix (fig.2) that NH and CO groups in this arrangement come into favourable position for the formation of hydrogen bonds, and the occurence of such hydrogen bonds in the α-helix has been shown experimentally by many techniques. If sterical hindrance by bulky side chains prevents the coordinates of the α-helix, in which the formation of hydrogen bonds is possible, as occurs in the case of poly-L-valine, no stable helix is formed at all. (For a review on the physical properties of polyamino acids see Fasman /8/).

We will now inquire about the contribution of hydrogen bonds to the stability of the helix. Biological molecules exist in aqueous solution. We must, therefore, look at the α-helix in an aqueous solution, where it is in equilibrium with the randomly coiled form of the peptide chain. The main stabilizing effects for both forms are shown in fig.3

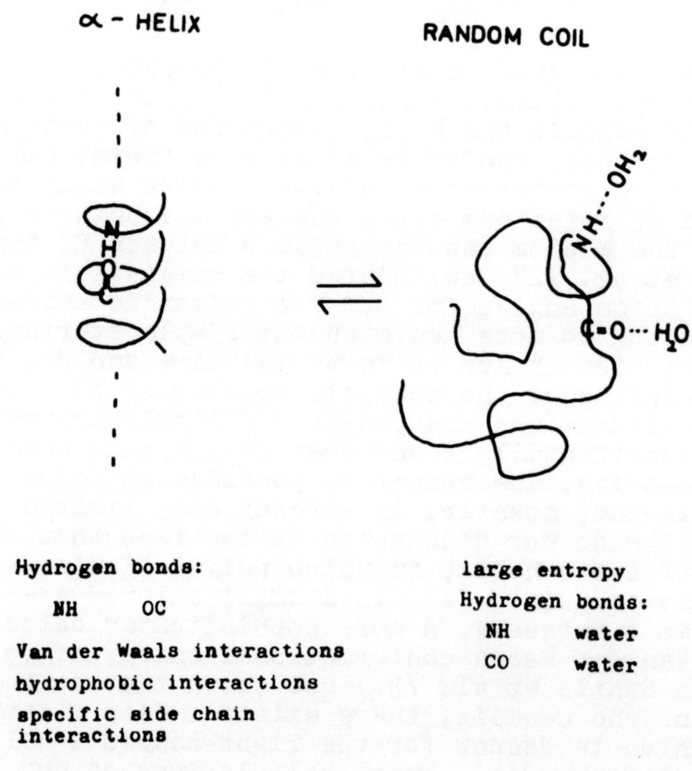

Fig.3 Main stabilizing effects for the α-helical and randomly coiled states of a peptide chain

The coiled state is favoured by strong hydrogen bonds with water or any other components of the solvent; the helical state by internal hydrogen bonds between the CO and NH groups. Thus, even when the internal bonds are quite strong, the balance may be zero or even in favour of the coil. If the stabilizing action of internal hydrogen bonds is compensated by hydrogen bonds with the solvent, side chain effects must help to stabilize the helix. An illuminating example is the homologous series of polyamino acids with the primary structure:

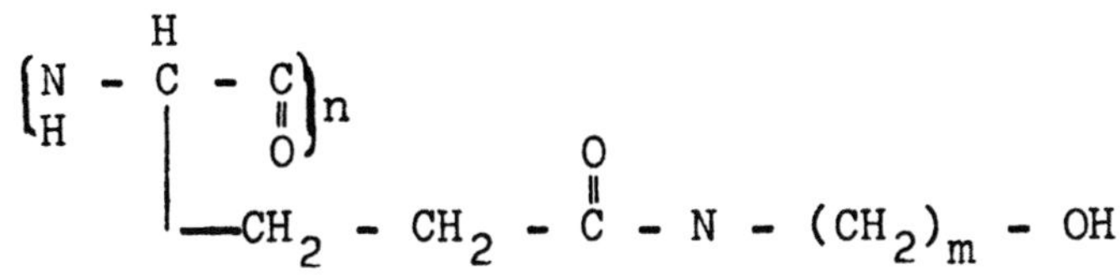

The member of this series with only one CH_2 group in the amino alcohol attached to the glutamic acid chain ($m = 1$) does not form an α-helix in water in the temperature range from 0 to 80°C (Lotan et al. /18/). Already one additional CH_2 group in the side chain ($m = 2$), however, gives rise to a 50% helix formation at 0°C. Three CH_2 groups give rise to a very stable helix. If one remembers that the backbone and the hydrogen bonds were not altered by these modifications, the stabilizing action of hydrophobic side chains becomes clear. It is illustrated experimentally that water competes with the internal hydrogen bonds by the following results. All members of the series form stable α-helices in methanol, but these helices are destroyed when water is added (Lotan et al. /18/).

At this point a few words will be added concerning the most important type of interactions between aliphatic or aromatic side chains, the so called hydrophobic interactions (Kauzmann /12/; Nemethy and Scheraga /20/; Nemethy /21/). Here hydrogen bonds between water molecules, surrounding the protein, play the dominant role. These hydrogen bonds are responsible for the high degree of structure which is preserved in liquid water (clusters, icebergs, or networks). The formation of hydrogen bonded, structured water is favoured at the boundary between water and a hydrophobic molecule. The immobilization of water molecules at this boundary causes a large decrease in entropy. The loss in entropy gives rise to a positive free energy for any process in which hydrophobic parts of a molecule have to be exposed to water. The reverse process, in which the boundary between hydrophobic side chains and water is decreased, is therefore favoured.

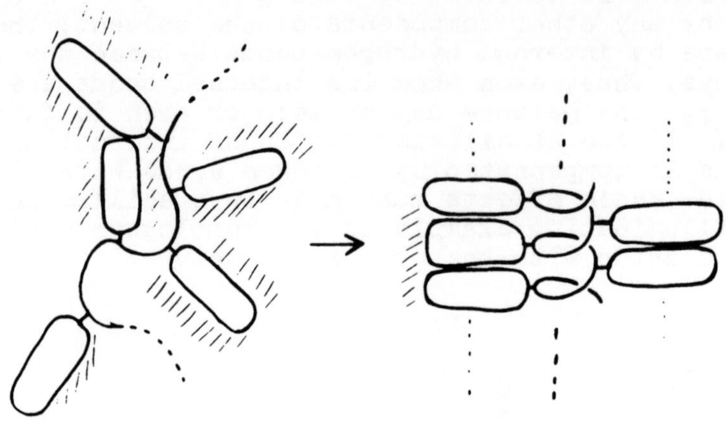

Fig.4 Hydrophobic side chain interactions: Aliphatic and aromatic side chains have a strong tendency to arrange in a regular way in order to make their region of contact with water (hatched) as small as possible

This leads to an apparent attraction force between the aliphatic or aromatic side chains of proteins (fig.4).

Hydrophobic interactions play a main role in the stabilization of the tertiary structure of proteins. The peptide chains of myoglobin and of hemoglobin are folded in such a way (fig.1) that most hydrophobic side chains are buried in the interior of the molecule (Kendrew /15/; Perutz et al. /24/; Perutz et al. /25/). Only side chains with hydrophilic groups are directed into the surrounding aqueaous solution. This principle has been found to be true for all other proteins whose structure has been solved. (lysozyme: Philips /26/; ribonuclease: Kartha et al. /11/; ribonuclease S: Wyckoff et al. /32/; carboxypeptidase A: Reeke et al./27/; α-chymotrysin: Matthews et al. /19/; papain: Drenth et al. /4/). There are many hemoglobins and other heme proteins which differ greatly in their amino acid sequence but have essentially identical structures. It is a fact that hydrophilic amino acid residues are rarely exchanged for hydrophobic residues or vice versa (Braunitzer /1/). The explanation is that if such a mutation occurs the new heme protein is unstable and is, therefore, eliminated during evolution. A good example of a hydrophobic interaction can be seen when the heme group is bound to the protein moiety of myoglobin or hemoglobin. The heme group rests with its hydrophilic side in a hydrophobic pocket of the protein. (see fig. 3.12 in Haggis /10/). The two propionic acid side chains, which make the other side of the heme group hydrophilic, protrude the surface. In the structure of sperm whale myoglobin, the hydrogen bonds between the propionic acid side chains and two amino acid side chains: Arg CD3 and His FG3, were observed (Watson /31/). (The symbols CD3 and FG3 indicate that the residues are located in the regions between the C- and the D-helix and between the F- and the G-helix respectively, see fig.1). It was first believed that these hydrogen bonds were essential for the binding of the heme to the protein. Later, however, hemoglobins were discovered in which the two amino acids were exchanged against aliphatic residues (summarized by Perutz /23/). This shows that the heme group is kept in its position even in the absence of hydrogen bonds by hydrophobic interactions. The importance of the hydrophobic interactions is shown directly by the fact that substitutions of residues in the hydrophobic pocket against hydrophilic ones have not been observed.

Very little accurate data concerning the absolute strength of hydrogen bonds, hydrophobic interactions and the many other stabilizing and destabilizing interactions are known for biological systems. We may only say that the energies are usually small, in most cases lower than 5 kcal/mole. We may, therefore, ask how it is possible that these weak and partially compensating interactions stabilize protein structures which are known to be extremely stable in broad regions of temperature, pH, etc. The answer is that we are dealing with cooperative systems in which the conversion of a segment of the molecule is dependent upon the state of the other parts of the molecule. By cooperativity is meant that the interactions between segments are such that there is a tendency for each segment to have the same conformational state as its neighbours. This phenomenon of cooperativity is the basis of an important property of all ordered macromolecules. The apparent equilibrium constant for a cooperative process is equal to the equilibrium constant of the elementary process, to the power of a large number. This number is the socalled cooperative length, which is the average number of residues in an uninterrupted sequence of identical conformational states. The cooperative length is about 100 for the α-helix (Zimm and Bragg /33/). For example, if all the stabilizing and destabilizing interactions sum up to a free energy which is only 0.2 kcal/mole in favour of the α-helix, then the equilibrium constant without cooperativity would be $K = 1.1$. This small free energy would usually be within the limits of error of any calculation which may be performed in order to determine the overall stability from the individual contributions of all stabilizing and destabilizing interactions. Indeed, for a noncooperative process, a value of 0.2 kcal could be ignored. Because of the high cooperativity, however, for the α-helix ⇄ coil transition (fig.3) an apparent equilibrium constant

$$K_{app} = 1.1^{100} \approx 10\ 000$$

results. The equilibrium is shifted completely towards the helical state in spite of the small difference in free energy between the helical and the coiled conformation! Cooperativity may be even larger in other systems (Engel /7/) and should always be remembered by all scientists who are trying to explain the stability of macromolecules by weak interatomic interactions like hydrogen bonds.

In the remaining part of this review, we shall have a quick glance on the dynamic aspects of hydrogen bonding and on hydrogen bonds in nucleic acids. In table 1 (Eigen /5/; Eigen /6/)the rate constants of hydrogen bond formation are compared with the rates of some important biological processes. Biological reactions must be extremely

Table 1

Rate Constants	
Formation of	
interionic interactions	$10^9 - 10^{11}$ M^{-1} sec^{-1}
H - bonds	$10^9 - 10^{10}$ M^{-1} sec^{-1}
hydrophobic interactions	$\approx 10^8$ M^{-1} sec^{-1}
enzyme - substrate complex antibody - hapten complex	$10^7 - 10^8$ M^{-1} sec^{-1}
Conformational transitions of	
polypeptides (n > 100)	$10^6 - 10^7$ sec^{-1}
polynucleotides (n > 100)	$< 10^6$ sec^{-1}
Turnover number of enzymes	
generally	$< 10^5$ sec^{-1}
in some cases	$< 10^6$ sec^{-1}

fast in order to meet the demands of the living world. Enzyme catalysis, for example, is a rapid process although it consists of a series of steps: the binding of the substrate, the chemical modifications done on the substrate, and the release of the products from the enzyme. This means that the elementary steps which are involved in the binding processes, namely the formation and the breakage of hydrogen bonds and hydrophobic interactions, have to be fast. The table shows that the rate constants of these interactions satisfy this need. We must remember that the formation of hydrophobic interactions is a process in which a change in water structure takes place. In this process the rate constant of hydrogen bonding again plays a role. An example of the binding of a substrate is the lysozyme-substrate complex whose structure is known from X-ray analysis (Philips /26/). The substrate, (a polysaccharide), is bound by about twelve specific hydrogen bonds and by a number of hydrophobic interactions in a cleft which runs through the enzyme molecule (Philips /26/). (At one specific point of the substrate, a bond is broken by the joint action of two COOH-groups. These belong to two amino acid side chains. In this catalytic cleavage, the formation of a hydrogen bond again plays an important role, but a detailed discussion of this important process is beyond the scope of this short introductionary review). In spite of the relatively weak interactions, a high specificity of binding of the substrate is achieved because of the participation of a large number of binding groups in a cooperative manner. The higher the cooperativity (specificity) of the process, the lower is its rate (Schwarz /30/). This again justifies the need for very fast elementary processes.

The reduplication of desoxyribonucleic acid (DNA) also requires high specificity of base pairing and a high rate of the elementary steps: The DNA-double helix is partially unfolded under the influence of an enzyme, (polymerase), and a new strand is synthesised which has the complementary sequence of basis. In this way, the genetic information layed down in the sequence of bases is transferred to a newly synthesised molecule. There are sets of hydrogen bonds between the base pairs in the DNA-double helix which are different for each of the various combinations of the bases.

This is quite different from the situation in the α-helix where the hydrogen bonds in the backbone are independent of the chemical nature of the individual amino acid residues. It has been shown that the specificity

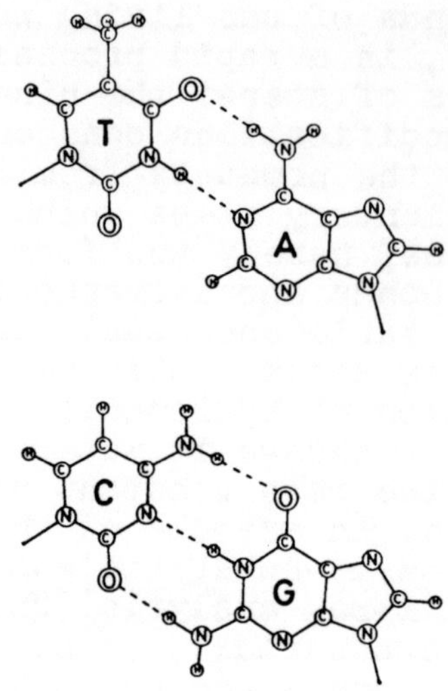

The complementary base pairs of DNA.

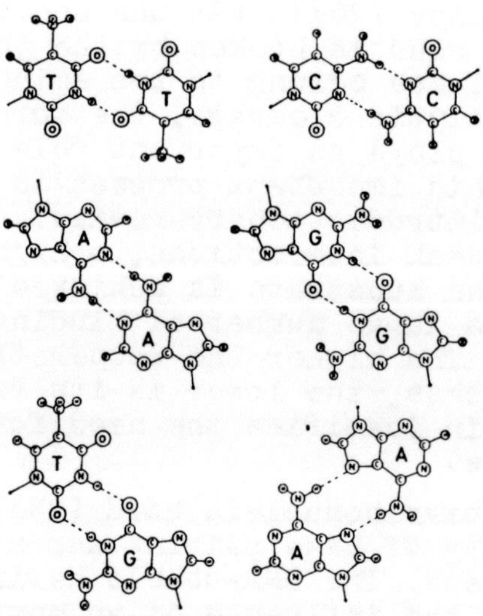

Other (noncomplementary) pair combinations.

Fig.5 Hydrogen bonds in complementary and non-complementary base pairs

which is seen in strand reduplication may be partially explained by the different stability of the possible sets of hydrogen bonds in different base pairs. The proper, that is, the complementary combinations are, by a factor of about 100, more stable than all others (Kyogoku et al. /17/; Küchler and Derkosch /16/; Funck /9/). It must be added that this specificity alone does not account for the enormous precision in base pairing during replication which is actually achieved in nature. The specificity of the process must be further increased by the enzyme polymerase. It must also be mentioned that, as in the case of proteins, hydrogen bonds are not the only stabilizing interactions in DNA.

Summary: Hydrogen bonds in biological macromolecules and in the surrounding water play an important role in biological systems. When discussing the contribution of hydrogen bonds inside of a biological molecule to the stability of its structure, we must consider the competition of water for the sites of hydrogen bonding. Therefore, the stabilizing action of hydrogen bonds in the folded state is more or less compensated by the interaction with water in the unfolded state. In most systems, however, small differences in free energy between the folded and unfolded state (or any other conformation) are sufficient for a complete stabilization of one of the two conformations, because of cooperativity. Differences in the strength of many possible sets of hydrogen bonds contribute to the specificity of base pairing in DNA. The rate of the formation and disruption of hydrogen bonds is very fast which is important for the dynamics of biological systems.

References

/1/ Braunitzer, G., Naturwissenschaften 54, 407 (1967)

/2/ Corey, R.B., L.Pauling, Proc.Roy.Soc. (London) B.141, 10 (1953)

/3/ De Santis, P., E.Giglio, A.M.Liquori, A.Ripamonti, Nature 206, 456 (1965)

/4/ Drenth, J., J.N.Jansonius, R.Koekoek, H.M.Swen, B.G.Wolthers, Nature 218, 929 (1968)

/5/ Eigen, M., Naturwissenschaften 12, 426 (1963)

/6/ Eigen, M., Berichte der Bunsenges. für Physikal. Chemie, 68, 889 (1964)

/7/ Engel, J., in "Conformation of Biopolymers, G.N. Ramchandran ed." Vol.2, Academic Press, New York (1968)

/8/ Fasman, G., "Poly-α-Amino Acids" Marcel Dekker, New York (1967)

/9/ Funck, T., cited by M.Eigen, Nobelsymposium (1967) 333

/10/ Haggis, G.H., in "Introduction to Molecular Biology, G.H.Haggis ed." p.60, Longmans, Green Co., London, (1964)

/11/ Kartha, G., J.Bello, D.Harker, Nature 213, 862 (1967)

/12/ Kauzmann, W., Adv.Protein Chem. 14, 1 (1959)

/13/ Kauzmann, W., in "Connective Tissue, Intercellular Macromolecules", Little Brown, Boston, p.43 (1964)

/14/ Kendrew, J.C., H.C.Watson, B.E.Strandberg, R.E. Dickerson, D.C.Philips, V.C.Shore, Nature 190, 663 (1961)

/15/ Kendrew, J.C., Brookhaven Symposium in Biology, 15, 216 (1962)

/16/ Küchler, E., J.Derkosch, Z.Naturforsch.21b, 209 (1966)

/17/ Kyogoku, Y., R.C.Lord, A.Rich, J.Am.Chem.Soc. 89, 496 (1967)

/18/ Lotan, N., A.Yaron, A.Berger, Biopolymers 4, 239 (1966)

/19/ Matthews, B.W., P.B.Sigler, R.Henderson, D.M.Blow, Nature 214, 652 (1967)

/20/ Nemethy, G., H.A.Scheraga, J.Chem.Phys. 36, 3401 (1962)

/21/ Nemethy, G., Angew.Chem. 79, 260 (1967)

/22/ Perutz, M.F., Proteins and Nucleic Acids, Elsevier, Amsterdam, London (1962)

/23/ Perutz,M.F.,J.Mol.Biol. 13, 646 (1965)

/24/ Perutz, M.F., J.C.Kendrew, H.C.Watson, J.Mol.Biol. 13, 669 (1965)

/25/ Perutz, M.F., H.Muirhead, J.M.Cox, L.C.G.Goaman, F.S.Mathews, E.L.McGandy, L.E.Webb, Nature 219, 29 (1968)

/26/ Philips, D.C., Scientific American 215, 78 (1966)

/27/ Reeke, G.N., J.A.Hartsuck, M.L.Ludwig, F.A.Quiocho, T.A.Steitz, W.N.Lipscomb, Proc.Nat.Acad.Sci. U.S. 58, 2220 (1967)

/28/ Scheraga, H.A., S.J.Leach, R.A.Scott, G.Nemethy, Disc.Faraday Soc. 40, 268 (1965). See also Poland, D., Scheraga in "Poly-α-Amino-Acids, G.Fasman ed." p.391, Marcel Decker, New York (1967)

/29/ Schroeder, W.A., R.T.Jones, in "Progress in the Chemistry of Organic Natural Products (L.Zechmeister ed.)" Springer, New York, Vol.23, 113 (1965)

/30/ Schwarz, G., Ber.d.Bunsengesellschaft 68, 843 (1964) J.Mol.Biol. 11, 64 (1965)

/31/ Watson, H.C., in "Heme and Hemoproteins, B.Chance, R.W.Estabrook, T.Yonetani, editors", 63, Academic Press, New York (1966)

/32/ Wykoff, H.W., K.D.Hardman, N.M.Allewell, T.Inagami, L.N.Johnson, J.M.Richards, J.Biol.Chem. 242, 3984 (1967)

/33/ Zimm, B.H., J.K.Bragg, J.Chem.Phys. 31, 526 (1959)

THE NATURE OF THE HYDROGEN BOND

G. Statz and E. Lippert

Iwan N. Stranski-Institute of Physical

Chemistry, Technische Universität Berlin

Abstract

The hydrogen bond is one of the most important forms of molecular interaction. It determines the physical and chemical properties of many liquids and crystals. Such specific interactions are mostly observed between the elements of the second line of the periodic table. The three-center-problem (donor, proton, acceptor) with four electrons can be treated theoretically with the aid of the electrostatic theory, the valence-bond theory, the charge-transfer theory or the SCF-MO theory. These electronic theories of hydrogen bonding are reviewed. From the experimental point of view the most important methods of investigation of the hydrogen bond are the diffraction methods (X-ray, electron and neutron), nuclear magnetic resonance and IR- and far IR-spectroscopy. The results of these experimental methods will be shown.

This paper is not a special contribution to the physics of ice but gives a general survey of some important aspects of hydrogen bonding as they are also observed in other systems. At first we intend to deal with the influence of hydrogen bonding to NMR, IR and far-IR-spectra and the possibilities of experimental determination of equilibrium constants and energies, and then to give a summary of the possibilities of theoretical descriptions of hydrogen bonding.

The hydrogen bond AH...B is a specific molecular interaction in which beside the proton atoms with non-bonding

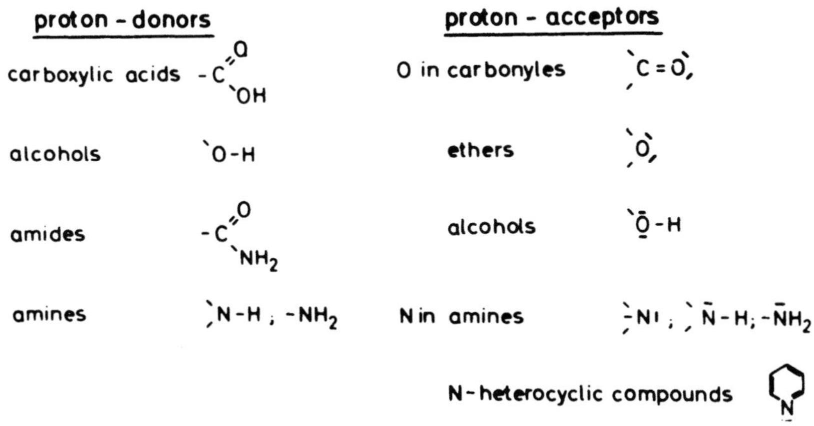

Figure 1

electrons take part. In such a bond the proton must protrude far enough out of the electron cloud of the donor atoms A and must penetrate the electron cloud of the acceptor atom B deep enough. Therefore, hydrogen bonds are mainly observed between the elements of the second row of the periodic table, i.e. between the elements B, C, N, O, and F. In addition weaker interactions occur to sulphur and to the halogenes. The functional groups acting most frequently as proton donors and acceptors in organic chemistry are summarized in fig.1. The energies resulting from the association of these compounds range from 3 to 10 kcal per mole and hydrogen bond. There are also stronger bases like sulphoxide and phosphine oxide which are forming much stronger hydrogen bonds with the corresponding acids /1/.

Because of the influence of these molecular interactions many physical and chemical properties of the associating substances are changed.

Beside the alteration of the dielectric and thermodynamic properties (e.g. dielectric constant, dipole moment, conductivity, vapour pressure, melting and boiling point, solubility) the proton signals in the proton resonance spectra are shifted to lower fields or to higher frequencies respectively. In the infrared spectra in frequency of the AH-stretching vibration is diminished and the integrated intensity of the corresponding absorption band is enlarged, whereby the association band shows anomalous width and in some cases a complicated structure. Finally new normal vibrations of the associated molecules arise with very low frequencies in the far-infrared. The

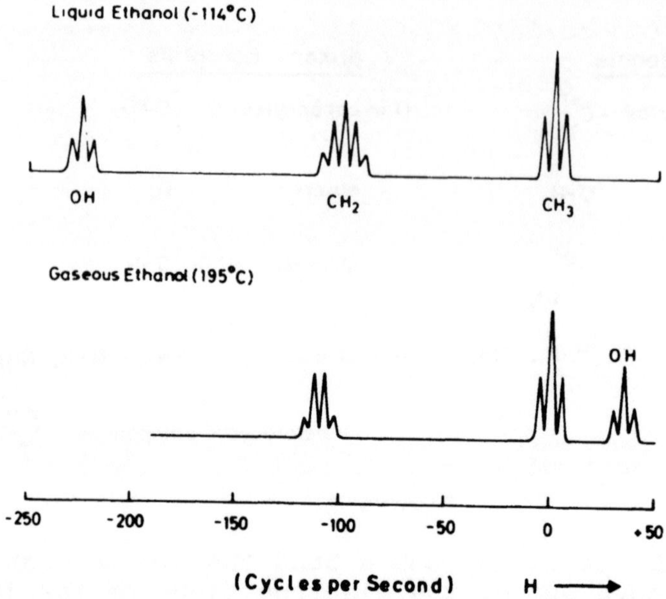

Figure 2

proton magnetic resonance and the infrared spectroscopy are the most popular methods for the investigation of hydrogen bonding. For that reason the results gained by these methods shall be described in detail by means of some examples.

Due to the association of two molecules by hydrogen bonding the resonance signal of the AH-proton is shifted to lower fields because of the diminuation of the electric shielding of the proton which is connected with the increase of the AH-distance as well as under the influence of diamagnetic anisotropy of the charge distribution. This is shown in fig.2 which contains the proton resonance spectra of ethanol as a vapour at high temperature without any association and in the liquid phase with association of the alcohol molecules /2/.

As the "time of observation" in nuclear resonance is large, compared with the exchange velocity of the AH-proton, it is not possible to observe discrete proton signals belonging to different association forms. One sharp signal, possibly split by coupling, results. Its spectral position is determined by the averaged contributions of all present species, that is, by the present association equilibrium. Therefore, the possibility exists to determine the nature of the equilibrium (e.g. monomer-dimer or monomer-trimer) as well as the equilibrium constant and finally from its temperature dependence the H-bond energy, by measuring the dependence of the chemical shift of the proton signal on the concentration of a substance dissolved in an inert solvent. E.g., for a monomer-dimer equilibrium in solution the chemical shift of a proton signal is described by

$$c\,\delta = c_M\,\delta_M + 2c_D\,\delta_D\,. \qquad (1)$$

Furthermore one finds for the overall concentration

$$c = c_M + 2c_D \qquad (2)$$

and for the assumed equilibrium

$$c_D = c_M^{\,2}\,K_c\,. \qquad (3)$$

From these three equations one obtains by rearranging

$$\delta = \left(\frac{\delta-\delta_M}{c}\right)^{1/2}\left(\frac{\delta_D-\delta_M}{2K_c}\right)^{1/2} + \delta_D\,. \qquad (4)$$

The chemical shift of the proton signal for the monomer δ_M is obtained by the extrapolation $c \rightarrow 0$. With the knowledge of this value one obtains the equilibrium constant and the chemical shift of the dimer from the diagram δ as a function of

$$\{(\delta-\delta_M)/c\}^{1/2}$$

for a large concentration range which gives a straight line. For a monomer-trimer equilibrium, for example, no straight line in this diagram is obtained; in this case one has to modify the above expression accordingly. If for a certain equilibrium an equilibrium constant could be determined, one obtains from its temperature dependence the H-bond energy from

$$\left(\frac{\partial \ln K_c}{\partial T}\right)_c = \frac{\Delta H}{RT^2}\,. \qquad (5)$$

Fig.3 contains the curves of the chemical shift against the logarithm of the concentration of t-butanol in CCl_4, that have been calculated assuming only one form of association for dimers, trimers, and tetramers respectively. The measured values give the best fit to the curve of the monomer-trimer equilibrium. If one plots these values in a diagram according to equation (4) for a monomer-trimer equilibrium (δ against $\{(\delta-\delta_M)/c\}^{1/2}$) the points

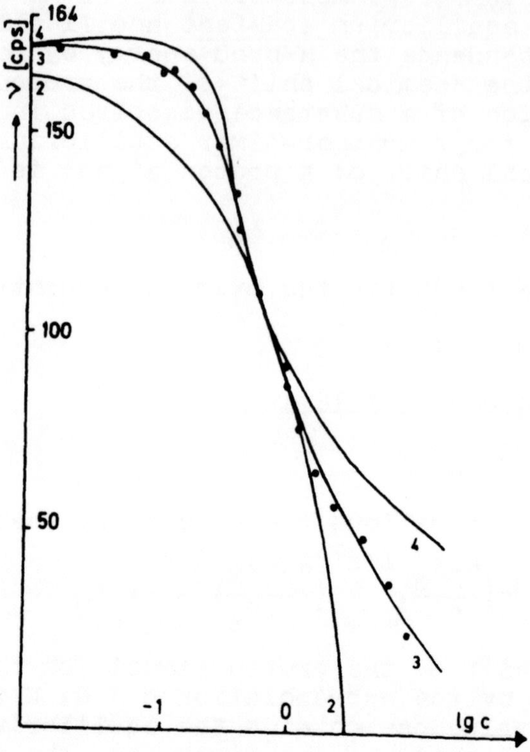

Figure 3

fit actually a straight line /4/. For smaller concentrations beyond 0.05 mole/l strong deviations from the straight line occur which apparently must be attributed to a dimerization of the t-butanol. Consequently the association of t-butanol to trimers, e.g., shown as ring trimers in fig.4 is proved for not too small concentrations. In alcohols with voluminous substituents the capacity for association may be cancelled by steric hindrance. This leads to a concentration and temperature independence of the position of the OH-signal as shown in fig.5 for triphenyl carbinol /5/.

Corresponding to the association shift of the proton signal in proton magnetic resonance one observes in infrared spectroscopy an association band at lower frequencies beside the AH stretching vibration. Since the vibrational frequency as well as the association shift increase with the strength of the hydrogen bond, a correlation diagram as in fig.6 results /6/. Similar correlations exist in addition, e.g., to the A...B distance and to other properties connected with the strength of

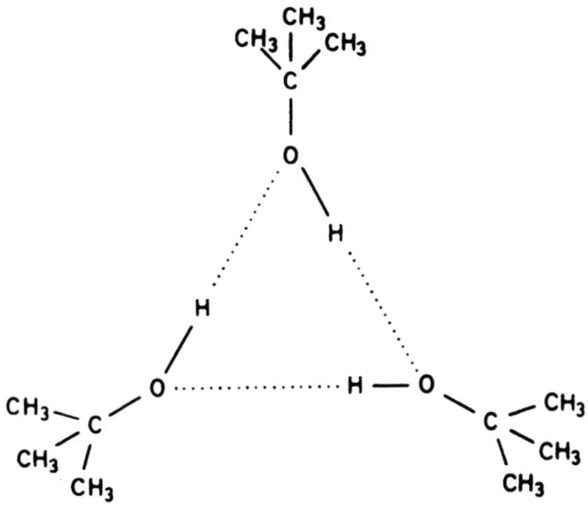

Figure 4

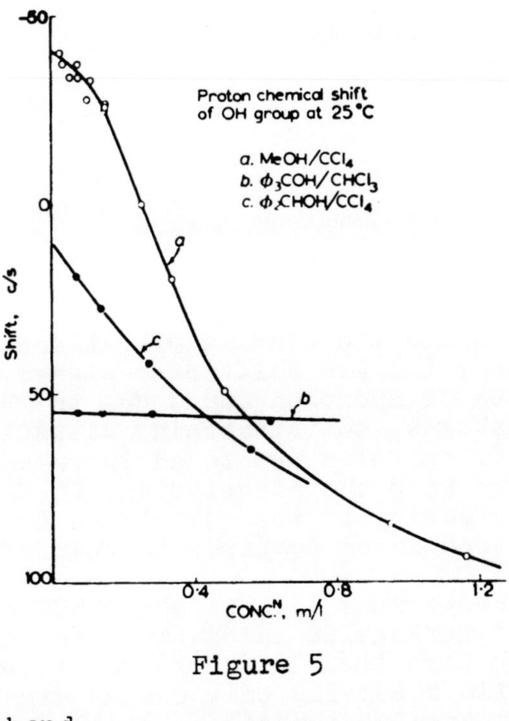

Figure 5

the hydrogen bond.

In the infrared spectra the formation of hydrogen bonds is detected most distinctly in the change of the proton

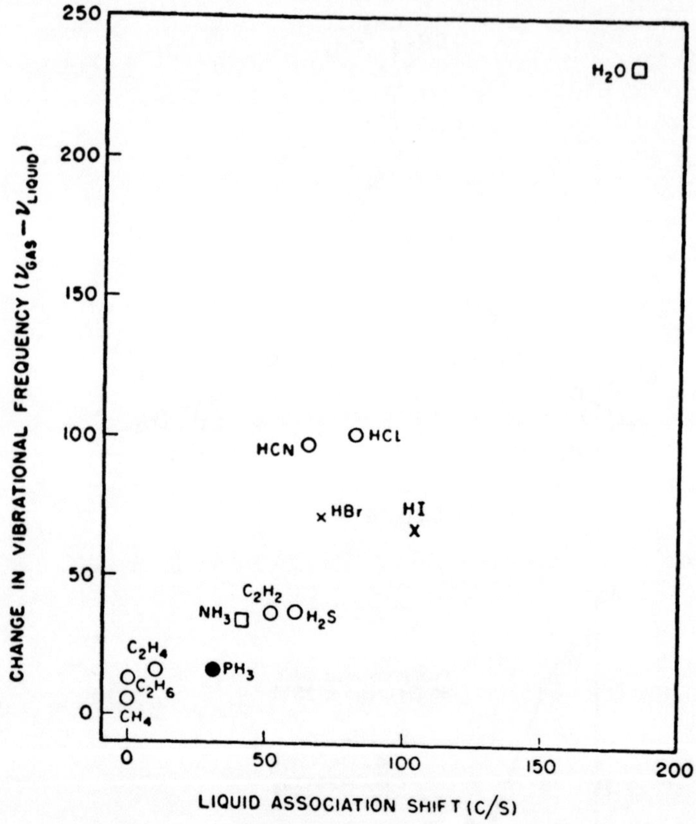

Figure 6

vibration bands. While the bending vibrations of the proton δ (AH) and γ (AH) are shifted to higher frequencies with the formation of hydrogen bonds due to the increase of the force constants, the stretching vibration ν (AH) shifts to lower frequencies due to an increase of the medium AH-distance with the association. At the same time the integrated intensity of the correlated absorption band increases strongly as compared to that of the monomer band.

This frequency change runs e.g. to about 500 cm^{-1} for the ring dimer of carboxylic acids; however, this value might increase to more than 1500 cm^{-1} for stronger acid-base adducts. While observing only one AH signal in proton magnetic resonance at a position resulting from the contributions of all present species, one can detect in infrared spectroscopy the monomer band as well as different association bands at the same time.

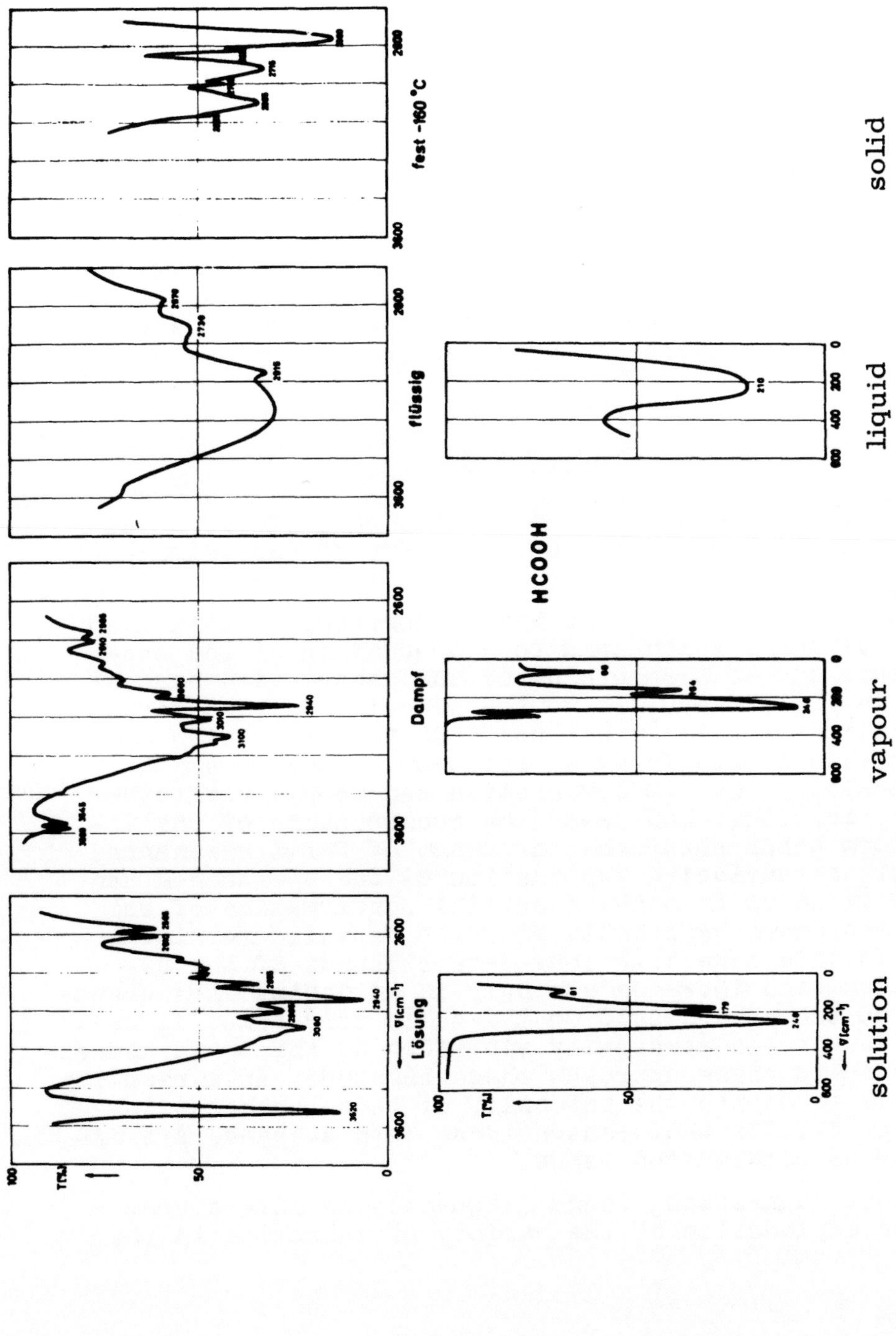
HCOOH
Lösung
Dampf
flüssig
fest -160 °C
solution
vapour
liquid
solid

Figure 7

The AH-proton in the free molecule vibrates at the highest frequency, while the corresponding absorption band is sharp. In addition to the monomer band a broader association band appears at frequencies that are the lower, the stronger the hydrogen bond. This band can show either an indistinct structure as in alcohols which results from the superposition of the stretching vibration bands in different present association forms like dimers, trimers, tetramers and so on, or the association band can be split into a lot of sharp subbands as in carboxylic acids (e.g. in HCOOH-vapour 17 subbands), see fig.7. Consequently the centre of the association band in alcohol solutions shifts to lower frequencies with increasing concentration, and the connected equilibrium shifts to higher association. Using the matrix isolation technique one succeeds in observing the different species in the same spectrum /7/. The carboxylic acids associate on the other side mainly to ring dimers of the structure

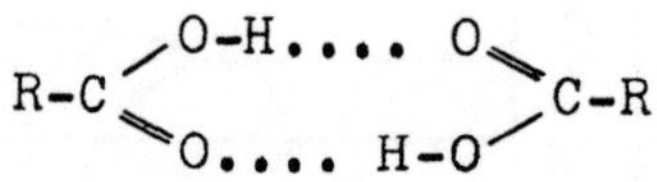

so that the association bands of the pure liquid acid, of the vapour, and of the solution are of very similar shape.

The pure formic and acetic acids, however, contain also chain multimers which lead to a broadening of the substructure and to a smoothing of the association band as shown in fig.7 and 9. As to the association band for the carboxylic acid, it is not possible to distinguish different association forms in solution. Owing to strong anharmonicity the ν(OH)-vibration can couple with other combination tones and cause the substructure of the association bands under the influence of Fermi resonance. However, a convincing explanation of the phenomenon cannot be given up to date. A partial confirmation of this Fermi resonance hypothesis seems to be valid for formic acid. In this case both long-wave subbands of the association band increase strongly in intensity when changing from the pure liquid to the solid acid (fig.7), because the proton stretching vibration of the chain adducts in the solid phase approach these subbands. As a result of Fermi resonance the intensity of these subbands increases /8/. For that reason these both subbands are explained as combination bands.

A further suggestion, found frequently in literature, for the explanation of the variety of submaxima in the association bands of the carboxylic acids is to assume a combination of the ν(OH)-vibration with the far-IR vibrations of the acid molecules in the dimer towards each other. This cannot be accepted after our intense investigations on these vibration bands.

By analogy with the proton magnetic resonance one can also prove the association equilibrium existing in solution from the concentration dependence of the extinction of the monomer band in the IR spectrum.

For the calculation of the corresponding equilibrium constant one has to use the following equations

$$E_M = \varepsilon_M c_M d \qquad (6)$$

$$K = c_M^2/c_D$$

$$c = c_M + 2c_D$$

on the simplified assumption of a monomer dimer equilibrium. Hence it follows

$$E_M/d = \frac{K}{2}\varepsilon_M^2(dc/E_M) - \frac{K}{2}\varepsilon_M. \qquad (7)$$

Plotting E_M/d as a function of dc/E_M one obtains a straight line from which one can calculate the equilibrium constant. Beside the alternations of the proton vibrations in consequence of association by hydrogen bonding new normal vibrations appear which can be described in a first approximation by vibrations of associated molecules towards each other. Because of the small force constants and the big vibrating masses these vibrations ensue with very low frequencies and are found in the far infrared, that means with frequencies lower than 250 cm^{-1}.

In the ring dimers of the carboxylic acids six of such vibrations are possible (as presented in fig.8) three of which are Raman-acitve and three IR-active.

The solution and vapourspectra of formic acid contain all three IR-active bands, among which the vibration ν(OH...O) has the highest frequency. The far-IR spectrum of the pure, liquid formic acid shows, however, only one very broad band from which the existence of long chains can be deduced. In the low temperature spectrum these chains are straightened and this effects again sharp absorption bands, as also observed in the chain adducts of solid acetic acid (fig.9). With increasing mass of the substituent bonded to the acid group the association vibrations shift further to longer wave-lengths

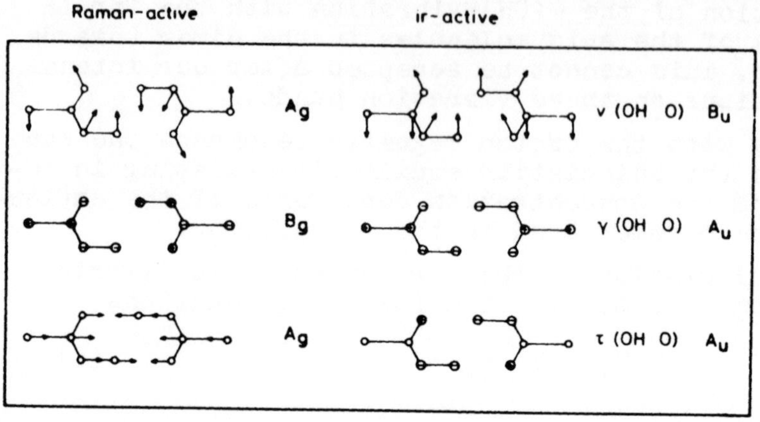

Figure 8

and often ν(OH...O) only is to be observed as a very weak band above 33 cm^{-1} /10/. Some fundamental data of the hydrogen bond, as the frequency of the ν(OH)-vibration, the bond energy and the force constant between A and B can be already gained from an empirical model for the double minimum potential influencing the proton in the hydrogen bond, as it was for instance determined by Lippincott and Schroeder /11/ (fig.10) using two Morse functions as well as one term for the van der Waals repulsion and one for the electrostatic attraction between atoms A and B employing 10 empirical constants or by McKinney and Barrow /12/ using a simple one-dimensional quantum mechanical model. Neutron diffraction experiments have confirmed the existence of two equilibrium distances for the proton. From the splitting of the ν(OH)-association band in the overtone region Bell and Barrow /13/ concluded the existence of a double-minimum potential in which the vibration level v=2 is split into the observed double band (fig.10).

The first theoretical treatment of the hydrogen bond based on Pauling's idea /14/ that in a hydrogen bond the hydrogen atom with its 1s-orbital cannot form more than one pure covalent bond. Therefore the hydrogen bonding has to be reduced to ionic forces. In the simpler electrostatic theory for the treatment of the three-centre-four-electron system of a hydrogen bond,i.e. OH...O', the two electrons of the OH-bond and the lone pair of O are arranged on a straight line between the three atomic cores assumed as positive point charges in such a way that the right values for the dipole moments of the OH-bond and

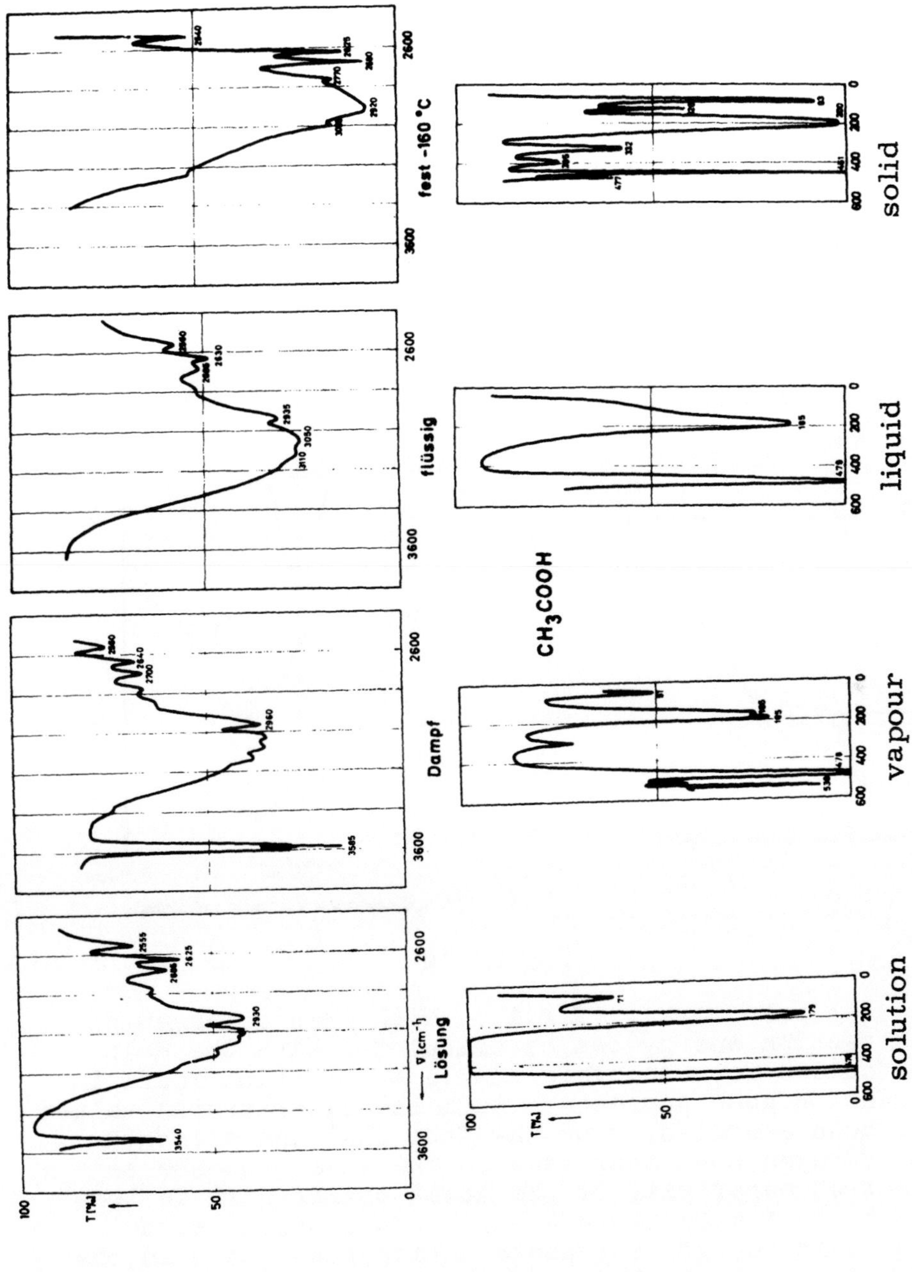
Lösung
Dampf
flüssig
fest -160 °C
CH3COOH
solution
vapour
liquid
solid
T[%]
ν [cm-1]
3600
2600
3540
2930
3585
2960
3110
3050
2935
2920

Figure 9

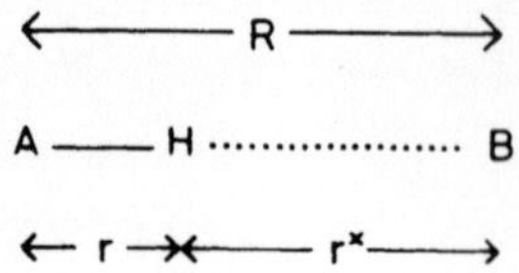

Lippincott + Schroeder :
(1957)

$$V = D_0\left(1 - e^{-\frac{n(r-r_{eq})^2}{2r}}\right) + D_0^*\left(1 - e^{-\frac{n^*(r^*-r_{eq}^*)^2}{2r^*}}\right) + Ae^{-bR} - BR^{-m}$$

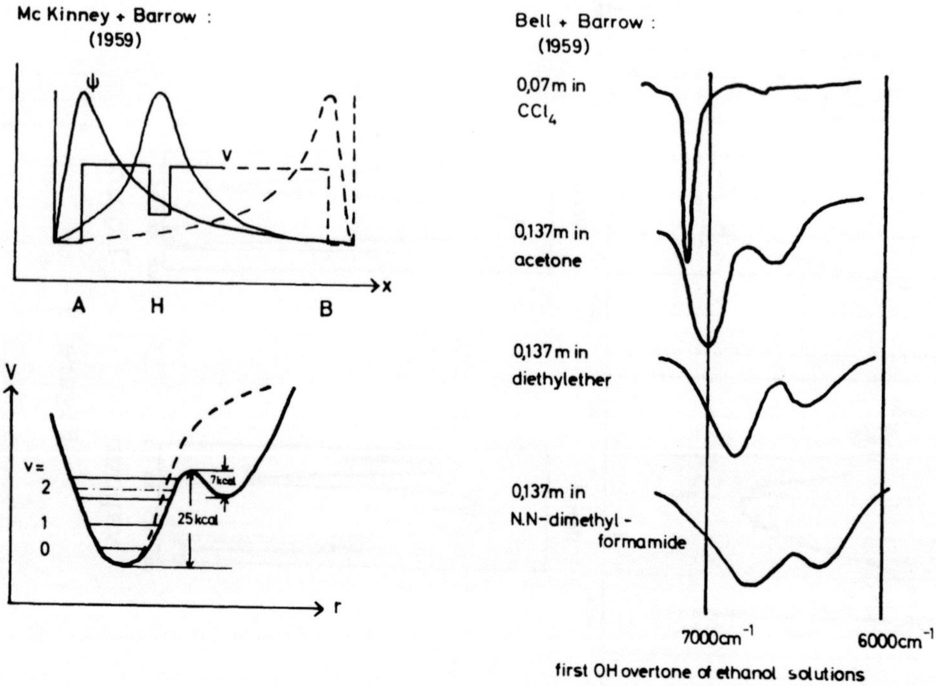

Figure 10

of the lone pair result (fig.11). With this arrangement the interaction energy can be calculated with the help of the Coulomb formula. This simple calculation leads in some cases to good agreements with the experimental values for the bond energies. From the fact that the strength of the hydrogen bond increases in the same sequence as the electron negativity of the atoms taking part in the bond, i.e. in the sequence N, O, F, follows, that the electrostatic energy represents an important part in the hydrogen bond energy. The charge on B exerts a force on the proton which opposes the force acting on the proton

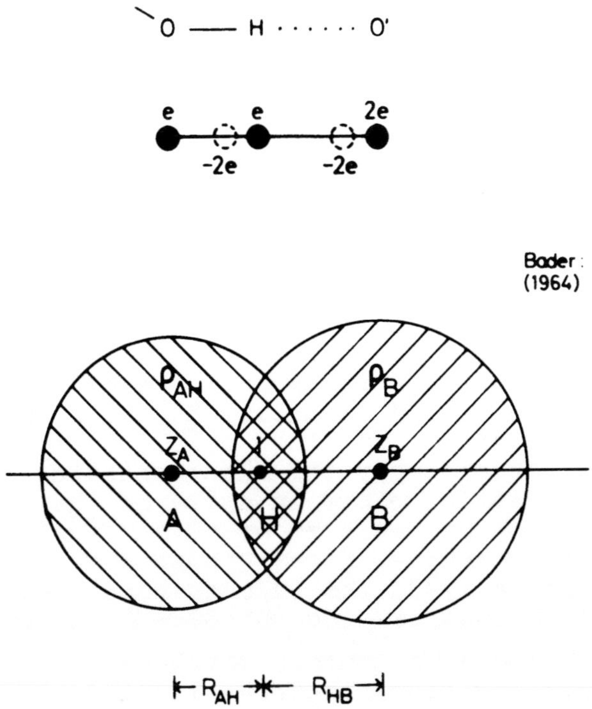

$$K_{\nu_{AH}} = \frac{2}{R^3_{AH}}(Z_A - Q_A) + \frac{2}{R^3_{HB}}(Z_B - Q_B) + 4\pi(\rho_{AH}(H) + \rho_B(H))$$

Figure 11

in the free molecule. The AH-distance is thus increased and the stretching-force constant for the proton is decreased. This leads to the observed low frequency shift of the ν(OH)-band.

In the elaborated electrostatic theories, for instance from Bader /15/, a homogeneous charge distribution is assumed by considering all electrons. The electron distribution of the AH-bond is described by an atom with z_A+1 electrons centered on A; the electron distribution of B is represented by a corresponding atom B. In addition there are three nuclei with the charge z_A, 1 and z_B (fig.11). From this model one obtains just the same results as in the before mentioned theory but, however, not more important data.

These electrostatic theories are all based on the

assumption that the hydrogen bond does not disturb the charge distributions of the components AH and B (e.g. by charge transfer). If this assumption would be valid the hydrogen bond could be described as an entirely ionic effect. The hydrogen bond is, however, only partially ionic - other non-ionic properties appear in addition. These are reflected by the following experimental facts: no correlation exists between the hydrogen bond strength and the dipole moment of the base B (e.g. phenol forms a strong H-bond with dioxane (0.3 D) and a weak H-bond with acetonitrile (3.5 D)). The A-B-distances are always much smaller than the sum of the van der Waals radii of both atoms. Thus a strong repulsion would result, consequently other short-range forces must be effective. Finally the increase of the intensity of the ν(OH)-stretching band cannot be explained solely by the electrostatic theory. Accordingly the electrostatic theories are not completely satisfactory in spite of their remarkable success and must be replaced by more effective quantum mechanical theories. Three groups were in principle developed since 1947: the valence-bond theories, the charge-transfer theories and the SCF-MO theories.

The valence-bond theory was mainly developed by Coulson /16/. In this case the four electrons involved in the hydrogen bond OH...O' are distributed over the atomic orbitals p_1', h' and p_2' (fig.12). The wave function for the hydrogen bond is constructed by linear combinations of these orbitals. Characteristic for this theory is the choice of the wave functions belonging to special mesomeric structures:

ϕ_1 describes the covalent bond between O and H and the lone pair belonging to O'.
ϕ_2 describes the ionic bond between O and H with the lone pair belonging to O'.
ϕ_3 deals, in contrast to the electrostatic effects described above, with a new phenomenon which has its source in quantum mechanics. This new phenomenon is the charge transfer from O' to O. This wave function thus refers to the covalent bond between H and O' containing now the lone pair at O. In this way the partially covalent character of the hydrogen bond is introduced.

For solving the secular equation $| \underline{H} - E\underline{S} | = 0$ the overlap matrix is calculated non-empirically using Slater's orbitals; for the diagonal elements of the Hamiltonian matrix four terms are taken into account:

The covalent bond energy expressed by a Morse term, the short range repulsion energy expressed by the

Valence bond

Coulson, Danielson
(1954)

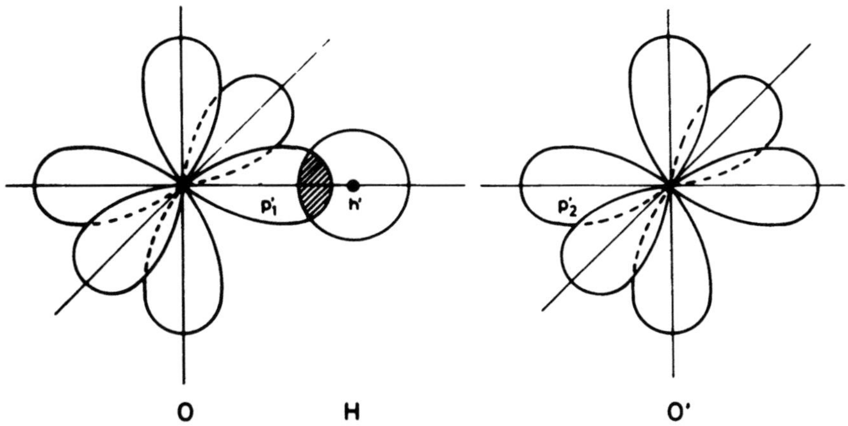

$$\Phi = C_1\Phi_1 + C_2\Phi_2 + C_3\Phi_3$$

$$\Phi_1 = N_1\left\{ \left| p_1 \bar{h} p_2 \bar{p}_3 \right| - \left| \bar{p}_1 h p_2 \bar{p}_3 \right| \right\} \quad \text{O} - \text{H} \quad \text{O}$$

$$\Phi_2 = N_2 \left| p_1 \bar{p}_1 p_2 \bar{p}_2 \right| \quad \bar{\text{O}} \quad \overset{+}{\text{H}} \quad \text{O}$$

$$\Phi_3 = N_3\left\{ \left| p_1 \bar{p}_1 h \bar{p}_2 \right| - \left| p_1 \bar{p}_1 \bar{h} p_2 \right| \right\} \quad \bar{\text{O}} \quad \text{H} - \overset{+}{\text{O}}$$

	Coulson (1959)
a) electrostatic	+6 kcal/mole
b) short range repulsion	-8,4
c) charge transfer	+8
d) polarization	+3
	+8,6
exper	+6,1 for ice

Figure 12

empirical formula Ke^{-br}, the energy connected with the transfer of an electron from B to A, as well as the polarization energy.

The non diagonal elements of the Hamiltonian matrix are calculated in such a way that the secular equation gives the exact value E_o for the bond energy of the free OH-group. By convenient choice of the parameters in the matrix elements the secular equation can be solved - the quality of the solution strongly depends on the choice of empirical parameters. The results given by the valence-bond

Charge transfer

Bratož
(1965)

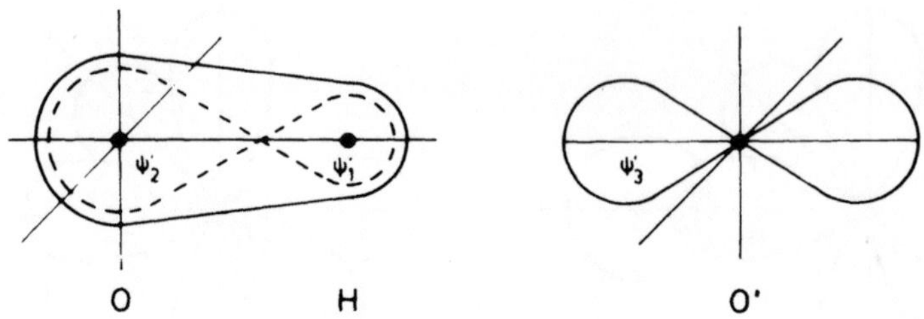

$$\Phi = c_1\Phi_1 + c_2\Phi_2$$
$$\Phi_1 = N_1 \left| \psi_1 \bar{\psi}_1 \quad \psi_3 \bar{\psi}_3 \right|$$
$$\Phi_2 = N_2 \left\{ \left| \psi_1 \bar{\psi}_1 \psi_2 \bar{\psi}_3 \right| - \left| \psi_1 \bar{\psi}_1 \bar{\psi}_2 \psi_3 \right| \right\}$$

Figure 13

theories can be summarized as follows:
The forces implied in hydrogen bonding are

1. electrostatic forces
2. short-range repulsion forces
3. charge-transfer forces
4. polarization forces.

The success of the electrostatic theories can be explained by the fact that the effects 2 and 3 frequently cancel each other (fig.12). The amount of charge transferred from O' to O is determined by c_3^2 (fig 12). This quantity can be neglected for long H-bonds, so that these are predominantly of electrostatic nature.

In the charge transfer theories, e.g. from Puranik and Kumar /17/ and from Bratoz /18/ (fig.13) the four electrons involved in the hydrogen bond OH...O' are distributed over three orbitals: the bonding and non-bonding orbitals ψ_1' and ψ_2' and the orbital ψ_3' belonging to the lone pair at O'. The wave function is composed in two parts:

ϕ_1 describes a situation in which two electrons are placed in the OH-bond and two ones at O'. Hence it reflects the electrostatic and short-range repulsion interactions.

ϕ_2 describes a situation in which charge transfer has

taken place from orbital ψ_3' of the lone pair to the non-bonding orbital ψ_2.

The diagonal elements of the Hamiltonian matrix are again determined semi-empirically from the energy of the free OH-group, the electrostatic energy by applying the Coulomb formula to formal charges on the atoms, the short-range repulsion energy determined by Ke^{-br}, and the charge-transfer energy. The non-diagonal elements of the Hamiltonian matrix and the overlap matrix are empirically determined. Also in the charge-transfer theory the hydrogen bond energy can be split into four parts as in the valence bond theory. The charge transfer component is larger if the ionization potential of the lone pair is smaller. This is in agreement with the experimental results. In the charge transfer process a fraction of the electron charge is transferred to the non-bonding OH-orbital. Hence the OH-bond is weakened, the bond length is increased and the stretching-force constant for the proton is decreased. This is just what is observed experimentally. Additionally the charge migration rises the polarity of the OH...O' system and thus increases the intensity of the ν(OH)-band .

Besides these already mentioned theories trials are recently made to consider in some simple examples AH...B as a single large molecule that can be calculated with the help of the SCF-MO theory using modified Slater's orbitals. Such attempts had been made by Fischer-Hjalmars /19/, Grahn /20/, Paoloni /21/, Hofacker /22/, and Rein and Harris /23/ for the systems $(H_2O)_2$, $(HF)_2$, $(H_2S)_2$, $(HCl)_2$, $(H_9O_4)^+$ and for guanine-cytosine base pairs.

In consequence of always necessary neglections all these theories can only supply approximate values for the hydrogen bond energies and the direction of some experimentally observed effects caused by hydrogen bond association. A complete quantitative description of the hydrogen bond cannot be given up to date.

References

/1/ Hadzi, D., N.Kobilarov, J.Chem.Soc. 4A, 439 (1966)

/2/ Schneider, W.G., Hydrogen Bonding, D.Hadzi, Ed., London (1959) p.55

/3/ Saunders, M., J.B.Hyne, J.Chem.Phys. 29, 1319 (1958)

/4/ Lippert, E., Ber.Bunsenges. 67, 267 (1963)

/5/ Reid, C., T.M.Connor, Hydrogen Bonding, D.Hadzi, Ed., London (1959) p.77

/6/ Schneider, W.G., H.J.Bernstein, J.A.Pople, J.Chem.Phys. 28, 601 (1958)

/7/ Pimentel, G.C., Hydrogen Bonding, D.Hadzi, Ed., London (1959) p.107

/8/ Mikawa, Y., J.W.Brasch, R.J.Jakobsen, J.Mol.Spec. 24, 314 (1967)

/9/ Harris, J.T., M.E.Hobbs, J.Am.Chem.Soc. 76 1419 (1954)

/10/ Statz, G., E.Lippert, Ber.Bunsenges. 71, 673 (1967)

/11/ Lippingcott, E.R., R.Schroeder, J.Chem.Phys. 23, 1099 (1955); J.Phys.Chem. 61, 921 (1957)

/12/ McKinney, P.C., G.M.Barrow, J.Chem.Phys. 31, 294 (1959)

/13/ Bell, C.L., G.M.Barrow, J.Chem.Phys. 31, 300 (1959)

/14/ Pauling, L., The Nature of the Chemical Bond and the Structure of Molecules and Crystals, Cornell Univ. Press, Ithaca, New York (1940)

/15/ Bader, R.F.W., Can.J.Phys. 42, 1822 (1964)

/16/ Coulson, C.A., Hydrogen Bonding, D.Hadzi, Ed., London (1953) p.339

/17/ Puranik, P.G., V.Kumar, Proc.In.Acad.Sci. 58, 29 (1963)

/18/ Bratoz, S., Symposium sur les forces intermolé-culaires, Bordeaux 1965

Bratoz, S., Electronic Theories of Hydrogen Bonding, Advances in Quantum Chemistry (1965)

/19/ Fischer-Hjalmars, I., R.Grahn, Acta Chem.Scand. 12, 584 (1958)

/20/ Grahn, R., Arkiv f.Fysik 15, 257 (1959)
21, 13 (1962)

/21/ Paolini, L., J.Chem.Phys. 30, 1045 (1959)

/22/ Hofacker, L., Hydrogen Bonding, D.Hadzi, Ed., London (1959) p.375

/23/ Rein, R., F.E.Harris, J.Chem.Phys. 41, 3393 (1964)

POSITRON ANNIHILATION IN THE WATER-ICE SYSTEM

O. Mogensen [x)]

Laboratory of Applied Physics II

Technical University of Denmark, Lyngby

A positron, entering a solid is usually thermalized or nearly thermalized before annihilating with an electron into photons. The properties of the photons depend on the electron-positron state on annihilation. We can therefore get information about the electronic structure of the solid by investigating the photons. Positron annihilation in water and ice gives quite anomalous results as compared to annihilation in other materials /1,2/. In water and ice about half the positrons form positronium or other bound states with electrons. Positronium is a hydrogen-like atom, which in the singlet state decays into two photons with a lifetime of about 0.1 nsec. The lifetime for the three-photon decay of the triplet state is about 100 nsc for free positronium. Triplet positronium in a solid usually decays through two-photon annihilation during collisions with electrons bound in the solid. The lifetime is about 2 nsec. The positrons not forming positronium or other bound states annihilate with electrons bound in the solid in about 0.2 nsec.

The experimental methods used belong to the techniques of γ-spectroscopy. In fig.1 are shown the two types of experiments usually performed. The standard positron life time technique is the following: A Na^{22} positron source is used, and the spectrum of time delays between the

---

x) Other members of the positron annihilation group: G. Trumpy, K. Petersen, and L. Smedskjaer.

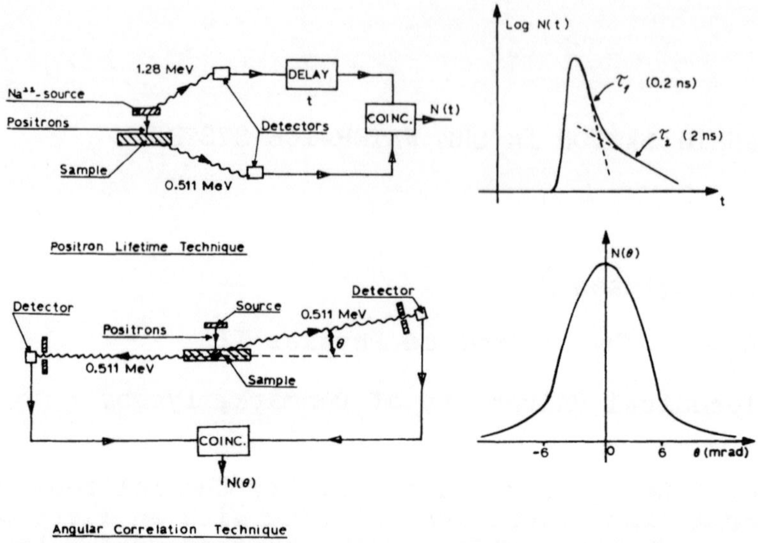

Fig.1 Diagram of typical positron annihilation experiments

1.27 MeV prompt nuclear photon from the decay of Na^{22} and one of the 0.51 MeV annihilation photons is recorded in a multichannel delayed-coincidence system. The typical decay curve for ice shows a short lifetime due to singlet positronium decay and annihilation of free positrons with electrons bound in the lattice and a longer lifetime due to triplet positronium pick-off annihilation with the lattice electrons. A diagram of a two-photon angular correlation set-up is also shown in fig.1. The two photons are emitted in nearly opposite directions. The apparatus counts the number of coincident photons hitting the two detectors as function of the angle Θ - the deviation from 180°. The angular correlation curve shown is a typical metal curve. The angle Θ is directly proportional to the center-of-mass momentum of the electron positron pair, if we neglect the influence of the lattice. In most cases the effective detector areas used are long linear slits. Hence the set-up determines approximately the distribution of a component of the center-of-mass momentum of the annihilating pairs.

We shall now review the most important results of positron annihilation experiments performed on the water-ice system. Fig.2 gives the result of lifetime experiments according to P.Jauho and M.Virnes /3/. We see that

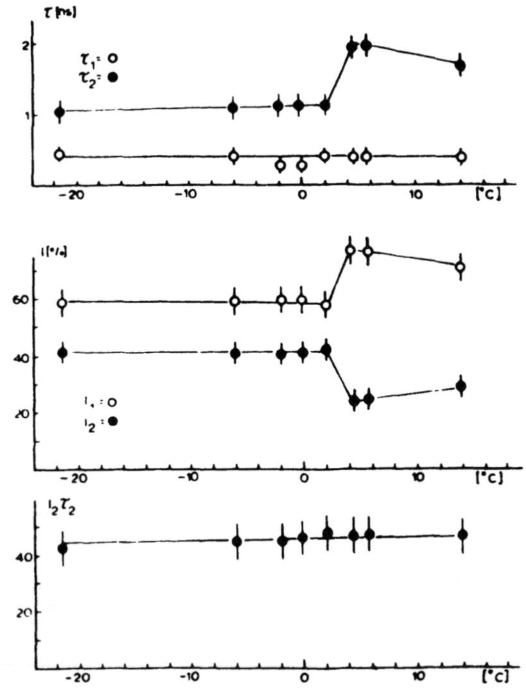

Fig.2 Lifetime τ_1 and τ_2, their intensities I_1 and I_2 and $I_2\tau_2$ as function of temperature in water and ice (P. Jauho and M. Virnes /3/)

the long lifetime τ_2 suddenly increases between 2°C and 4°C and no change is observed at the melting point. Also the relative intensities change abruptly near 4°C. This is probably the only discontinuity in the properties of water occuring near the maximum density point.

Some two-photon angular correlation curves for water and ice as a function of the temperature are shown in fig.3. The measurements are due to P.Colombino, B.Fiscella and L. Trossi /4/. The -144°C curve consists mainly of two components: The narrow component, corresponding to a small pair center-of-mass momentum, is interpreted as due to the presence of thermalized or nearly thermalized positronium atoms with typical energies of about 10^{-1} to 10^{-2} eV. If the positron annihilates with a bound electron, which has a typical energy of about 5 eV, either through free positron annihilation or due to triplet positronium pick-off annihilation, the broad component arises. The narrow component is completely absent in the room temperature and the 4°C curve. At the melting point part of

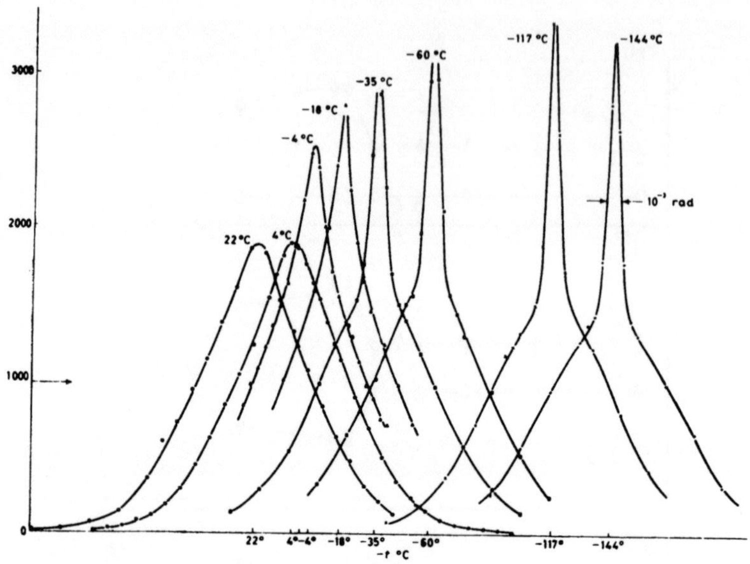

Fig.3 Two-photon angular correlation curves for water and ice as function of temperature (Colombino et al. /4/)

the narrow component is present, but a sudden decrease of the breadth of this component appears between -18°C and -25°C, as shown in fig.4.

We have also performed angular correlation measurements on the water-ice system. We remeasured the 22°C, -18°C, -25°C, and -40°C curves in good agreement with the results shown in fig.3 and fig.4. Our positron source was $Na^{22}Cl$ dissolved in the water or ice samples in contradiction to Colombino et al., who used a positron source placed outside the samples. In some preliminary measurements the strength of the solution was about $0.02 \cdot 10^{-3}$ mole/ℓ, and in the latest measurement we used a solution of about $0.6 \cdot 10^{-3}$ mole/ℓ. The results were independent of the solution strength. This indicates that there are no detectable effects due to the rather strong radiation of the samples or due to the NaCl impurity. This is remarkable, considering the fact that most other properties of ice are strongly impurity or defect dependent.

We have also measured an angular correlation curve at -22°C. This curve lies between the -18°C and the -25°C curves, shown in fig.4. Hence we must conclude that the narrowing does not occur abruptly at a certain temperature.

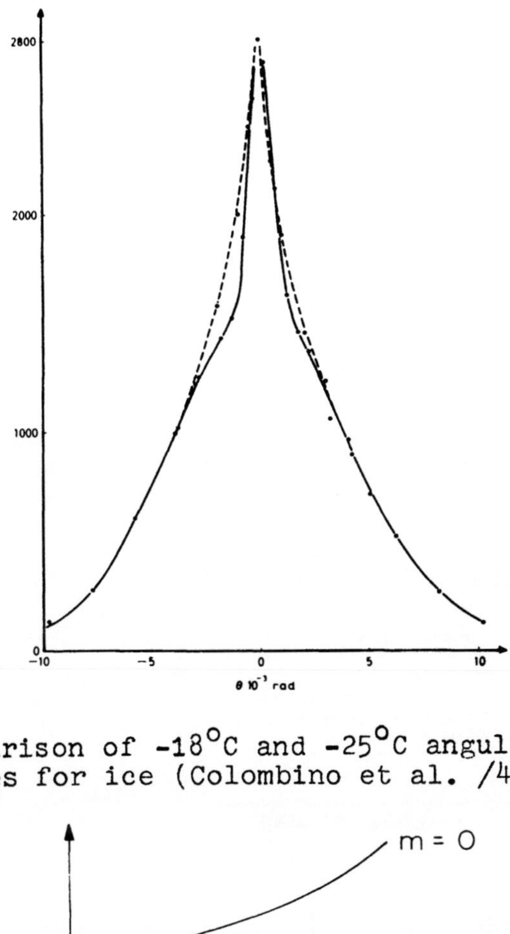

Fig.4 Comparison of $-18^{\circ}C$ and $-25^{\circ}C$ angular correlation curves for ice (Colombino et al. /4/)

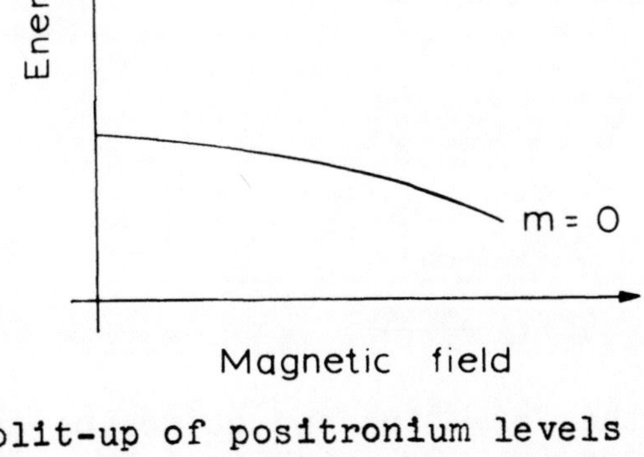

Fig.5 The split-up of positronium levels in a magnetic field

Ice also shows anomalous magnetic quenching effects /1,2/. In fig.5 is shown the effect of a magnetic field on the ground states of positronium. The triplet m = o state couples with the singlet m = o state and this reduces the number of positronium atoms decaying through three-photon annihilation. This means, e.g., that the relative intensity of the long lifetime component decreases for increasing magnetic field. Water shows this magnetic effect which, however, is not found in ice.

In fig.6 is shown a diagram /1,2/ illustrating the mostly used model for interpretation of formation of positronium in solids. The positrons are slowed down to about 15 eV during a time in the order of 10^{-16} sec. V is the ionization energy, E the lowest excitation energy, and 6.8 eV the binding energy of positronium. Below (V - 6.8) eV positronium formation is not allowed because of energy conservation. Above E excitation of the molecules is the most probable process. Hence it is usually assumed that positronium is only formed in the energy interval

$$\Delta = E_{max} - E_{min} = E - (V\text{-}6.8)\ \text{eV},$$

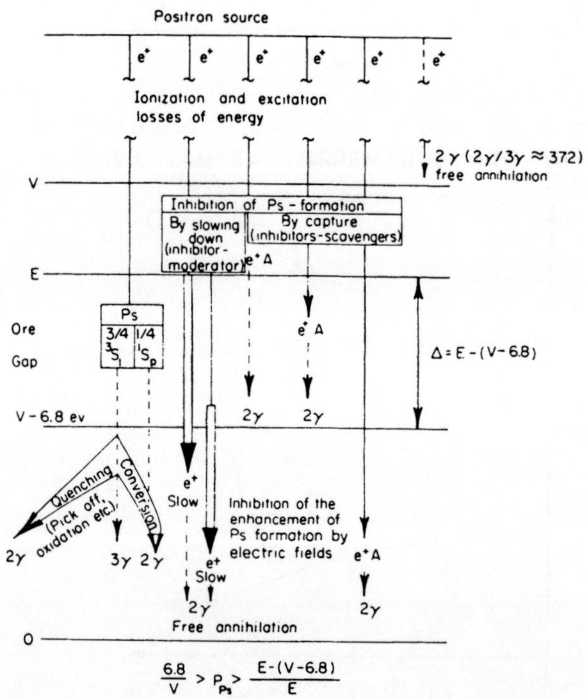

Fig.6 Diagram of the formation and quenching of positronium in solids (V.I.Goldanskii /1,2/)

also called the Ore gap. The model has been used in the gas case with great success and it can also be used as a good qualitative model for those solids, where positronium is formed /1,2/.

The anomalous positron annihilation properties of ice and water have not been explained yet. The different models /1,2,4/, which have been proposed, are not unique for the water-ice structure. I feel that it is necessary to include the peculiar behaviour of hydrogen in ice in the explanations, and one could conjecture that the positrons or positronium atoms get bound to or replace hydrogen in the structure. Further experimental work is needed to find the right explanation. We have planned to continue the measurements and we consider the following possibilities: Measurements on D_2O-ice, on single crystals of ice, and of phase transitions in ice. It is also important to check the measurements, which have been done until now, under different sample conditions, in order to see whether impurities or defects are of importance. Especially a lifetime magnetic quenching experiment on ice would be valuable.

References

/1/ Stewart, A.T., L.O.Roellig, editors, Positron Annihilation, Academic Press, New York (1967)

/2/ Goldanskii, V.I., Atomic Energy Review, VI (1), IAEA, Vienna (1968)

/3/ Jauho, P., M.Virnes, Phys.Letters 26A. 208 (1968)

/4/ Colombino, P., B.Fiscella, L.Trossi, Nuovo Cim. 38, 707 (1965)

MODELS FOR THE WATER MOLECULE AND RELATED IONS

Reinhards Siksna

Institutet för Högspänningsforskning

Uppsala Universitet, Uppsala, Sweden

Positive, so-called atmospheric, ions in the ambient atmosphere have been studied for more than 70 years, only recently has it become generally accepted that they consist mostly of oxonium (H_3O^+) ions and the hydrates $(H_3O^+)(H_2O)_n$ /2,4,6,11/.

I have designed models of these ions which show the orbitals of the lone-pair electrons and other electrons as well as (in most models) the atomic nuclei. Such models are useful for understanding many properties of the ions under consideration.

In designing the models the tetrahedral configuration of the atoms of the water molecule in ice was taken into account. The elements of the models are shown in figs. 1 and 2.

It is known that some simple molecules with hydrogen atoms centered about atoms of the second row of the

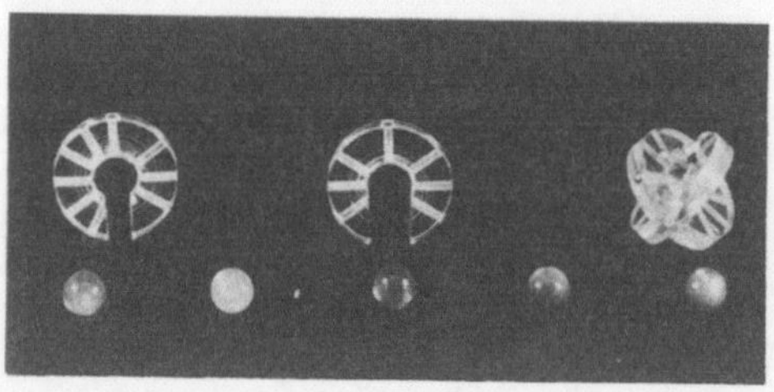

Fig.1 Elements of the model core, various atomic nuclei of different colours, and the composite core

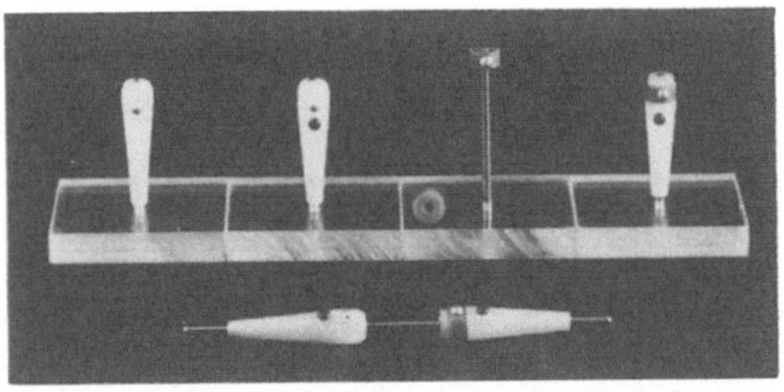

Fig.2 One-electron orbital, lone-pair electron orbital, proton, hydrogen atom, one-electron orbital with added hydrogen atom. On the front, below: hydrogen bond. The spins of the electrons are denoted with white and black colored rivets.

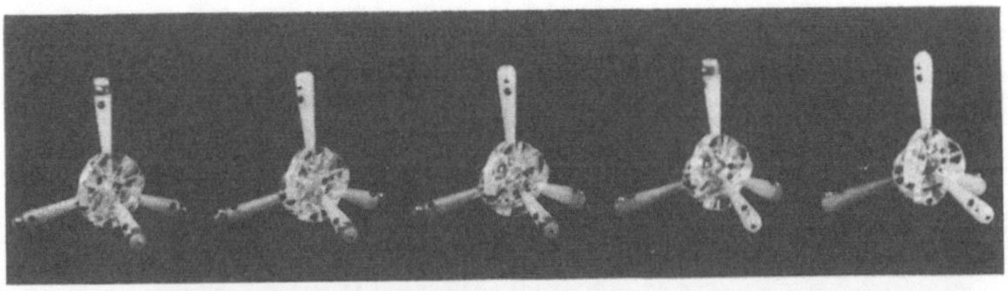

Fig.3 Tetrahedral structure of molecules with hydrogen centered about the atoms of the second row of the periodic system: CH_4, NH_3, H_2O, HF, Ne, with the angles: H-C-H, 109.5°; H-N-H. 108°; H-O-H, 104.5°

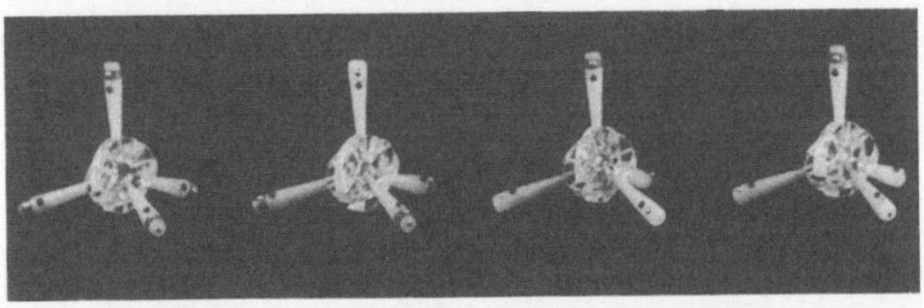

Fig.4 The structure of ions derived from the water molecule: H_4O^{++}, H_3O^+, H_2O. OH^-.

periodic system can also be considered tetrahedral or almost tetrahedral /1,5/,as shown in fig.3. Similar structures may be obtained from the water molecule by adding or removing protons (fig.4). H_4O^{++} has not been observed experimentally; however, for the sake of completeness it is considered theoretically, e.g., by Rosenfeld /9/.

In aggregates of water molecules, hydrates of the oxonium ion and derivates of these structures, the hydrogen bond plays an important role which is illustrated by the models of fig.5. A model of a more complicated structure

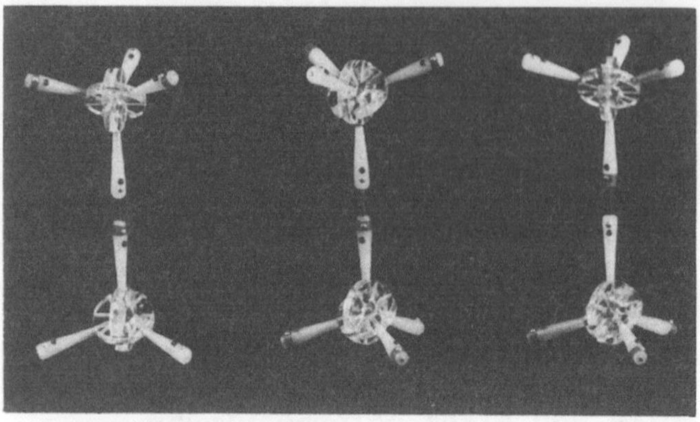

Fig.5 Hydrogen bond in the structure of H_2O-H_2O, H_3O^+ - H_2O, and H_3O^+ -OH^-

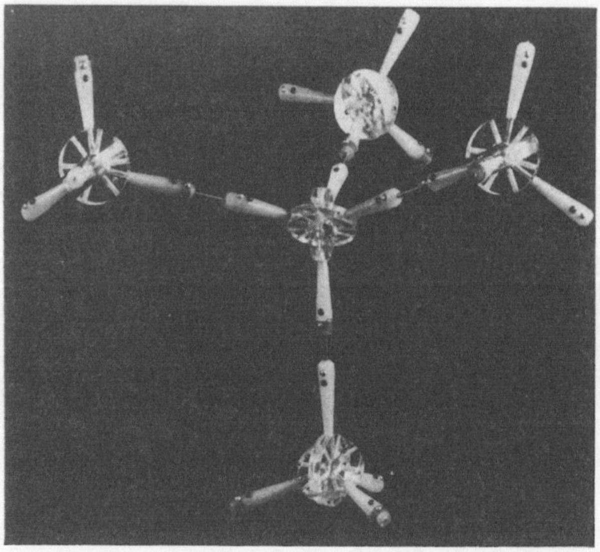

Fig.6 Hydrated hydroxyl ion, $(OH^-)(H_2O)_4$

with hydrogen bonds, the fourfold hydrated negative hydroxyl ion (OH^-) $(H_2O)_4$, is presented in fig.6. This ion has been observed experimentally in the mass spectrometer /7/. A similar model can be constructed for the fourfold hydrated positive oxonium ion (H_3O^+) $(H_2O)_4$ which has also been observed in the mass spectrometer /10/. By continuing the aggregation of water molecules with hydrogen bonds, the ice structure may be obtained in fig.7. The concept of a hydrogen bond as a lone-pair electron

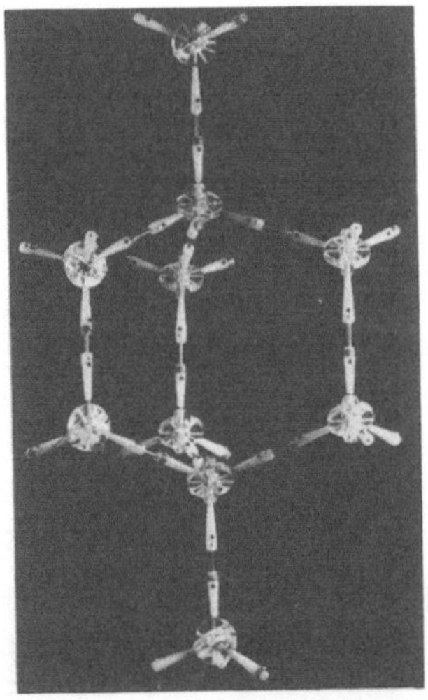

Fig.7 Ice structure

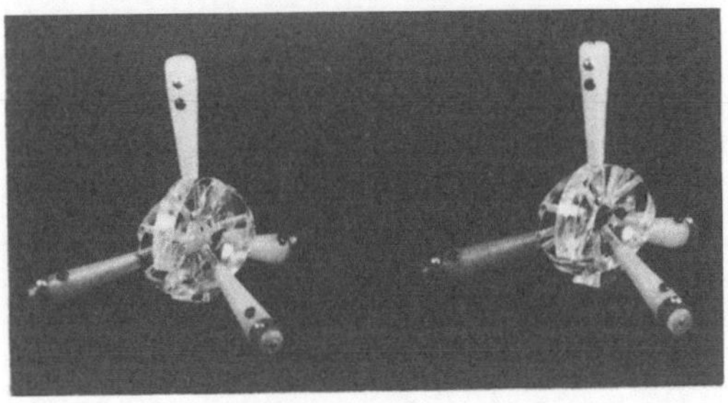

Fig.8 The structure of NH_3 and H_3O^+

orbital of one oxygen atom and a one-electron orbital of the neighboring oxygen atom overlapped with the hydrogen s-orbital is portrayed to some extent by the model used.

Concerning the ice structure, the difficulties for realization of Bjerrum D-defects begin to show in the model, fig.7. Dunitz /3/ has considered this point. On the other hand when we take the models of NH_3 and H_3O^+ (fig.8) and the models of OH^- and HF (fig.9), reflections may be made to obtain the structure of ice doped with ammonium or hydrogen fluoride.

It is evident that the models presented, although they are only crude pictures of the real molecules and ions, are nevertheless useful visual aids. It may not be easy to construct better classical models of something as

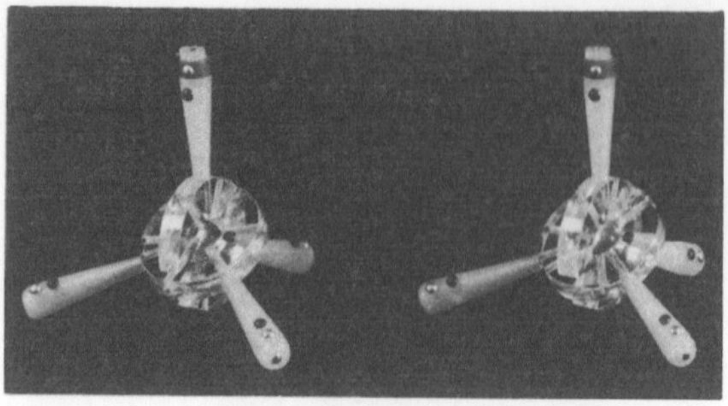

Fig.9 The structure of HF and OH^-

subtle as the orbital concept. Mulliken, in his Nobel lecture /8/ delivered in Stockholm, 1966, has given the following definition of an orbital: " An orbital means, roughly, something like an orbit; or, more precisely, something as much like an orbit as is possible in quantum mechanics. Still more precisely, the term "orbital" is simply an abbreviation for one-electron orbital wave function, or, preferably, for one-electron orbital eigenfunction."

In conclusion I should like to point out that while the models are not means for quantitative solution of problems, they may well be useful tools in formulating questions concerning problems under consideration.

References

/1/ Dickens, P.G., J.W. Linnet, Quart.Rev. (London), 11, 291-312 (1957)

/2/ Dolezalek, H., IUGG XIVth General Assembly Lucerne, September-October 1967; IAMAP Report of Proceedings, pp.238-240. Toronto IAMAP April 1968

/3/ Dunitz, J.D., Nature 197, 860-862 (1963)

/4/ Eichmeier, J., Z.Geophys. 34, 297-322 (1968)

/5/ Linnett, J.W., A.J.Poö, Trans.Faraday Soc. 47, 1035-1041 (1951)

/6/ Mohnen, V., IVth International Conference on the Universal Aspects of Atmospheric Electricity, Tokyo, May 12 - 18, 1968

/7/ Moruzzi, J.P., A.V.Phelps, J.Chem.Phys. 45, 4617-4627 (1966)

/8/ Mulliken, R.S., Science 157, 13-24 (1967)

/9/ Rosenfeld, J.L.J., J.Chem.Phys. 40, 384-389 (1964)

/10/ Shahin, M.M., J.Chem.Phys. 45, 2600-2605, (1966)

/11/ Siksna, R., IVth International Conference on the Universal Aspects of Atmospheric Electricity, Tokyo, May 12 - 18, 1968.

IRRADIATION-PRODUCED SOLVATED ELECTRONS IN ICE

Klaus Eiben

Institute of Radiation Chemistry

Nuclear Research Center, Karlsruhe, Germany

Abstract

Influences of the structure and the temperature on production and stability of solvated electrons formed as one of three reactive intermediates during the radiolysis of ice have been studied by several groups. Experimental evidence for the presence of irradiation-produced solvated electrons was established utilizing the techniques of ESR and optical spectroscopy as well as pulse-radiolysis. From the results it is evident that the stability of solvated electrons in ice is increased as the lattice of the ice structure is progressively destroyed. The production of solvated electrons in crystalline ice shows a remarkable temperature dependence. Their stability is small compared to those in alkaline ice glasses. This is explained by the very low concentration of preexisting trapping sites for electrons in crystalline ice. A model of the solvation mechanism derived from the experimental results is discussed.

1. Introduction

In recent years investigations of the radiation-chemical decomposition of water have been extended to ice /1-3/. The irradiation-induced ionisation and dissoziation of the water molecule in the gaseous, liquid and solid state results in the formation of H-atoms, hydroxylradicals, electrons, and $H_2O^{\oplus}$-radicals.

$$\begin{array}{lll} \text{Primary Radicals} & H_2O \rightsquigarrow H;\ OH;\ e^{\ominus};\ H_2O^{\oplus}. & \\ & & (1) \\ \text{Products} & H_2;\ H_2O_2\,. & \end{array}$$

These primary radicals are extremly reactive and thus shortlived. Some of them, H-atoms /4/, OH-radicals /5, 6, 40/and electrons /7, 40/ can be stabilized and trapped at suitable sites in ice at low temperatures. They have been detected by optical and/or ESR spectroscopy. In the absence of any special radical scavenger the only detectable decomposition products of the ice radiolysis are hydrogen and hydrogenperoxid/6/. The yields of primary radicals and of their stable products depend in a very complicated way on the temperature and on the degree of order of the solid state.

Among the above mentioned primary radicals, the electron first observed in glassy frozen alkaline aqueous solution /8a/, has become whether hydrated or solvated in the liquid state /8b/ or stabilized or trapped in the solid state the most widely studied entity in radiation chemistry since its discovery six years ago /2, 8c/. Basically the various electrons are described as extra electrons bound to a group of water molecules. (The term of $e_{aq}^{\ominus}$ is used for a hydrated electron in liquid water; $e_s^{\ominus}$ is introduced for shorttime stabilized electrons in the solid state, i.e. ice, while these are distinguished from trapped electrons $e_t^{\ominus}$, which are indefinitely stable in glassy or pure ice at low temperatures. Mobile electrons, $e_m^{\ominus}$, are free or thermalized electrons. Mobile electrons are in a dynamical equilibrium with the various types of electrons:

$$e_m^{\ominus} \rightleftharpoons e_t^{\ominus};\ e_s^{\ominus};\ e_{aq}^{\ominus} \qquad (2)$$

2. Method of investigation

For the investigation of the irradiation effects on ice essentially the three methods compiled in fig.1 have been applied. In pulseradiolytical experiments /9/ an optical absorption of the transient radical is produced by the application of an intense irradiation pulse of micro- or nanosecond duration. The light absorption, its wavelength dependence and the concomitant or delayed decay are studied by fast electronic detection. Compared to Co-60-γ-radiolysis, which normally involved the technic

Pulseradiolysis (transparent ice-crystal; hexagonal)	Temperature	286°K →	138°K
	Yield; $G_{e_s\theta}$	0,3 →	3·10
	λ_{max} (nm)	680 →	630
	$T_{1/2}\cdot 10^{-6}$sec	20 →	> 10^5
	$\frac{\Delta\lambda max}{dT}$ =	$-1,2\cdot 10^{-3}$ (eV/deg)	

		H_2O	D_2O
Co-60-γ-Radiolysis (transparent ice-crystal; hexagonal)	Temperature	77°K	77°K
	Yield; $G_{e_s\theta}$	$\sim 3\cdot 10^{-4}$	$\sim 8\cdot 10^{-4}$
	λ_{max} (nm)	640±5	625±5
	Lifetime	∞	∞

Crossed Beam Technic ("polycrystalline" condensed water vapour)	Temperature	77°K	77°K
	Color	blue	blue
	ESR Linewidth	10,6	2,7
	ESR g-factor	2,0009	2,0007

Fig.1 Temperature dependent behaviour of irradiation produced solvated electrons in crystalline and "polycrystalline" ice

of matrix isolation at temperatures below 80°K, the pulseradiolytical method has advantages in detecting short lived intermediates, even at low temperatures /12, 13/. In the crossed beam technic /10,11/ a water vapor jet is transmitted through an ionizing beam on its way to a surface at which the irradiated vapor rapidly is cooled to 77°K. Some of the transient radicals formed in the gasphase get inbedded into the ice of the condensate. The effects observed with this technic thus are not necessarily comparable with the events occuring during the irradiation of the solid phase. This matter will be discussed later.

The transparent ice crystals for the optical studies were formed by slow immersion into a cooling bath of -6 °C of vacuum degased, triply distilled water, sealed into test-tubes. The mechanical shaped cylindrical crystals were cooled to the desired temperature. The structure of the so formed ice crystals was not analysed, but is assumed to be hexagonal throughout the temperature range studied.

3. Results

The results concerning the formation of stabilized electrons in irradiated ice crystals are shown in fig.1. The important characteristics that should be emphasized are these:

With decreasing temperature of ice one observes /12, 13/:

a) An overall decrease in the yield of stabilized electrons, $e_s^\ominus$, by several orders of magnitude - dropping markedly from -5° to -40° but only slowly thereafter. (This observation does not exclude, that the yield of irradiation-produced mobile electrons, $e_m^\ominus$, as precursors of $e_s^\ominus$, is temperature dependent; instead a temperature dependent competition reaction seems to be involved).

b) A remarkable blue shift of the absorption maximum, the temperature-gradient being similar to the one observed in the liquid state for $e_{aq}^\ominus$ /14/.

c) The increase in lifetime by several orders of magnitude .

d) A small but remarkable isotope effect of the absorption maximum and the ESR singlet /11/ for heavy ice crystals.

In principle the observed temperature dependent behavior of stabilized electrons in ice does not differ from that of hydrated electrons in water /14/ and solvated electrons in other polar liquids /15/. Remarkable is the fact that apparently no sudden change in the radiation chemical behaviour is observed at the phase transition point, solid-liquid; thus strongly suggesting that $e^\ominus_{aq}$ and $e_s^\ominus$ both are one and the same species. The pulse-radiolytical investigations have been confirmed by A. Nilsson et al./16/.

The results of the crossed beam technic as studied by Marx et al./10/, and Smith et al./11/ apparently show, that during the collision of water molecules with ionizing irradiation, electrons are formed in the gas-phase which afterwards are trapped in the matrix of the rapidly condensed water. Although the authors do not indicate absolute yields, the number of trapped electrons at 77^oK

is many orders of magnitude higher than in γ- or pulse-irradiated ice crystals. From this the presence of an important structural effect seems to be evident.

The very small radiation-chemical yield of trapped electrons in crystalline ice at 77°K had been overlooked for many years and has not been observed till the intensive irradiation of optically long ice samples /6/. On fig.2 the absorption maximum described to trapped electrons is seen at 640 nm which according to the predicted value of Ershov et al./33/ in vitreous neutral ice is in close agreement. Fig.2 shows also the very strong UV-absorption with peaks at 280 & 230 nm, which is several orders of magnitude greater as the absorption in the visible.

This UV-absorption, attributed to trapped hydroxyl- and hydroperoxy-radicals, was also observed in pulse radiolytical experiments of ice in the temperature range between -30°C and -130°C /12/. The very complex decay kinetics of these radicals have been studied, and it has been evaluated that they are more or less independent from those of stabilized electrons /12/. Therefore the appearance of these radicals in ice will not be considered further in the discussion of stabilized electrons /17/.

The temperature dependent decay kinetics of stabilized electrons in ice are complex. First or second order behaviour is observed, which indicates that the decay

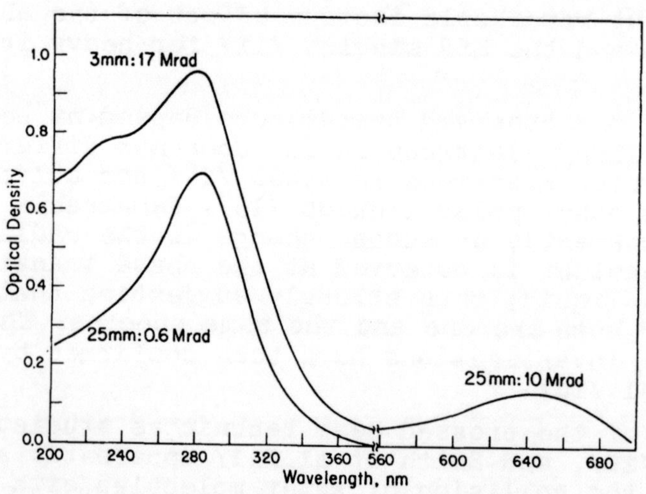

Fig. 2 Absorption spectra of ice crystals after Co-60-γ-irradiation at 77°K. Optical pathlength and total absorbed dose for each sample are indicated

mechanism is ruled by competing reactions (such as eq. 2, 5 and 6). An activation energy observed from first order rate constants is calculated to be E_{act} = 9 ± 2 kcal/mole /12/ and has to be compared with results of Shubin et al. /13/, who claimed to find overall first order decay rates (E_{act} = 6 ± 1 kcal/mole).

4. Discussion

Mechanism of solvation and stabilization of electrons in water. Whereas the chemical behaviour of the solvated electron in water is well established the mechanism of stabilization and solvation in the various polar media still remains uncertain. Among several models those are briefly mentioned here which have been applicated to trapped electrons in the solid state.

In the model established by Jortner & Sharf /18/ the electron in alkaline glassy ice is assumed to be stabilized in an expanded s-type orbital near the alkali cation. The lack of an ESR spin coupling between electron and alkali nuclear spin and the absence of a cation dependence on the position of the absorption maximum are in contrast to this model.

In a model postulated by Moorthy and Weiss /19/ it is assumed that the electron occupies a 3 s-orbital in the environment of the $O^{\ominus}$ radical-anion - a radical formed in strongly alkaline ice via a proton transfer from hydroxyl radicals according to eq. (3)

$$HO + OH^{\ominus} \rightleftharpoons O^{\ominus} + H_2O. \qquad (3)$$

The following observations are in contrast to the proposed model:

a) Electrons are also stabilized in glassy frozen alcohols, which do not contain any centers corresponding to the $O^{\ominus}$-radicalanion /20/.

b) In water alkali-metal condensates electrons are stable up to -130°C /21/ although it has been proved that no $O^{\ominus}$-radicalanions are formed in this system. A similar result is observed in the above mentioned crossed beam experiments.

While these two models were developed to describe trapped electrons in aqueous alkali hydroxide glasses, the anion vacancy theory put forward by Symons et al. /22/ for the same system, can also be applicated to other polar solvents. Originally it had been assumed that the electron may be solvated or stabilized in an irradiation produced anion vacany. However, this assumption is contrary to experimental results of Barzynski and Schulte-Frohlinde /23/ who were able to demonstrate that a significant structural effect exists in the irradiation of frozen polar solutions. Rapidly frozen and thus glassy solids of aqueous alkalihydroxides and of alcohols are capable of stabilization of irradiation-produced electrons /25/ whereas the crystals, slowly grown from the same liquids, do not show the ability of electron stabilization. Since water does not readily form a glass, a comparison of this structural effect in ice is currently not possible. From this structural dependence the conclusion is drawn that at least in the solid state suitable sites capable to form potential wells for electron trapping must have been formed prior to irradiation /25,26/.

But the results of the crossed beam experiments indicate already that, under certain conditions even in pure water, electrons are trapped at 77°K with a high yield /10, 11/. However, the observed stabilization will be due rather to the circumstance that the specific arrangement of water molecules, which provide a trap for electrons, is formed in the gas-phase and both, the electron together with its polarized solvation shell is condensed rapidly on the cold surface. This view is supported by the observation that water condensates obtained without irradiation +) are incapable of electron trapping during subsequent irradiation /11/.

On the other side the polaron theory derived from solid state physics has been used successfully to describe solvated electrons in neutral ammonia solutions and was also applied to hydrated electrons in water /27/. Since this model also favours the assumption that the electron "digs its own hole" in the polar system, it seems applicable only, if at all, to the liquid state, where orientational together with electrical polarization may occur rapidly enough to surround a free electron with a stabilizing solvation shell. In this model the electron is bound through long range interactions with polarized solvent molecules which can be described as a selfinduced stable quantum state of the polarized dielectric.

The light absorption of the solvated electron thus is a 1s→2p transition and the position of the absorption maximum depends on the mean radius of the cavity as well as on the static and optical dielectric constants of the

+) These unirradiated "polycrystalline" condensates seemingly do not consist of amorphous ice according to the description in /24/.

medium. According to this modified polaron model the cavity radius of the electron in water is assumed to be zero at room temperature. A decisive difficulty of the application of this model is the dipol relaxation time of ice. Platzman /28/ has pointed out that the relaxation time of water molecules in ice at low temperatures could be several years and hence the irradiation-produced electron, $e_m^{\ominus}$, moves rather freely. On its free path it reacts with an excess-proton which also is formed during an early radiation-chemical stage according to eq.4.

$$H_2O^{\oplus} + H_2O \xrightarrow{10^{-14} sec} H_3O^{\oplus} + OH, \quad (4)$$

$$H_3O^{\oplus} + e_m^{\ominus} \longrightarrow H + H_2O, \quad (5)$$

$$H_3O^{\oplus} + e_s^{\ominus} \longrightarrow H + H_2O. \quad (6)$$

The mobility of excess protons in ice is extremely high and is assumed to occur via reorientation of hydrogen bridges, as has been ascertained by Eigen et al. /29/ and recently by Engelhardt and Riehl /30/. Thus charge combination according to eq. 5 and 6 should influence the yield and the stabilization process in ice and it seems to be evident that mobile electrons disappear (eq.5) for lack of suitable potential wells. The few, which give rise to trapped electrons at 77°K, are frozen-in defect sites or crystal distortions. That charge separation effectively occurs during irradiation of ice is evaluated from scavenger experiments /31/.

The recently proposed vacancy model /1,26,27,35,38/ describes qualitatively most of the experimental results. It is suggested that the potential well of the electron is based on a tetrahedral arrangement /36,37/ of four water molecules of which the ice lattice and the liquid water is assumed to be composed. The electron is bound inside by electrical and orientational polarization of the water molecules at the four corners of the tetrahedron, the size of this arrangement not being changed relative to a non occupied tetrahedron. The orientational polarization seemingly involves hydrogen bond breakage. The electron moves with a high probability by tunneling as long as the water molecules rotate rapidly and the tetrahedral structure changes frequently. With decreasing temperature less tetrahedra will exist with an already preformed orientation to provide the probability of forming deep potential wells for electrons. On the other side tunneling is increasingly hindered at lower temperatures, thus the few electrons seem stabilized indefinitely. Walker /2/ has suggested the presence of different deep potential wells to explain the temperature dependent shift of the absorption maximum. This has been ascertained by Ershov et al. /32,33/ who found that λ_{max} shifts towards shorter wavelengths if photo-bleaching with light $\lambda > \lambda_{max}$ is applied to samples with trapped electrons. The number of differently sized traps is increased as the degree of disorder in the lattice is increased by suitable additives such as neutral salts /38/ and alkalihydroxides /33/. This is supported by the observed increase of the bandwidth of the light absorption in alkalihydroxide and alcohol glasses compared to ice /32,33/. On the other side solvent /15/ as well as temperature /12,14/ dependent shifts of the absorption maximum can be explained by changes of the vacancy radius which in turn depends on changes of the static dielectric constant. If the dipoles are frozen-in at low temperatures, the dielectric constant and the vacancy radius will become constant. The temperature dependent shift of the absorption maximum thus reaches a plateau in alaklihydroxide-glasses at -105°C /34/ and in pure ice at -140°C /12/.

Weiss /39/ recently considered that the electron in water and ice is a bound electron hole pair, which is described as an exiton of small radius. The calculated optical transition energy which according to this model is proportional to E_o^{-2} agrees fairly well with the experimentally observed.

Finally, the conclusion can be drawn that the hydrated electron in the liquid state and the stabilized electron in the solid state seem to be similar entities. Since the presently accepted models account only for most but not all of the experimental observations of the electron stabilization and solvation in a polar medium, considerably more work is needed to find an all encompassing description of the simplest reducing species known in chemistry.

References

/1/ Kevan, L., in Proceedings of the 19th Farkas Memorial Symposium, Jerusalem, Dec. 1967

/2/ Walker, D.C., Quarterly Reviews 21, 79 (1967)

/3/ Schulte-Frohlinde, D., K.Vacek, in: M.Ebert and A.Howard, Current Topics in Radiation Research V, in press.

/4/ Piette, L.H., R.C.Rempel, H.E.Weaver, J.M.Flournoy, J.Chem.Phys. 30, 1623 (1959)

/5/ McMillan, J.A., M.S.Matheson, B.Smaller, J.Chem. Phys. 33, 609 (1960)

J.M. Flournoy, L.H.Baum, S.Siegel, J.Chem.Phys. 30, 2229 (1962)

/6/ Ghormley, J.A., A.O.Stewart, J.Am.Chem.Soc. 28, 2934 (1956)

/7/ Eiben, K., I.A.Taub, Nature 216, 782 (1967)

/8a/ Schulte-Frohlinde, D., K.Eiben, Z.Naturforsch. 17a, 445 (1962)

/8b/ Hart, E.J., J.W.Boag, J.Am.Chem.Soc. 84, 4090 (1962)

/8c/ Solvated Electron; Advances in Chemistry, Series 50, 1965, American Chemical Society.

/9/ Baxendale, J.H., M.Ebert, J.P.Keen, A.J.Swallow, Proceedings of the Manchester Symposium on Pulse Radiolysis 1965; Academic Press, London

/10/ Marx, R., S.Leach, M.Horani, J.Chim.Phys. 60, 726 (1963)

/11/ Seddon, W.A., D.R.Smith, P.E.Bindner, Canad.J. Chem. 46, 1747 (1968)

/12/ Taub, I.A., K.Eiben, to be published in J.Chem.Phys.

/13/ Shubin, V.N., V.A.Zhigunov, V.I.Zoltarevsky, P.I. Dolin, Nature 212, 1002 (1966)

/14/ Gottschall, C., E.J.Hart, J.Phys.Chem. 71, 2102 (1967)

/15/ Arai, S., M.C.Sauer, J.Chem.Phys. 44, 2297 (1966)

/16/ Nilsson, G., H.C.Christensen, J.Fenger, P.Pagsberg and S.O.Nielson, Abstracts Eigth Intern.Free Radical Symposium, Novosibirsk, USSR 1967 and in: Advances in Chemistry, Series 81, American Chemical Society 1968, pp. 71

/17/ For detailed studies of the OH-radical in single crystals of ice see:

Siegel, S., J.Chem.Phys. 39, 390 (1963)

Brivati, J.A., M.C.R.Symons, D.J.Tinling, H.W. Wardale, D.O. Williams, Chem.Communications, 1965, 402 and J.Chem.Soc.Ser.A 1967, 2112,

Dibbin A.H., Nature 209, 394 (1966)

Gunter, T.E., J.Chem.Phys. 46, 3818 (1967)

/18/ Jortner, J., B.Sharf, J.Chem.Phys. 37,2506 (1962)

/19/ Moorthy, P.N., J.J.Weiss, Phil.Mag. 10,659 (1964)

/20/ Dainton, F.S., J.P.Keen, J.T.Kemp, G.A.Salmon, J.Teply, Proc.Chem.Soc. 265 (1964)

/21/ Bennett, J.E., B.Mile, A.Thomas, Nature 201, 919 (1964) and J.Chem.Soc.London, Ser.A 1967, 1394

/22/ Blandamer, M.J., C.Shields, M.C.R.Symons, Nature 199, 902 (1963) and J.Chem.Soc. 4352 (1964)

/23/ Barzynski, H., D.Schulte-Frohlinde, Z.Naturforsch. 22a, 2131 (1967)

/24/ Ghormley, J.A., J.Chem.Phys. 48, 503 (1968)

/25/ Eiben, K., D.Schulte-Frohlinde, Z.Phys.Chem.N.F. 45, 20 (1965)

/26/ Schulte-Frohlinde, D., Radiation Research 1966, Proceedings of the Cortina Conference (1966), North Holland Publ.Comp. (1967), pp. 251

/27/ Jortner, J., Radiat.Res.Suppl., 4 , 24 (1964)

/28/ Platzman, R.L., N.A.S. - N.R.C. Reports No. 305 p.22 (1953)

/29/ Eigen, M., L.De Maeyer, H.C.Spatz, Ber.dtsch. Bunsengesell. 68, 19 (1964)

/30/ Engelhardt, H., N.Riehl, Phys.Kondens.Mat. 5, 73 (1966)

/31/ Moorthy, P.N., J.J.Weiss, Nature 201, 1318 (1964)

/32/ Ershov, B.G., I.E.Makarov, A.K.Pikaev, Khimiya Vysokikh Energii 1, 404 (1967)

/33/ Ershov, B.G., I.E.Makarov, A.K.Pikaev, Khimiya Vysokikh Energii 1, 472 (1967)

/34/ Taub, I.A., K.Eiben, to be published

/35/ Kevan, L., in: "Progress in Solid State Chemistry", Ed.H.Reiss, 2, 304 (1965)

/36/ Fueki, K., J.Chem.Phys. 45, 183 (1966)

/37/ Natori, M., T.Watanabe, J.Phys.Soc. Japan 21, 1573 (1966)

/38/ Ershov, B.G., A.K.Pikaev, Khimia Vysokikh Energii 1, 29 (1967)

/39/ Weiss, J.J., Nature 215, 150 (1967)

/40/ Khodzhaev, O.F., B.G.Ershov, A.K.Pikaev, Jsv. Akad.Nauk SSSR, Ser.Khim 1968, 246

FORMATION AND STRUCTURE OF COLOUR CENTRES IN IRRADIATED ICE *

J.J. Weiss

School of Chemistry, University of Newcastle-upon-Tyne, England

I believe that the situation in water and also the work with ionising radiations in water vapour referred to by Eiben is now fairly clear, at least in principle. One must add here that irradiation of water with electron pulses in the nanosecond region by Thomas and Bensasson /1/ has shown that there is also a much shorter living highly reactive species (lifetime ∿70 n sec) which, as was suggested previously could be a radiation-produced electron-hole pair (charge-transfer exciton) /2/ which absorbs light in a similar region as the trapped electron although its absorption spectrum is not yet fully known.

However, the problem of special interest is concerned with the formation of radiation-produced trapped electrons (colour centres) in ice. It was really some sort of breakthrough when it was shown first by Shubin et al. /3/ that the absorption spectrum of the trapped electron in ice is very similar to that of the hydrated electron in water. Moreover, one knows also from the recent work of Smaller /4/ that the ESR spectrum of the electron in water is very similar to that of the electron in alkaline ice. I think, however, that there is no longer any difficulty to understand the formation of trapping sites for the electron in irradiated ice: in short, the traps are radiation-produced, basically by the radiation-induced decomposition of some water molecules in

---

* Discussion remarks on "Irradiation-produced solvated electrons in ice" given by K.Eiben

the ice structure. Without going into the radiation chemical details here it is known that irradiation of ice leads to decomposition of water molecules.

To visualise the trapping of an electron, let us consider a water molecule in the ice surrounded by its four water molecules in the more or less perfect Bernal-Fowler structure. If such a water molecule is decomposed, a vacancy is created in its place where an electron can be trapped. In the intact ice structure two protons would be pointing towards the centre of the vacancy. If now an electron occupies this vacancy, diffusion of Bjerrum D-defects to the vacancy should be induced by the field of the electron and would result in what would amount to a rotation of two water molecules to point their protons towards the centre of the vacancy. This would create a more efficient trap for the electrons as now the protons of all the four water molecules would be directed towards the vacancy. The radiolytic decomposition of a water molecule is a necessary but not a sufficient condition for the creation of a vacancy for an electron in the ice matrix. It is not only necessary for a water molecule to be decomposed but also that the decomposition products can be displaced, away from the vacancy, into some suitable interstitial positions.

This latter process is likely to be more efficient in glassy than in crystalline materials. This would thus explain the increased yields which have been found in glassy phases by various authors.

The mechanism outlined above has a certain similarity to that proposed for the formation of F-centres in alkali halides under the influence of radiation which also involves the formation of radiation-produced anion vacancies.

References

/1/ Thomas, J.K., R.V.Bensasson, J.Chem.Phys. 46, 4147 (1967)

/2/ Weiss, J.J., Nature 215, 150 (1967)

/3/ Shubin, V.N., V.A.Zhigunov, V.I.Zoloratevsky, P.I.Dolin, Nature 212, 1002 (1966)

/4/ Smaller, B., J.Chem.Phys., (in press).

MECHANICAL PROPERTIES OF ICE SINGLE CRYSTALS

Akira Higashi

Department of Applied Physics, Faculty of Engineering, Hokkaido University, Sapporo/Japan

Abstract

The most notable attainments of the last ten years of research on the mechanical properties of ice single crystals are reviewed with especial reference to the work of the author and his colleagues. Of primary importance is the fact that the equation $\dot{\gamma} = C \cdot \tau^m \cdot \exp(-Q/RT)$ holds for both creep and the stress-strain relation in wide ranges of strain rate, stress, and temperature ($\dot{\gamma}$, τ and T). Although the stress exponent, m, differs in basal (1.6) and non-basal glide (6.5), the activation energy is the same in both cases ($Q \approx 16$ kcal/mole).

Johnston's theory of the dislocation mechanism for the plastic deformation of crystalline materials has been found to apply very well in the case of ice. Anisotropy, both in yield stress and in the stress exponent of the strain rate (rate of dislocation motion), has been explained by the characteristic dispositions as revealed by X-ray diffraction topographs.

Further progress and prospects are discussed concerning deformation studies which have now been extended to correlate with the effects of hydrogen disorder or of proton transfer on the dislocation motion.

1. Introduction

Many papers have been published concerning the mechanical properties of ice single crystals since Glen /1/ reviewed the subject in Advances in Physics ten years ago. At that time, there were very few reports of experiments with single crystals and those few were somewhat rudimentary. It had been established that the basal plane is the preferential glide plane in ice and that the glide direction is <11$\bar{2}$0>. Steinemann /2/ had derived the power law, $\dot{\gamma} = \tau^n$, relating the creep rate, $\dot{\gamma}$, to the applied shear stress, τ, from his experiments with artificial ice single crystals. Other important achievements occuring up to about 1955 were included in Glen's review /1/.

The rate of advancement of knowledge was slower for ice than for other materials, probably because of the paucity of good crystals, many of which were needed, and also because of the difficulties encountered in carrying out experiments in temperatures below 0 °C. Nakaya's very extensive work on bending creep, which he did at SIPRE using many natural glacier ice single crystals, is probably the first mile stone from which many further developments rapidly evolved. His work contains much reliable, reproducible data which may be analysed from the viewpoint of the modern aspects of the mechanical properties of ice. However, Nakaya only analysed his data with respect to the geometry of deformation in basal glide planes and to his idea of elementary layers /3/. After his rather sudden death, quantitative analyses of his creep curves under various conditions were made by this author and the work is now nearly complete /4/.

There have been several experiments reported which used artificial ice single crystals: Butkovich and Landauer /5/ obtained from their tension creep experiments very nearly the same power law as that obtained for polycrystalline ice; from their stress relaxation experiments, Readey and Kingery /6/ determined that the value for the stress exponent in the power law is about 2.5 for small strains and decreases to about 1.5 for large ones; later, Glen and Jones /7/, studying uniaxial tension creep at temperatures of about -60°C, obtained a value of about 4 for this exponent. Such differences in values may be attributed in part to differences in the ice single crystals used by the individual investigators and in part to the different ranges of strain and the testing methods.

In this review, the author will begin with the results of deformation experiments in natural glacier ice single crystals, conducted by his colleagues and himself, which were initially stimulated by Prof. Nakaya and his work and which have been carried out over the past several years. Single crystals from Alaska's Mendenhall Glacier are still the best crystals for such work in terms of experimental reproducibility as well as purity. As is shown in table 1, recently analysed melt water from

Table 1 Impurity concentration in melt water from natural ice single crystals (analysed by Dr. and Mrs. S. Kanamori, Nagoya University, Japan)

Ion impurity	Cl^-	Ca^{++}	Mg^{++}	SO_4^{--}
Concentration (ppm)	0.2	1.5	0.02	0.8

these crystals compares very favorably with the best obtainable distilled water. The large number of Mendenhall crystals brought back by expeditions to Alaska in 1960 and 1964, made it possible to carry out extensive deformation experiments, studying both creep and stress-strain relations, in a very wide range of stress, temperature, and strain rate. The results obtained have been interpreted using dislocation theory and the dislocation characteristics involved have been further examined using X-ray diffraction topographic techniques. Developments in the research field which correlate dislocation motion with hydrogen disorder or proton transfer will be discussed in a later section.

2. Plastic deformation of ice single crystals

The results of the above mentioned deformation experiments /8,9,10/ will be discussed first. Figure 1 shows a typical creep curve for basal glide in ice single crystals as obtained from bending experiments. The shape of this curve

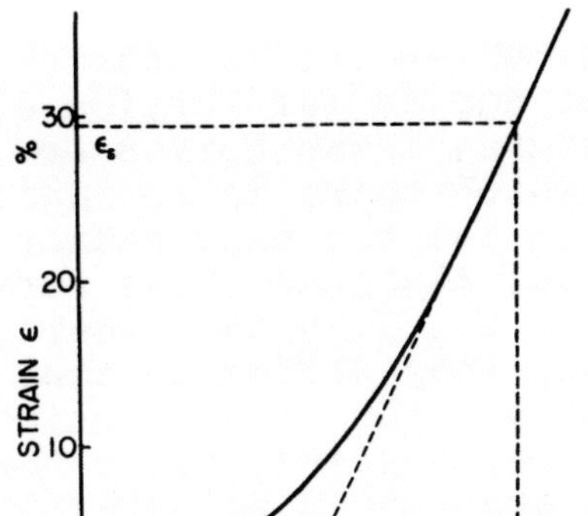

Fig.1 Typical creep curve (Higashi et al./8/)

is very similar to the shape of those curves obtained for Ge by Penning and De Wind /11/ and for InSb by Peissker, Haasen, and Alexander /12/. The curve is characterized by a gently initial slope followed by a steeper, more constant slope and its shape may be expressed by the following parameters:

ε_s: strain at the point where the slope becomes constant,
t_s: time when the ε_s is attained,
t_i: time at which the extrapolated straight line of the constant creep portion intersects the time axis,
$\dot{\varepsilon}_s$: creep rate in the region of constant creep.

The relation, $\varepsilon_s = \dot{\varepsilon}_s(t_s - t_i)$, is naturally established and the creep rate, $\dot{\varepsilon}_s$, is a major parameter which will be hereinafter used to describe experimental results.

The temperature and stress dependence of the stationary creep rate, $\dot{\varepsilon}_s$, is expressed by:

$$\dot{\varepsilon}_s = K_1 \tau^m \exp(-Q_1/RT) \quad (1)$$

where T is the absolute temperature and τ the applied stress. The exponent, m, equals 1.58 and the activation energy, Q_1, about 15.8 kcal/mole. Note should be taken of the small value for m.

The stress-strain relation was obtained for basal glide in ice single crystals /9/ from tension experiments with ice cylinders in which the tensile axes were oriented so as to make an angle of 45° with the basal plane of the crystal. The rate of extension of the specimen was varied from 4×10^{-4} to 2×10^{-2} by changing the gear system of the tensile testing machine. The testing temperature varied from -40 to -15 °C.

Typical stress-strain curves are illustrated in fig. 2 which shows a group of curves for varying strain rates at a constant temperature. These curves are characterized by large yield drops which there is no indication of recovery. This means that ice crystals cannot be work-hardened in basal glide. Although these curves resemble those obtained for LiF, Ge, InSb and others, which also show large yield drops, they differ in that the slope of the initial, linear portion is a function of the strain rate and the temperature. However, since the magnitude of the yield drop cannot accurately be determined, the maximum shear stress, τ_{max}, will be used as the parameter by which the curves are characterized. A similar group of stress-strain curves was obtained from experiments with varying temperatures and a constant strain rate.

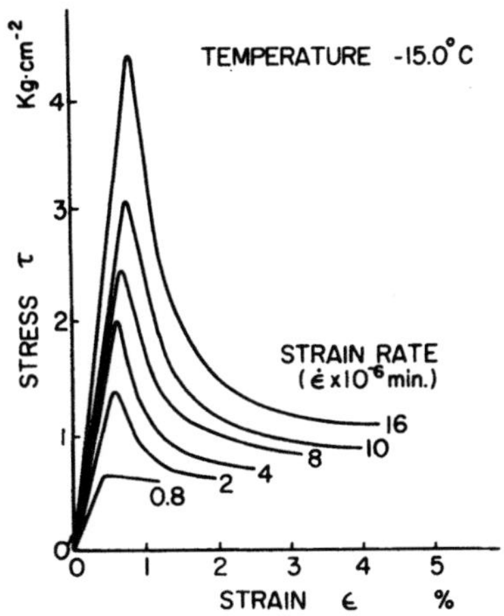

Fig.2 Stress-strain curves for various strain rates at -15 °C (Higashi et al. /9/)

From these curves, the strain-rate and temperature dependence of τ_{max} was obtained as follows:

$$\tau_{max} = C_1 \cdot \dot{\varepsilon}^{\frac{1}{m}} \cdot \exp\left(\frac{E_2}{RT}\right). \qquad (2)$$

This equation may be converted to:

$$\dot{\varepsilon} = C_2 \cdot (\tau_{max})^m \cdot \exp\left(-\frac{Q_2}{RT}\right) \qquad (3)$$

where $Q_2 = mE_2$. The empirical value for m, as derived from these curves, is approximately 1.53 which agrees well with the value for creep in equation (1). The value for Q_2 is 15.9 kcal/mole which is the activation energy of the strain rate as obtained from the stress-strain relation and which also agrees well with that obtained from the creep experiments.

Tensile tests in non-basal glide were also carried out /10/ using the same tensile machine described above in which the specimen was oriented so that the c-axis was perpendicular to the tension axis. The specimens were thin, rectangular bars, approximately 15 x 10 x 60 mm, and were prepared from saw-cut large bars by chemical polishing with ethyl alcohol. They were cut so that the c-axis of the specimen was perpendicular to the 15x60 mm face. The angle θ, between the a-axis and the specimen's side, 10 x 60 mm face, was taken as a variable parameter with which to define the inclination of the principal glide plane, $(10\bar{1}0)$, in non-basal glide against the tensile axis. There is some evidence that $(10\bar{1}0)\langle 11\bar{2}0\rangle$ is the principal glide system in non-basal glide. Etch pit studies of deformed ice single crystals have also revealed the existence of a $(10\bar{1}0)\langle 0001\rangle$ glide system, however, this latter system does not apply in these experiments since there was no applied stress acting upon $\langle 0001\rangle$ dislocations.

Figure 3 illustrates typical load-elongation curves for specimens with varying values for θ at a testing temperature of -19 °C and an axial strain rate of $3 \times 10^{-6}\ sec^{-1}$.

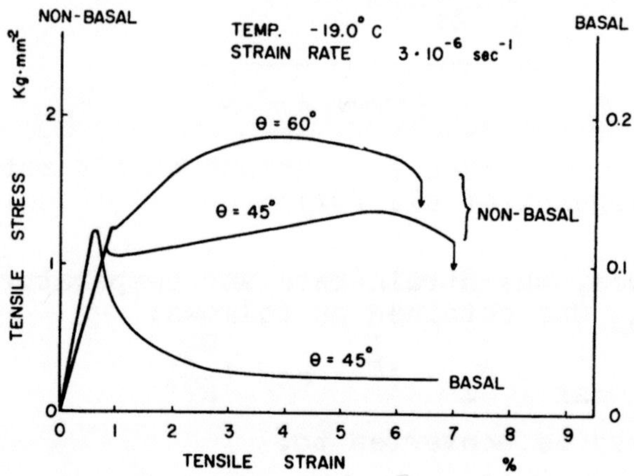

Fig.3 Typical stress-strain curves in basal and non-basal glide. θ is angle between principal glide plane and the tensile axis. (Higashi et al. /10/)

The shape of these curves differs greatly from the shape obtained in basal glide which is given in the figure for comparison, and more nearly resembles that of metals and other substances which show work-hardening. The tensile yield stress in non-basal glide, σ_y, is much higher than that observed in basal glide, as is illustrated by the difference in the ordinate scales for basal and non-basal glide in this figure. The tensile yield stress depends upon the strain rate, temperature and the angle, θ. When the load and elongation rates are converted to resolved stress and strain in the primary glide plane, $(10\bar{1}0)$, according to the following equations, the dependence upon angle θ vanishes. In these equations

$$\tau = \left(\frac{\sigma}{2}\right) \sin 2\theta \quad (4)$$

and

$$\dot{\gamma} = \dot{\varepsilon}/\cos\theta \quad (5)$$

τ is the resolved shear stress, σ the applied load, $\dot{\gamma}$ the resolved shear strain rate, and $\dot{\varepsilon}$ the elongation rate. The dependence of the resolved shear yield stress, τ_y, upon the resolved shear strain rate, $\dot{\gamma}$, and the temperature, T, is expressed as follows:

$$\tau_y = A \cdot \gamma^{\frac{1}{m}} \exp (E_3/RT) . \qquad (6)$$

When this is converted in the following style as was done with equation (2):

$$\dot{\gamma} = B \cdot (\tau_y)^m \exp (-Q_3/RT) \qquad (7)$$

the value for the exponent, m, is 6.5 and the activation energy, $Q_3 = mE_3$, is approximately 16.5 kcal/mole.

The work-hardening rate depends very little upon the temperature or strain rate, but it does depend upon crystal orientation and the angle θ. The experimental results shown in table 2 indicate that the work-hardening rate decreases rapidly as θ approaches 45°. This may be related to the relative differences in the resolved shear stresses in the three $(10\bar{1}0)$ planes with variation in the angle θ and this will be discussed later.

Table 2 Orientation dependence of the work-hardening rate in ice single crystals deformed in non-basal glide

Orientation of primary non-basal glide plane (angle θ, degrees)	Work-hardening rate ($kg/mm^2 sec$)
30	12.2
37	8.9
45	2.7
60	9.7

3. Dislocation mechanisms in plastic deformation

As was previously stated, creep curves in basal glide in ice single crystals strongly resemble those of Ge and InSb and others which several investigators (Van Bueren /13/, Peissker et al. /12/, and Haasen /14/) have interpreted on the basis of dislocation motion in the crystals. Johnston's /15/ curves for non-work-hardening material calculated from his dislocation theory of yielding are quite similar to those obtained by the author and his colleagues. Therefore, Johnston's theory was adopted to explain basal glide, i.e.:

$$\dot{\varepsilon} = 2bnv \tag{8}$$

where b is the Burgers vector, n the mobile screw dislocation density, and v the screw dislocation velocity. If we assume appropriate dependences for n upon $\dot{\varepsilon}$, and v upon τ, which correspond to those which etch pit studies demonstrate to exist in LiF, Ge, or Si, the stress dependence may be obtained for $\dot{\varepsilon}_s$ as is expressed by equation (1). The simplest expressions for these dependences are:

$$n = \alpha\varepsilon \tag{9}$$

and

$$v = (\frac{\tau}{D})^m \tag{10}$$

where the numerical factors α, m, and D may be determined by etch pit observations of dislocation multiplication and screw dislocation velocity under varying stress.

Since technical difficulties prevent reliable, direct observation of dislocation multiplication and velocity in ice crystals with the etch pit method, this interpretation would appear to be very hypothetical but there is an indirect way of proving that the mechanism is really operating. Johnston's theory predicts a large yield drop in basal glide in ice single crystals because of the comparatively small numerical value for m, and this is clearly demonstrated by the stress-strain curves shown in fig. 2. It was therefore concluded that Johnston's theory of yielding and creep might also be applied to basal glide deformation in ice single crystals. In this theory, the incubation time, t_i, is understood to be the time during which the low dislocation density ($\sim 10^4$ cm^{-2}), present before stress is generated in the crystal, gradually increases with stress to the point where mobile dislocation velocity is greatly decreased by virtue of mutual interaction, at which point yield drop, or a stationary creep rate, is established.

The remarkable anisotropy of the mechanical properties of ice single crystals is clearly illustrated in fig. 3 which shows the yield stress in non-basal glide to be about ten times that in basal glide. Since the same equation may be used to express the experimental results of the stress and temperature dependence of the yield stress in both basal and non-basal glide (eq. 6), non-basal deformation should also be subject to interpretation by dislocation motion and multiplication. If this is true, the stress exponent of the dislocation velocity should be 6.5 - 2, or 4.5, because the stress dependence of the strain rate includes a term expressing dislocation multiplication. When the crystal is work-hardened, this term should be proportional to the square of the stress.

The reader may get some idea of the comparative strength of ice as a material from table 3 in which the yield stress for ice in basal and non-basal glide is compared with that in other materials.

Table 3 Yield strength in ice and other materials

Material	Upper limit yield stress (kg/mm^2)	Experimental conditions	Investigator
Fe	40	50 mm/min, at -78 °C	Takeuchi, Ikeda /16/
Ge	5	at +500°C	Patel, Chaudhuli /17/
LiF	0.5	at room temp.	Johnston, Gilman /18/
Ice (non-basal)	0.5	$1.8 \cdot 10^{-4}$/min, at -14.5 °C	Higashi, Mae, Fukuda /10/
Ice (basal)	0.04	$1.6 \cdot 10^{-4}$/min, at -15.0 °C	Higashi, Koinuma, Mae /9/

The activation energy for the strain rate is almost identical in both basal and non-basal glide, approximately 16 kcal/mole, which is also the same as that of the self-diffusion in ice crystals as measured either by tritium /19/ or by O^{18} /20/. This value should also be adopted for the diffusion of vacancies in ice crystals in quantitative explanations of void formation during non-basal glide /21/. The fact that the activation energy is the same for all of these different processes means that the self-diffusion of intact water molecules in ice crystals takes place by vacancy migration. It may therefore be concluded, that the rate of dislocation movement is thermally controlled in the same way in both basal and non-basal glide by the diffusion of molecular vacancies to jogs that may be formed thermally or by the intersection of screw dislocation.

Many attempts have been made to observe dislocation behaviour in ice crystals under tension using methods previously developed for metals and other materials. Several workers /22-28/ used etching techniques to determine slip systems and dislocation Burgers vectors, despite the limitations of such techniques. One example of etch channels from which non-basal slip systems were determined is given in fig.4 (see /25/ for details). In 1965, Hayes and Webb /29/ first demonstrated the usefulness of X-ray diffraction topography as a method of direct observation of dislocations in ice crystals and the results of their extensive work have only recently been published /30/. The work of the author and his colleagues has followed this first report of Hayes and Webb and they have used X-ray diffraction techniques to examine both natural /31/, and artificial ice single crystals, in the latter case studying both deformation mechanisms /10,32/ and crystal growth /33/.

Dislocation Burgers vectors of $\frac{a}{3}\langle 11\bar{2}0\rangle$ have been determined from topographs taken with the $(10\bar{1}0)$ diffracting planes and of $c\langle 0001\rangle$ from those taken with (0002) diffracting planes /3,22,24/. Topographs taken on these planes indicate that all observed dislocations lie on the basal planes of the crystal. Fig.5, taken with the $(11\bar{2}0)$ diffracting and $(10\bar{1}0)$ scanning planes, shows that the dislocations lie parallel to the basal planes. The lines indicating dislocations show short perpendicular jogs from one basal plane to another. Topographs taken with the (0002) diffracting planes also show dislocations lying parallel to the basal plane. It is possible to conclude from these observations therefore, that dislocations with $c\langle 0001\rangle$ Burgers vectors are edge dislocations and those with $\frac{a}{3}\langle 11\bar{2}0\rangle$ vectors are edge, screw or 60° dislocations. Since in non-basal glide there is no stress acting on the $\langle 0001\rangle$ dislocations, the anisotropy of the yield stress may be attributed to the difference in mobility of $\frac{a}{3}\langle 11\bar{2}0\rangle$ dislocations on the basal (0001) and non-basal $(10\bar{1}0)$ plane. This difference in mobility, or in resistance to dislocation motion, may be explained by dislocation intersections which occur, not on the basal planes, but on the prism planes where dislocations intersect others on neighbouring prism planes. These intersections result from the relative difference in the velocities of dislocations moving on different prism planes, on which different resolved shear stress acts. The fact that work-hardening occurs in non-basal glide and that its rate depends upon crystal orientation with reference to the tensil axis (angle θ dependence shown in table 2), may also be interpreted by dislocation motion on the prism planes. When angle θ is 45° between one of the $10\bar{1}0)$ planes and the tensile axis, there is maximum resolved shear stress on this plane (see eq.4). Dislocations on

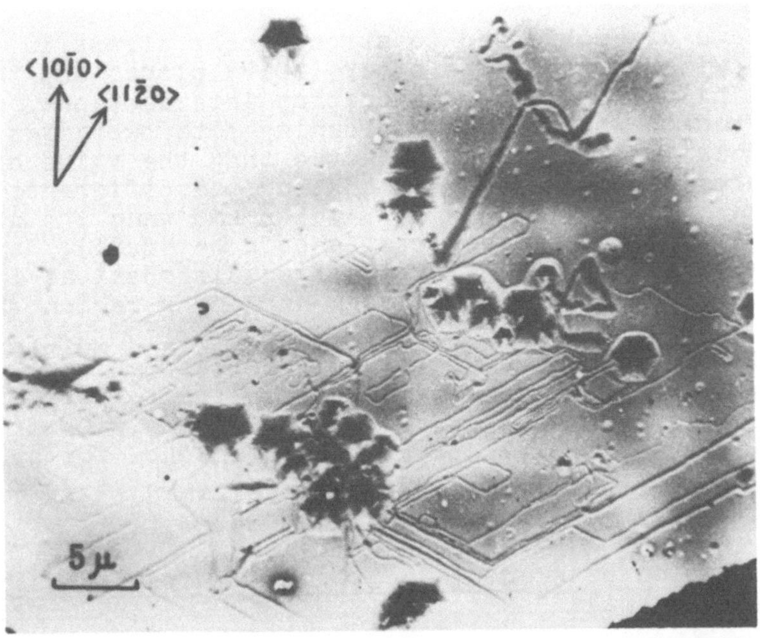

Fig.4 Etch channels on the (0001) plane of an ice crystal sheared in non-basal glide. Electron microscope, x3000. (Muguruma and Higashi /25/)

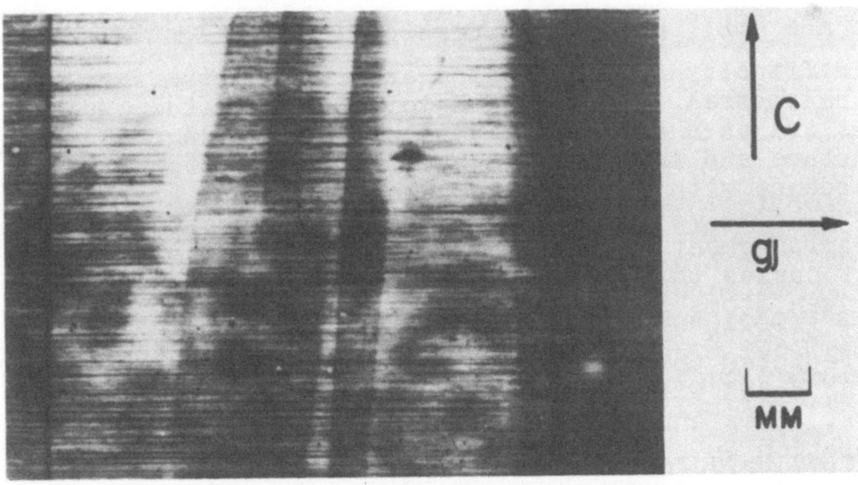

Fig.5 X-ray topograph of natural ice single crystal, (11$\bar{2}$0) diffracting planes scanned on (10$\bar{1}$0) surface. (Fukuda and Higashi /32/)

this plane are subjected to strong shear stress in comparison with those on the other $(10\bar{1}0)$ planes and there are a minimum number of dislocation intersections. This is the reason why the work-hardening rate decreases as θ approaches 45°. It should be noted that the value of the work-hardening rate when $\theta = 45^\circ$, or the difference in the work-hardening rate when $\theta = 30^\circ$ and when $\theta = 60^\circ$ (which, in all other respects, should be equal), may be understood as resulting in part from dislocations intersecting voids formed in the work-hardening region /21/.

Precise data are needed for the movement and multiplication of dislocations in ice single crystals under stress in both basal and non-basal glide to confirm the mechanisms stated above. Some observations have been made for basal glide using X-ray diffraction topography and the results will be reported in a paper to be published /32/, but there is still much to be done.

4. Correlation between dislocations and hydrogen disorder in ice crystals

Although it has been confirmed that dislocations with $\frac{a}{3}\langle 11\bar{2}0\rangle$ Burgers vectors are of primary importance in plastic deformation of ice single crystals, the details of their structure within the crystal lattice and the behaviour of the hydrogen bonds associated with their movement are still not well understood.

The difficulty of the problem arises from the fact that an ice crystal is a compound in which only the oxygen atoms are on the crystal lattice of its hexagonal wurtzite structure and in which the hydrogen atoms are arranged in accordance with the Bernal-Fowler rules but are otherwise random. Glen /34/ has discussed the changes occuring in the arrangement of hydrogen atoms surrounding a dislocation as it passes through a crystal and has concluded that the hydrogen atoms must be reoriented by thermally activated point defects ahead of the moving dislocation. Accordingly, he predicted that ice would soften at low temperatures if it was doped with HF which produces excess L-Bjerrum defects. Jones and Glen /35,36/ seem to have proven this prediction in both creep and stress-strain relation experiments with HF doped ice single crystals. As is shown in fig. 6, when ice single crystals are doped with very weak HF solutions, the tensile creep rate increases and the maximum shear stress is reduced. However, these effects were not observed when NH_3, which produces D-defects, was used as a dopant.

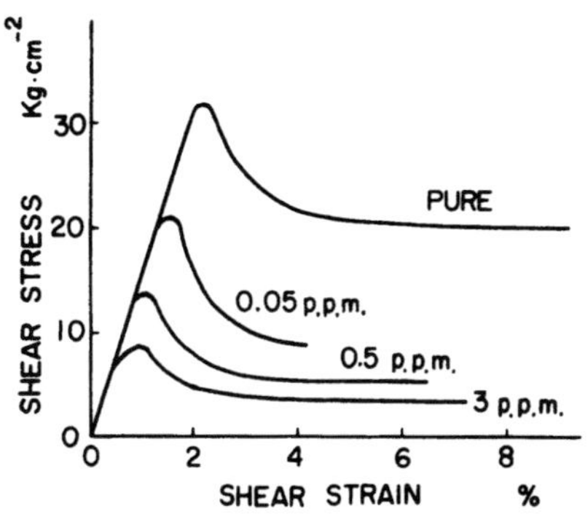

Fig.6 Stress-strain curves at -70 °C for pure ice and ice containing various concentrations of HF. (Jones and Glen /36/)

The author and his colleagues have taken a different approach to the problem. Just as the passage of dislocations may produce reorientation of hydrogen bonds through changes in the number or disposition of Bjerrum defects, so may plastic deformation change the dielectric constant or the conductivity of a crystal. Experimental studies of such effects are now under way and, as the preliminary results given in fig. 7 show, when ice single crystals are subjected to plastic deformation, their dielectric constant increases in the higher temperature range, their conductivity increases, and the temperature at which conductivity is a maximum increases with compression. These results may possibly indicate that dislocation motion results in changes in the number and/or disposition of Bjerrum defects which then produces changes in the electrical properties.

However, the author does not believe that reorientation of hydrogen atoms is the sole controlling mechanism for dislocation motion. Glen's theory does not appear to explain the remarkable anisotropy of the mechanical properties of ice. Consideration of the discussion of activation energy given in the preceding section requires examination of the possibility that proton ring or molecular diffusion is of primary importance. In any case, the new informations and ideas which will be presented in this symposium may offer a solution, or at least help immeasurably in the task of combining what is known about the mechanical properties of ice with other properties and structures.

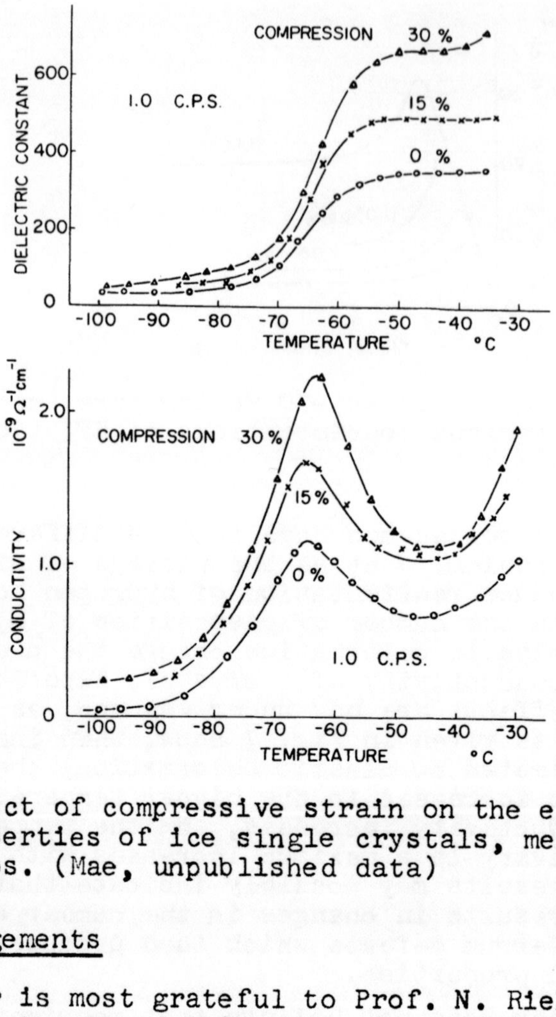

Fig.7 Effect of compressive strain on the electric properties of ice single crystals, measured at 1 cps. (Mae, unpublished data)

Acknowledgements

The author is most grateful to Prof. N. Riehl for his invitation to attend this symposium and to present this review. He is also indebted to Dr. J.W. Glen for very helpful discussion of hydrogen disorder and to Dr. S. Mae for supplying unpublished data.

References

/1/ Glen, J.W., Advances in Physics 7, 254-265 (1958)

/2/ Steinemann, S., J. Glaciol. 2, 404-412 (1954) and Beiträge zur Geologie der Schweiz, Hydrologie No.10 (1958)

/3/ Nakaya, U., SIPRE Res.Report 28 (1958)

/4/ Nakaya, U., to be published as TSC Res.Report

/5/ Butkovich, T.R., J.K. Landauer, U.G.G.I., Association Int.d'Hydrologie Scientifique, Symposium de Chamonix, publication No.47, 318-325 (1958)

/6/ Readey, D.W., W.D. Kingery, Acta Metal. 12, 171-178 (1964)

/7/ Glen, J.W., S.J. Jones, Physics of Snow and Ice, Proc.Int.Conf.on Low Temperature Science, 1966, 267-275 (1967)

/8/ Higashi, A., S. Koinuma, S. Mae, Jap.J.Appl.Phys. 4, 575-582 (1965)

/9/ Higashi, A., S. Koinuma, S. Mae, Jap.J.Appl.Phys. 3, 610-616 (1964)

/10/ Higashi, A., S. Mae, A. Fukuda, Proc.Int.Conf.on Strength of Metals and Alloys (Sept. 1967, Tokyo), 784-789 (1968)

/11/ Penning, P., G. De Wind, Physica 25, 765-774 (1959)

/12/ Peissker, E., P. Haasen, H. Alexander, Phil.Mag. 71, 1279-1303 (1962)

/13/ Van Bueren, H.G., Physica 25, 775-791 (1959)

/14/ Haasen, P., Proc.Conf.National Physical Laboratory, 3-8 (1963)

/15/ Johnston, W.G., J.Appl.Phys. 33, 2716-2730 (1962)

/16/ Takeuchi, T., S. Ikeda, J.Phys.Soc. Japan 18, 488-495 (1963)

/17/ Patel, J.R., A.R. Chaudhuli, J.Appl.Phys. 34, 2788-2799 (1963)

/18/ Johnston, W.G., J.J. Gilman, J.Appl.Phys. 30, 129-144 (1959)

/19/ Itagaki, K., J.Phys.Soc. Japan 19, 1081 (1964)

/20/ Delibaltas, P., O. Dengel, D. Helmreich, N.Riehl, H. Simon, Phys.Kondens.Materie 5, 166-170 (1966)

/21/ Mae, S., Phil.Mag. 18, 101-114 (1968)

/22/ Muguruma, J., J.Electromicro. 10, 246-250 (1961)

/23/ Muguruma, J., Nature 190, 37-38 (1961)

/24/ Kuroiwa, D., W.L. Hamilton, Ice and Snow, 34-55, MIT Press (1962)

/25/ Muguruma, J., A. Higashi, J.Phys.Soc. Japan 18, 1261-1269 (1963)

/26/ Muguruma, J., A. Higashi, Nature 198, 573 (1963)

/27/ Muguruma, J., J.Fac.Sci. , Hokkaido Univ. Ser.II, 5, 11-22 (1963)

/28/ Levi, L., E.M. de Achaval, E. Suraski, J.Glaciol. 5, 691-699 (1965)

/29/ Hayes, C.E., W.W. Webb, Science 147, 44-45 (1965)

/30/ Webb, W.W., C.E. Hayes, Phil.Mag. 16, 909-925 (1967)

/31/ Fukuda, A., A. Higashi, to be published

/32/ Fukuda, A., A. Higashi, Physics of Ice, Proc.Int. Symp., Munich, Plenum Press

/33/ Higashi, A., M. Oguro, A. Fukuda, J.Crystal Growth (in press)

/34/ Glen, J.W., Phys.kondens.Materie 7, 43-51 (1968)

/35/ Jones, S.J., Phys. Letters 25A, 366-367 (1967)

/36/ Jones, S.J., J.W. Glen, Phil.Mag. (in press, 1968)

INFLUENCE OF THE SURFACE LAYER ON THE PLASTIC DEFORMATION OF ICE SINGLE CRYSTALS

Jiro Muguruma ♦

Snow and Ice Section, Division of Building Research, National Research Council of Canada

Ottawa / Canada

Abstract

The removal of the mechanically disturbed surface layer of ice single crystals greatly increases the maximum stress for basal glide in constant strain rate tests. This shows that surface dislocation sources have a pronounced effect upon the deformation.

The results of various studies of the deformation behaviour of ice single crystals in constant strain rate tests /1-4/ show a wide variation in the shape of the stress-strain curve. This variation has been considered to be due to the differences between crystals in the dislocation density.

In this note, evidence is presented to show that not only the quality of the crystals, but also the method of specimen preparation, strongly influences the yield behaviour of single crystals of ice. The results are interpreted in terms of surface sources of dislocations.

The specimens used were rectangular plates cut from single crystals in such a way that the basal plane was 45°

♦ On leave from the Department of Applied Physics, Hokkaido University, now at the Cavendish Laboratory, Cambridge, England.

to the long axis of the specimen and normal to its largest face. These plates were prepared either by planing with a fine blade or by polishing with fine emery paper until they were 5 mm in thickness, 20 mm in width and 25 mm gauge length. In some specimens, additional surface layer was removed either by a chemical method /5/ or careful slight melting. Deformation in compression and at a constant rate of crosshead movement was achieved with a Hounsfield tensometer driven by a synchronous motor and equipped with a load cell and gear train for varying the strain rate.

Curves A and B (figure 1) are typical results for specimens with and without chemical polishing. It can be clearly seen that the maximum stress for the specimen polished chemically (curve A) is about twice as great as the one that has only been polished mechanically. In order to increase the maximum stress, it was found necessary to remove a layer of approximately 200 μ in thickness by chemical polishing.

A characteristic increase in the maximum stress could be obtained only for crystals that would have had a fairly high maximum stress without additional removal of

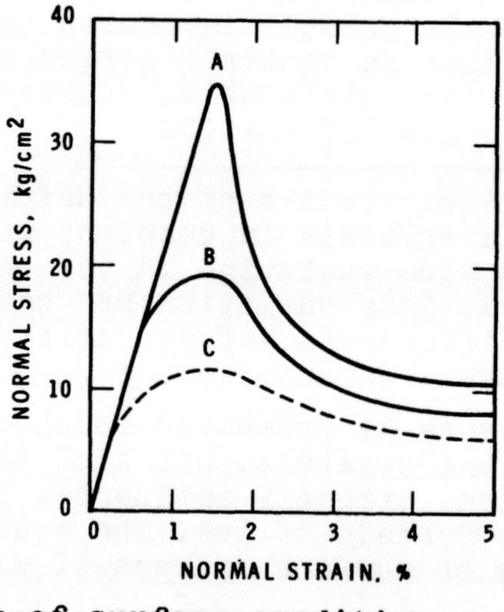

Fig.1 The effect of surface condition on the compressive stress-displacement curve of ice single crystals in basal glide at -10°C. (Crosshead speed = 0.063 mm/min; specimen length = 25 mm).
Curve A: Result for chemically polished crystal.
Curve B: Result for mechanically polished crystal.
Curve C: Result for crystal that appeared to have quite a high initial density of dislocations

surface, as exhibited by curve B. According to the general relationship for the plastic strain rate /6/, crystals with high maximum stress have initially a low density of mobile dislocations; those with low maximum stress have a high density. Curve C is an example of the behaviour of crystals that appeared to have had quite a high initial density of dislocation according to the above relationship. For these crystals, the removal of surface did not have any effect upon the maximum stress.

In order to establish that surface sources of dislocations are of primary importance in the deformation behaviour, specimens without additional surface removed were annealed in a kerosine bath at -3°C for 120 hours. The maximum stress of some of the annealed specimens was almost the same as that obtained for specimens with surface removed by chemical polishing. On the other hand, crystals with low maximum stress (curve C) had the same maximum stress after annealing as similar crystals without annealing. This indicates that the dislocations generated at the surface layer due to mechanical polishing are more easily removed from the crystal by image forces than internal dislocations.

If the surface sources are a dominating factor, it would be expected that slip bands in a specimen without surface removed would be faint and more closely spaced because many surface sources would be active and the displacement per slip band would not need to be so large to produce a given strain. The slip bands of the specimen with surface removed should be clear and more widely spaced because fewer dislocation sources would operate and a larger displacement per slip band would be required to produce the same strain. Observations proved the validity of these assumptions and provided additional evidence that surface sources of dislocations play a very important role in the deformation process of ice crystals.

The results show that the surface layer of ice crystals does not appear to act as a strong barrier to dislocation movement, but does act as a source of dislocations. This emphasizes that surface treatment is of primary importance when carrying out studies on the structure-sensitive properties of ice crystals.

Detailed results will be published.

References

/1/ Higashi, A., S.Koinuma, S.Mae, Jap.Jour.Appl.Phys. 3, 610 (1964)

/2/ Readey, D.W., W.D.Kingery, Act.Met. 12, 171 (1964)

/3/ Glen, J.W., S.J. Jones, Proc.Int.Conf. on Low Temp.Sci., Sapporo Vol.1, Part 1, 267-275 (1966)

/4/ Jones, S.J., Phys.Letters 25A, 366 (1967)

/5/ Muguruma J., S.Mae, A.Higashi, Phil.Mag. 13, 625 (1966)

/6/ Johnston, W.G., J.J.Gilman, J.Appl.Phys. 30, 140 (1959)

IMPURITY EFFECTS ON THE PLASTICITY OF ICE AND THEIR EXPLANATION IN TERMS OF HYDROGEN REORIENTATION

S. J. Jones* and J.W. Glen

Department of Physics, University of Birmingham

Birmingham, England

Abstract

Experiments on the plastic deformation of single crystals of ice at low temperatures show that a few parts per million of HF dissolved in the ice produce very marked softening. NH_3 produces a slight hardening while NH_4F has no apparent effect. The softening can be explained in terms of the ability of L- defects to reorient hydrogen bonds thus allowing dislocations to move despite hydrogen disorder in ice; the failure of NH_3 to produce a similar effect then implies either that D-defects are not formed, or that they are unable to reorient the bonds in question.

1. Introduction

During the course of experiments on the mechanical properties of ice it was thought that the dislocations in the crystal would very likely interact with the point defects (ionic and Bjerrum) which existed in the lattice. It was decided therefore to dope ice single crystals with various impurities known to vary the number of point defects, and to observe the effect on the mechanical properties. It was discovered that HF had a considerable softening effect, while NH_3 produced a slight hardening and NH_4F had no apparent effect. This discovery led to

* Now at: Inland Waters Branch, Department of Energy, Mines and Resources, Ottawa, Canada.

the realisation (Glen /6/) that the hydrogen disorder in ice must impede the movement of dislocations in the lattice, and this theory can explain satisfactorily the experimental results on doped ice.

The paper that describes this work in detail will be published in Philosophical Magazine (Jones and Glen /1/); the present paper summarizes the main results for the International Symposium on the Physics of Ice.

2. Apparatus and Specimens

Two types of tests have been used at various temperatures down to -70°C: creep tests in tension under a constant load, and compressive constant strain-rate tests all of which were conducted at a nominal strain-rate of 2.7×10^{-7} sec^{-1}. Details of these testing machines can be found in Glen and Jones /2/ and Jones and Glen /3/. The method used to grow the crystals is described in Glen and Jones /2/; cylindrical single crystals were produced about 0.5 cm in diameter and 3 cm in length for the tensile tests, and about 1.5 cm in diameter and 3 cm in length for the compressive tests. They were doped with either HF, NH_3 or NH_4F. The impurity concentrations were determined by melting the crystals after they had been mechanically tested and analyzing the melt water by colorimetric methods.

3. Results for HF- doped crystals.

Creep curves obtained at -60°C for HF- doped crystals are shown compared with two pure crystals in fig.1 and in fig.2, results obtained in the compressive constant strain-rate tests at -70°C are shown. In both cases it can be clearly seen that the HF doping has considerably softened the crystals: in the creep tests a greater strain-rate is observed, and in the constant strain-rate tests a smaller stress is required to maintain the same strain-rate. A similar result was obtained, as shown in fig.3, if HF was allowed to diffuse into the crystal in the middle of a test as described in Jones and Glen /1/.

The dependence of the flow stress on HF concentration was determined and it was shown that, at a constant stress, the strain-rate, $\dot{\varepsilon}$, depended on HF concentration C as

$$\dot{\varepsilon} \propto C^{+0\cdot62} .$$

This dependence of strain-rate on HF concentration is similar to the HF dependence of the dielectric

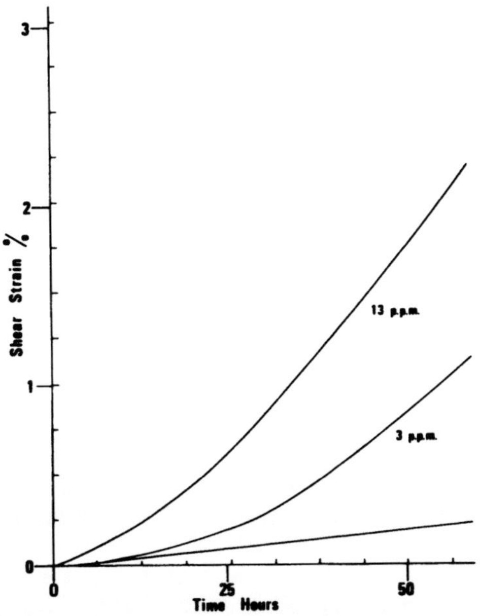

Fig.1 Creep curves obtained at -60°C for HF doped ice, compared with a pure specimen

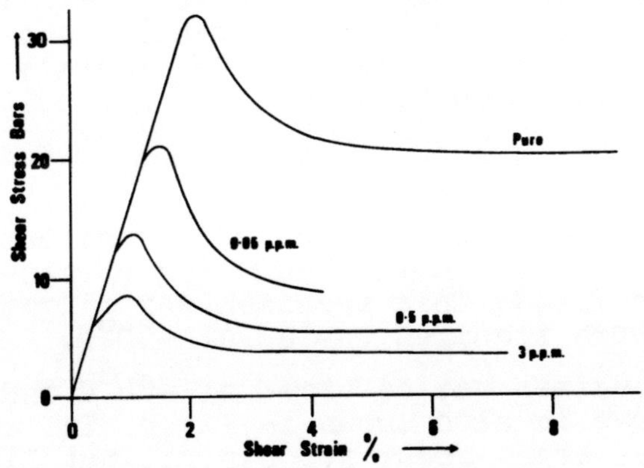

Fig.2 Stress-strain curves obtained at -70°C for ice crystals of various concentrations of HF, compared with a pure specimen

relaxation time, t_D,

$$1/t_D \propto C^m$$

with m = 0.5 at low HF concentrations, increasing to m = 1.0 at high concentrations (Gränicher /4/).

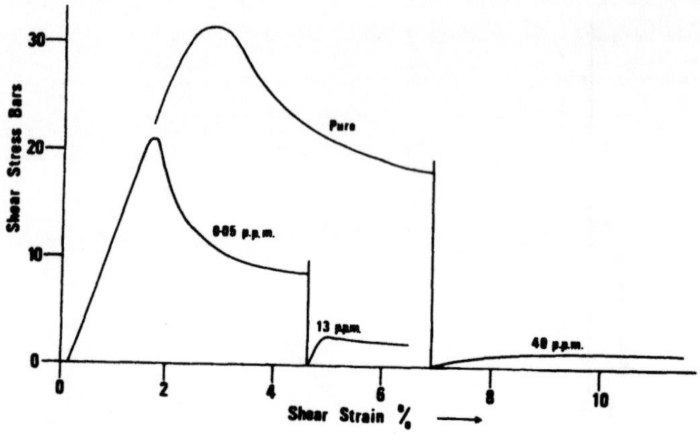

Fig.3 Stress-strain curves obtained at -70°C. The as-grown crystals were deformed, doped with HF by diffusion, and returned to the testing machine

This softening effect of HF was also investigated at a higher temperature, -20°C, and was found to be a much smaller effect at this temperature. This is an indication that the activation energy of the creep process for HF doped ice is less than the activation energy for pure ice. A direct determination of this activation energy was attempted and although the results showed an appreciable scatter, it appeared the HF crystals did have a lower activation energy.

4. Other dopings

One NH_3-doped specimen was deformed at a constant rate in compression at -60°C and the resulting stress-strain curve is shown in fig.4. This specimen was slightly harder than any pure specimen tested at -60°C.

One NH_4F-doped specimen was deformed at -60°C and its stress-strain curve is also shown in fig.4. The amount of doping is very large, about 100 p.p.m., but little effect on the stress-strain curve is noticed.

5. Discussion of results

The relative softness of HF-doped ice is thought to be due to an increase in dislocation velocity, and the hardness of NH_3-doped ice to be due to a decrease in this velocity. When impurities such as these are dissolved in ice they form point defects of the ionic and Bjerrum types (Kröger /5/). HF dissolved in ice greatly increases the number of L-and H_3O^+- defects at the expense of the

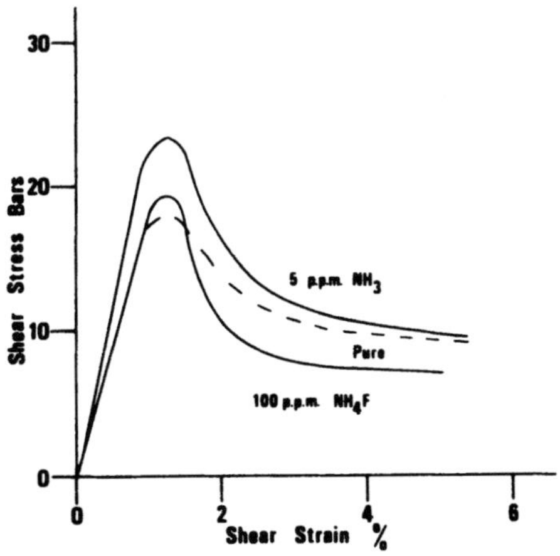

Fig.4 Stress-strain curves of an NH_3 and an NH_4F doped crystal deformed at -60°C, compared with a pure specimen

D and OH^- ones, while NH_3, if it enters the ice lattice substitutionally, increases the number of D- and OH^--defects at the expense of the L and H_3O^+.

Using the ideas of Glen /6/ it is easy to explain qualitatively the softness of HF doped ice crystals: the number of point defects has been greatly increased by the HF and hence the probability of reorientation of a bond which is holding up a dislocation is greatly increased. The dislocation will therefore move faster and the crystal will be softer. However, the inclusion of NH_3 in the ice lattice should also greatly increase the number of point defects and hence the reorientation of bonds, giving the same effect as HF but the experimental evidence is in contradiction with this - if anything NH_3 hardens ice single crystals. Two possible explanations for this have been offered by Jones and Glen /1/. The first assumes D-and OH^--defects are formed by NH_3, but that they are unable to reorient the bonds in question due to the attraction of D-defects by the dislocation core. The second questions whether NH_3 does in fact produce the mobile D-and OH^--defects.

The equilibrium concentration of the different types of defect present in NH_4F doped ice is rather complicated. All four types of defect may be expected to form giving large numbers of H_3O^+-and OH^--ions and L-and D-Bjerrum defects. The law of mass action must apply to these reactions which leads to a cancelling out of the L-and D-defects to give perfect bonds, and of the ions to give H_2O molecules. The work of Levi, Milman and Suraski /7/

has shown that the cancelling process is not complete. It is probable, then, that the majority defects in NH_4F-doped ice are D-defects and that the slight hardening expected was not observed. Clearly further experiments need to be done.

The smaller activation energy of HF doped specimens can also be explained with this theory. In the HF doped ice the activation energy should correspond to the energy of movement of L-defects while in pure ice it should correspond to the combined energies of formation and movement of the defects.

Further support for these ideas comes from the correlation mentioned earlier between the dependence of the strain-rate and dielectric relaxation time, on HF concentration.

Acknowledgements

We would like to thank the Royal Society for the provision of the cold laboratory used in the preparation and handling of specimens, and the Science Research Council for a research studentship held by one of us (S.J.J.).

References

/1/ Jones, S.J., J.W.Glen, Phil.Mag. (1968), (in press)

/2/ Glen, J.W., S.J.Jones, Proc.Int.Conf. Low Temperature Science, Sapporo 1966, Vol.1 Part 1,267-275

/3/ Jones, S.J., J.W.Glen, Union Géodésique et Géophysique Internationale, Association Internationale d'hydrologie Scientifique, Assemblée Generale de Berne 25.9-7.10, (1967), Commission des Neiges et Glaces, (in press).

/4/ Gränicher, H., Phys.kondens.Materie 1, 1-12 (1963)

/5/ Kröger, F.A., Chemistry of Imperfect Crystals. North-Holland Publishing Company, Amsterdam (1964)

/6/ Glen, J.W., Phys.kondens.Materie 7, 43-51 (1968)

/7/ Levi, L., O.Milman, E.Suraski, Trans.Farad.Soc. 59, 2064-75 (1963).

ELASTIC MODULI OF ICE

G. Dantl

I. Physikalisches Institut

Universität Stuttgart / Germany

Abstract

The velocities of longitudinal and transversal sound in ice single crystals have been determined for different orientations from the melting point down to -140 °C using a supersonic pulse-echo method and a double-pulse interference technique. From these data the five elastic moduli, the five elastic constants and the compressibility respectively have been calculated. A model of an aging effect is proposed which explains the differences of earlier results for the elastic moduli, the density and the piezoelectricity.

Introduction

Due to its character of bonding ice occupies a special position among solids: It lies between a molecular and an ionic crystal. The interaction between next but one neighbouring molecules in these open type crystals is small, so that they are especially suited to a theoretical calculation of their elastic properties. This is probably why Born applied his lattice theory first to the diamond which also has this open type structure. His co-worker Penny /9/ applied this theory to ice, and it is interesting to see to what extent the assumptions made prove true.

The elastic moduli and the elastic coefficients of ice have been measured repeatedly. Although single moduli have been measured over a wide temperature range /5, 6/, the complete set has only been measured for a small temperature range near the melting point /1,2,3,4/.

To determine the full set of the five elastic moduli, at least five independent measurements (e.g. of the velocity of sound) had to be made. Supersonic methods in the Mc. range are especially advantageous because, due to the small wave length (at 30 Mc. the wave length is almost 10^{-2} cm), one approaches what can be called an infinitely extended medium. One also has the possibility of making two independent measurements at one orientation of the crystal. These two measurements are made with longitudinal and transversal waves.

Experiments

a) The pulse-echo method.
Short supersonic pulses traverse the ice sample and the corresponding time is measured. The elastic moduli can be calculated from the velocity of sound considering the mode of vibration and the crystallographic orientation. The samples were cylindrical and had plane-parallel surfaces. Their length was about 26 mm and their diameter 30 mm. A transmitting-receiving quartz transducer was attached to one of the plane surfaces. The pulse is reflected at the end of the sample and returns to the quartz, where it is reflected again etc. Every time the pulse returns to the quartz it produces an electrical signal which is proportional to the amplitude of the sound wave. This signal is made visible on an oscilloscope. The transit time can be determined by means of a calibrated electrical delay line. The accuracy is better than 1 %. The supersonic transmitter covers a frequency band from 5 to 200 Mc. Pulse duration is 1 to 2 μs. This method allows the reliable measurement of dependence of sound velocity on temperature.
Variations of pulse shape due to multiple reflections cause relatively large errors in the determination of the absolute value of the moduli. Therefore the double-pulse interference method was also used on the same ice samples.

b) Double-pulse interference method /7/.
Here also supersonic pulses are transmitted through the sample. As soon as the first echo has arrived at the transducer at the front of the sample, a second pulse is sent /19/. If these pulses are coherent one gets interference. In this arrangement, the specimen has the function of an acoustic delay line. As the degree of interference varies with frequency, maximum resp. minimum amplitudes are found when the total path, 2d, is an exact multiple of one half the wavelength. When the frequency is raised from a value at which an amplitude minimum is

observed to that where the next minimum is obtained, then exactly one more wavelength fits into the distance 2d. The formula relating the velocity of sound v and the length d of the sample with the two neighbouring frequencies f_n and f_{n+1}, at which the amplitudes have a minimum, is

$$v = 2d\,(f_{n+1} - f_n)\ .$$

In this way the velocity of sound can be found by one determination of length and two measurements of frequency. In practice, the above formula has to be corrected for the change in phase shift at the reflecting surface /7/, because the quartz transmitter does not oscillate at its resonance frequency.

c) The growing and preparation of the ice specimen. Purest ice single crystals were used. The crystals, grown upward from an oriented seed crystal, have a diameter of 10 cm and a length of 25 cm. The growth velocity was 0.2 µm/s. The water used was extremely pure and was produced in a quartz multiple-distillation apparatus. At 25 °C the water of the melted crystals had a specific conductivity of $0.30 \cdot 10^{-6}$ $(\mathrm{Ohm}\cdot\mathrm{cm})^{-1}$, and a pH-value of 6.80 ± 0.15. The perfectly clear crystals contained no grain boundaries. They were oriented parallel to the crystallographic c-axis to an accuracy better than ±0.6° by means of a conoscope. With appropriate mechanical methods and interferometric tests we succeeded in getting the end planes parallel to within 10^{-5} cm/cm /7/. The coupling medium for the temperature range between 0 °C and -60 °C was silicone oil DC 200/20. For lower temperatures silicone oil AS 50 was used. The velocity of sound was measured in the following five orientations to determine the five elastic moduli of ice:

modulus	mode	orientation of the wave-front
c_{11}	long.	$(1\ 0\ \bar{1}\ 0)$
c_{33}	long.	$(0\ 0\ 0\ 1)$
c_{44}	transv.	$(1\ 0\ \bar{1}\ 0)$ $(0\ 0\ 0\ 1)$
c_{66}	transv.	$(1\ 0\ \bar{1}\ 0)$
c_{13}	quasi-long.	$(1\ 1\ \bar{2}\ 1)$

Results

The velocities of sound were measured by means of the pulse-echo method. From these the values of the elastic moduli of monocrystalline ice were determined in the frequency range from 5 to 190 Mc. No frequency dependence of the elastic moduli (fig. 1) could be observed within the limits of experimental accuracy. The dependence on temperature was measured by the pulse-echo method at 30 Mc. (longitudinal waves) and at 28.75 Mc. (transversal waves) in intervals of 5 °C. The temperature range of the experiment was traversed monotonously by either increasing or decreasing the temperature. The deviation of the temperature from the desired value was less than 0.05 °C.

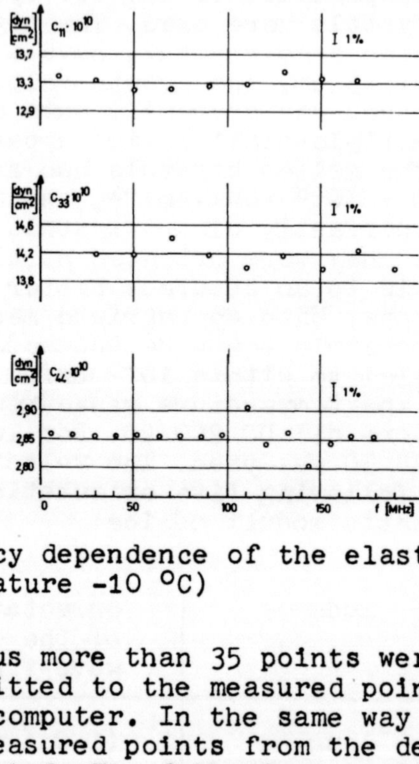

Fig.1 Frequency dependence of the elastic moduli (temperature -10 °C)

For each modulus more than 35 points were measured. Parabolas were fitted to the measured points (fig. 2) with the help of a computer. In the same way the mean square error of the measured points from the determined parabolas was calculated. The following equations describe the temperature dependence of the elastic moduli:

$$c_{11}(T) = 12.904\ (1-1.489\cdot10^{-3}\cdot T-1.85\cdot10^{-6}\cdot T^2) \pm 0.3\ \%$$

$$c_{33}(T) = 14.075\ (1-1.629\cdot10^{-3}\cdot T-2.93\cdot10^{-6}\cdot T^2) \pm 0.4\ \%$$

$$c_{44}(T) = 2.819\ (1-1.601\cdot10^{-3}\cdot T-3.62\cdot10^{-6}\cdot T^2) \pm 0.7\ \%$$

$$c_{12}(T) = 6.487\ (1-2.072\cdot10^{-3}\cdot T-3.62\cdot10^{-6}\cdot T^2)\ \pm\ 2\ \%$$

$$c_{13}(T) = 5.622\ (1-1.874\cdot10^{-3}\cdot T \qquad \pm\ 7\ \%$$

all $c_{ik}(T)$ in 10^{10} (dyn/cm^2), T in ($^\circ$C).

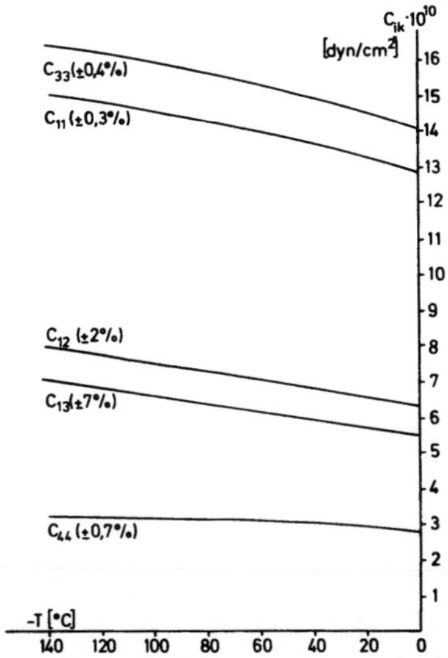

Fig.2 Elastic moduli of aged ice

The absolute values of the moduli c_{11}, c_{33} and c_{44} were determined by means of the double-pulse method and with different samples. Both methods exhibited the same dependence on the temperature. Using the known relations between the elastic moduli c_{ik}, the compressibility κ, and the elastic coefficients s_{ik} (fig.3), leads to the following equations:

$$\kappa(T) = 11.94\ (1+1.653\cdot10^{-3}\cdot T+3.12\cdot10^{-6}\cdot T^2)\ \pm\ 15\ \%$$

$$s_{11}(T) = 10.40\ (1+1.070\cdot10^{-3}\cdot T+1.87\cdot10^{-6}\cdot T^2)\ \pm\ 1\ \%$$

$$s_{33}(T) = 8.48\ (1+1.405\cdot10^{-3}\cdot T+4.66\cdot10^{-6}\cdot T^2)\ \pm\ 1\ \%$$

$$s_{44}(T) = 33.42\ (1+1.505\cdot10^{-3}\cdot T+4.04\cdot10^{-6}\cdot T^2)\ \pm\ 1\ \%$$

$$s_{12}(T) = -4.42\ (1+0.463\cdot10^{-3}\cdot T-2.06\cdot10^{-6}\cdot T^2)\ \pm\ 6\ \%$$

$$s_{13}(T) = -1.89\ (1+1.209\cdot10^{-3}\cdot T+6.15\cdot10^{-6}\cdot T^2)\ \pm\ 20\ \%$$

all $s_{ik}(T)$ in 10^{-12} (cm^2/dyn), T in ($^\circ$C).

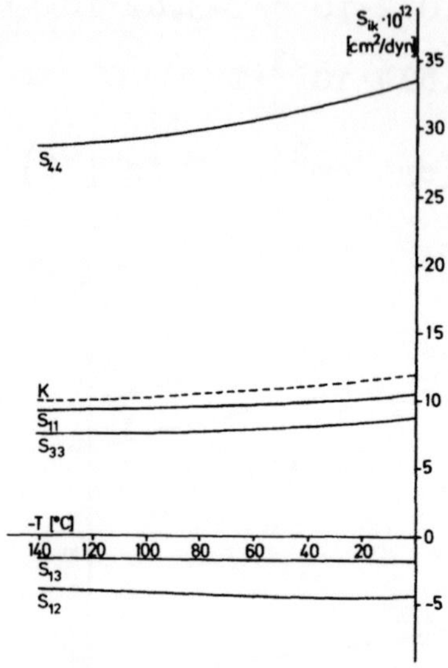

Fig.3 Elastic coefficients and compressibility of aged ice

Discussion

The distance between oxygen atoms is rather large and the hydrogen bond is strong and rigid in ice. The interaction between the next but one nearest neighbours can be neglected when applying Born's dynamic lattice theory to ice (Penny /9/). The calculation, therefore, neglects the hexagonal cell and is based only on the tetrahedral sub-cell. One thus obtains three equations relating the five macroscopic elastic moduli. The measured c_{ik} values agreed very well with these equations, better than those measured by other authors /1,6/.
The values of the moduli mentioned above deviate up to 5.6% from those of other authors, and these differ also among each other. These deviations exceed the experimental limits of error. The same is true for measurements of density and piezoelectricity. The differences may be interpreted by the assumption of an aging process in ice. We assume that an orientation of the H_2O dipoles, due to the growing process, changes into disorientation within several days. To compare the data of different authors, age and previous treatment of the samples have to be known. This hypothesis is the result of the following measurements and considerations:

Density. Since the end of the last century /13/ it has been known that density measurements of ice differ by up

to 0.22 %. The results depend on the conditions of growth and on the age of the ice. Recent measurements /14/ confirm that fluctuations of the density can reach more than 0.3%. The smallest density measured is that of aged ice.
The decrease in density by 0.18 % corresponds to an increase of the O-O distance of 0.06 % and to a decrease of the power constants of the Morse potential by 2 % /15, 16/. The power constant is proportional to the elastic modulus.

Piezoelectricity. The assumption that the dipole orientation disappears with time is confirmed by the piezoelectric properties of ice. The hexagonal structure of the oxygen lattice cannot cause piezoelectricity. Piezoelectricity has therefore to be due to a polar structure, i.e. in certain regions of the crystals the H_2O dipoles have to be similarily oriented. Reports of piezoelectricity in the literature /10,11/, show both positive and negative results. A rather strong piezoeffect was found on crystals which Deubner et al. /10/ obtained from super-cooled water. This effect vanished within several days.

Measurements of the mechanical damping in ice /7/ indicated no piezoelectricity comparable in magnitude to that mentioned by Deubner et al. for the ice sample used in this work.

Behaviour near the melting point. The values of the moduli measured by Brockamp and Querfurth /3/ are rather large. Therefore, one can assume a polar structure in the ice used. These authors found that the moduli decrease by about 5% when measured between -0.2 °C and the melting point. This result can be interpreted by assuming that the existing polar structure can be removed close to the melting point in an especially short period of time.

Electrical potential during growth. Workman and Reynolds /18/ have shown that ice develops a potential during growth (in pure ice crystals it is almost 100 volts) in the direction of the c-axis. One assumes that polar regions due to this potential arise which produce a dipole interaction effective at long ranges. This interaction leads to a greater density of the ice. The ice used for this work had been grown slowly and stored for at least 8 months near the melting point. If the aging hypothesis is correct then the polar structure had been removed, the O-O distance enlarged, and the elastic moduli reached their smallest value. The values of other authors, measured with 'younger' ice, were indeed always larger.

References

/1/ Jona, F., P. Scherrer, Helv.Phys.Acta 25, 35 (1952)

/2/ Bass, R., D. Rossberg, G. Ziegler, Z.Phys. 149, 199 (1957)

/3/ Brockamp, B., H. Querfurth, Z.Polarforsch. 5, 253 (1964)

/4/ Bogorodskii, V.V., Soviet Physics Acoustics 10, 124 (1964)

/5/ Zarembowitch, A., Bull.Soc.Franc.Minéral.Crist. LXXXVIII, 17 (1965)

/6/ Proctor, M., jr., JASA 39, 972 (1966)

/7/ Dantl, G.,Thesis,TH Stuttgart (1967)

/8/ Cady, W.G., Piezoelectricity, New York, Dover Publ. Inc. (1964)

/9/ Penny, A.H.A., Proc.Cambridge Phil.Soc. 44, 423 (1948)

/10/ Deubner, A., R. Heise, K. Wenzel, Naturwissenschaften 47, 600 (1960)

/11/ Teichmann, I., G. Schmidt, phys.stat.sol. 8, K145 (1965)

/12/ Rossmann, F., Experimentia 6, 182 (1950)

/13/ Dorsey, N.E., Properties of Ordinary Water-Substance. New York: Reinhold Publ.Corp. (1957)

/14/ Dantl, G., I. Gregora, Naturwissenschaften 55, 176 (1968)

/15/ Lippincott, E.R., R. Schroeder, J.Chem.Phys. 23, 1099 (1955)

/16/ Hadži, D. (ed.), Hydrogen Bonding, London, Pergamon Press (1959)

/17/ Dantl, G., Z.Phys. 166, 115 (1962); Z.Phys. 169, 466 (1962)

/18/ Workman, E.J., S.E. Reynolds, Phys.Rev. 78, 254 (1950)

/19/ McSkimin, H.J., in: Mason (ed.), Physical Acoustics IA;New York, Academic Press (1964)

ELASTIC ANOMALIES OF ICE AT LOW TEMPERATURES

D. Helmreich

Physik-Department

Technische Hochschule München, Germany

Abstract

The temperature dependence of the longitudinal elastic constants of ice single crystals was measured by an ultrasonic pulse echo method. With heating respectively cooling rates less than a critical value, deviations to lower values of the elastic constants were observed.

The time dependence of the anomaly was studied at fixed temperatures. After sufficiently long time, the elastic constants reach limiting values; the maximum of these values was observed near 105°K.

The results are interpreted as representing a second order phase transition from a disordered proton configuration to a more ordered state with decreasing temperature. The increased ordering is obtained by migration of orientational faults.

1. Introduction

When calorimetric and electric measurements were first made on ice at lower temperatures it was found that deviations from a normal behaviour occur between 85°K and 150°K.

In 1936 Giauque and Stout /1/ found such an abnormal behaviour of the specific heat c_p in polycrystalline ice. They discovered that .."between 85°K and 100°K, the attainment of temperature equilibrium in the solid was less rapid than at other temperatures." Süßmann /2/

confirmed that anomaly with measurements on ice single crystals. Recent results of specific heat measurements on pure and doped ice have been reported by Pick /3/. His findings show also an abnormal behaviour.

Eckener /4/ found that the static dielectric constant attains a maximal value of about 300 at 100°K. This value is very small - it means that only a small part of the crystal will be affected.

With the aid of Gibb's electrical free energy G_2 these findings can be interpreted under a common point of view. If only small forces are applied and if a transition of the second order is assumed, G_2 is given by

$$G_2(x_{ij}, E_i, dT) = 1/2\ \Sigma\chi_{ij}E_iE_j + 1/2\ \Sigma c_{ijkl}x_{ij}x_{kl} +$$

$$+ 1/2\times c/T\times(dT)^2 + \text{mixed terms of 2 nd order.}$$

For a supercritical phase transition the coefficients of second order terms attain a maximal value at the transition point. Correspondingly, their reciprocals - the stability coefficients - attain a minimal value, thus reflecting the lowered stability of the whole system at the transition point.

The dielectric constant and the specific heat of ice show such instabilities. The purpose of this work was to search for a similar behaviour of the elastic constants.

2. Experimental set-up

For that reason, the two longitudinal elastic constants c_{11} and c_{33} have been measured by an ultrasonic pulse echo method. The time lapse between the first und eleventh reflected pulses - $10\times\tau_{ij}$ - has been determined with a time mark generator. Since the directions of particle velocity and sound propagation coincide, one only needs to know the transit time τ_{ij} and the length of the sample in order to compute the sound velocity v_{ij} and the elastic constant c_{ij}. Dantl's thermal expansion coefficient /5/ was used for the calculation.

The ice samples have either been cut from monocristalline seed crystals - or from the central portion of crystals having been grown according to a modified Bridgeman method. These crystals measured 58 mm in diameter and about 200 mm in length.

At the time of experiment the ice crystals had an age between two weeks and ten months; the melting water of the samples had a conductivity of 3 to 70×10^{-6} $(\text{ohm.cm})^{-1}$ for pure ice and HF doped ice respectively. No precautions against impurities were taken because the anomalies at low temperatures occur only in ice crystals which are not too pure.

The desired crystal orientation (c-axis normal and parallel to the direction of sound wave propagation was accomplished by controlled melting and checked with a conoscope. From these oriented crystals the samples were cut with a lathe. After polishing their reflecting surfaces they were adjusted in a sample container. The final sample shape was cylindrical with 16 mm diameter and 30 mm length.

The expansion coefficient of ice ($\alpha = 51\times10^{-6}/^{o}C$) is much greater than that of the quartz resonator used ($\alpha = 8\times10^{-6}/^{o}C$). This great difference results in considerable mechanical strain. The bonding agent, therefore, must be ductile to very low temperatures in order to compensate these strains and to prevent the decoupling of the resonator from the sample.

4-methyl-1-penten was found to be the best of all bonding materials tested. But at $-10^{o}C$ - where the sample is installed - 4 methyl-1 penten is still rather volatile. In order to avoid the evaporation,a sample container was constructed which operates on the principle of piston and cylinder. The ice sample is adjusted in the cylinder. The piston which carries the quartz oscillator shuts the cylinder airtight. Consequently the bonding agent is prevented from escape.

3. Results and discussion

The measured elastic constants have been fitted to a second order equation by a least square fit. Equations of higher degree do not improve the approximation.

Fig.1 shows such curves for the elastic constants c_{11} and c_{33}. In the following, these curves will be called "norm curves". They are compared with the results of Zarembovitch, Kahane /6/, Proctor /7/ and Dantl /8/. One sees clearly that all curves have the same shape but their absolute values differ by an amount greater than the reported errors.

Dantl /8/ discussed these findings in detail.

At cooling and heating rates less than 0.7 respectively 0.3 deg/min, deviations to lower values of the elastic constants have been observed. Fig.2 shows experimental

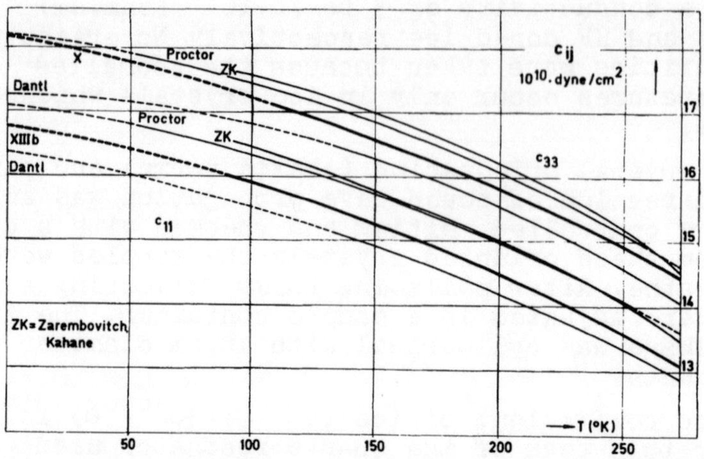

Fig.1 Temperature dependence of longitudinal elastic constants c_{11} and c_{33} of ice Ih /12/

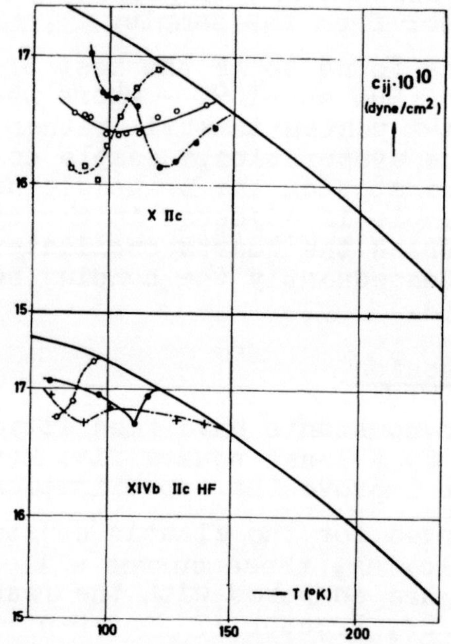

Fig.2 Deviations from the norm curve for different heating and cooling rates /12/

values of the elastic constants of two different crystals, each with different heating and cooling rates. The full line represents the norm curve. Larger cooling rates correspond to a lower temperature of maximum deviation T_a; larger heating rates correspond to a higher T_a.

The deviations of the elastic constants to lower values at low temperatures as well as the relation between temperature T_a and cooling resp. heating rate can be interpreted by the transition of ice from a phase 1 into a phase 2. When cooling a crystal-being in phase 1-at a sufficiently low rate, phase 2 will soon begin to develop.

Fast cooling causes the freezing in of phase 1. In consequence, one obtains the norm curves mentioned above.Phase 2 is then suppressed. Analogous considerations apply to the case of heating.

The observed softening of the crystal at the temperature T_a indicates that a disordered configuration is removed with decreasing temperature. As the oxygen lattice remains unaltered down to liquid nitrogen temperature it is assumed that the ordering affects the proton configuration.
The measured effects are very small; thus, only a small part of the crystal will be ordered. A finite number of ordered domains will be imbedded in a disordered continuum.

Subsequent experiments were performed to study the time dependence of that anomaly at fixed temperatures ranging from 85°K to 150°K.
In order to take advantage of the high relative accuracy (5×10^{-4}), each norm curve was remeasured at the time of experiment. Fig.3 shows such a cycle of measurements. The crystal is cooled at a rate greater than 0.7 deg/min to a certain fixed temperature (106°K in fig.3). After reaching the maximal deviation from the norm curve at that fixed temperature the crystal is cooled to liquid nitrogen temperature. The maximal deviation attained is preserved by further fast cooling. That means, it is possible to undercool the new phase.

The differences between the saturation values and the norm curve are plotted in a log $\Delta\tau$ vs. $1/T$ - plot for various temperatures (fig.4). The maximum difference is reached at a temperature between 100°K and 110°K.

From the time dependence of the deviations a mechanism of this ordering can be deduced . The concentration of ordered domains increases from zero to a constant value c_2 with decreasing temperature. If one assumes the transit time τ_{ij} to be proportional to that concentration, one should find a similar increase in the transit time with decreasing temperature. Experimentally, however, one finds a maximum for the difference between the saturation values and the norm curve. This observation indicates a hindered transition.

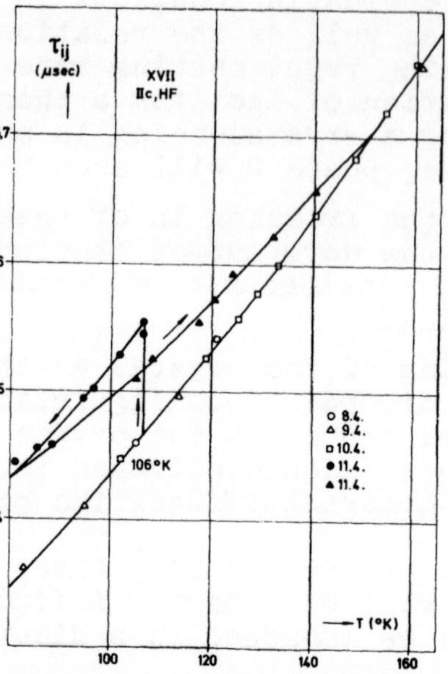

Fig.3 A typical cycle of measurements performed in order to find the time dependence of the deviations /12/

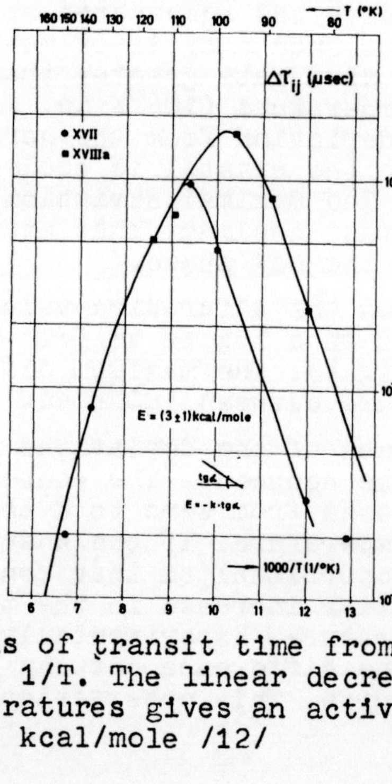

Fig.4 Deviations of transit time from the norm curve $\Delta\tau_{ij}$ vs. 1/T. The linear decrease of $\Delta\tau_{ij}$ at low temperatures gives an activation energy of E = (3±1) kcal/mole /12/

Small imperfections in the crystal act as nuclei for ordered regions. Such a nucleus can grow by adding correctly ordered molecules to it. Usually, such molecules will not be found in the vicinity of a nucleus. They have to be transferred from distant regions to the nucleus by a diffusion-like mechanism.

Normally, two activation energies are involved in the diffusion: E_o for freeing the particles to diffuse and E_b for their mobility. The total energy can be found in our case when plotting the measured transit time τ_{ij} in a log $\Delta\tau_{ij}$ vs. 1/T-plot (fig.4). In this representation the transit time decreases linearly with decreasing temperature. The slope of the straight line gives the energy.

This activation energy was found to be 3±1 kcal/mole. This is a value which is usually assumed to be valid for the mobility of Bjerrum faults. Bjerrum /9/ estimated 2.5 kcal/mole on theoretical grounds. Jaccard /10/ and Engelhardt /11/ found experimental values of 5.2 kcal/mole and 4.0 kcal/mole resp.

Therefore, the motion of Bjerrum faults is responsible for the ordering process. If the ordered regions are to be supplied with additional correctly ordered water molecules, the Bjerrum faults have to reach them. There they orient the water molecules correctly. Then these molecules can easily be attached to the ordered region. As the transition and the mobility freeze in according to the same rule, one can say that the concentration of Bjerrum faults remains constant with decreasing temperature. Therefore, the attainment of an equilibrium concentration requires much more time than available in the experiment.

References

/1/ Giauque, W.F., J.W. Stout, J.Amer.Chem.Soc. 58, 1144 (1936)

/2/ Süßmann, W., Thesis, TH München (1967)

/3/ Pick, M., Physics of Ice, Proc.Int.Symp., Munich, Plenum Press, New York (1969)

/4/ Dengel, O., U.Eckener, H.Plitz, N.Riehl, Phys.Letters 9, 291 (1964)

/5/ Dantl, G., Z.Phys. 166, 115 (1962); ibid.169, 466 (1962)

/6/ Zarembovitch, A., A.Kahane, C.R.Acad.Sc. Paris, 258, 2529 (1964)

/7/ Proctor, Th.M., J. Acoust.Soc.Am.39, 972 (1966)
/8/ Dantl, G., Phys.kond.Mat. 7, 390-397 (1968)
/9/ Bjerrum, N., Danske Mat.Phys.Medd. 27, 32 (1951)
/10/ Jaccard, C., Helv.Phys.Acta 32, 89 (1959)
/11/ Engelhardt, H., N.Riehl, Phys.kond.Mat. 5, 73 (1966)
/12/ Helmreich, D., B.Bullemer, Phys.kond.Mat., submitted.

X-RAY DIFFRACTION TOPOGRAPHIC STUDIES OF THE DEFORMATION BEHAVIOUR OF ICE SINGLE CRYSTALS

Akeharu Fukuda and Akira Higashi

Department of Applied Physics, Faculty of Engineering, Hokkaido University, Sapporo, Japan

Abstract

The dynamic behaviour of dislocations in basal glide in ice single crystals was investigated using a special apparatus which served both to hold the specimen on the goniometer head of a Lang camera and to bend the crystal in it. Topographs taken in the early stages of creep showed that new dislocations generated at or near loading sites, inclusions, or stress concentrations and that they were primarily screw dislocations with Burgers vectors of a/3 <11$\bar{2}$0>. At -22°C, the dislocation velocity was estimated to be approximately 10^{-3}cm/sec.

1. Introduction

During the last few years the mechanical properties of ice single crystals have been extensively investigated. /1, 2, 3, 4/.
Dislocation theories have been used to interpret the experimental results obtained in studies of creep, relaxation, and stress-strain relations and Johnston's mechanisms for motion and multiplication of dislocations under stress have been applied to basal glide in ice single crystals /5/. Although many investigators have attempted to confirm the predicted behaviour of dislocations in ice crystals, the majority of these investigations utilized etching techniques which were both limited and uncertain. No direct observation of dislocations had been attempted until Hayes and Webb /6/ applied X-ray

diffraction topography techniques to thin, dendritic ice crystals.

In the experiments reported here, X-ray topography was used to observe the distribution and structure of dislocations in bulk ice single crystals. The natural glacier single crystals which had been used in previous plastic deformation experiments were examined first /7/. The dislocation structure was clearly revealed by topographs taken with the $(10\bar{1}0)$ diffracting planes scanned over the (0001) plane and also by those taken with the $(1\bar{1}20)$ diffracting planes on the $(\bar{1}100)$ plane. Dislocations with Burgers vectors of $\frac{a}{3}\langle 11\bar{2}0\rangle$ were lying only on the basal plane. Topographs taken with the (0002) diffracting plane scanned over the $(\bar{1}100)$ plane revealed that there also were edge dislocations with Burgers vectors of $c\langle 0001\rangle$ on the basal plane. The fact that all of the dislocations appear to be bound in the basal plane shows that it would be difficult for dislocations to move from one basal plane to another and explains the strong anisotropy of yield stress and the stress dependence of glide in ice single crystals which has been reported /3/.

The work has been extended from the static results described above to include dynamic behaviour of dislocations.
This paper describes the results obtained from topographs taken of specimens subjected to repeated deformation to allow observation of generation, movement, and multiplication of dislocations in ice single crystals.

2. Experimental Procedures

The ice single crystals used in these experiments were artificially grown from water using a modified Czochralski method /8/. The crystals were grown in the direction of the c-axis and had an initial dislocation density of the order of 10^4 cm^{-2} or less. With such low dislocation densities, it is easy to distinguish those dislocations generated during deformation from those initially existing in the crystal. Parallelepipeds about 1 x 1 x 3 cm were cut from the crystals with a band-saw and chemically polished in alcohol to about 1 x 8 x 22mm. Final surface treatment consisted of rinsing with n-hexane to remove any residual alcohol and then allowing the surface to sublimate slightly. This method produced platelets with no artificially introduced surface defects.

The scanning surface for the X-ray topographs was 8 x 22 mm and perpendicular to the c-axis of the crystal. The long direction of the platelet was either the

$\langle 1\bar{1}00 \rangle$ or the $\langle 2\bar{1}\bar{1}0 \rangle$ to allow observation of the effect of direction in maximum bending shear stress.

A special specimen holder was designed in which the specimen could be deformed without disturbing the precise alignment required to fix it on the goniometer head of the Lang camera. Figure 1 (a) illustrates this device: the specimen (A) is set in an aluminium mount (B) which is covered on both sides by Mylar diaphragms (M) which prevent evaporation of the ice surface. The mount is sandwiched between two casings (C and D), one of which (D) is attached to a circular disk. This disk fits into a ring (E) which allows it to rotate 60° in two directions in order to change the diffracting plane.

When the specimen was deformed, the specimen holder was removed from the ring, placed in a horizontal position and supported at both ends as is shown in fig. 1(b). As this figure also illustrates, loading was obtained by applying a wedge to the top surface of the specimen. This method of straining is a kind of three-point-bending which had also been used in previous creep experiments but, in this case, the load was deliberately applied in an off-center position to allow easier observation of dislocation behaviour in the center of the specimen. There was no accurate measure of the bending but the loading time was used as a measure of the strain. A 560 g weight was used and if an elastic equation for bending /1/ was used, this weight produced a maximum shear stress in the crystal of approximately 100 kg/cm^2.

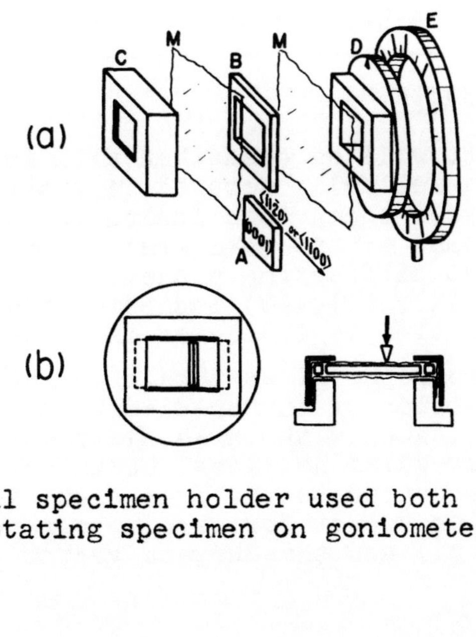

Fig.1 Special specimen holder used both for bending and rotating specimen on goniometer head

The first set of topographs was taken on three different $(10\bar{1}0)$ diffracting planes with an undeformed specimen. The specimen holder was then removed from the camera and the specimen was strained for periods of 30 to 120 seconds and replaced in the camera for another set of topographs. This procedure was repeated to allow observation of the processes of dislocation movement and multiplication with increasing strain. All of these experiments were carried out in a cold room at -22°C. Table 1 describes the experimental procedure followed out for two specimens. The results obtained for these two specimens are discussed in the following section. The corresponding figure numbers are also given in the table so that the reader may determine at what stage the topographs shown were taken.

Table 1 Experimental Procedures

Specimen No. 1: Long axis ∥ $<1\bar{1}00>$											
Process	L	S	L	S	L						
Topographs	0		1		2						
Loading time (seconds)		90		90							
Figure No.	2		3								
Specimen No. 2: Long axis ∥ $<2\bar{1}\bar{1}0>$											
Process	L	S	L	S	L	S	L	S	L	S	L
Topographs	0		1		2		3		4		5
Loading time (seconds)		30		60		90		90		120	
Figure No.	4		5						6		
							7(1)		7(2)		7(3)
	8(1)		8(2)				8(3)				

3. Experimental Results

Experimental results will be described with every topographs shown in the following several figures. In figures from 2 to 6, three topographs indicated (a), (b) and (c) are the topographs of the same area of crystal taken with three different diffracting planes. In figures 7 and 8, sub-numbers (1), (2), (3) indicate time series of the topographs. The time is indicated in table 1.

Figure 2: Three topographs of an undeformed specimen in which the long direction of the specimens was $<1\bar{1}00>$. The topographs were taken with three different diffracting planes. The irregular hexagonal areas outlined on the photographs correspond to the same areas in the crystal. The diffraction vector for each topograph is indicated by an arrow ($\underline{g}|$) and the Burgers vector ($|\underline{b}$) was

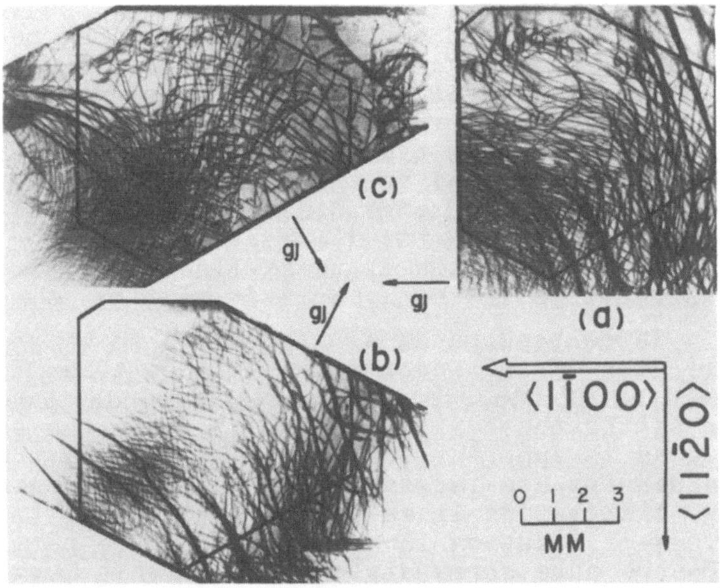

Fig.2 Topographs of undeformed specimen No. 1 taken with different diffracting planes; diffraction vectors denoted by **g**

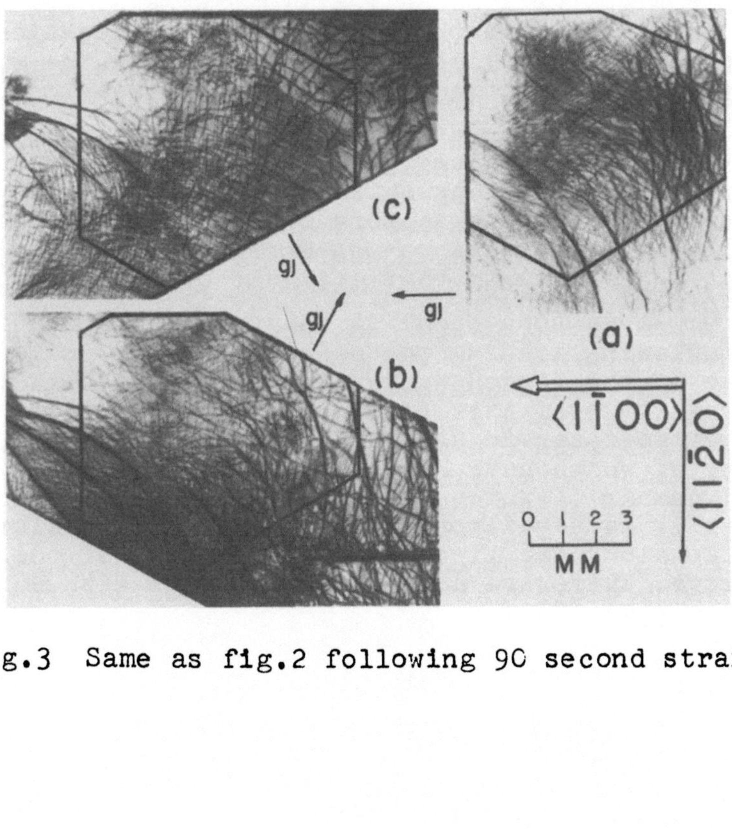

Fig.3 Same as fig.2 following 90 second strain

determined by assuming that $\underline{g}| \cdot |\underline{b} = 0$ for a topograph on which the image vanished. In this figure the fine, vertical lines in b and c and the tilted lines near the right side of the hexagon in a and c are screw dislocations. The horizontal lines appeared in a and b are 60° dislocations with Burgers vectors of $\frac{a}{3}\langle 2\bar{1}\bar{1}0\rangle$ and those in a and c are the same type with Burgers vectors of $\frac{a}{3}\langle 1\bar{2}10\rangle$. The thick lines at the bottom of the figures a and b are polygonized edge dislocations because they vanish in figure c when the diffraction vector, $\underline{g}|$, is parallel to the direction of these lines. There are helical dislocations in the upper left of figs. a and c.

Figure 3: Three topographs of a specimen in which the long direction of the specimen was $\langle 1\bar{1}00\rangle$ taken following straining of the specimen for 90 seconds. The disposition of the photographs, a, b and c are the same as in fig.2. As is apparent from the photographs, the dislocation density has increased, primarily from generation of new dislocation lines parallel to $\langle 2\bar{1}\bar{1}0\rangle$ (a and b). The new dislocation lines in this figure parallel to $\langle 2\bar{1}\bar{1}0\rangle$ are pure screw dislocations with a Burgers vector of $\frac{a}{3}\langle 2\bar{1}\bar{1}0\rangle$. The small group of dislocations in the upper left appear to have generated close to the helical dislocation and they have a Burgers vector of $\frac{a}{3}\langle \bar{1}2\bar{1}0\rangle$. Another group of screw dislocations of the same Burgers vector may be seen on the right side of the hexagon. In this figure there are no new dislocations with Burgers vectors parallel to $\langle \bar{1}\bar{1}20\rangle$ which is the direction perpendicular to the long axis of the specimen.

Figure 4: Three topographs of an undeformed specimen (No.2) in which the long axis is parallel to $\langle 11\bar{2}0\rangle$. The primary dislocations shown are polygonized edge dislocations with Burgers vectors of $\frac{a}{3}\langle 2\bar{1}\bar{1}0\rangle$ in b and c which are not present in a and screw dislocations which run almost vertical in the left half of the hexagons.

Figure 5: Three topographs of the specimen shown in fig.4 following loading on the specimen for 30 seconds. The edge of the loaded wedge fell on a line between the two arrows in b of this figure. The new dislocations which may be observed are pure screw dislocations, the majority of which have Burgers vectors of $\frac{a}{3}\langle 2\bar{1}\bar{1}0\rangle$ and a small number of which have Burgers vectors of $\frac{a}{3}\langle 1\bar{2}10\rangle$. It is very apparent that these new dislocations generated along the line denoting the edge of the loading wedge. There are no new dislocations with Burgers vectors of $\frac{a}{3}\langle 11\bar{2}0\rangle$.

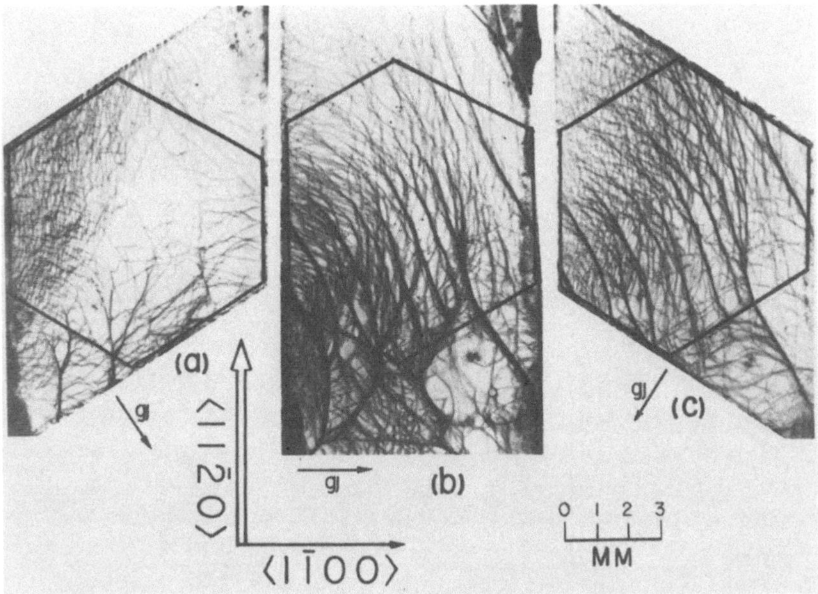

Fig.4 Topographs of specimen No.2, undeformed

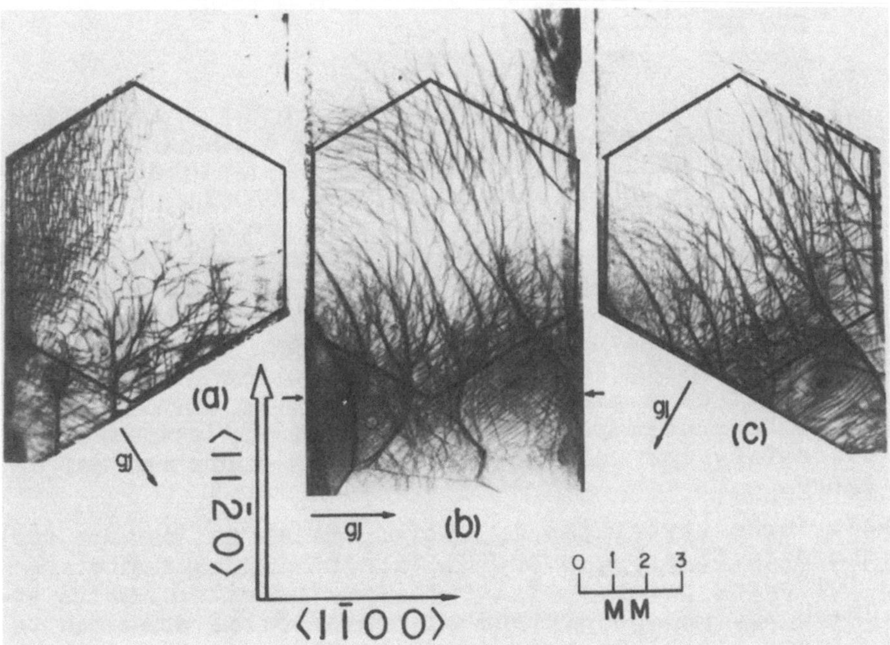

Fig.5 Same as fig.4 following 30 second strain

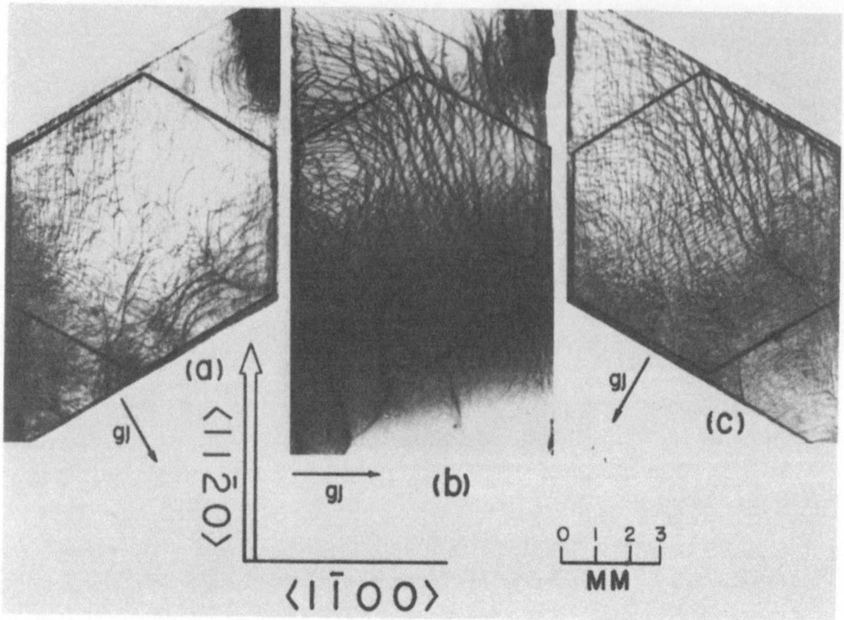

Fig.6 Topographs of specimen No.2 following fourth period of strain

Figure 6:
Three topographs of the specimen shown in fig.4 following four different periods of loading to a total of 270 seconds (see table 1). The density of dislocations has increased so much that it is very difficult to distinguish individual dislocation lines. A group of new screw dislocations with Burgers vectors of $\frac{a}{3}$<2$\bar{1}\bar{1}$0> have moved from their originating site under the line of the wedge to the upper portion of the topographs. Those few screw dislocations with Burgers vectors of $\frac{a}{3}$<1$\bar{2}$10> which were present in fig.5 have moved to the lower left. The polygonized edge dislocations present in figs.4 and 5 appear to have decomposed into separate dislocations as is evident by the thinner, clearer lines present in this figure.

Figure 7: Three topographs in a time series of the specimen shown in figs. 4, 5 and 6. Time of taking topographs following periods of loading is indicated in table 1. These topographs are enlargements of the area to the right and above the hexagon delineated in previous figures. The lower right corner of the topographs shows an area of the edge of the specimen which was accidentally frozen to the frame of the holding device. When this specimen was stressed a stress concentration

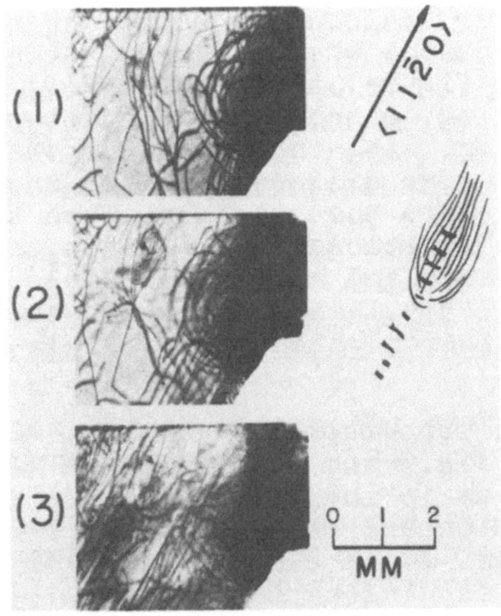

Fig.7 Topographs showing generation and movement of dislocation loops

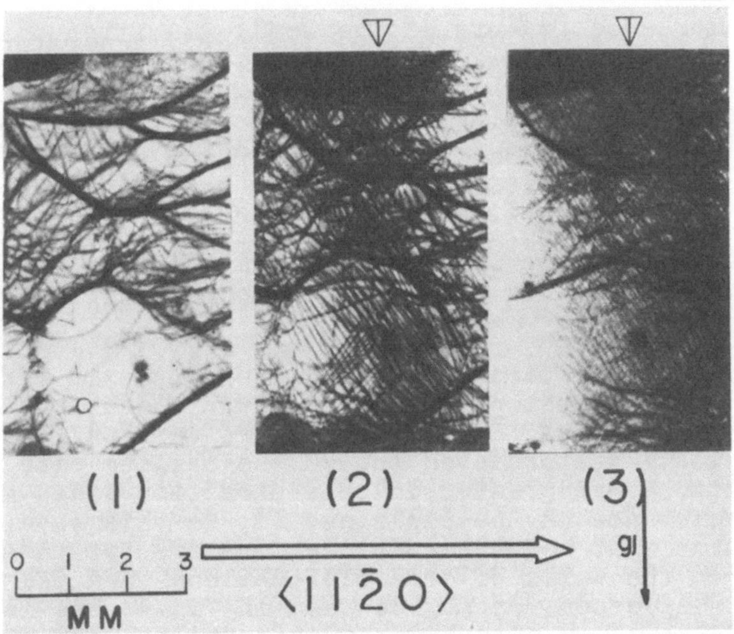

Fig.8 Topographs showing generation and movement of screw dislocations from an edge wall near the loading wedge

occurred in this area and dislocation loops were generated here. Another dislocation source is shown at the top of the topographs where edge and screw dislocations (see drawing on figure) appeared in 7(2) following 270 seconds of loading. Since they are not present in the topographs in fig. 6 b, they have Burgers vectors of $\frac{a}{3}\langle 11\bar{2}0\rangle$ and the edge dislocations are short segments which penetrate into the specimen. When the specimen was strained for an additional 120 seconds, these edge dislocations disappeared and many screw dislocations appeared as is shown in 7(3). This dislocation source may be a very minute inclusion which is invisible in the topographs.

Figure 8: The three topographs in the figure are similar series to those in fig.7 and were taken in the early stages of deformation as is indicated in table 1. They show dislocation generation and multiplication near the edge of the loading wedge. As is clearly shown in the figure, some of the new screw dislocations originated from polygonized edge dislocations which were also being decomposed, deformed, and removed.

4. Discussion and Conclusions

The most remarkable feature of the newly generated dislocations was that they were primarily straigth lines of screw dislocations which only occasionally formed loops. These new screw dislocations had preferential directions. In the first specimen in which the long axis was parallel to $\langle 1\bar{1}00\rangle$, there were many dislocations with $\frac{a}{3}\langle 2\bar{1}\bar{1}0\rangle$ and $\frac{a}{3}\langle \bar{1}2\bar{1}0\rangle$ Burgers vectors but none with $\frac{a}{3}\langle \bar{1}\bar{1}20\rangle$ Burgers vectors. This preference may be explained by the fact that the shear stress component is zero for this specimen in the $\langle \bar{1}\bar{1}20\rangle$ direction(see fig.9a, $|b_3$).

In the second specimen, in which the long axis was parallel to $\langle 11\bar{2}0\rangle$, most of the new screw dislocations had Burgers vectors of $\frac{a}{3}\langle 2\bar{1}10\rangle$ and there were only a few with $\frac{a}{3}\langle 11\bar{2}0\rangle$ vectors even though the maximum resolved shear stress was greater in this direction (fig.9b, $|b_2$). This preference may be explained if, in this case, it is assumed that new screw dislocations are apt to originate from the walls of edge dislocations when the crystal is stressed. As may be seen in fig.4, the majority of dislocations in this specimen are polygonized edge dislocations with $\frac{a}{3}\langle 2\bar{1}\bar{1}0\rangle$ Burgers vectors which seemed to be sources of new screw dislocations with the same vectors (see also fig.8 (2)).

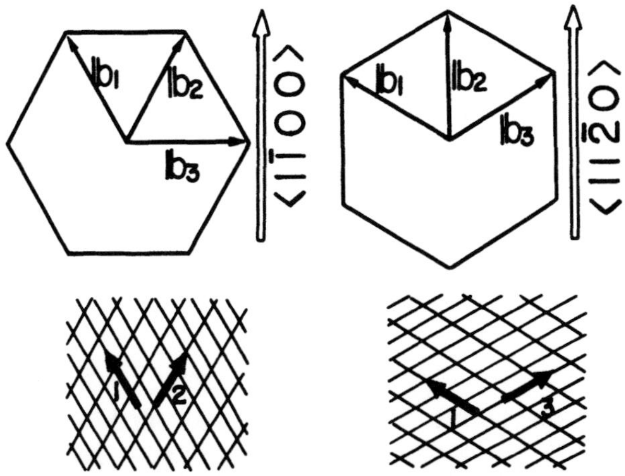

(a) (b)

Fig.9 Preferential direction of Burgers vectors of newly generated dislocations with respect to the long axis of the bent crystal

Inclusions or other minute defects which may cause stress concentrations may also be sources of dislocations. The type and direction of dislocations generated under stress primarily depend upon the type and direction of the original dislocations in the crystal and also upon the direction of maximum shear stress with respect to the crystal axis.

Although the amount of strain was not measured in the deformation tests, a rough estimate was made, based upon previous bending creep experiments /1/. In the experiments reported herein, the maximum shear stress thus calculated was approximately 100 Kg/cm^2, indicating that deformation was still in the incubation stage in the creep process even after the specimens had been strained for several hundred seconds and, therefore, that the topographs taken show the behaviour of dislocations in a very early stage of deformation. The topographs taken of the second specimen (figs.5, 6, 7,8), show a bunch of screw dislocations with $\frac{a}{3}\langle 2\bar{1}\bar{1}0\rangle$ Burgers vectors. If it is assumed that the upper front of a group of dislocation of this kind is a moving front of dislocations generated under the edge of the loading wedge and further assumed that this front moves only when the specimen is stressed, the velocity of dislocation motion can be evaluated. This may also be done with the movement of dislocations from the source of stress concentration illustrated by fig.7. This velocity was calculated to be of the order of 10^{-3} cm/sec at a temperature of -22°C and a stress of 100 kg/cm^2. If Johnston's equation, $\dot{\varepsilon} = 2bnv$, is used, this value gives a reasonable strain rate at this stage of creep. The velocity of the original dislocations in the crystal appears to be much slower.

More precise observations and measurements are now under way using better ice single crystal specimens which will make it easier to distinguish the behaviour of new dislocations generated under stress.

Acknowledgements

The authors wish to thank Prof.K.Kohra, University of Tokyo, for his kind comments on the X-ray apparatus and topography. This research was supported by the Scientific Research Fund of Ministry of Education, Government of Japan.

References

/1/ Higashi A., S.Koinuma, S.Mae, Jap.J.Appl.Phys. 4, 575-582 (1965)

/2/ Higashi A., S.Koinuma, S.Mae, Jap.J.Appl.Phys. 3, 610-616 (1964)

/3/ Higashi A., S.Mae, A.Fukuda: Proc.Int.Conf. on Strength of Metals and Alloys, (Sept. 1967, Tokyo), The Japan Inst. of Metals, 784-789 (1968)

/4/ Glen J.W., S.J.Jones, Physics of Snow and Ice (Proc.Int.Conf. on Low Temperature Science), Vol.1, Part 1, 267-275 (1968)

/5/ Higashi A., Physics of Snow and Ice, Vol.1. Part 1, 276-289 (1968)

/6/ Hayes C.E., W.W.Webb, Science 147, 44-45 (1965)

/7/ Fukuda A., A.Higashi: To be published in Jap.J. Appl.Phys.

/8/ Higashi A., M.Oguro, Oyo Buturi 36, 988-994 (1967) (Text in Japanese with English summary)

TENSILE AND FLEXURE PROPERTIES OF SALINE ICE

J.E. Dykins

U.S. Naval Civil Engineering Laboratory

Port Hueneme, California 93041 / USA

Abstract

Laboratory tests were conducted on axially loaded tensile specimens to define the boundary curves for the strength envelope spanning the salinity range of normal seawater to brackish-water (4- to 5-parts per thousand) ice for both horizontally and vertically oriented specimens. These tests covered the temperature range from -4°C to -27°C. It was found that the grain (crystal) size had little effect on the strength; however, the orientation of the grain and subgrain (dendrite) structure in relation to the stress orientation had an appreciable effect. A good correlation exists for the effect of temperature and salinity by treating the strength as a function of the brine volume to the one-half power.

The flexure strength tests of the annual sea ice at McMurdo, Antarctica, has been investigated for seasonal change (period of increasing thermal activity) for November through mid-February. The accumulated data includes results of large in-situ, and medium and small beams covering a temperature range of -7.2°C to -2°C. Analysis of the field data for flexure strength as a function of the brine volume to the one-half power results in a best line fit which has an almost identical slope as found for the horizontal tensile data. The flexure strength-zero brine intercept was 2.1 kg/cm^2 higher than the horizontal tensile strength-zero brine intercept.

1. Introduction

An essential to the development of analytical methods for predicting the structural capacity of an ice system is a qualitative understanding of the solid continuum behavior of the material for a given set of boundary conditions. The tensile and rupture modulus (flexure strength) are thus two of the several mechanical properties that are of considerable interest. This paper covers research investigating these properties as a function of temperature and salinity.

Laboratory tests were conducted on axially loaded tensile specimens to define the boundary curves for the strength envelope spanning the salinity range of normal seawater to brackish-water (4- to 5-parts per thousand) ice (figure 1). The tests cover the temperature range from -4 °C to -27 °C. These results have been compared with laboratory and field obtained flexure strength of natural sea ice.

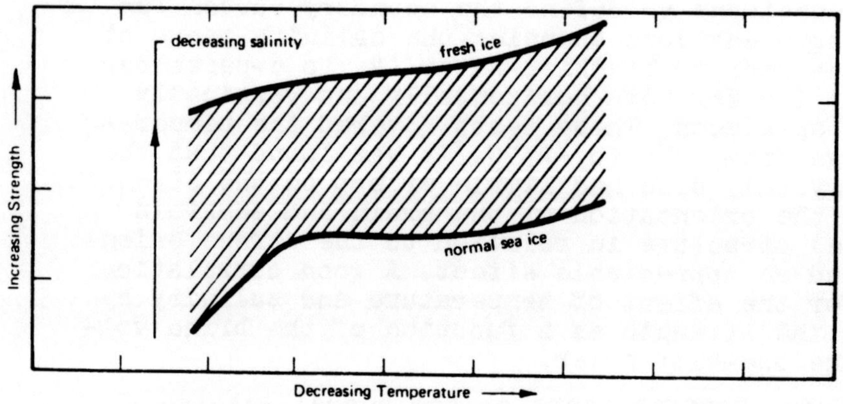

Fig.1 Envelope of tensile strength as function of temperature and salinity

Due to the lengthy nature of this research, it has been desirable from time to time to report work completed to date. Thus some of the material contained in this writing has appeared in previous publications, i.e., the tensile data pertaining to polycrystalline ice frozen from natural seawater, NCEL Technical Report R-415 of January 1966, and in the Proceedings of the International Conference on Low Temperature Science, 1966, Vol. 1, Part 1, Hokkaido University, Sapporo, Japan.

2. Experimental Procedure

The chamber used for the laboratory experimental work maintained a cyclic test temperature within -20 to -15°C, however, specimen temperature was little affected due to the cycle frequency. Air velocities caused by the refrigeration fans measured 60 cm above the surface of the freezing tank approximately 180 cm above the floor, ranged from 1.4 to 6.1 km/h. The surface exposure of the freezing tank was 163 by 257 cm. The tank walls and floor were insulated to produce uni-directional ice growth from the surface, duplicating natural conditions. The thermal activity through the water column, ice layer, and air immediately above the ice was automatically recorded at hourly intervals. The thermocouples for this purpose were at 7.62 cm spacing.

When the ice growth approached 56 cm, the freezing cycle was interrupted, and the chamber temperature was adjusted to the test schedule. The ice block was divided into quadrants designated for extraction of either horizontal or vertical specimens. Obtaining horizontal specimens required sawing the quarter block free and elevating it from the main block.

Test temperatures. Selection of the test temperatures was based on determining the strength of the ice in relation to the eutectic temperature of the two major salts in normal seawater; these being sodium sulfate ($Na_2SO_4 \cdot 10H_2O$) with a eutectic point of -8.2°C, and the major salt, sodium chloride ($NaCl \cdot 2H_2O$), with a eutectic point of -23.3°C. The test temperatures were -4, -10, -20 and -27°C.

Salinity and density. Salinity of the ice blocks and the test specimens was determined by measuring the specific conductance of the melted sample. Density of the ice was determined by two methods. Because of the irregular shape of the broken tensile specimens, volume was determined by fluid displacement. For samples taken specifically for density, the volume was determined both by fluid displacement and measured dimension. All specimens were weighed on an optical precision balance scale having an accuracy of ± 0.03 g.

Tensile specimens. Sampling of the ice for tensile specimens for each of the test temperatures was programmed to provide a random method for selecting the time for extracting the specimen and the area of the block from which it was to be extracted. The sample population for each group of data represents ice from more than one tank.

Following a temperature change, the ice block temperature was permitted to become nearly stable before specimens were extracted. Further temperature stabilization of the specimen was provided by an overnight soak, generally ranging from 16 to 24 hours.

Tensile specimens were machined on a regular metal lathe for the bellshaped ends needed for gripping the specimen for the axially applied load. This work was done in a chilled room adjoining the chamber where the testing was performed. The temperature of this room approximated the test temperatures of -4 °C and -10 °C but for the two lower test temperatures of -20 °C and -27 °C, it generally ran 8 to 10 degrees warmer. After machining the specimen, which generally took under 10 minutes, it was returned to the test chamber where it temperature-stabilized for 3 to 4 hours before testing (figure 2).

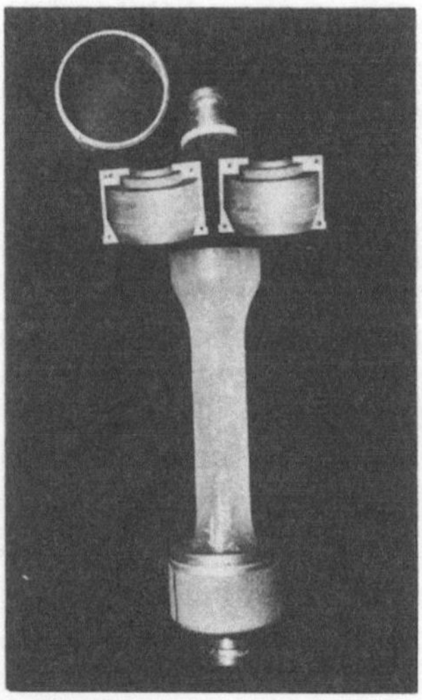

Fig.2 Tensile specimen and gripping heads

A gage-section area of 2 square inches was adopted after excessive breakage of the specimen in the test head was found when using a gage-section of 3 square inches. The majority of the specimens had gage lengths of 1 to 2.33 times the gage diameter; however, no discernible strength difference was found in a small group of specimens tested with shorter gage lengths.

All tensile specimens were tested on a universal 0- to 10,000-pound testing machine. The load frame of the machine was inside the chamber so that the specimen would not have to be removed to a new temperature environment. Axial alignment of the test load was maintained by a standard universal knuckle in the linkage. The load rate used was 0.5 in./min, and it was found necessary to preload the specimen to approximately 10 psi (0.7 kg/cm^2) and hold for a few seconds to remove all slack in the system before applying the test load.

3. Ice Block Characteristics

Natural seawater with a mean salinity of 32.54 ppt was used to determine the tensile values of the 7- to 9-ppt ice (lower boundary curve). A mixture of distilled and natural seawater producing a solution of 4.4-ppt salinity was used to determine the tensile strength of 1- to 2-ppt brackish ice (upper boundary curve). Since the freezing tank was not equipped to maintain a constant water salinity, each of the ice types was a product of increasing water salinity as the growth progressed. This is defined as a closed, or confined, freezing system. Thus the top 20 inches of natural seawater ice was the product of water for which salinity ranged from approximately 33 to 53 ppt, while the brackish-water ice was a product of water with salinity ranging from approximately 4 to 11 ppt.

Salinity. Prior to extracting strength specimens, salinity specimens were taken at 1-inch increments to determine the vertical and horizontal salinity profiles. Representative vertical profiles for both types of ice are shown in figure 3. Each curve represents the average of three cores from individual tanks of ice. A consistent distinguishing feature between the profiles of the two ice types, aside from difference in actual salinity values, was the general uniformity of the salinity distribution below the upper 3-inch strata of the brackish ice as opposed to the ragged-appearing, multi-spiked profile of the natural seawater ice. This leads to a tentative assumption that the macro-structure of the natural seawater ice sufficiently differs from that of brackish-water ice to permit instability in brine suspension. However, as pointed out in the later discussion of petrographic features, this is not readily discernable. The 8- to 9-ppt salinity of the seawater ice beginning at about the 4-inch (10-cm) depth is fairly typical of salinities reported for young natural sea ice (Weeks and Lee /1/). The high-surface salinities which are converse to the low-

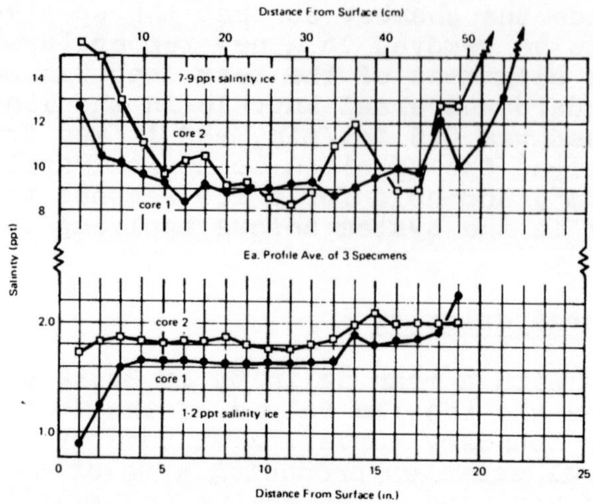

Fig.3 Vertical salinity profile

surface salinities found for brackish ice, are also in the general range of values found for young sea ice. No literature has been found to compare the bottom layer with a comparable thickness of natural sea ice.

The horizontal salinity profiles, though not included, showed a fairly uniform distribution and were in good agreement with the intersection of the vertical profile.

Density. Vertical density profiles of the ice of each tank were determined using 3-inch-diameter (7.62-cm) by 3-inch-long specimens taken prior to strength sampling. The curves shown in figure 4 are a 3-specimen average. Density of the natural seawater ice was found to center around 0.920 to 0.925 gm/cm^3, while the brackish ice

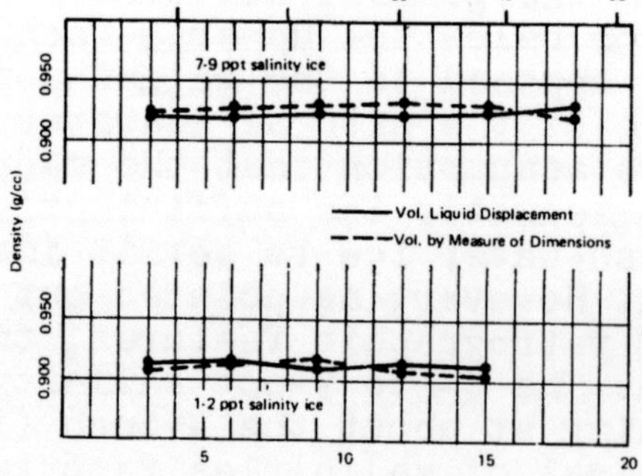

Fig.4 Vertical density profiles

was found to be slightly less dense, with mean values ranging from about 0.900 to 0.915 gm/cm^3.

Petrographic Structure. To determine the petrographic structure, horizontal thin sections, approximately 0.12 mm thick, were examined and photographed between crossed polaroids for grain structure (crystal) and under the microscope to observe the inclusions. Specimens with well-defined grain (crystal) and subgrain (dendrite) structure are shown in fig. 5 (natural seawater specimen) and fig. 6 (brackish-water specimen).

Fig.5 Horizontal thin section of natural seawater ice enlarged to show grain and subgrain structure. The grid is 1 cm on a side

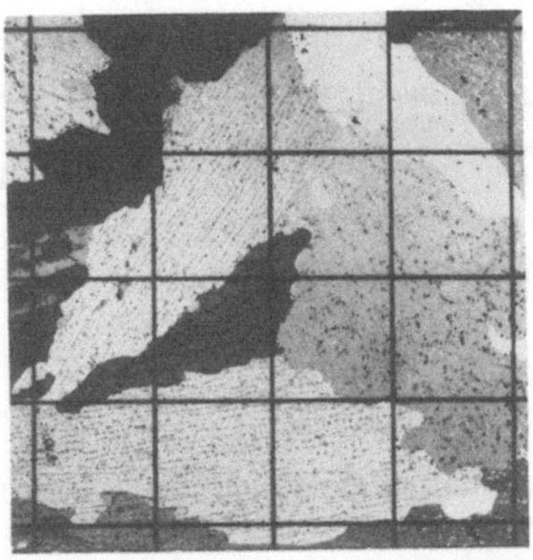

Fig.6 Horizontal thin section of brackish ice enlarged to show grain and subgrain structure. The grid is 1 cm on a side

A review of the literature on investigations of young sea ice (Bennington /2/; Weeks and Hamilton /3/; Anderson /4/) indicated that the natural seawater ice grown in the laboratory was essentially typical of naturally formed sea ice. As would be expected for the laboratory ice, there was the absence of the normal surface zone of slush ice resulting from agitation and snow pollution.

The quantity of samples necessary to make a statistical analysis prevents other than the general assessment of the difference in petrographic structure between the natural and brackish-water ice types. Less twining of the grains with less irregularity of the boundaries were identifying characteristics of the brackish ice. Grain size variation with depth was also much less evident for the brackish ice (figure 7). Variation in the center-to-center spacing of the dendrite (platelet width) with depth was not notable in either ice type. The platelet spacing throughout the 4- to 40-cm strata for natural seawater ice ranged from about 0.3 to 0.5 mm, while for this same thickness of brackish ice, the platelet spacing ranged from about 0.25 to 1.0 mm.

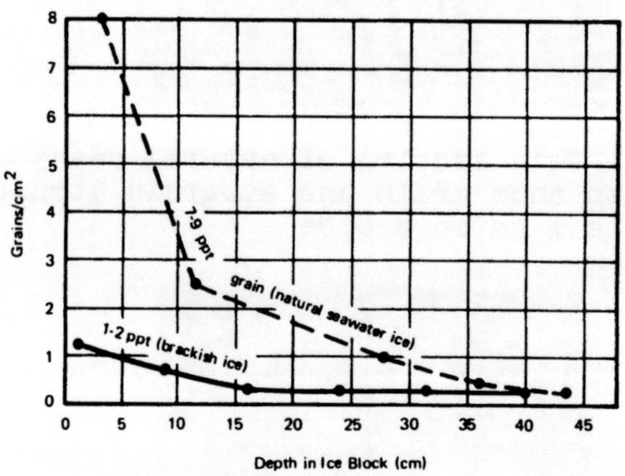

Fig.7 Increase in grain (crystal) size with depth

4. Tensile-Strength Analysis

The tensile strength was investigated for dependency of strength on temperature, grain (crystal) orientation and subgrain (dendrite) structure, brine volume, and depth. The tests were conducted to determine the strength with the stress field applied across the dendrites (specimen extracted from horizontal plane of the ice block), and with the stress field applied along the longitudinal axis of the dendrite (specimen extracted from vertical plane). The direction of applied stress on the horizontal specimens represents the relationship of bending stress to grain orientation in surface loading of an ice sheet.

The tensile data are summarized in table 1. The table gives the mean tensile strength, a summary of statistical analysis, data dispersion, mean density, and mean salinity for each sample in relation to orientation and temperature. The analysis is based on normal distribution; however, most of the data groups only approximate the normal and are actually skewed to the right. Analysis by normal distribution was selected after finding no other transformation technique that would improve the fit. It is believed that the standard deviation by normal distribution with the 95% confidence limit reasonably represents the dispersion about the mean strength.

In figure 8, it is seen that the mean tensile strength representing horizontal orientation of natural seawater ice is 41 psi (2.9 kg/cm^2) for -4 °C, 67 psi (4.7 kg/cm^2) for both -10°C and -20°C, and 78 psi (5.5 kg/cm^2) for -27°C. From the same figure, we find the mean tensile strength of brackish-water ice as follows: 64 psi (4.5 kg/cm^2) for -4 °C, 94 psi (6.5 kg/cm^2) for -10 °C, 98 psi (6.9 kg/cm^2) for -20 °C, and 112 psi (7.9 kg/cm^2) for -27 °C.

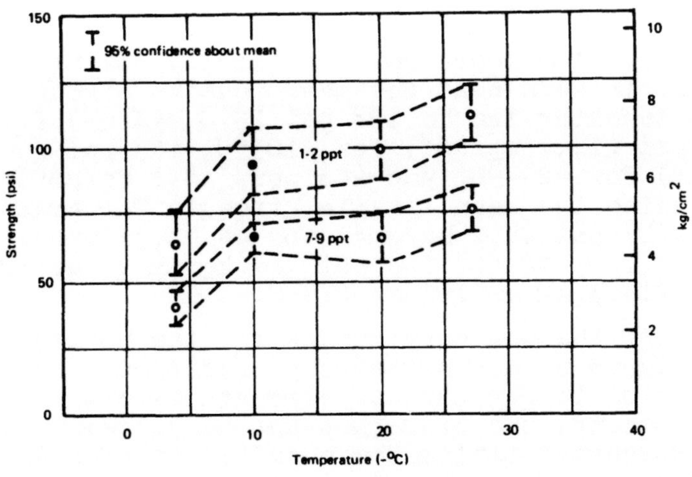

Fig.8 Mean horizontal tensile strength versus temperature, showing the effect of salinity

Table 1. Summary of Tensile Strength Data

Spec. Orientation w/Surface	Ice Temp. (°C)	No. of Spec.	Mean Tensile Strength (psi)	Standard Error of Mean (psi)
			(Natural Seawater	
Horizontal	-4	27	41	2.96
"	-10	47	67.4	2.92
"	-20	47	66.9	4.44
"	-27	47	77.8	4.45
Vertical	-4	86	128	4.0
"	-10	42	152.3	5.60
"	-20	62	163.1	5.76
"	-27	21	208	12.39
			(Brackish-Water	
Horizontal	-4	20	64	5.48
"	-10	26	94	6.27
"	-20	32	98	6.08
•• "	-27	38	112	5.74
Vertical	-4	41	87	3.32
"	-10	58	188	5.52
"	-20	61	198	7.00
"	-27	47	221	4.75

• Numbers in parentheses indicate number of specimens.
•• Data not yet published.

From figure 9, which represents vertically oriented specimens, it is seen that the mean tensile strength for natural seawater ice is 128 psi (9.0 kg/cm^2) for -4 °C, 152 psi (10.7 kg/cm^2) for -10 °C, 163 psi (11.5 kg/cm^2) for -20 °C, and 208 psi (14.6 kg/cm^2) for -27 °C. Likewise the mean tensile strength for brackish-water ice is 87 psi (6.1 kg/cm^2) for -4 °C, 188 psi (13.2 kg/cm^2) for -10 °C, 198 psi (14 kg/cm^2) for -20 °C, and 221 psi (15.5 kg/cm^2) for -27 °C.

Two features of the boundary curves for the horizontal specimens (figure 8) are noteworthy. First the non-linear effect of temperature on strength is clearly evident and second, the near-constant displacement between the two curves implies the possibility of a linear salinity effect on strength. If it were not for two anomalies found in the boundary curves for the vertically oriented specimens (figure 9), they would have

95% Confidence Limit on (mean)		Range of Observations		Mean	Mean
Lower (psi)	Upper (psi)	Lower (psi)	Upper (psi)	Density (gm/cm³)·	Salinity (ppt)·
Ice 7-9% Salinity)					
35	47	10	62	0.876(60)	5.30(27)
62	73	30	109	0.919(42)	7.59(37)
58	76	22	148	0.927(47)	8.47(45)
69	87	33	164	0.926(47)	8.44(41)
120	136	82	187	0.892(98)	5.87(86)
141	164	67	253	0.922(41)	6.86(41)
152	175	65	247	0.924(48)	8.32(45)
182	234	98	302	0.936(21)	8.85(19)
Ice 1-2% Salinity)					
52	76	19	106	0.914(19)	1.67(20)
81	107	19	150	0.911(26)	1.60(26)
87	110	56	167	0.926(31)	1.76(32)
100	124	46	184	0.928(38)	1.63(38)
80	94	35	128	0.907(41)	1.46(41)
177	199	85	271	0.915(58)	1.53(58)
184	212	68	325	0.920(61)	1.66(60)
211	230	109	340	0.916(47)	1.54(47)

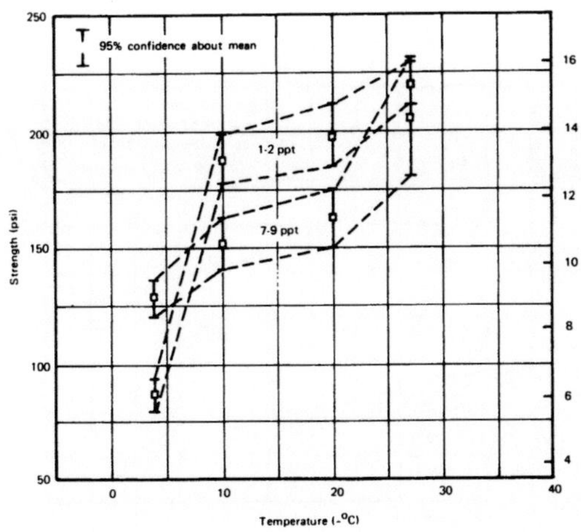

Fig.9 Mean vertical tensile strength versus temperature, showing the effect of salinity

much the same characteristic as the horizontal curves. These anomalies are: (1) the low-strength value of the vertical specimens found for the 1- to 2-ppt salinity tested at -4 °C, which presently has no explanation, and (2) the high-strength value of the 7- to 9-ppt-salinity vertical specimens tested at -27 °C, which from the statistical point, however, is not real if the data is viewed at the 95% confidence level.

A method of relating the effect of salinity on strength is by determining the brine volume present in the ice as a function of salinity and temperature (Assur /5/). The data in figure 10 has been plotted, using average strength versus average brine volume for the specimens at each test temperature. A significant point of interest occurs by converting the temperature-salinity effect into a single function, brine volume, and plotting the two distinct data groups, 7- to 9-ppt and 1- to 2-ppt ice; the horizontal data configure into a continuous curve of almost constant slope. A least squares line fit of the horizontal data results in an equation of the following form:

$$\text{Tensile Strength} = 120.68 - 10.214 \sqrt{\text{Brine Volume} \times 10^3}$$

The fitted line, together with the 95% confidence level, are shown in figure 10. The high absolute value of the correlation coefficient, 0.9219, indicates a close relationship between the horizontal tensile strength and the square root of the brine volume. The intercept of the best fit line predicts that the maximum horizontal tensile strength of ice at zero brine content to be 120.68 psi (8.5 kg/cm^2).

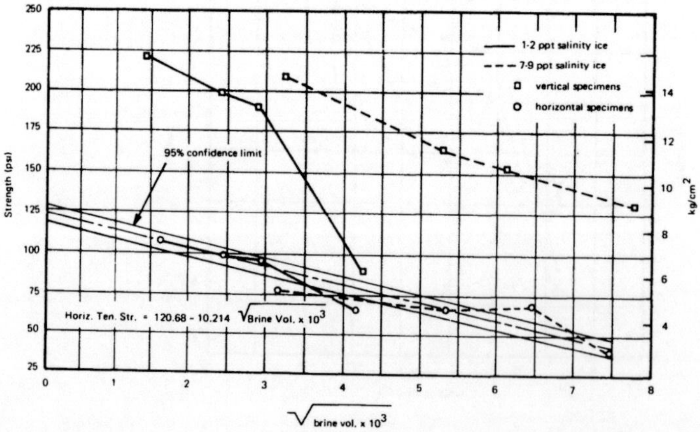

Fig.10 Mean tensile strength versus brine volume

For record and information, the calculated best fit line for vertical specimens (not plotted in figure 10) was made by first including the data for -4 °C, 1- to 2-ppt ice, and then excluding the point. The equation resulting from including the point was:

$$\text{Vertical Tensile Strength} = 228.0 - 14.308 \cdot \sqrt{\text{Brine Volume} \times 10^3}.$$

The equation resulting from excluding the point was:

$$\text{Vertical Tensile Strength} = 238.53 - 14.090 \cdot \sqrt{\text{Brine Volume} \times 10^3}.$$

An excellent correlation between strength and brine volume is indicated by high order of the correlation coefficient, 0.9708, when the point is excluded. A much lower correlation coefficient, 0.6748, results from including the data point. In comparing the two equations, however, the difference is quite insignificant, amounting to only about a 4.5% strength difference at the zero brine intercept. The slope difference between the two lines can also be considered insignificant.

To examine the tensile strength for correlation with grain (crystal) structure vertical profiles of the strength of the ice block were plotted (figures 11 and 12). The profiles relate the break face of the specimen to its in-situ position by 4-inch strata.

Since the profiles indicate a fairly uniform strength throughout the observed strata, and we are aware from examining the petrographic structure (figure 7) that the grain size of the natural seawater ice increased appreciably with depth while the grain size of the

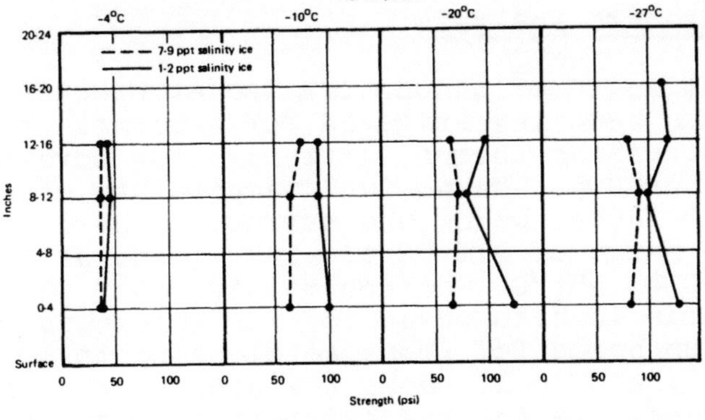

Fig.11 Horizontal tensile strength versus depth strata

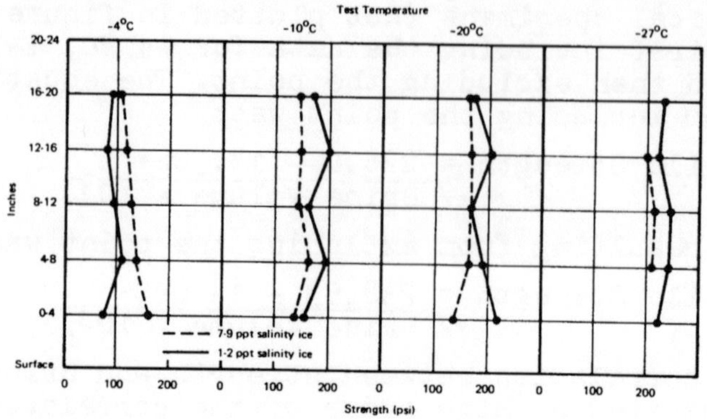

Fig.12 Vertical tensile strength versus depth strata

brackish ice was reasonably constant with depth, it is thus fairly conclusive that grain size had no appreciable effect on the strength.

The tensile strength relationship to grain (crystal) growth axis orientation with the stress field, however, is appreciable as indicated by the difference in horizontal and vertical strength.

It was also observed during testing of horizontal specimens that failure strength was commonly associated with orientation of the subgrain (ice platelets within crystal), e.g., the tendency was for higher strengths to be associated with subgrain orientations perpendicular to the failure plane, and conversely lower strength when parallel to failure plane of specimen. This was discussed in some detail in publications noted in the introduction section.

5. Flexure Strength Analysis

The flexure strength of annual sea ice at McMurdo, Antarctica, has been investigated for seasonal change (period of increasing thermal activity) for November through mid-February. The nominal temperature of a stratum 2 to 3 inches below the surface in early November is in the range of -12 °C to -14 °C, and at mid-February -3 °C to -5 °C. The accumulated data includes results of large in-situ beams (of ice thickness up to 8 feet), medium beams (of cross sections up to 60 square inches), and small beams (of cross sections 4 square inches). The apparatus for cutting and testing large in-situ beams is shown in figure 13.

Fig.13 Apparatus for cutting and testing large in-situ beams

The ice saw is capable of cutting ice up to 10 feet thick, and the load frame of applying loads in excess of 200,000 pounds. Load measurement is made with a strain-gage load cell.

The flexure strength data is summarized in table 2. All beams were failed by a center-point load applied at a rate which, in general, produced a stress rate in excess of 14 psi/sec (1 $kg/cm^2/sec$). Temperature effect on the strength is shown in figure 14. In addition to beam size, the data represents two conditions of temperature.

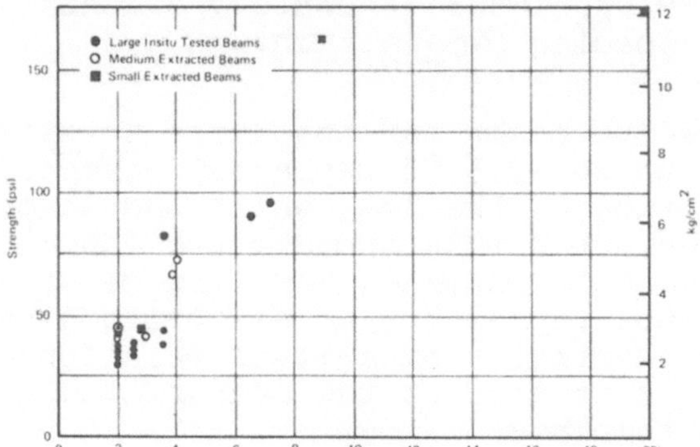

Fig.14 Flexure strength versus temperature

Table 2. Summary of Flexure Strength Data

Beam Type w/Center-Point Load	Beam Span (in.)	Beam Width (in.)	Beam Depth (in.)
Fixed-End	734.4	36.0	73.0
"	722.4	35.5	73.0
"	1072.8	40.4	84.3
"	1000.8	40.5	88.0
"	892.8	39.8	94.0
"	918.0	39.7	93.0
"	970.8	39.0	94.0
Simple-Supported	312.0	36.0	73.0
"	314.0	34.0	73.0
"	193.0	35.5	72.0
"	456.0	40.5	84.3
"	492.0	40.5	84.7
"	428.4	40.5	88.0
"	372.0	40.5	88.0
"	86.6	7.0	6.9
"	82.3	7.3	8.6
"	60.8	7.0	5.9
"	29.5	7.3	3.8
"	28.5	7.0	3.9
• "	16.0	2.0	2.0
•• "	17.0	2.0	2.0
••• "	17.0	2.0	2.0

• Average value for 28 specimens annual sea ice, McMurdo, Antarctica.

•• Average for 19 specimens laboratory brackish-water ice (rate of loading 0.5 in./min).

The small and medium beams and the large <u>in-situ</u> beams at temperatures up to -2.5 °C represent essentially an isothermal condition of the ice, while the large in-situ beams tested above this temperature (absolute value) have a thermal gradient. The flexure strength for this group thus represents an effective strength and not an exact strength-temperature relationship since the temperature is treated as the mean of the gradient across the beam thickness.

Relating the salinity effect on the flexure strength through conversion to brine volume is shown in fig.15.

Mean Temp. (-°C)	Mean Salinity (ppt)	Flexure Strength (psi)	$\sqrt{\text{Brine Vol.}}$(ppt)
2.5	5.47	36.5	10.30
3.5	5.47	38.5	8.75
2.0	5.11	38.0	11.09
2.0	5.11	33.0	11.09
3.6	5.5	82.0	8.65
6.5	5.5	90.0	6.65
7.2	5.5	95.0	6.36
2.5	5.47	39.2	10.30
2.5	5.47	34.4	10.30
3.5	5.47	44.4	8.75
2.0	5.11	33.8	11.09
2.0	5.11	30.5	11.09
2.0	5.11	46.0	11.09
2.0	5.11	40.5	11.09
2.0	5.2	42.0	11.18
2.0	5.9	46.0	11.92
3.0	5.2	41.0	9.18
3.9	5.2	66.2	8.12
4.0	5.2	77.6	8.03
2.8	5.2	43.4	9.49
9.5	1.84	161.0	3.26
20.0	1.98	173.0	2.59

***Average for 18 specimens laboratory brackish-water ice (rate of loading 0.5 in./min).

For this data, the equation for best line fit (by least squares) was:

$$\text{Flexure Strength} = 194.74 - 14.734\sqrt{\text{Brine Volume} \times 10^3}\ .$$

The high absolute value of the correlation coefficient, 0.9389, indicates a close relationship between the flexure strength and the square root of the brine volume. Figure 16 shows the interrelationship between the flexure strength and the horizontal tensile strength.

A point of interest at this early stage of developing the flexure strength property is that by eliminating the two laboratory obtained data points and considering only the field flexure data, the equation for the best line fit was:

$$\text{Flexure Strength} = 151.38 - 10.434\sqrt{\text{Brine Volume} \times 10^3}\,.$$

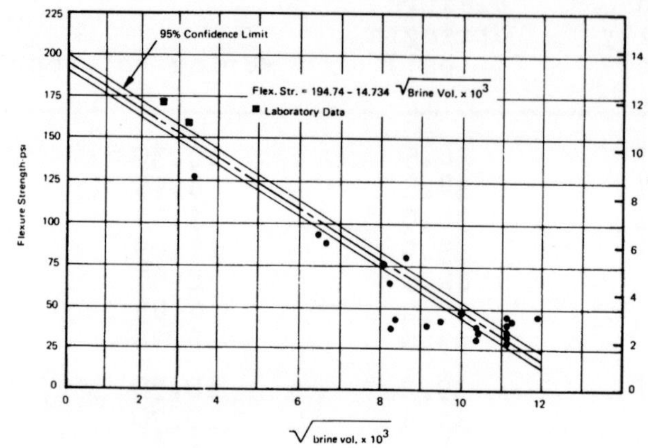

Fig.15 Flexure strength versus brine volume

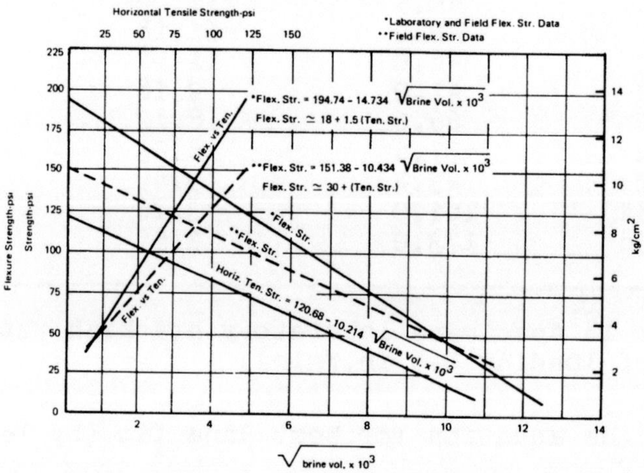

Fig.16 Flexure-tensile strength relationship

The correlation coefficient was 0.8266. It is noteworthy, however, to observe that the line of this equation has almost identically the same slope as the horizontal tensile strength, i.e., -10.434 and -10.214 respectively. Comparing the monograph data of Weeks and Assur /6/ (figure 3) for natural sea ice cantilever beams with the above flexure data which includes both fixed-end and simple beams, we observe a consistent strength difference amounting to about 50% greater values for the fixed-end and simple beam data of comparable brine volume.

6. Summary

The characteristics which included salinity, density, and petrographic structure of the laboratory grown natural seawater ice did not show an appreciable difference from reported values for young natural sea ice. When comparing the laboratory grown brackish-water ice with the characteristics of the natural seawater ice, aside from the lower salinity, the principal difference was the predominance of larger crystal size throughout the thickness and less twining, or interlocking, of the crystal boundaries.

Neither the grain size nor the center-to-center spacing of the subgrain structure were identified to have much influence on the tensile strength. By contrast, however, the orientation of the grain structure in relation to the stress field had a significant influence, i.e., the vertical tensile strength, in general, was from 2 to 3 magnitudes greater than the horizontal strength. A significant influence on the horizontal tensile strength was also found to result from the orientation of the subgrain structure in relation to the failure plane of the specimen. When alignment of the subgrains was parallel with the failure plane, a low tensile strength resulted, but as the alignment of the subgrains approached 90 degrees to the failure plane, a high strength value was observed. This observed behavior for the horizontal specimens explains the reasons for much of the scatter found in the horizontal strength data. Although this had a pronounced effect on specimen strength, it would be of lesser consequence in causing strength variation of a large ice mass because of the nonpreferred orientation of the subgrain structure, and thus, the random orientation would tend to normalize this effect for a large failure surface. The data indicates that neither the horizontal nor vertical tensile strength is a linear function of the temperature; however, there appears to be a very good linear correlation between the horizontal tensile strength and the brine volume to the one-half power. The vertical tensile strength correlation with brine volume was less apparent due to the divergency of two of the data points, which clearly need additional investigation for verification.

The flexure strength data, though presently of limited quantity, shows the same strong trend as the tensile strength to behave as a linear function of the brine volume. By this analysis technique, the data definitely indicates the value of the flexure strength property for a given set of boundary conditions will be higher than the horizontal tensile strength property. This is not surprising, however, as other materials have a similar characteristic, e.g., concrete.

Though the brine volume technique appears to be a worthy method of correlating strength for the effect of temperature and salinity, there is presently a limitation that must be recognized. This author believes, however, this may only be of consequence when considering the strength properties of the ice following the period it first reaches thermal saturation at the melting temperature. After it reaches this condition, the degree of internal melting would be a time-dependent process, with the liquid phase theory becoming increasingly less valid. For example, the annual ice sheet in the bay at McMurdo, Antarctica, reaches near-isothermal state generally around mid-December and maintains this temperature condition until mid-February; at this time, the ice generally breaks up. The net heat flux into the McMurdo ice, however, is probably less than for many arctic regions, based on the observations that very little surface melt or deterioration occur during this isothermal period.

Our future plans are to continue with research on the mechanical properties of ice as an essential to the development of analytical methods for predicting the structural capacity of an ice system for any given set of boundary conditions.

References

/1/ Weeks, W.F., O.S. Lee, J. Arctic Inst. North Amer. 15, 93-108 (1962)

/2/ Bennington, K.O., J. Glaciol. 4, 669-688 (1963)

/3/ Weeks, W.F., W.L. Hamilton, CRREL Res. Rept. 101, 1-11 (1962)

/4/ Anderson, D.L., Research 13, 310-318 (1960)

/5/ Assur, A., Arctic Sea Ice. Nat. Acad. Sci.-Nat. Res. Council U.S.A. Publ. 598, 106-138 (1958)

/6/ Weeks, W.F., A. Assur, CRREL Res. Rept. 11-C3, 1-80 (1967)

INFRARED SPECTRUM OF ICE Ih IN THE RANGE 4000 TO 15 CM^{-1} *

E. Whalley

Division of Applied Chemistry

National Research Council Ottawa, Canada

Abstract

The first part of this contribution concerns the origin of the well-known difference between the optical and microwave refractive indices of ice, which are 1.31 and 1.78 respectively, corresponding to permittivities of 1.72 and 3.1. To determine the origin of this difference, J.E.Bertie and I have measured the absorptivity of ice Ih over the range 4000 to 30 cm^{-1}. The contributions of the various bands to the microwave refractive index have been evaluated by means of the Kramers-Kronig relations. About three quarters of the infrared contribution to the refractive index arises from the translational lattice vibrations below about 325 cm^{-1}, and about 15% arises from the rotational vibrations whose absorption is centred near 850 cm^{-1} but extends from about 1040 to 525 cm^{-1}. An approximate theory of the infrared contribution to the permittivity, analogous to Szigetti's theory for ionic crystals, suggests that the effective charge of a water molecule is about 0.3 e when e is the electronic charge. The second part of this contribution concerns the absorption of light by the sound waves of ice. Since ice is on orientationally disordered crystal, all its vibrations are infrared active. Hence, the spectrum of the band of translational vibrations below 325 cm^{-1} is closely related to the density of vibrational states. The

---

* N.R.C. No. 10507

maximum in the transverse acoustic branch is at about 65 cm^{-1}, so that below about 50 cm^{-1} are the sound waves. These are infrared active. A theoretical discussion and experimental measurements will be presented.

1. Introduction

The work described in this paper will be published in full in two papers /1,2/ and only a summary will be given here.

In a perfect crystal, the lattice excitations can be expanded as plane waves characterized by wave vectors. The excitations can then be described as quasi-particles known as phonons. The interaction of the nuclear vibrations with light is strictly subject to the selection rule that the sum of the wave vectors before and after the interaction is zero. For absorption then the wave vector of the photon absorbed is equal to the wave vector of the phonon produced. Thus, out of the 10^{25} or so normal vibrations of a mole of crystal, only a handful can interact with light as fundamentals.

Ice is not a perfect crystal. On the contrary, the molecules are orientationally disordered, and this means that the lattice excitations cannot be expanded in plane waves. In general, therefore, the excitations cannot be characterized by wave vectors, and so wave vector cannot be conserved. Instead, all 10^{25} or so normal vibrations can and do interact with light. The interpretation of the infrared spectrum of ice is therefore quite different from that of a perfect crystal.

In general terms the fundamental spectrum is reasonably well understood. The O-H stretching vibrations are centred about 3300 cm^{-1} and cover the range approximately 3550 to 3000 cm^{-1}. The band is so wide, no doubt, because of both intramolecular and intermolecular coupling of individual O-H vibrations. There is at present no satisfactory detailed interpretation of the band /3/.

The bending vibrations of the water molecule absorb near 1600 cm^{-1}, and little else is known about them. The IR-band of the rotational vibrations is centred about 850 cm^{-1} at 100°K and occupies the range 1040 to about 525 cm^{-1}. Here again there is clearly very strong coupling of the motion of different molecules, but a detailed understanding is lacking.

Finally, below about 325 cm^{-1} the translational vibrations of water molecules absorb. In contrast with the bands at higher frequencies, the origin of the features of the translational band is rather well understood in detail /4,5/. The main reason for this is that the orientational disorder probably plays only a minor role in determining the form of the normal vibrations. For example, if a hydrogen bonded O---O bond is stretched, the potential energy depends hardly at all on which oxygen atom the hydrogen is attached to. The vibrations can therefore be expanded in plane waves, and the concept of a phonon and a wave vector is again useful. Much information can thus be transferred from the known density of phonon states in crystals of similar structure.

Although the phonons are related to those of crystals with similar structure, the interaction of the phonons with light is quite different /4/. The reason for this is that while the potential energy of stretching an O---O bond depends little on which oxygen atom the hydrogen is attached to, the dipole moment change depends strongly. For the two bonds

O-H---O

and

O----H-O

the change of dipole moment on stretching must, by symmetry, be of the same magnitude but of opposite sign.

Since crystallographically equivalent pairs of oxygen atoms have either of these hydrogen atom configurations essentially at random, the dipole moment density induced by the phonons cannot be expanded in plane waves. This means that wave vector is not conserved, and all the normal vibrations of the crystal can absorb light. The infrared spectrum is therefore closely related to the density of vibrational states. Theory /4/ suggests that the density of states is more closely related to the absorptivity divided by the frequency squared than to the absorptivity itself, and features that can be associated with features in the density of states such as the TO, LO, LA, and TA maxima, the minimum between the LO and LA branches, and so on can be clearly recognized and can be fitted to a simple force field /5/.

The object of the work reported in this paper is to contribute towards the understanding of the detailed electrical properties of the vibrations of ice in two ways. The main object of the first was to determine the origin of the rather large difference of 0.47 between the optical (1.31) and microwave /6/ (1.78) refractive indexes. This has been accomplished /1/ by measuring the absorptivity of ice in the range 4000 to 30 cm^{-1} and calculating the contribution Δn of the various regions to the low-

-frequency refractive index by means of the Kramers-Kronig relations. It is, in fact, possible to determine the origin of the polarization in qualitative terms by rather simple considerations, as is shown in the next section.

The object of the second investigation/2/ was to measure the absorption of light below 50 cm^{-1} where the excitations are high-frequency sound waves. The measured absorptivity is compared with a theory /2/ which relates it to the difference of limiting permittivities at high and low frequency of the translational band, and the Debye temperature determined from heat capacities.

2. A simple analysis of the infrared contributions to the refractive index

A rough idea of the origin of the infrared contributions to the refractive index can be obtained from the following consideration of a very simple model. Each spectral region of ice in which absorption is strongis represented by a set of N_i uncoupled diatomic oscillators, each having wave number ν_i. In the actual crystal they are of course coupled, but the coupling will have only a slight effect on the refractive index at frequencies much smaller than the band maximum. The reduced mass of the oscillator is m_i and the effective charge is $f_i e$, defined in terms of the dipole moment μ induced by a displacement x of the oscillator by the relation

$$\mu = f_i \, ex ,$$

where e is the electronic charge.

According to the Kramers-Kronig relations the contribution $\Delta n(i)$ of an absorption band i to the low-frequency refractive index is

$$\Delta n(i) = \frac{1}{2\pi^2} \int_{\text{band } i} \frac{K(\nu)}{\nu_i^2} \, d\nu . \qquad (1)$$

For approximate purposes the band width can be assumed to be much smaller than the frequency, so that equation (1) becomes

$$\Delta n(i) = \frac{1}{2\pi^2 \nu_i^2} \int_{\text{band } i} K(\nu) \, d\nu . \qquad (2)$$

If for the present approximate purposes we neglect the effects of the internal field, the integrated absorptivity of one oscillator of the class i is

$$A_i = \frac{1}{N_i} \int_{\text{band } i} K(\nu)\, d\nu ,$$

$$= \frac{\pi}{3c^2} \left(\frac{\partial \mu}{\partial Q_i} \right)^2 , \qquad (3)$$

where Q_i is the normal coordinate, c the speed of light in vacuum, and N_i the number densitiy of oscillators. From equations (2) and (3)

$$\Delta n(i) = \frac{N_i}{6\pi c^2 \nu_i^2} \left(\frac{\partial \mu}{\partial Q_i} \right)^2 . \qquad (4)$$

The normal coordinate Q_i is

$$Q_i = \sqrt{m_i}\, x.$$

Hence equation (4) becomes

$$\Delta n_i = \frac{N_i f_i^2 e^2}{6\pi c^2 \nu_i^2 m_i} ,$$

which for ease of computation can be written for ice as

$$\Delta n_i = 0.25 \frac{v_i f_i^2}{M_i (10^{-3} \nu_i)^2} , \qquad (5)$$

where v_i is the number of oscillators per molecule contributing to the absorption band i and M_i is the reduced atomic weight of the oscillator.

The values of v_i, M_i, and ν_i are listed in table 1 for the O-H stretching, the rotational, and the translational bands, together with the square f_i^2 of the fractional charge that would have to move if all the infrared contribution to the low frequency refractive index were caused by the band i. Since f_i must certainly be much less than 1, only the translational band can contribute a large part of Δn. In spite of the crude model, the conclusion appears to be definite enough to be reliable.

Table 1 Effective charges of the oscillators as fractions f'_i of the electronic charge if an absorption region is to cause all the infrared polarization

Band i	v_i	M_i	ν_i/cm^{-1}	f'^2_i
O-H stretch	2	1	3300	10
Rotational	2	2	850	1.4
Translational	3	9	200	0.2

3. The absorptivity of ice

The maximum and minimum absorptivities of ice in the range 4000 to 30 cm^{-1} are in the ratio about 10^4. Consequently, several different thicknesses must be used to measure the absorptivity. No direct measurement of thickness has been made in this work, but instead the thickness was determined indirectly as follows.

The absorbances κ(= $\log_{10} I_o/I$ where I_o and I are the intensities before and after transversing the thickness of the sample) of several evaporated and carefully annealed films were measured in overlapping regions. All the curves were scaled to one particular curve, so the absorbance of one (as yet unknown) thickness was determined. The infrared contribution to the low-frequency refractive index is, according to the Kramers-Kronig relations

$$\Delta n = \frac{1}{2\pi^2} \int_{\text{infrared}} \frac{K(\nu)}{\nu^2} d\nu . \qquad (6)$$

In terms of the absorbance κ of the film of unknown thickness ℓ this can be written

$$\Delta n = \frac{2.303}{2\pi^2 \ell} \int \frac{\kappa(\nu)}{\nu^2} d\nu . \qquad (7)$$

All terms in this equation are known except ℓ, and consequently the thickness can be determined. The thickness was also determined by a second method in which the absorbance at the peak at 4940 cm^{-1} was compared with Ockman's /7/ directly measured value. The two methods yielded thicknesses of 0.86 and 0.88 μm respectively. The absorptivity so obtained is shown in fig.1.

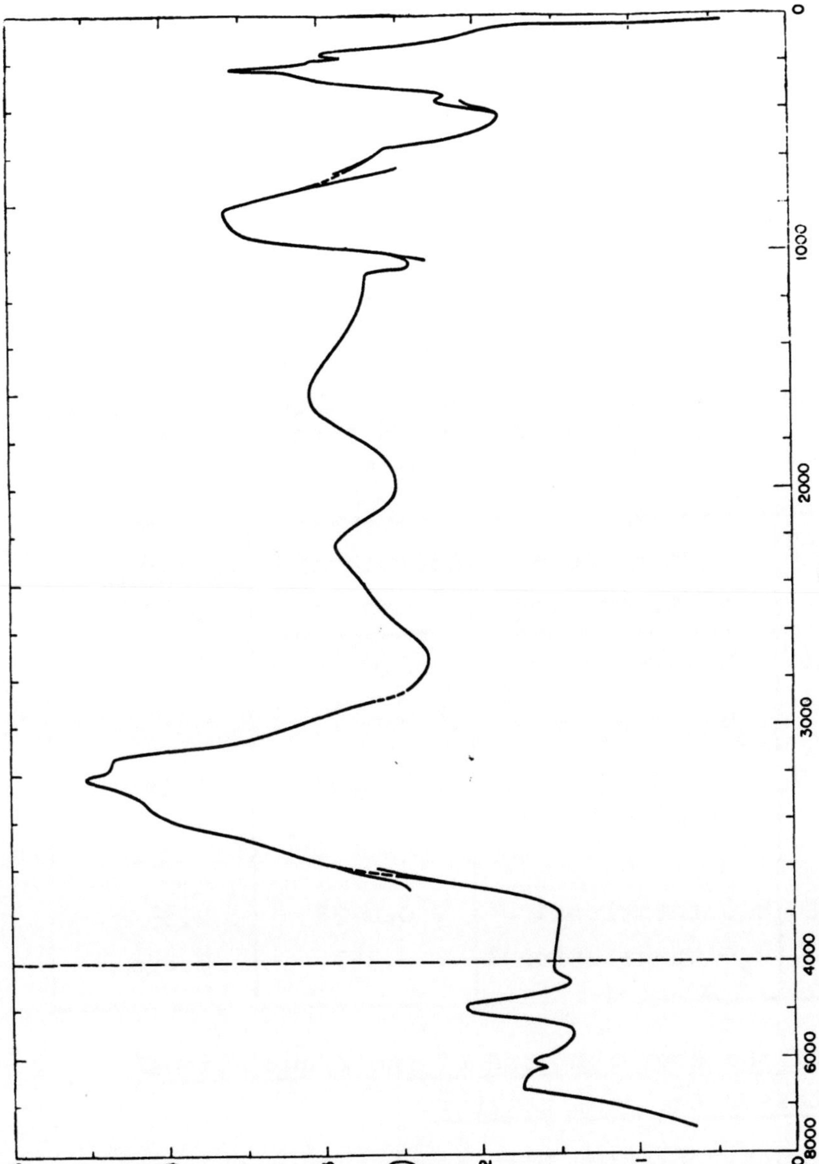

Fig.1 Absorptivity of H_2O ice Ih in the range 7000 to 30 cm^{-1}

The exact origin of the infrared contribution to the low-frequency refractive index can be most easily shown graphically by rewriting equation (6) to give

$$\Delta n = \frac{1}{2\pi^2} \int \frac{K(\nu)}{\nu} \, d \ln \nu . \tag{8}$$

The contribution of any region to Δn is then proportional to the area in that region under the curve of $K(\nu)/\nu$ against log ν. This is shown in fig.2. The region of the translational lattice vibrations below 325 cm^{-1} clearly contributes the most part of Δn. A somewhat more quantitative analysis is given in table 2. About 74% is caused by the translational band, although its integrated absorptivity, which is given in column 4 of table 2, is quite low compared with that of the O-H stretching band.

Table 2 Contributions of the various absorption regions to the microwave refractive index compared with the integrated absorptivities

Region, cm^{-1}	Main cause of absorption	Contribution to Δn	$10^6 \int K d\nu$ / cm^{-2}
3600 to 2900	O-H stretching	0.0309	6.86
2900 to 1000	H-O-H bending and combinations	0.0198	1.14
1000 to 320	H_2O rotational	0.0728	0.85
320 to 17	H_2O translational	0.347	0.238

4. Theory of the contributions of the translational vibrations to the permittivity

The experimental relation between resonance absorption and polarization in a static field is most easily and directly based on the absorptivity and the contribution to the refractive index as equation (1) shows: both sides can be measured directly. On the other hand, the most direct theoretical relationship between the low-

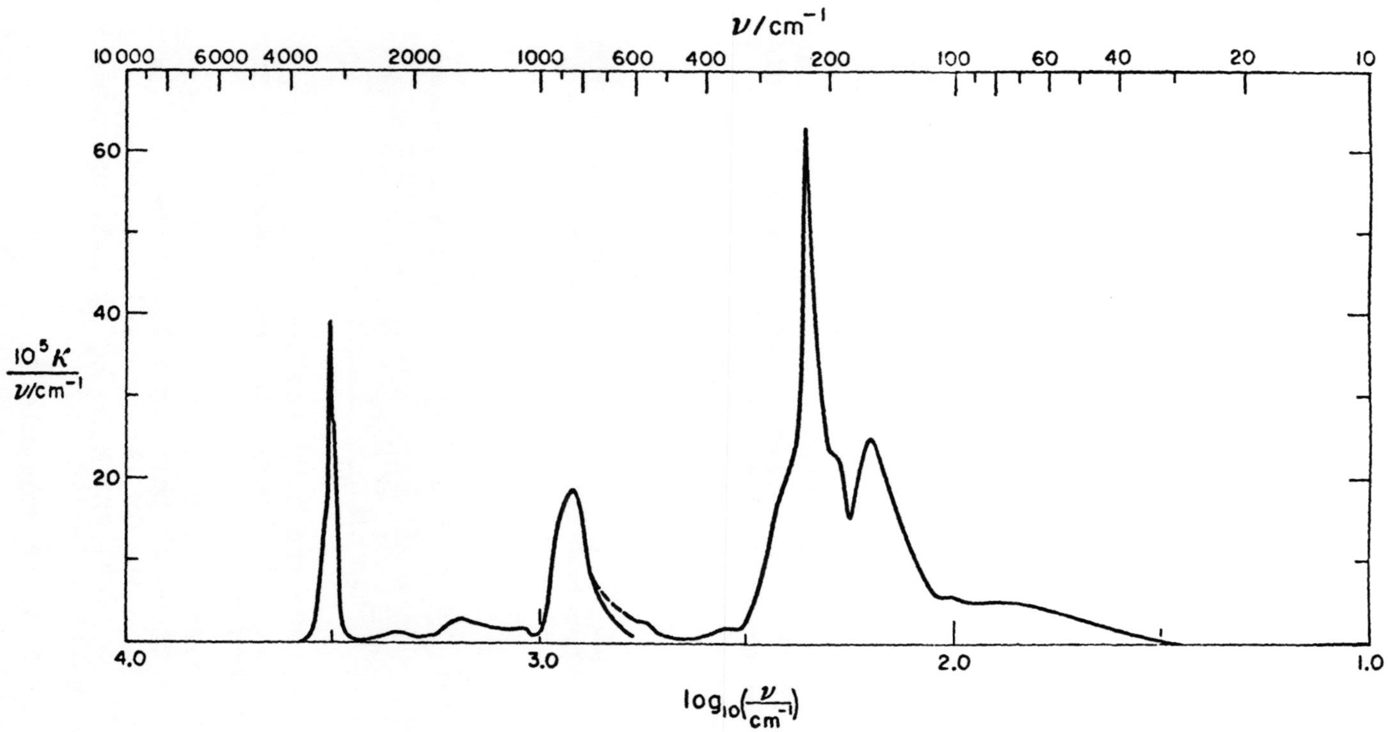

Fig.2 Graph of $\kappa(\nu)/\nu$ against $\log \nu$ for a film of undetermined thickness. According to eq. (8), the relative areas of bands measure their relative contribution to the low frequency refractive index

-frequency polarization and the molecular properties causing absorption is based on the permittivity. Szigetti /8/ and earlier workers have discussed the corresponding relation for harmonic cubic diatomic crystals in which only one vibration contributes to the infrared polarization. Here, we are dealing with an orientationally disordered crystal in which all vibrations are very weakly absorbing, and none is strongly absorbing.

The model assumed is that of a randomly oriented polycrystal that is isotropic on a macroscopic scale. The symmetry of the crystal is arbitrary. The dynamics and the electrical properties of the vibrations can be described in terms of the force constants ϕ_i for the displacements of the internal coordinates s_i, and the corresponding dipol moment derivates $\underline{M}_i$. On the basis of the remarks in the introduction, the values of ϕ_i are the same for the same kind of internal coordinate in different unit cells, in spite of the orientational disorder. The values of $\underline{M}_i$ will however not be the same in all unit cells. We will assume that the magnitudes are the same, but that they are in a number of directions whose unit vectors sum to zero. Any particular internal coordinate has one of the allowed directions of $\underline{M}_i$ at random. It is further assumed for simplicity that

$$R = M_i^2/\phi_i$$

is independent of i. It is not unreasonable that the stronger the force constant the larger the square of the dipole moment derivative, but it is not of course necessary that it be so. If only one force constant is important, this equation is trivially satisfied.

The development, which will not be given here, leads to the relation

$$\varepsilon_o - \varepsilon_\infty = 4\pi N \frac{\varepsilon_o + 2}{3} \frac{\varepsilon_\infty + 2}{3} R \qquad (9)$$

where ε_o and ε_∞ are the limiting low- and high-frequency permittivities of the absorption band, and N the number density of molecules.

The microwave permittivity of ice is 3.1 /6/, and the limiting high-frequency permittivity of the translational band is, from the data in the preceding section, $(1.78 - 0.35)^2 = 2.04$, and $\varepsilon_o - \varepsilon_\infty$ is 1.1. By using these values with equation (9) we find

$$R = 1.29\ \text{Å}^3\ \text{molecule}^{-1}.$$

The O---O stretching force constant in ice /5,1/ is about 0.16 mdyn Å^{-1}, so that the modulus of the dipole moment derivative M for the stretching of the O---O bond is

$$M = 1.4 \text{ Debye } Å^{-1}$$
$$= 0.3 \text{ e.}$$

The effective charge of the water molecules in ice for the O--O stretching vibration is therefore about 0.3 electrons, which is rather high for a crystal composed of neutral molecules. The value is within a factor 1.5 of that given in table 1 as a result of a very simple and crude theory. The molecules do not of course carry a real charge, and the effective charge measures the flow of charge when molecules are displaced relative to one another.

5. Absorption of light by the sound waves of ice /2/

A brief summary of the absorption of light by the translational lattice bands of ice I was given in the introduction. All the vibrations are active as fundamentals, and not only the zero-wave-vector vibrations that are the only active fundamentals in perfect crystals.

The maximum in the transverse acoustic branch occurs /5,1/ at about 65 cm^{-1}, and at frequencies well below this, the normal vibrations can be described as sound waves. The sound waves of ice should also absorb light, and the theory of the intensity should be particularly simple.

It is assumed that the potential energy V can be written as the sum of square terms of the internal coordinates

$$2V = \sum_{\ell,\kappa} \phi(\kappa)\, s^2(\ell,\kappa), \qquad (10)$$

where $s(\ell,\kappa)$ is the κth internal coordinate associated with unit cell ℓ and $\phi(\kappa)$ is the corresponding force constant. The change $\Delta\underline{\mu}$ of the dipole moment of the crystal for the change of the internal coordinates is

$$\Delta\underline{\mu} = \sum_{\ell,\kappa} \underline{M}(\ell,\kappa)\, s(\ell,\kappa), \qquad (11)$$

where $\underline{M}(\ell,\kappa)$ is the dipole moment derivative with respect to coordinate $s(\ell,\kappa)$. In an ordered crystal $\underline{M}(\ell,\kappa)$ is independent of the unit cell ℓ, but in a disordered crystal it is different for different ℓ because of the disorder.

$\underline{M}(\ell,\kappa)$ can be divided into a regular part $\underline{M}'(\kappa)$ that has the diffraction symmetry of the crystal. It is independent of ℓ, and is equal to the average value of $\underline{M}(\ell,\kappa)$ over all unit cells,

$$\underline{M}'(\kappa) = \frac{1}{N} \sum_{\ell} \underline{M}(\ell,\kappa), \tag{12}$$

where N is the number of unit cells. The difference $\underline{M}''(\ell,\kappa)$ between $\underline{M}(\ell,\kappa)$ and $\underline{M}'(\kappa)$,

$$\underline{M}''(\ell,\kappa) = \underline{M}(\ell,\kappa) - \underline{M}'(\kappa), \tag{13}$$

describes the electrical irregularity of the crystal. A non-zero value of $\underline{M}'(\kappa)$ causes the usual activity of low-wave-vector vibrations, whereas a non-zero value of $\underline{M}''(\ell,\kappa)$ causes all vibrations to be active. In a perfect crystal of course $\underline{M}''(\ell,\kappa)$ is zero.

If only one coordinate $s(\kappa)$ contributes to the potential energy and if $\underline{M}''^2(\ell,\kappa)$ has only one value $\underline{M}''^2(\kappa)$, then the integrated reduced absorptivity $A(\underline{k},j)$ of a normal vibration of wave vector $\underline{k}$ in branch j

$$A(\underline{k},j) = V \int_{\substack{\text{band} \\ \underline{k},j}} K(\nu)\, d\nu, \tag{14}$$

where V is the molar volume, $K(\nu)$ the absorptivity due to the vibration $\underline{k},j$, and ν the vacuum wave number of the light, is

$$A(\underline{k},j) = \frac{\pi}{3c^2} \frac{(n^2+2)^2}{9n} R(\kappa)\, \omega^2(\underline{k},j). \tag{15}$$

In the equation

$$R(\kappa) = M''^2(\kappa)/\phi(\kappa), \tag{16}$$

$\omega^2(k,j)$ is the circular frequency of vibration $\underline{k},j$, and the refractive index n allows for the effects of the material on the field. Equation (15) is also obtained if several coordinates contribute appreciably to the potential energy and the dipole moment, providing that $R(\kappa)$ is the same for each one.

By summing over all vibrations, assuming that each normal vibration has a band width small compared to the frequency interval over which the absorptivity changes appreciably, it can be shown that the absorptivity $K(\omega)$ at frequency ω is given by

$$K(\omega) = \frac{2\pi^2}{3Vc} \frac{(n^2+2)^2}{9n} R\omega^2 g(\omega), \tag{17}$$

where c is the velocity of light and g (ω) is the density of vibrational states at frequency ω.

If the frequency is low enough that the Debye distribution holds, then

$$g(\omega) = 9N \frac{\omega^2}{\omega_D^3}, \tag{18}$$

where ω_D is the Debye frequency.
The quantity R can be obtained in various ways. One is from the contribution of the translational band to the static permittivity according to equation (9). From equations (9), (17), and (18), in terms of the vacuum wave number ν of the light absorbed,

$$K(\nu) = 3\pi^2 f\ (\varepsilon_0 - \varepsilon_\infty)\ \nu_D \left(\frac{\nu}{\nu_D}\right)^4, \tag{19}$$

where

$$f = \frac{\varepsilon_0 + 2}{\varepsilon_\infty + 2} \frac{1}{\sqrt{\varepsilon_0}}$$

arises from the effects of the internal field, etc. The absorptivity is proportional to the fourth power of the wave number.

The preceding theory gives of course only the fundamental absorption. Szigetti's /9/ analysis suggests that the absorptivity of regular crystals due to difference bands at frequencies well below the Debye frequency and at low temperatures is given approximately by an expression of the kind

$$K(\nu) = \beta\ T\ \nu^2 \tag{20}$$

where β is a constant and T is the temperature. Although transitions other than those considered by Szigetti are active in orientationally disordered crystals, it is likely that the order-allowed transitions dominate the disorder-allowed transitions, and eq. (20) will be an adequate first approximation. At low enough frequencies, therefore, order-allowed difference bands will dominate the disorder-allowed fundamentals and the fourth-power dependence will change to a square dependence. The absorptivity in the square region will be approximately proportional to the temperature. For ice,

$$\varepsilon_0 = 3.1,\ \varepsilon_0 - \varepsilon_\infty = 1.1,\ \text{and}\ \nu_D = 157\ \text{cm}^{-1}.$$

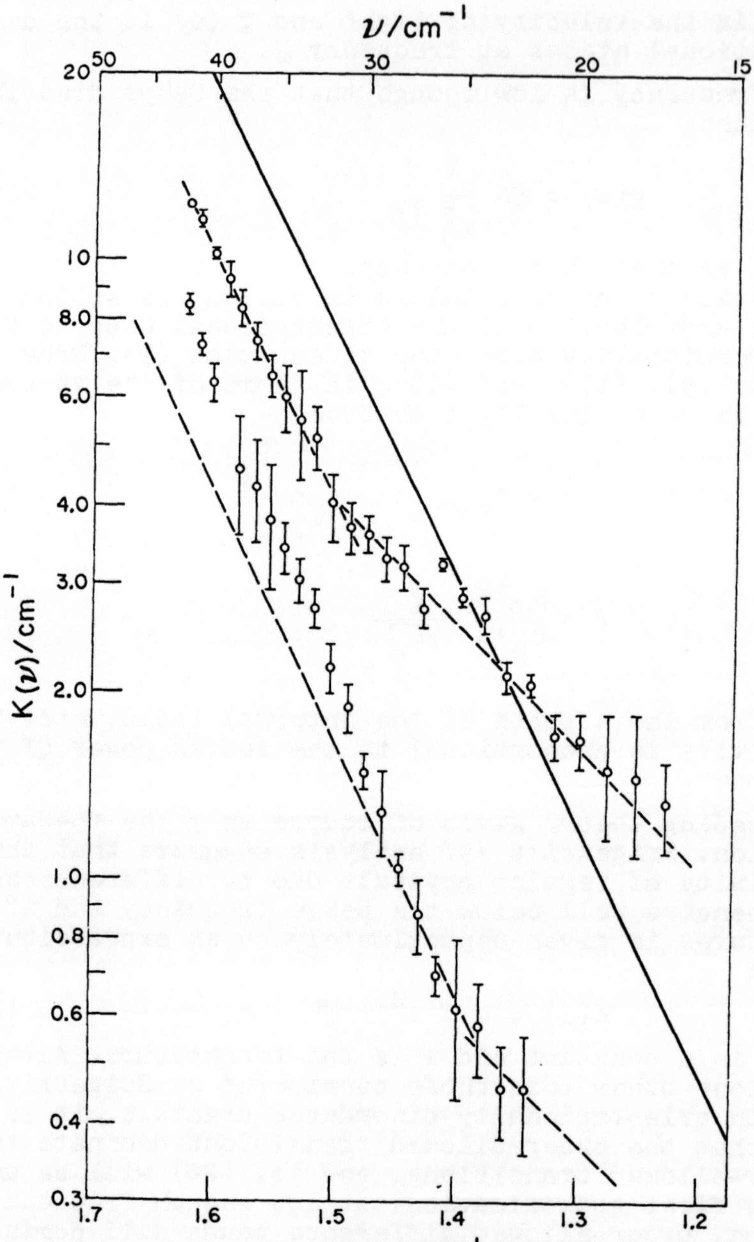

Fig.3 Absorptivity of ice in the range 44 to 17 cm^{-1} at 100°K (lower points) and 200°K (upper points). The solid line in equation (21), and the dashed lines have slopes of either 4 or 2

Hence from equation (19)

$$K(\nu) = 4.6 \times 10^3 (\nu/\nu_D)^4 \text{ cm}^{-1}. \quad (21)$$

Experimental measurements of the absorptivity of ice in the range 50 to 17 cm^{-1} are shown in fig.3 which is a log-log plot of $K(\nu)$ against ν at 100°K and at 200°K. The line of equation (21) is also drawn for comparison.

The theory clearly agrees well qualitatively with experiment. The absorptivity at high frequencies at both 100 and 200°K is approximately proportional to the fourth power of the wave number, and is no doubt due to the absorption of light by the high-frequency (∿1 THz) sound waves in one-quantum transitions. At lower frequencies the absorptivity is approximately proportional to the square of the wave number and is strongly dependent on temperature, as expected for two-quantum difference bands according to equation (20).

Quantitatively the theory is less satisfactory, at least partly because the approach to the TA maximum at 65 cm^{-1} causes the absorptivity to rise more steeply than the fourth power of the wave number; because the simple theory does not allow for the correlation in the orientation of neighboring molecules and so predicts a higher absorptivity than is observed; and because of other approximations in the theory.

References

/1/ Bertie, J.E., H.J.Labbé, E.Whalley, Absorptivity of Ice I in the range 4000 to 30 cm^{-1}, J.Chem. Phys. submitted

/2/ Whalley, E., H.J.Labbé, Optical Spectra of Orientationally Disordered Crystals. III. Infrared and Raman Spectra of the Sound Waves and Application to Ice Ih, J.Chem.Phys. submitted

/3/ Whalley, E., Developments in Applied Spectroscopy 6, 277 (1968)

/4/ Whalley, J.E.Bertie, J.Chem.Phys.46, 1264 (1967)

/5/ Bertie, J.E., E.Whalley, J.Chem.Phys.46, 1271 (1967)

/6/ Lamb, J., A.Turney, Proc.Phys.Soc.B 62, 272 (1949)

/7/ Ockman, N., Advan.Phys. 1, 199, (1958)

/8/ Szigetti, B., Trans.Faraday Soc. 45, 155 (1949); see also H.Fröhlich, Theory of Dielectrics, 2nd Edition , Oxford University Press 1958, p. 152

/9/ Szigetti, B., Proc.Roy.Soc. A 258, 377 (1960)

LATTICE DYNAMICS OF ICE

B. Renker and P. v. Blanckenhagen

Kernforschungszentrum Karlsruhe

Institut für Angewandte Kernphysik

Abstract

The frequency distribution function of ice Ih has been calculated in extension to the work of Forslind. Additionally it was derived by an analysis of inelastic neutron data corrected for multi-phonon processes and multiple scattering. A comparison is performed including IR-results. The temperature dependence of the specific heat was calculated from both frequency distribution functions and compared with $C_v(T)$ calculated from the measured $C_p(T)$.

1. Introduction

The density of states function contains detailed information about the dynamics of a solid and therefore its determination is of manifold interest. It may be obtained from neutron scattering and IR experiments and can be calculated theoretically from a force constant model of the solid. Neutron measurements on ice have been performed by several authors / 1, 2, 3/.
But until now no effort has been made to correct for multiple scattering and multi-phonon processes in the derivation of the generalized frequency distribution function ρ_{gen} from the measured spectra. For the hexagonal lattice symmetry no complete theoretical calculation has been published so far. In extension of the dispersion relations for selected directions by Forslind /4/ in the present work the frequency distribution of the lattice vibrations was calculated from the

elastic moduli. The results are compared with ρ_{gen} derived by inelastic neutron scattering and with the optical density function divided by ν^2 taken from the work of Bertie and Whalley /5/.

2. Lattice dynamics

The position of an atom j with mass M_j in the unit cell m may be written

$$\underline{R} = \underline{r}_j^m + \underline{u}_j^m \tag{2.1}$$

where $\underline{u}_j^m$ is the displacement of the atom from its mean position. The potential energy of the crystal in harmonic approximation is

$$V = -\frac{1}{2} \sum_{mm'} \sum_{jj'} \sum_{\alpha\beta} \Phi_{\alpha\beta}\binom{mm'}{jj'} u_\alpha\binom{m}{j} u_\beta\binom{m'}{j'} \tag{2.2}$$

where α,β indicates the cartesian components. $\Phi_{\alpha\beta}$ is the force constant matrix and according to the translational invariance it depends only on the relative position m-m' of any two unit cells. Using the dynamical matrix

$$C_{\alpha\beta}\left(\frac{\underline{k}}{jj'}\right) = \frac{1}{\sqrt{M_j M_{j'}}} \sum_{mm'} \Phi_{\alpha\beta}\binom{m-m'}{j\ j'} e^{i\underline{k}\underline{r}\binom{m-m'}{j\ j'}} \tag{2.3}$$

the equation of motion can be written as

$$\omega^2 P_\alpha(j) + \sum_{j'} \sum_{\beta} C_{\alpha\beta}\left(\frac{\underline{k}}{jj'}\right) P_\beta(j') = 0 \tag{2.4}$$

where $\underline{k}$ is the wave vector and the $\underline{P}(j)$ are the polarisation vectors. Now ω^2, P and $C(\underline{k})$ are assumed to be functions of the wave vector and they may be expanded in terms of $\underline{k}$. In the long wave length limit one may drop all terms of higher order than two and the equation of motion can be written in a form containing only amplitudes U for the motion of the unit cell as a whole.

$$\omega^2 U_\alpha + \sum_\beta C'_{\alpha\beta}(\underline{k})\, U_\beta = 0. \tag{2.5}$$

A comparison between eq. (2.5) and the equations for elastic waves in an anisotropic continuum gives relations between the macroscopic elastic constants and the $C'_{\alpha\beta}$ depending on atomic constants /4, 6/. These relations are used to derive numerical values for the atomic constants, which are necessary for a solution of eq. (2.4).

2.1 Calculation of the frequency spectrum

Fig. 1 shows the position of the oxygen atoms in the hexagonal ice lattice. For the position of the hydrogen atoms the Pauling model assumes a statistical distribution for the two possible positions of the proton in the H-bridge. It seems to be a good approximation to assume a lattice having point-masses 18 on the position of the oxygen atoms. Further we assume that no forces of long range are important, that a model with nearest neighbour interaction describes all modes where the H_2O molecules move as a whole. The following results have been taken from Forslind's paper /4/.

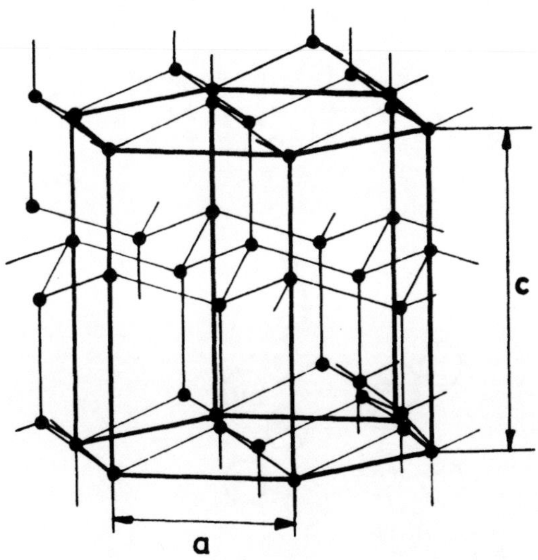

Fig. 1 Structure of ice Ih

By application of symmetry operations the number of coefficients $C_{\alpha\beta}$ in eq. (2.4) are reduced and finally the following dynamical equation is derived.

$$\begin{vmatrix}
X^{11}_{11} & C^{12}_{11} & 0 & C^{14}_{11} & 0 & 0 & 0 & C^{14}_{12} & 0 & 0 & 0 & C^{14}_{13} \\
C^{12}_{11} & X^{22}_{11} & C^{23}_{11} & 0 & 0 & 0 & C^{23}_{12} & 0 & 0 & 0 & C^{23}_{13} & 0 \\
0 & C^{32}_{11} & X^{33}_{11} & C^{34}_{11} & 0 & C^{32}_{12} & 0 & 0 & 0 & C^{32}_{13} & 0 & 0 \\
C^{41}_{11} & 0 & C^{43}_{11} & X^{44}_{11} & C^{41}_{12} & 0 & 0 & 0 & C^{41}_{13} & 0 & 0 & 0 \\
0 & 0 & 0 & C^{14}_{21} & X^{11}_{22} & C^{12}_{22} & 0 & C^{14}_{22} & 0 & 0 & 0 & C^{14}_{23} \\
0 & 0 & C^{23}_{21} & 0 & C^{21}_{22} & X^{22}_{22} & C^{23}_{22} & 0 & 0 & 0 & C^{23}_{23} & 0 \\
0 & C^{32}_{21} & 0 & 0 & 0 & C^{32}_{22} & X^{33}_{22} & C^{34}_{22} & 0 & C^{32}_{23} & 0 & 0 \\
C^{41}_{21} & 0 & 0 & 0 & C^{41}_{22} & 0 & C^{43}_{22} & X^{44}_{22} & C^{41}_{23} & 0 & 0 & 0 \\
0 & 0 & 0 & C^{14}_{31} & 0 & 0 & 0 & C^{14}_{32} & X^{11}_{33} & C^{12}_{33} & 0 & C^{14}_{33} \\
0 & 0 & C^{23}_{31} & 0 & 0 & 0 & C^{23}_{32} & 0 & C^{21}_{33} & X^{22}_{33} & C^{23}_{33} & 0 \\
0 & C^{32}_{31} & 0 & 0 & 0 & C^{32}_{32} & 0 & 0 & 0 & C^{32}_{33} & X^{33}_{33} & C^{34}_{33} \\
C^{41}_{31} & 0 & 0 & 0 & C^{41}_{32} & 0 & 0 & 0 & C^{41}_{33} & 0 & C^{43}_{33} & X^{44}_{33}
\end{vmatrix} = (E\omega + C) = 0 \qquad (2.6)$$

The elements of the determinant depend on $\underline{k}$, the atomic force constants (α, β, γ, δ, Θ, ε, κ), and the nearest neighbour distance d according to fig. 2.

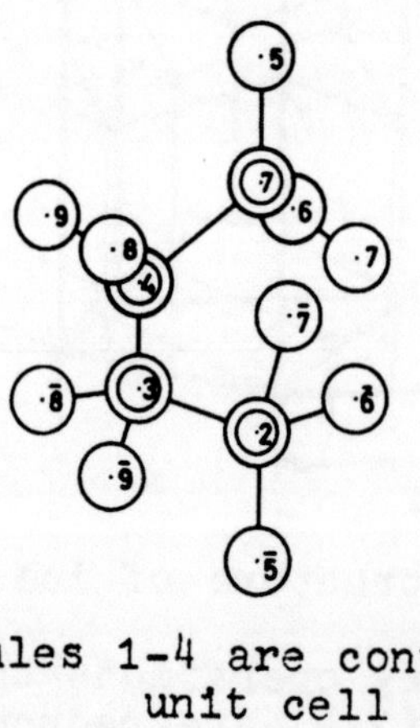

Fig.2 Molecules 1-4 are contained in the unit cell /4/

The equations are given in the appendix.

The numerical values of the atomic constants derived in the long wave length limit from the macroscopic elastic moduli /6/ are listed below:

α = 379 dyne/cm	θ = 2830 dyne/cm
β = 4445 dyne/cm	ε = 1616 dyne/cm
γ = 3600 dyne/cm	κ = 21549 dyne/cm
δ = 2830 dyne/cm	

2.2 The sampling method

A possible way to derive the frequency spectrum from (2.4) is the "root sampling method". Eq. (2.6) is solved for a large number of $\underline{k}$-vectors and the resulting frequencies ν (k)are distributed over a field of frequency channels of widths $\Delta\nu$. The resolution depends on the widths $\Delta\nu$ and the number of frequencies. Good resolution requires a large number of diagonalisations which normally cannot be obtained with reasonable computing times especially for complicated lattices. An improved sampling method proposed by Gilat and Dolling /7/ was used to calculate the frequency spectrum of ice.
The main idea of this procedure is to solve the eigenvalue problem only for a small number of k-vectors spaced on a regulary crude "mesh" in the irreducible part of the Brillouin zone. The normal modes for a fine "mesh" are obtained by linear extrapolation. According to the Jacobi-method a diagonalisation of $C(\underline{k})$ is derived by a unitary transformation.

$$A = U^{+} C^{o} U \quad \text{with} \quad A_{ij} = \begin{cases} 4\gamma^{2}{}_{o}(k) & i = j \\ 0 & i \neq j \end{cases} \tag{2.7}$$

The matrix D describes the change in C due to a change δk_i in the i-th component of $\underline{k}$.

$$D^{i}_{jl}(\underline{k}) = C^{i}_{jl}(\underline{k} + \underline{e}_i\ \delta k_i) - C^{o}_{jl}(\underline{k}). \tag{2.8}$$

The change of the j-th eigenvalue due to δk_i is given in first order approximation by

$$\Delta^{i}_{jj}(\underline{k}) = \{U^{+}(\underline{k})\ D^{i}(\underline{k})\ U(\underline{k})\}_{jj} \tag{2.9}$$

and

$$(\text{grad } \nu_j)^i \cong \frac{\delta \nu_j}{\delta k_i} \cong \frac{1}{8\pi^2 \nu_{oj}(\underline{k})} \frac{\Delta^i{}_{jj}(k)}{\delta k_i} . \qquad (2.10)$$

$\nu_j(\underline{k} + \Delta \underline{k})$ is given by the extrapolation

$$\nu_j(\underline{k} + \Delta \underline{k}) = \nu_{oj}(\underline{k}) + (\text{grad } \nu_j \Delta \underline{k}). \qquad (2.11)$$

As the number of calculations of U which takes most of the time is substantially reduced compared to the normal sampling method, the improved method saves about a factor of 100 in computing time for equal resolution /8/.
The sampling region is the irreducible part of the Brillouin zone shown in fig.3.

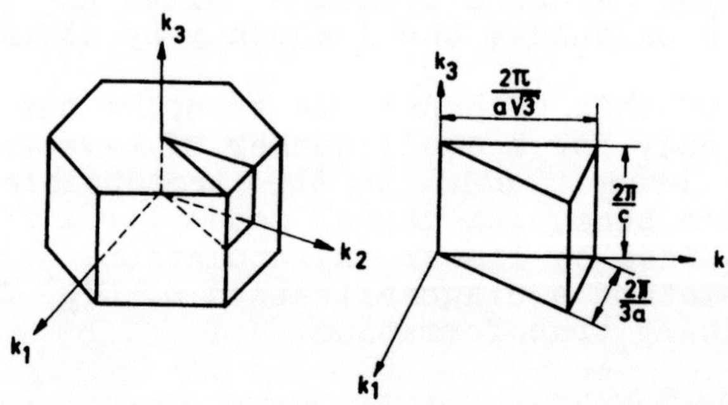

Fig.3 Brillouin zone and its irreducible part

For the sampling process this volume **was** composed of rectangular and triangular elements. With this method it was possible to fill up the volume exactly even with a very crude "mesh" of $\underline{k}$ vectors. For the numerical calculations on the IBM 7074 a Fortranprogram was available /9/.

2.3 Inelastic neutron scattering

In the case of one atom per cell the double differential cross-section for neutron scattering can be written

$$\frac{d^2\sigma}{d\Omega dE} = \frac{1}{k_BT}\frac{k'}{k_o} e^{-\beta/2} a_b^2 S(\alpha,\beta) \tag{2.12}$$

where

$S(\alpha,\beta)$ is the scattering law

$\alpha = \frac{\hbar^2\kappa^2}{2Mk_BT}$ is proportional to the square of momentum transfer $\hbar(\underline{k}-\underline{k}_o) = \hbar\underline{\kappa}$

$\beta = \frac{\hbar\omega}{k_BT}$ is proportional to the energy transfer $\hbar\omega$

M is the mass of the primary scattering atom

a_b is the bonded scattering amplitude.

According to Egelstaff and Schoffield /10/ $S(\alpha,\beta)$ may be expanded in a phonon series.

$$S(\alpha,\beta) = e^{-2W} \sum_{n=1}^{\infty} \frac{1}{n!} (B\alpha)^n T_n(\beta) \tag{2.13}$$

$$S(\alpha,\beta)^1 = e^{-2W} \alpha \frac{\rho(\beta)}{2\beta \sinh(\beta/2)} \tag{2.14}$$

$$F = \frac{B\alpha\, T_1(\beta)}{e^{2W} S(\alpha,\beta)} \tag{2.15}$$

$S(\alpha,\beta)^1$ is the one phonon term and F describes the contribution of the one phonon process to the total scattering for the fixed variables α, β.

If there are z atoms per unit cell eq. (2.12) can be written

$$\frac{d^2\sigma}{d\Omega dE} = \frac{1}{k_BT}\frac{k}{k_o} e^{-\beta/2} \sum_{l=1}^{z} a_{bl}^2 S_l(\alpha,\beta). \tag{2.16}$$

If we define a generalized frequency distribution function $\rho_{gen}(\beta)$ the phonon series formalism may be generalized:

$$\frac{d^2\sigma}{d\Omega dE} = \frac{1}{k_B T} \frac{k}{k_o} \frac{\hbar\kappa^2}{4\omega \sinh(\beta/2)} \sum_{l=1}^{z} a_{bl}^2 \frac{1}{M_l} \rho_{gen}^{l}(\beta)\, e^{-2W_l} + \text{multiphonon terms.} \qquad (2.17)$$

Further we will have 3z dispersion curves with polarisation vectors $\underline{P}_i^l$, describing the amplitude of the i-th nucleus in the branch i as a function of wave vector $\underline{k}$. $\rho_{gen}^{l}(\beta)$ is then defined as

$$\rho_{gen}^{l}(\beta) = \frac{1}{3N} \sum_{i=1}^{3z} |P_i^l(\underline{k})|^2 \, \delta(\beta-\beta_i) \qquad (2.18)$$

where N is the number of unit cells in the crystal. For a fixed frequency the $|P_i^l|^2$ may be simply expressed as $|P_i^l|^2 = M_l/M_i$, where M_i is the effective mass corresponding to the mode in the i-th branch. In this formalism librations may be treated as oscillations.

3. Experimental method

A neutron beam from the "cold neutron source" at the reactor FR2 was filtered by a polycrystalline Be-block for background reduction. The experiments were performed with the rotating crystal time of flight technique. The incident neutron energy was 5 meV and the primary energy resolution was ≈ 4%. The sample was 0.5 mm thick and cooled by Peltier-elements to 261 °K. Inelastic neutron spectra could be simultaneously obtained for scattering angles between ≈ 10° and ≈ 100°. The flight path for the smallest angle (250 cm), the collimator and the sample container were He-filled for a reduction of air scattering. The observed spectra were corrected for background and counter efficiency.

3.1 Neutron scattering results

Fig. 4 shows a typical time of flight spectrum obtained at scattering angle of 11.5°; similar spectra were obtained for larger scattering angles. The general tendency is an increase of the inelastic scattering intensity and of the "smoothing" due to multi-phonon

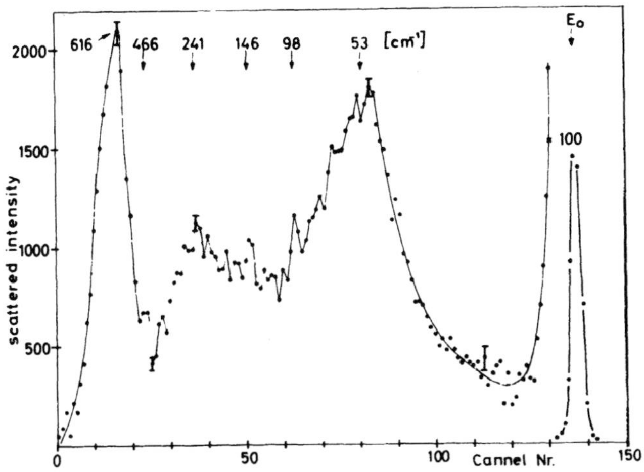

Fig.4 Time of flight spectrum, $\Delta t = 16\ \mu s$, $E_o = 5$ meV, $\theta = 11.5^o$, $T = 261^oK$

processes with increasing momentum transfer.

This spectrum was converted into the double differential cross-section which is necessary to calculate the frequency distribution function $\rho_{gen}(\beta)$ by eq. (2.12 and 2.14) in the one-phonon-approximation. Starting with approximately correct $\rho_{gen}(\beta)$ the multi-phonon contribution was calculated by eq. (2.15) using the LEAP-code which bases on the phonon expansion. Then corrections for multiple scattering were calculated with an additional code /11/ which bases on the theory of Lane and Miller /12/ and uses $S(\alpha,\beta)$ calculated with LEAP as input data. Both corrections were applied to the initial $\rho_{gen}(\beta)$. Using this corrected function $\rho_{gen}(\beta)$ the LEAP and multiple scattering calculation was repeated until reasonable agreement between the calculated and the measured scattering law was obtained.

Fig.6 shows the final $\rho_{gen}(\beta)$ function, in fig.5 curves for multiple scattering are drawn. It is seen that multiple scattering is an important correction, especially for small scattering angles and energy transfers, where it becomes very large.

The statistical error in fig.6 is reasonable but additional uncertainties arise for high β from the multiphonon correction. It was impossible to derive the slope of the $\rho_{gen}(\beta)$ function at high β with sufficient accuracy therefore the curve was broken off at a value which gave the correct weight.

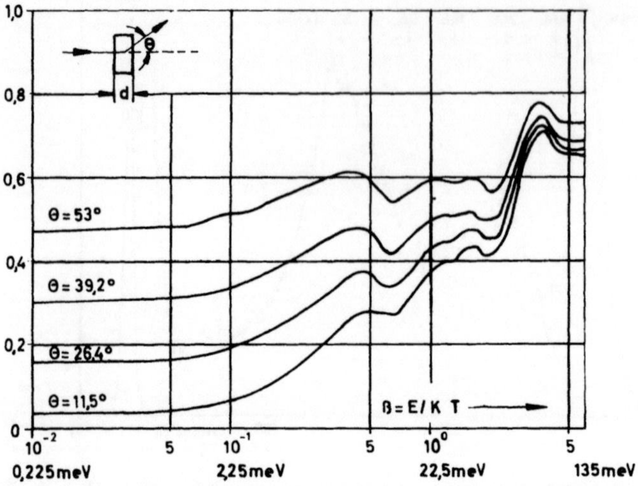

Fig.5 Multiple-scattering : correction-factor for E_o = 5 meV and d = 0.5 mm

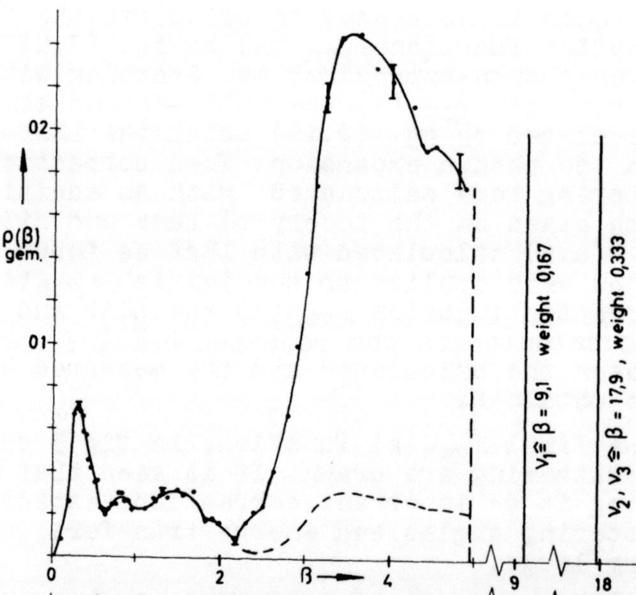

Fig.6 Generalized frequency distribution T = $261^{O}K$

According to the fact that the molecular vibration frequencies are nearly the same for the free and the bonded molecule /13/ one may conclude that the structure of the molecules in both phases is not very different. Therefore, with the effective rotational mass $M_r = 2.32$ derived by Nelkin /14/ for the free molecule one gets a relation

$$\frac{M_T}{M_{Libr}} = 7.76 \text{ (sec. 2.4)}$$

for the areas of translational and librational part in the $\rho_{gen}(\beta)$ function. This factor is also used to calculate the true density function $\rho(\beta)$ (dashed curve in fig.6. A frequency distribution function - derived by the extrapolation method - which gives the higher β region was derived by Harling /15/ using higher incident neutron energies (101 to 616 meV).

3.2 Comparison with optical and theoretical results

Fig.7 shows the translational part of $\rho(\beta)$, as derived from neutron measurement (fig.7a), from optical data (fig.7b) and from the theoretical calculation based on the elastic moduli (fig.7c). The dotted lines in fig.7 show the initial curves folded by the energy dependent resolution function employed in the neutron measurement. The optical curve has been taken from Bertie and Whalley /5/. They have shown that in the case of orientationally disordered crystals all frequencies should be infrared and Raman active. In simple cases with no coupling between translational and librational part the optical density devided by ν^2 should show the same features as the frequency distribution function /16/. The lower energy part is not resolved. The theoretical curve may be discussed on the basis of the dispersion relations calculated by Forslind (fig.8).
The branches with ω_1 represent the acoustical and ω_{2-4} the optical vibrations. The indices 11 and 33 refer to transversal and 22 to longitudinal polarisation.

A comparison of the different curves for equal resolution in fig.7 shows good agreement in the position of maxima and minima. An obvious difference in its shape shows the peak at $\sim 230\ cm^{-1}$ in the neutron curve, whilst the optical and theoretical curve are in fairly good agreement. The remaining difference between fig.7b and 7c may be explained by the fact, that our calculation based on the elastic moduli-derived at long wavelength - may fail to describe the high optical vibrations correctly.

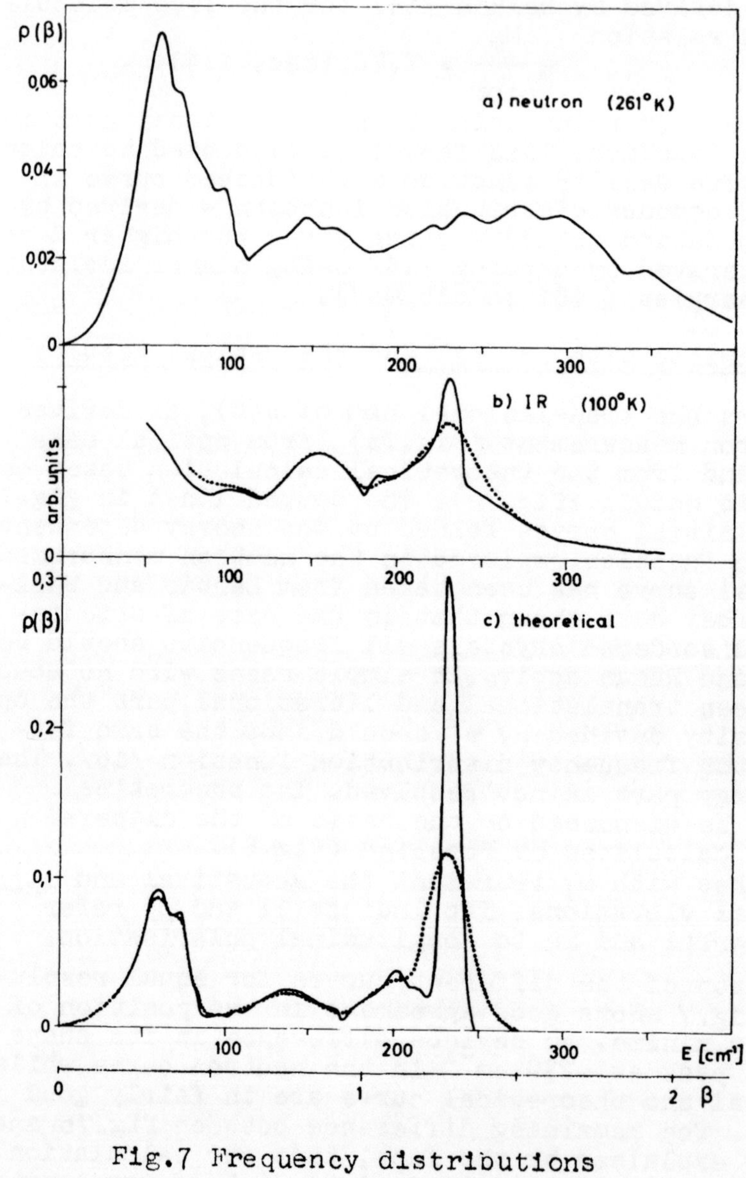

Fig.7 Frequency distributions

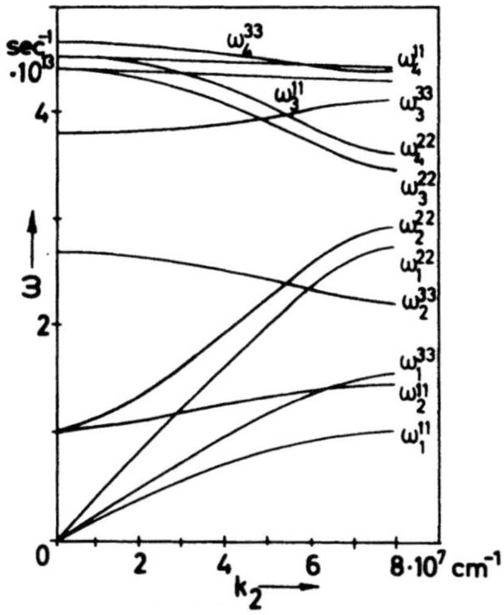

Fig.8 Dispersion curves along the k_2-axis

Table 1 Observed and calculated peaks in the spectrum of ICE Ih

Raman (R) and infrared (IR) data			**neutron data···**	**theoretical···**
compiled by Ockman·	168°K··	100°K··	261°K	
53 (R)				
97 (R)				
122 (R)	100 (IR)	103 (IR)	60	60
160 (IR)	160 (IR	164 (IR)	70	70
177 (R)				
193 (R)	184 (IR)	190 (IR)	145	140
212 (R)	222 (IR)	229 (IR)		202
232 (R)				
252 (R)	260 (IR)	275 (IR)		
272 (R)	300 (IR)	300-310 (IR)	225-300	230
294 (R)				
457 (R)			660	
516 (R)				
540 (IR)				
600 (IR)				
660 (IR)				
800 (IR)				

·see ref. /18/, ··Bertie and Whalley /5/, ···taken from ρ(β), data in cm^{-1}

According to the fact that the neutron measurement is only sensitive to the motions of the proton, whilst the IR measurement should be sensitive to all normal modes of the crystal a possible explanation for the difference between the neutron and the IR $\rho(\beta)$ might be that the protons are not rigidly coupled to the oxygen atoms.

From the function $\rho_{gen}(\beta)$ in fig. 6 we derive a mean amplitude u = 0.22 Å which is comparable to X-ray results from Owston /17/ u = 0.26 Å at 268°K.

A comparison between peaks observed in Raman and IR-spectra and the functions $\rho(\beta)$ in fig. 7a and 7c are given in table 1.

3.3 Temperature dependence of the specific heat.

Fig. 9 shows curves for the specific heat $C_v(T)$ derived from the different $\rho(\beta)$ functions. The hindered rotations have been treated as linear harmonic vibrations which is a good approximation for low exitation energies.

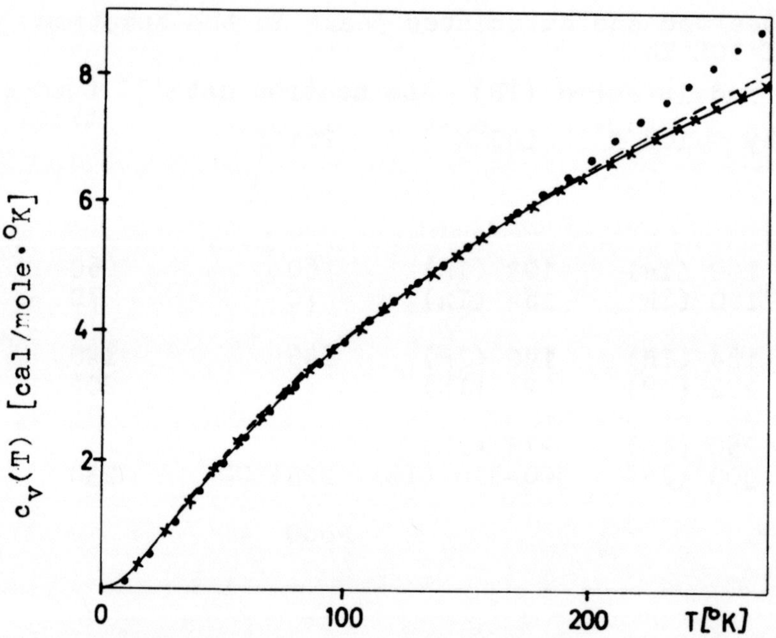

Fig.9 Specific heat of ice I h
·· measured
— corrected for anharmonicity
xx derived from neutron $\rho(\beta)$
-- derived from neutron $\rho_{theor}(\beta)$

The points in fig.9 are values of $C_v(T)$ which have been calculated from the measured $C_p(T)$ /19/ values using the constants for compressibility and cubic expansion mentioned by Leadbetter /20/. Both curves derived from $\rho(\beta)$ in fig.7a and 7b show good agreement with the measured values for temperatures $\lesssim 180°K$ Leadbetter has shown that there exists a contribution of anharmonicity (solid line in fig.9) and the $C_v(T)$ values calculated from the theoretical $\rho(\beta)$ should be compared with this curve.

Conclusion

The translational and librational frequency distribution function $\rho(\beta)$ ($E \lesssim 100$ meV) was derived from neutron scattering data regarding multiple scattering and multi-phonon corrections. The specific heat calculated with this $\rho(\beta)$ shows good agreement with the measured values for temperatures $T \lesssim 180°K$. $\rho(\beta)$ was also calculated using the atomic constants obtained in a long wave length approximation from the elastic moduli by Forslind. The theoretical $\rho(\beta)$ agrees with the measured $\rho(\beta)$ in the main features especially in the low energy part. To improve the atomic force model it would be necessary to determine additional interatomic force constants by a measurement of the dispersion relations of ice.

Acknowledgements

For many helpful discussions we want to thank the Drs. O.Abeln, G.Ehret and W.Gläser.

Appendix

$$M\,C_{11}^{14} = 4\alpha\, e^{i\underline{k}\underline{d}_{14}} + (\alpha+3\beta)\,(e^{i\underline{k}\underline{d}_{17}} + e^{i\underline{k}\underline{d}_{16}})$$

$$M\,C_{12}^{14} = M\,C_{21}^{14} = \sqrt{3}\,(\beta-\alpha)\,(e^{i\underline{k}\underline{d}_{17}} - e^{i\underline{k}\underline{d}_{16}})$$

$$M\,C_{13}^{14} = -\sqrt{3}\;\delta(e^{i\underline{k}\underline{d}_{17}} - e^{i\underline{k}\underline{d}_{16}})$$

$$M\,C_{31}^{14} = -\sqrt{3}\;\theta(e^{i\underline{k}\underline{d}_{17}} - e^{i\underline{k}\underline{d}_{16}})$$

$$M\,C_{22}^{14} = 4\beta\, e^{i\underline{k}\underline{d}_{14}} + (3\alpha+\beta)\,(e^{i\underline{k}\underline{d}_{17}} + e^{i\underline{k}\underline{d}_{16}})$$

$$M\,C_{23}^{14} = \delta(2e^{i\underline{k}\underline{d}_{14}} - e^{i\underline{k}\underline{d}_{17}} - e^{i\underline{k}\underline{d}_{16}})$$

$$M\,C_{32}^{14} = \theta(2e^{i\underline{k}\underline{d}_{14}} - e^{i\underline{k}\underline{d}_{17}} - e^{i\underline{k}\underline{d}_{16}})$$

$$M\,C_{33}^{14} = \gamma(e^{i\underline{k}\underline{d}_{14}} + e^{i\underline{k}\underline{d}_{17}} + e^{i\underline{k}\underline{d}_{16}})$$

$M\,C^{41}_{\alpha\beta}$ is obtained from $C^{14}_{\alpha\beta}$ by substituting

$\underline{d}_{41}$ for $\underline{d}_{14}$

$\underline{d}_{48}$ for $\underline{d}_{16}$

$\underline{d}_{49}$ for $\underline{d}_{17}$

$$M\,C^{23}_{11} = 4\alpha\, e^{i\underline{kd}_{23}} + (\alpha+3\beta)\,(e^{i\underline{kd}_{2\bar{6}}} + e^{i\underline{kd}_{2\bar{7}}})$$

$$M\,C^{23}_{12} = M\,C^{23}_{21} = \sqrt{3}\,(\beta-\alpha)\,(e^{i\underline{kd}_{2\bar{6}}} - e^{i\underline{kd}_{2\bar{7}}})$$

$$M\,C^{23}_{13} = \frac{\delta}{\theta}\,M\,C^{23}_{31} = \sqrt{3}\,(e^{i\underline{kd}_{2\bar{6}}} - e^{i\underline{kd}_{2\bar{7}}})$$

$$M\,C^{23}_{22} = 4\beta\, e^{i\underline{kd}_{23}} + (3\alpha+\beta)\,(e^{i\underline{kd}_{2\bar{6}}} + e^{i\underline{kd}_{2\bar{7}}})$$

$$M\,C^{23}_{33} = \gamma(e^{i\underline{kd}_{23}} + e^{i\underline{kd}_{2\bar{6}}} + e^{i\underline{kd}_{2\bar{7}}})$$

$M\,C^{32}_{\alpha\beta}$ is obtained from $M\,C^{23}_{\alpha\beta}$ by substituting

$\underline{d}_{32}$ for $\underline{d}_{23}$

$\underline{d}_{3\bar{8}}$ for $\underline{d}_{2\bar{6}}$

$\underline{d}_{3\bar{9}}$ for $\underline{d}_{2\bar{7}}$

$$M\,C^{23}_{23} = M\,\frac{\delta}{\theta}\,C^{23}_{32} = -\,\delta(2e^{i\underline{kd}_{23}} - e^{i\underline{kd}_{2\bar{6}}} - e^{i\underline{kd}_{2\bar{7}}})$$

$$M\,C^{12}_{\alpha} = \begin{bmatrix} \varepsilon & 0 & \\ 0 & \varepsilon & 0 \\ 0 & 0 & \kappa \end{bmatrix} e^{i\underline{kd}_{15}}$$

Replacing d_{15} by $\underline{d}_{2\bar{5}}$ gives $M\,C^{21}$

$\underline{d}_{34}$ gives $M\,C^{34}$

$\underline{d}_{43}$ gives $M\,C^{43}$

$$X^{jj}_{\alpha\alpha} = \omega^2 + C^{jj}_{\alpha\alpha}$$

$$C^{jj}_{\alpha\alpha} = -\frac{1}{M}\begin{bmatrix} 6(\alpha+\beta)+\varepsilon & 0 & 0 \\ 0 & 6(\alpha+\beta)+\varepsilon & 0 \\ 0 & 0 & 3\gamma+\kappa \end{bmatrix}$$

References

/1/ Larsson, K.E., U. Dahlborg, Reactor Science and Technology 16, 81 (1961)

/2/ Golikov, V.V., I. Zukovskaja, F.L. Sapiro, A. Skatula, E. Janik, Proceedings of the Symposium on Inelastic Scattering of Neutrons, Bombay 1964, Vol.II p. 201

/3/ Prask, H., H. Boutin, S. Yip, Journ.Chem.Phys. 48 3367 (1968)

/4/ Forslind, E., Svenska Forskningsinst. Cement Betong vid Kgl.Tek.Högskol. Stockholm Handl. No. 21, 1 (1954)

/5/ Bertie, J.E., E. Whalley, Journ. Chem. Phys. 46, 1271 (1966)

/6/ Penny, A., Proc. of the Cambr. phil. Soc. 44, 423 (1948)

/7/ Gilat, G., G.Dolling, Physics Letters 8, 304 (1964)

/8/ Gilat, G., L.J. Raubenheimer, Phys. Rev. 144, 390 (1966)

/9/ Abeln, O., ROSA and FOURI zwei Programme zur Berechnung der Frequenzverteilung eines Festkörpers, IAK-Arbeitsbericht 51/67

/10/ Egelstaff, P.A., P. Schoffield, Nucl. Sci. Eng.12 260 (1962)

/11/ Ehret, G., F. Carvalho, Korrekturen für die Vielfachstreuung langsamer Neutronen in unendlich ausgedehnten Platten, IAK-Arbeitsbericht 46/67

/12/ Lane, R.O., W.F. Miller, Nucl. Inst. Meth.16 1 (1962)

/13/ Springer, T., Nucleonik 3, 110 (1961)

/14/ Nelkin, M., Phys. Rev. 119, 741 (1960)

/15/ Harling, O.K., Conference on Inelastic Scattering of Neutrons, Copenhagen, 1968 Report SM 104/80

/16/ Whalley, E., J.E. Bertie, Journ. Chem. Phys. 46, 1264 (1967)

/17/ Owston, P.G., Quart.Rev.Chem.Soc.Lond. 4, 344 (1950)

/18/ Ockmann, N., Adv. Phys. 8, 199 (1958)

/19/ Giauque, W.F., J.W. Stout, J. Am. Chem. Soc. 58, 1144 (1936)

/20/ Leadbetter, Proc. Roy. Soc. Lond. 287, 403 (1965)

THE VAPOR PRESSURE ISOTOPE EFFECT OF ICE AND ITS ISOMERS

H. Wolff

Physikalisch-Chemisches Institut

Universität Heidelberg / Germany

Abstract

The vapor pressure isotope effect of compounds forming hydrogen bonds can be explained by the super-position of a normal effect of the intermolecular vibrations and of an inverse effect of the intramolecular vibrations. Its calculation for ice faces the difficulty that the vibrations interact with each other. As far as the valency vibrations are concerned, this difficulty is eliminated by calculating with the uncoupled vibrations. By reasonable choice of the intermolecular vibrations it is then possible to calculate the vapor pressure isotope effect of H_2O- and D_2O-ice as well as the effect of H_2O- and HDO-ice. The possibility of obtaining the temperature dependence of the vapor pressure isotope effect of the liquid isomers in the same way confirms the explanation given for the solid compounds. Because it is sensitive against deviations of the wavenumbers the calculation can serve to ensure the assignment.

Since the ratio of vapor pressures of heavy ice and light ice is smaller than unity at all temperatures, the vapor pressure isotope effect of ice is always normal. Even the ratio of vapor pressures of liquid D_2O and H_2O is smaller than unity up to a temperature of about 220°C. Only above this temperature this ratio attains a value greater than one, corresponding to an inverse vapor pressure isotope effect.

Other substances which form hydrogen bonds also show the normal effect, which changes to the inverse effect with increasing temperature, provided the substances do not show association in the gaseous state. The crossover is effected not only by increase in temperature but also by dilution with an appropriate solvent, as is shown in fig. 1 for dimethylamine in hexane /1/. The effect has also been shown for methylamine /2/, methanol /3/ and ammonia /4/. These experimental observations suggest that the normal effect is related to hydrogen, deuterium or tritium bonding of the compounds and that it can be explained by the existence of intermolecular vibrations which occur with association or condensation /5/.

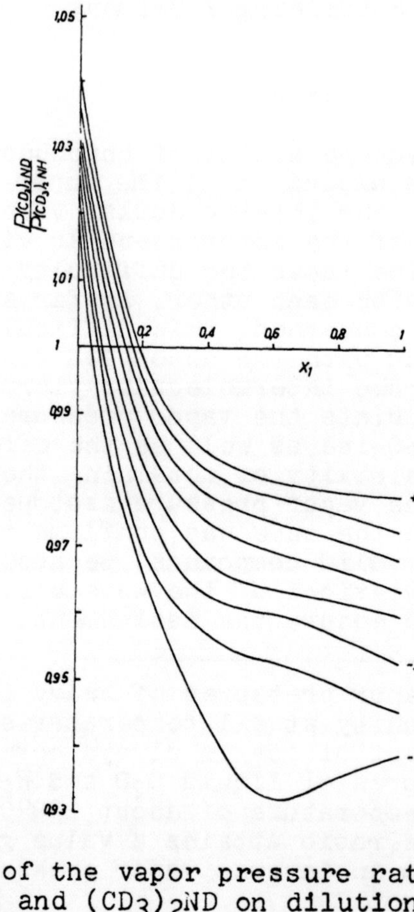

Fig.1 Changes of the vapor pressure ratios p^D/p^H of $(CD_3)_2NH$ and $(CD_3)_2ND$ on dilution with n-Hexane at temperatures ranging from +20 to -50 °C, x_1 = mole fraction dimethylamine (taken from reference /1/)

Bigeleisen /6/ has given a general formula for the calculation of the vapor pressure isotope effect. It may be reduced for water and ice to the equation /5,7,8/

$$\frac{p^D}{p^H} = \prod_{i=1}^{3N} q_i = \prod_{i=1}^{6} \left\{ \frac{\Theta^H_{i\ kond}}{\Theta^D_{i\ kond}} \; \frac{\sinh \frac{\Theta^D_{i\ kond}}{2T}}{\sinh \frac{\Theta^H_{i\ kond}}{2T}} \right\} \cdot$$

$$\cdot \exp \left(\sum_{i=7}^{3N} \{(\Theta^H_{i\ gas} - \Theta^H_{i\ kond}) - (\Theta^D_{i\ gas} - \Theta^D_{i\ kond})\}\right)/2T \qquad (1)$$

In the expression, when i ranges from 1 to 6, $\Theta_{i\ kond}$ gives the characteristic temperatures of the intermolecular vibrations of the condensed phase. With i ranging from 7 to 9 $\Theta_{i\ gas}$ and $\Theta_{i\ kond}$ give the characteristic temperatures of the intramolecular vibrations in the gaseous and in the condensed state.

As we have shown previously /5/, equation (1), originally derived by Bigeleisen's accurate method, may be obtained more simply if the pressures in the vapor pressure ratio are expressed by

$$RT \ln p = -(G_o - G_{kond}) \qquad (2)$$

where G_o, the Gibbs free energy of the gas at one atmosphere, and G_{kond}, the Gibbs free energy of the condensate, are calculated by the relations of statistical thermodynamics for systems without interaction, taking into account the total temperature function for the intermolecular vibrations and the zero point contribution for the intramolecular vibrations alone. Assuming that the translational and rotational contributions to G_o differ only in the terms related to the molecular weight M and the moment of inertia I, the expression

$$RT \ln \frac{p^D}{p^H} = \frac{3}{2} RT \ln \frac{M^D}{M^H} + \frac{3}{2} RT \ln \frac{I^D}{I^H} - \frac{R}{2} \sum_{i=7}^{3N} (\Theta^D_{i\,gas} - \Theta^H_{i\,gas})$$

$$+ \sum_{i=1}^{6} RT \ln \frac{\sinh \frac{\Theta^D_{ikond}}{2T}}{\sinh \frac{\Theta^H_{ikond}}{2T}} + \frac{R}{2} \sum_{i=7}^{3N} (\Theta^D_{i\,kond} - \Theta^H_{i\,kond}) \qquad (3)$$

results. If the square root of the mass ratio is replaced by the reciprocal ratio of translational vibrations and the square root of the moments of inertia by the reciprocal ratio of librations, formula (1) is obtained.

Neglecting anharmonicity and coupling of the vibrations, the first factor of equation (1) represents the influence of the translational and librational vibrations originating from the translations and rotations of the gas. As shown in fig. 2 (taken from /5/) this factor is smaller than one for a single intermolecular vibration and also of course smaller than one for the sum of all intermolecular vibrations. Therefore this factor decreases the vapor pressure ratio.

The second factor accounts for the changes of the intramolecular vibrations upon condensation. Since the changes are greater for hydrogen vibrations than for deuterium vibrations and are usually positive in the sum, this factor is greater than one and therefore increases the vapor pressure ratio. Consequently the vapor pressure isotope effect is normal when the first term is predominant and inverse if the second term predominates /5,9/.

Several difficulties are encountered if the vapor pressure ratio of the ices is calculated according to equation (1). There is no large error for the translational factor, because the ratio of translations is only about 1.034 for H_2O/D_2O and about 1.017 for H_2O/HDO. In addition the wavenumber of the largest translation is 229 cm^{-1} for H_2O-ice and 222 cm^{-1} for D_2O-ice /10/, corresponding to the region of T/Θ^H of about 1.0 of fig. 2, in which the factor q_i of a single vibration is about 0.998. The remaining translations occur at still lower frequencies and the factor q_i of a single translation is closer to one. Therefore factor $\prod_{i=1}^{3} q_i$ of all translations is assumed to be 0.998 - -0.997. A similar result is obtained by calculating that factor for all translations from the following equation/7/

$$\exp\{-\frac{1}{24}\cdot\left(\frac{3}{5}\right)^2\left(\Theta^{H^2}_{Debye}-\Theta^{D^2}_{Debye}\right)/T^2\}=\exp\{-92/T^2\} \quad (4)$$

choosing the value of 192 $^{\circ}K$ for Θ_{Debye} of H_2O-ice as determined by Giauque and Stout /11/ and this value reduced by a factor of $\sqrt{18/20}$ (square root of mass ratio) for Θ_{Debye} of D_2O-ice. Considering the other uncertainties of the calculation it seems best to neglect the influence of the translations and to set factor $\prod_{i=1}^{3} q_i$ equal to one.

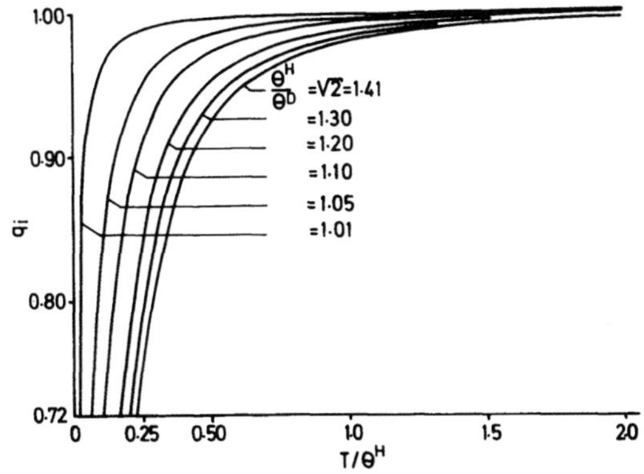

Fig.2 $$q_i = \frac{\Theta^H_{ikond}}{\Theta^D_{ikond}} \frac{\sinh \frac{\Theta^D_{ikond}}{2T}}{\sinh \frac{\Theta^H_{ikond}}{2T}}$$

as a function of T/Θ^H for different ratios Θ^H/Θ^D, ranging from 1.01 to $\sqrt{2}$ (taken from reference /5/)

The difficulties are more severe for the librations. The ratios of the librations are about 1.34 for H_2O/D_2O and about 1.17 for H_2O/HDO, and,based on the considerably higher wavenumbers the values of T/Θ_H are significantly lower than 1.0 (fig. 2). Correspondingly the factor q_i of a single vibration is essentially smaller than one and therefore the factor $\prod_{i=4}^{6} q_i$ of the librations has to be calculated. Unfortunately, however, librations give rise to broad and unresolved bands /12,16/. For the time being the maximum of the librational band can only be considered a good mean value and be used for each of the three librations. As the calculations are made for the temperature range between 0 and -40 °C, the wavenumbers which have been measured in this range by Giguère and Harvey /12/ should be used (table 1): measured at -15 °C these values are 800 cm^{-1} for H_2O-ice and 590 cm^{-1} for D_2O-ice. Similar values are also obtained by neutron scattering measurements (table 1). For HDO-ice a value of 683 cm^{-1} has been calculated from the ratio of the moments of inertia, choosing the above values for H_2O-ice and D_2O-ice and the mean of the values obtained from the following equations

Table 1 Recent measurements of librations of H_2O and D_2O, water and ice, given in cm^{-1}

Method	Authors	Temperature	Ice	Water	Ice	Water
Infrared	Giguère, Harvey /12/	9 °C		710		530
		-15 °C	800		590	
	Draeger, Williams /13/	28 °C		680		500
	Swain, Bader /14/	45 °C		667		
	Hornig, White, Reding /15/	-190 °C	812		605	
	Bertie, Whalley /16/	100 °K	840		640	
Raman	Walrafen /17/	10-95 °C		439 538 717		
	Weston /18/			(450) 800		
Neutron scattering	Prask, Boutin, Yip /19/	150 °K	650 (900)			
	Harling /20/	299 °K		625		480
		268 °K	750			

$$\nu^{HDO}/\nu^{H_2O} = \{(I_1 I_2 I_3)^{H_2O}/(I_1 I_2 I_3)^{HDO}\}^{1/6} = 0.845 \quad (5a)$$

$$\nu^{HDO}/\nu^{D_2O} = \{(I_1 I_2 I_3)^{D_2O}/(I_1 I_2 I_3)^{HDO}\}^{1/6} = 1.167 \quad (5b)$$

(I_1, I_2 and I_3 = principal moments of inertia).

Similar difficulties, if not greater ones, are encountered in the case of the intramolecular vibrations. The stretching vibrations of H_2O- and D_2O-ice are strongly perturbed, possibly by Fermi-resonance /15,21/ which occurs mainly between the symmetrical mode of stretching and the first overtone of the deformation, by coupling with the lattice vibrations /16/ or by both. To eliminate these difficulties we have chosen the stretching vibrations of HDO instead of those of H_2O and D_2O. For, as has been concluded from the small bandwidth, these vibrations of HDO-ice do not couple with the lattice /16/. Besides there is no possibility of Fermi-resonance, because the unperturbed stretching vibrations of HDO-ice do not coincide with the unperturbed overtone of the HDO-deformation vibrations. As a result of reducing both the valence force constants by the same quantities (both the H- or both the D-atoms of a particular molecule form hydrogen bonds) the OH-vibration of HDO-ice is the mean value of the symmetrical and the antisymmetrical stretching of H_2O-ice and the OD-vibration of HDO-ice is the mean value of both vibrations of D_2O-ice. Therefore by choosing the corresponding vibration of HDO the result is the same as if the calculation were made with the wavenumber of the symmetrical and the antisymmetrical stretching vibrations split by equal quantities due to the coupling resonance.

The wavenumbers used in the calculations are cited in table 2. Those used for calculating the ratio of vapor pressures of H_2O/D_2O are underlined and those for calculating the ratio of H_2O/HDO are overlined. The stretching vibrations cited for HDO-ice are those measured at 0 °C by Falk and Ford /21/, differing from those measured by Bertie and Whalley /16/ at 100 °K (in table 2 in brackets) by the temperature shift only. The values of the deformation vibrations to be used in every case are those of the compound in question.

The ratios calculated for temperatures between 0 and -40°C are given by the corresponding curves shown in fig. 3. As can be seen, the measurements made by other authors coincide satisfactorily with the calculated curves. Trying to use the values of the valency vibrations of H_2O- and D_2O-ice from table 2 instead of those of the stretching vibrations of HDO-ice the result of the calculations is not unequivocal because it is difficult to choose the proper frequencies, and in addition there is no agreement with the measurements.

Table 2 Wavenumbers of the intra- and the intermolecular vibrations used for the calculation of the vapor pressure ratios for the solid state. Values used for calculating values of H_2O/D_2O underlined, those of H_2O/HDO overlined

As-signment	H_2O Gas /22/	H_2O Solid /12,16/	D_2O Gas /22/	D_2O Solid /12,16/	HDO Gas /22/	HDO Solid /15,16,23/	$(\nu^H_{Gas}-\nu^H_{Ice}) - (\nu^D_{Gas}-\nu^D_{Ice})$ H_2O/D_2O	H_2O/HDO
ν_3	3756	3330-80	2788	2545	$\overline{3707}$	$\overline{3300}$ •	+ 120	0
				2485		(3277)		
		3220		2425				
ν_1	3652		2671	2360-95	$\overline{2727}$	$\overline{2440}$ •	+ 120	+ 120
		3110-50		2332		(2421)		
				2240				
ν_2	$\overline{\underline{1595}}$	$\overline{\underline{1650}}$	$\underline{1178}$	$\underline{1210}$	$\overline{1402}$	$\overline{1480}$ ••	− 23	+ 23
			$\sum\{(\nu^H_{Gas}-\nu^H_{Ice})-(\nu^D_{Gas}-\nu^D_{Ice})\}$			=	+ 217 cm^{-1}	+ 143 cm^{-1}
$\nu_{Libr.}$		800		590		683		

• Wavenumbers measured at 0 °C /21/

•• Mean value of the wavenumbers measured at 1470 and 1490 cm^{-1} /15,23/

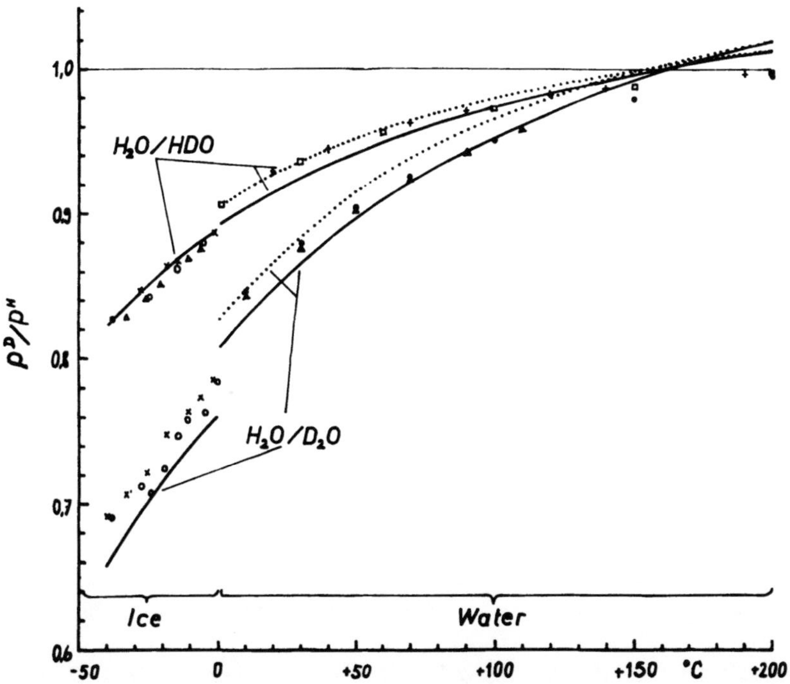

Fig.3 Calculated and measured vapor pressure ratios p^D/p^H of the H_2O/D_2O and H_2O/HDO pairs in the solid and liquid state.

——— Quotients calculated without the assumption of an equilibrium.

······· Quotients calculated with the assumption of an equilibrium, on the basis of mole-fractions of free groups /33/.

Measured values: H_2O/HDO-Ice: x/37/, o /38/, △ /39/;
H_2O/HDO-Water: +/40,41/, □ /42/;
H_2O/D_2O-Ice: x /37/, o /38/;
H_2O/D_2O-Water: ● /40,41/, ▲ /7/

Figure 4 shows besides the calculated ratios the partial factors as function of temperature. As can be seen, the partial factor of the intramolecular vibrations converges from higher values less rapidly toward 1.0 than the partial factor of the intermolecular vibrations increases toward 1.0, therefore the resulting ratio rises with increase in temperature. This explains the temperature dependence of the vapor pressure ratios.

In fig. 3 and 4 the curves calculated for the higher temperatures show the vapor pressure ratios and the partial factors of the liquid compounds as a function of temperature.

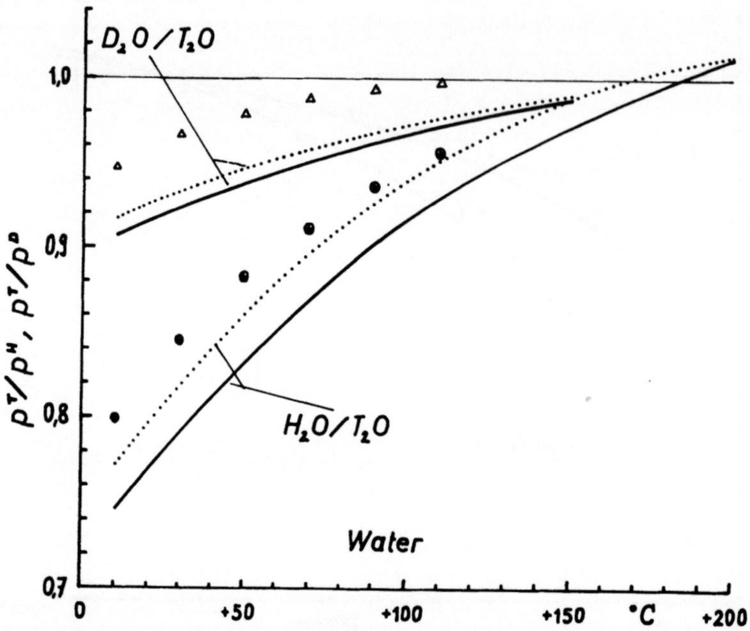

Fig.4 Calculated and measured quotients p^T/p^H and p^T/p^D of the H_2O/T_2O and D_2O/T_2O pairs in the liquid state.
——— Quotients calculated without the assumption of an equilibrium.
······ Quotients calculated with the assumption of an equilibrium, on the bases of mole-fractions of free groups /33/.
⊕ Measured quotients of the H_2O/T_2O pair.
△ Values of p^{T2O}/p^{D2O} determined from the measured values /7/ of p^{H2O}/p^{T2O} and p^{H2O}/p^{D2O}

Figure 5 also shows the curves for H_2O/T_2O and D_2O/T_2O in the liquid state. As is described more exactly in reference /24/ the ratios are calculated as follows:

1. Under the assumption of the continuum model of water structure, according to which liquid water forms an "essentially complete hydrogen bonded network with a distribution of hydrogen bond energies and geometries" /21/, loosening with increase of temperature only.

2. Under the assumption of a mixture model of water structure /21/, according to which water represents a temperature-dependent equilibrium of free and bonded OH- (or OD- or OT-) groups.

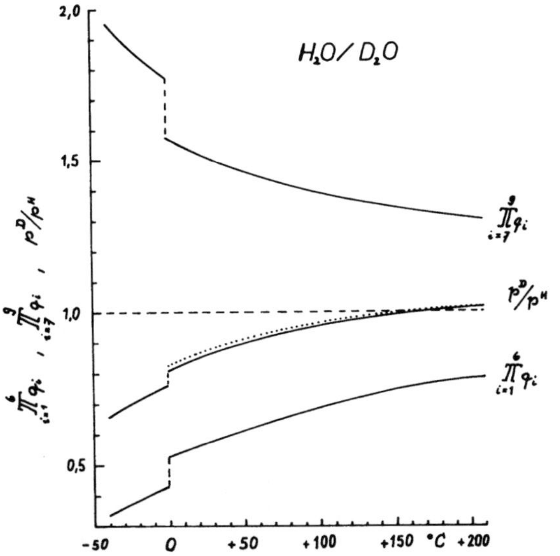

Fig.5 Partial factors of the H_2O/D_2O pair and the resulting vapor pressure ratios, $\prod_{i=1}^{6} q_i$ = partial factor of the intermolecular vibrations, $\prod_{i=7}^{3N} q_i$ = partial factor of the intramolecular vibrations, p^D/p^H = resulting vapor pressure ratios, all shown with solid lines. The ratios calculated assuming an equilibrium on the basis of free groups /33/ are shown with dotted lines

Assuming the continuum model, water can be treated as a crystalline system, the ratios can therefore be calculated with equation (1) again. Assuming the mixture model equation (1) has to be replaced by the relation

$$\frac{p^D}{p^H} = \left(\prod_{i=1}^{3N} q_i\right)_{\text{bonded}}^{1-x} \left(\prod_{i=1}^{3N} q_i\right)_{\text{free}}^{x} \qquad (6)$$

where x is the mole fraction of the free OH-groups, $\left(\prod_{i=1}^{3N} q_i\right)_{\text{bonded}}$ represents the ratio calculated with equation (1) from the vibrations of the bonded OH or OD, $\left(\prod_{i=1}^{3N} q_i\right)_{\text{free}}$ the ratio calculated with equation (1) from vibrations of free OH- or OD-groups. This relation will be obtained in a way analogous to the simplified derivation of equation (1), if it is taken into account, that with the existence of an equilibrium the Gibbs free energy of the condensed phase is defined by the relation

$$G_{Kond.} = x\, G^{free}_{Kond.} + (1-x)\, G^{bonded}_{Kond.} \quad .$$

The wavenumbers used are cited in table 3. The values tentatively used for x are those derived from dielectric measurements made by Haggis, Hasted and Buchanan /33/ (0.09 at 0 °C, 0.15 at 50 °C, 0.2 at 100 °C, 0.26 at 150 °C and 0.34 at 200 °C). Similar values have been obtained by other authors by other methods /34,35/.

As can be seen the ratios calculated with equation (1) (solid line) as well as those calculated with equation (6) (dotted line) are in satisfactory agreement with the measurements. The differences between the curves calculated with and without the assumption of an equilibrium are still within the limits of error, according to the simplifications of the calculation and to the uncertainties of the spectroscopic data which are largest in the case of tritium compounds. Besides, the temperature dependence of the ratios can be explained in the same way as that of the pairs of the solid compounds. Therefore, the calculation with the uncoupled vibrations seems justified with the pairs from the liquid compounds as well as with the pairs from the solid compounds.

Using symmetrical and antisymmetrical stretching vibrations for calculating the vapor pressure ratios both Van Hook /8/ and Jones /7/ chose a value above 3600 cm^{-1} for the antisymmetrical stretching vibrations of liquid H_2O. Since then it could be shown that this value has to be discarded and may not be used for calculating the vapor pressure isotope effect caused by the vibrations of the bonded OH-groups /17,36/. The correct value of the corresponding vibration of the bonded OH-group is supposed to be about 180 wavenumbers lower, which is a quantity approximately as large as the quantity which enters the calculation due to all internal vibrations. The agreement with the measurements seems to have been reached mainly by choosing very low values for the librations, 490-500 cm^{-1} /8/ respectively 439, 538 and 717 cm^{-1} /7/ (mean value: 565 cm^{-1}) for H_2O and corresponding values for D_2O. Furthermore, if a force field is used for the calculations/8/, agreement with the measurements may have in part also been obtained because the interaction force constants act as adjustable parameters. Due to its sensitivity against deviations of wavenumbers calculation of the vapor pressure isotope effect can therefore serve to ensure the assignment of the wavenumbers of the vibrations.

Table 3 Wavenumbers used for calculating the vapor pressure ratios of H_2O/D_2O, H_2O/HDO, H_2O/T_2O, and D_2O/T_2O for the liquid state

		H_2O	D_2O	T_2O	HDO	HTO	DTO
ν_3	Gas	3756 /22/	2788 /22/	2365 /25/	3707 /22/	3703 /25/	2733 /25/
	Liquid bonded	—	—	—	3405 /26/	3402 **	2528 **
	Liquid free *	3725 *	2765 **	2345 **	3676 **	—	—
ν_1	Gas	3652 /22/	2671 /22/	2233 /25/	2727 /22/	2295 /25/	2290 /25/
	Liquid bonded	—	—	—	2520 /26/	2120 **	2118 **
	Liquid free *	3627 *	2653 **	2218 **	2708 **	—	—
ν_2	Gas	1595 /22/	1178 /22/	998 /25/	1402 /22/	—	—
	Liquid bonded	1640 /28,29/	1205 /28/	1024 **	1455 /26/	—	—
	Liquid free *	1600 *	1181 **	1001 **	1406 **	—	—
$\nu_{Libr.}$	Liquid bonded	710 /12/	530 /12/	436 ***	610 ***	—	—
	Liquid free	<< $\nu_{Libr. Liquid bonded}$ (approximately full excited)					

* As has been shown /24/, in the calculation with equation (6) the vibrations of free groups can be replaced by the vibrations of the monomer molecules. The cited values of H_2O are those obtained for monomer molecules by measurements with matrix technique /27/.

** Values of D_2O, T_2O and HDO are calculated under the assumption that for corresponding vibrations of the pairs of isotopic compounds the relative solvent shift $(\nu_{gas} - \nu_{solute})/\nu_{gas}$ is approximately constant /30–32/.

*** calculated with equations (5).

References

/1/ Wolff, H., R. Würtz, Ber.Bunsenges.physik.Chem., to be published

/2/ Wolff, H., A. Höpfner, Ber.Bunsenges.physik.Chem. 71, 461 (1967)

/3/ Wolff, H., H.-E. Höppel, Ber.Bunsenges.physik.Chem. 72, 722 (1968)

/4/ Wolff, H., A. Höpfner, Ber.Bunsenges.physik.Chem. to be published

/5/ Wolff, H., A. Höpfner, Ber.Bunsenges.physik.Chem. 71, 730 (1967)

/6/ Bigeleisen, J., J.Chem.Phys. 34, 1485 (1961)

/7/ Jones, W.M., J.Chem.Phys. 48, 207 (1968)

/8/ Van Hook, W.A., J.Phys.Chem. 72, 1234 (1968)

/9/ Wolff, H., E. Wolff, H.-E. Höppel, Ber.Bunsenges. physik.Chem. 72, 644 (1968)

/10/ Bertie, J.E., E. Whalley, J.Chem.Phys. 46, 1271 (1967)

/11/ Giauque, W.F., J.W. Stout, J.Am.Chem.Soc. 58, 1144 (1936)

/12/ Giguère, B.A., K.B. Harvey, Canad.J.Chem. 34, 798 (1956)

/13/ Draegert, D.A., D. Williams, J.Chem.Phys. 48, 401 (1968)

/14/ Swain, C.G., R.W.F. Bader, Tetrahedron 10, 182 (1962)

/15/ Hornig, D.F., H.F. White, F.P. Reding, Spectrochim. Acta 12, 338 (1956)

/16/ Bertie, J.E., E. Whalley, J.Chem.Phys. 40, 1637 (1964)

/17/ Walrafen, G.E., J.Chem.Phys. 47, 114 (1967)

/18/ Weston, R.E. Jr., Spectrochim.Acta 18, 1237 (1962)

/19/ Prask, H., H. Boutin, S. Yip, J.Chem.Phys. 48, 3367 (1968)

/20/ Harling, O.K., The dynamics of liquid H_2O and D_2O and solid H_2O from the inelastic scattering of epithermal neutrons. Symposium on Inelastic Neutron Scattering, Copenhagen 1968 (Nuclear Science Abstracts 22, 26199 (1968)

/21/ Falk, M., T.A. Ford, J.Chem.Phys. 48, 244 (1968)

/22/ Benedict, W.S., N. Gailar, E.K. Plyler, J.Chem.Phys. 24, 1139 (1956)

/23/ Haas, C., D.F. Hornig, J.Chem.Phys. 32, 1763 (1960)

/24/ Wolff, H., E. Wolff, Ber.Bunsenges.physik.Chem., in press

/25/ Friedman, S., L. Haar, J.Chem.Phys. 22, 2031 (1954)

/26/ Waldron, R.D., J.Chem.Phys. 26, 809 (1957)

/27/ Van Thiel, M., E.D. Becker, G.C. Pimentel, J.Chem. Phys. 27, 486 (1957)

/28/ Thompson, W.K., Trans.Farad.Soc. 61, 2635 (1965)

/29/ Greinacher, E., W. Lüttke, R. Mecke, Z.Elektrochem., Ber.Bunsenges.physik.Chem. 59, 23 (1955)

/30/ Hallam, H.E. in M.Davies, Infrared Spectroscopy and Molecular Structure, Elsevier Publishing Co., Amsterdam (1963) p.405

/31/ Buckingham, A.D., Proc.Roy.Soc. (London) 248A, 169 (1958)

/32/ Buckingham, A.D., Trans.Farad.Soc. 56, 753 (1966)

/33/ Haggis, G., J.B. Hasted, J. Buchanan, J.Chem.Phys. 20, 1452 (1952)

/34/ Luck, W.A.P., Ber.Bunsenges.physik.Chem. 69, 626 (1965)

/35/ Pauling, L., Die Natur der Chemischen Bindung, Verlag Chemie Weinheim/Bergstrasse 1962, p.434

/36/ Walrafen, G.E., J.Chem.Phys. 48, 244 (1968)

/37/ Kiss, I., H. Jakly, H. Illy, Acta Chim.Acad.Sci. Hung. 47, 379 (1966)

/38/ Matsuo, S., H. Kuniyoshi, Y. Miyake, Science 145, 1454 (1964)

/39/ Merlivat, L., G. Nief, Tellus 19, 122 (1967)

/40/ Kirshenbaum, I., Physical Properties of Heavy Water, McGraw Hill Book Co., New York (1951)

/41/ Landolt-Börnstein, 6. Aufl. II, 2a, p.63

/42/ Merlivat, L., R. Botter, G. Nief, J.Chim.Phys. 60, 56 (1963)

SOME EXPERIMENTS ON THE REGELATION OF ICE

E. Hahne and U. Grigull

Institut für Technische Thermodynamik

Technische Hochschule München/Germany

Abstract

This classical experiment with a wire cutting through a block of ice was performed with variations in wire material, wire size and effective bearing pressure. As material, copper and "Perlon" were used with diameters ranging from 0.125 to 1 mm. Pressures of 5, 8 and 10 bar were applied. In order to investigate surface effects two different metallic materials, copper, and stainless steel (V2A), of markedly different thermal conductivity were coated with silver and insulating varnish. The experimental results presented in terms of speed of the wire cutting through the ice show good agreement with results predicted by a model theory based on heat conduction.

Experimental setup

The classical experiment for the regelation- or refreezing- phenomenon is performed with a wire cutting through a block of ice and nevertheless leaving the block undivided.

It was the object of our experiments to learn about the governing mechanism of this phenomenon.

For this reason different wires were used to penetrate through a block of ice 40 cm long and 12 cm in square. A sketch of the experimental set up is given in figure 1. Three wires of different diameters were used at a time

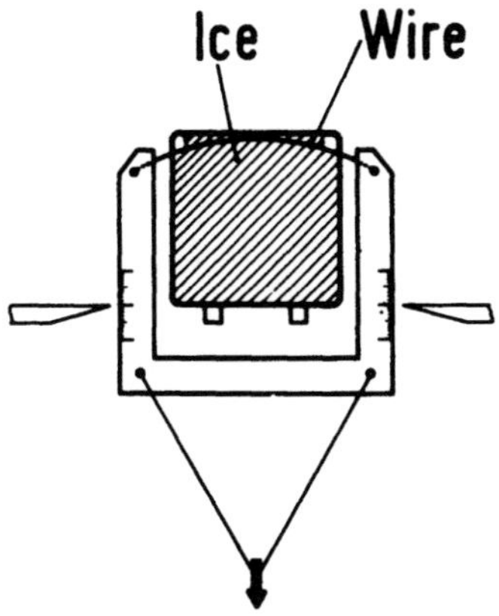

Fig.1 Experimental set up

in one run. Each wire was held in a rigid wooden frame which was loaded with such a weight that the bearing pressure was the same for each wire.

This pressure was calculated from the weight, the diameter and the bearing length of the wire. Since the experiments were performed in a room with a rather constant temperature of 22°C, heat was conducted along the wire into the ice by the so-called fin-effect. The wire length influenced by this effect could be calculated as well as measured. Since the surplus heat from the room prevented the refreezing of the water above the wire, a wedge-shaped mark showed the influenced length on either side of the block. This influenced, heated length was not considered as bearing.

The penetrating wires were always placed some distance away from the ends of the block, so that influences by edge effects or structural differences in the ice are kept small.

The descent of the wires within certain time intervals was observed from a millimeter scale on the frames and from this the penetration speed was calculated.

Preparation of ice

The first experiments had shown that the penetration speed is quite sensitive to impurities in the ice, especially to air.So it had to be a main object to produce constant quality ice. Clear transparent blocks were obtained from distilled water which had been kept boiling for at least 30 minutes in order to drive out the air. The freezing of the water took about 8 hours in a rectangular shaped container which floated in a cold bath. One long side of the container was open and the water was stirred continuously. Several times the liquid part with the remaining impurities and reabsorbed air was exchanged for fresh treated water.

Still remaining impurities collected in the upper part, in the section which froze last.

In figure 2 two blocks of ice are shown, one with air and the other treated as described.

In the test runs the blocks were turned upside down, so that the contaminated section formed the bottom part and could not affect the experiment.

Fig.2 Ice blocks

Experiments

In the experiments different wire diameters, different bearing pressures and different materials were applied.

The materials were selected with respect to their thermal conductivity, copper, with a thermal conductivity $k = 370$ W/m·deg, spring steel with $k = 30$ W/m·deg and Perlon with $k = 0.3$ W/m·deg were used.

The results of the experiments are presented in figure 3. The penetration speed is given as a function of wire diameter, with the mean pressure as parameter. There is a distinct difference in the behaviour of the two materials. The Perlon string is moving considerably slower than the copper wire. For both materials an increase in diameter causes a decrease in penetration speed, while

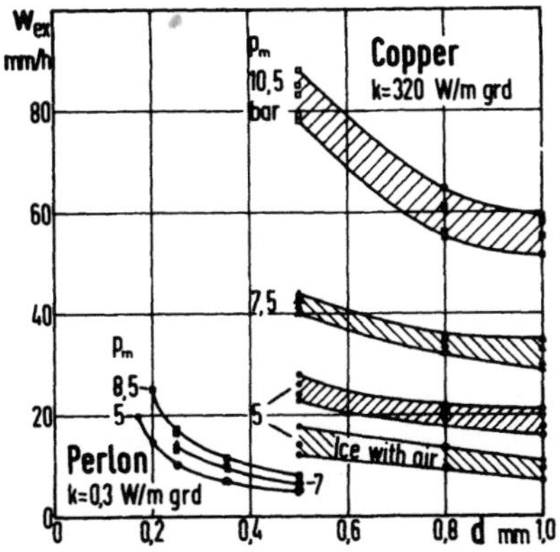

Fig.3 Penetration speed of copper and Perlon wires

an increase in pressure also increases the speed. There is a striking difference in the scatter of the results. The reproducibility of Perlon tests is much better than that of copper. Reasons for this will be discussed.

The plot also shows some results for ice with air inclusions. The penetration speed in this case is always below that for air-free ice.

In order to learn about possible surface effects /1/ copper wires and spring steel wires were either silver plated or coated with insulating varnish. Again these coating materials were chosen for their great difference in thermal conductivity. The results of these tests are presented in figure 4.

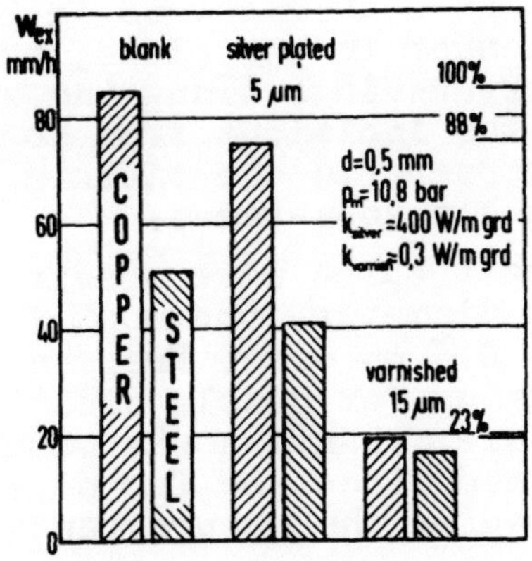

Fig.4 Penetration speed of silver-plated and varnished wires

The penetration speed is shown for blank, silver-plated and varnished wires. In all runs the diameter of the wires was constant 0.5 mm, and the applied bearing pressure 10.8 bar. The silver-plating was 10 μm thick, the varnish coating 15 μm.

The difference in speed between the two different materials is maintained even though the surfaces of the probes were the same: silver or insulating varnish.

The plot shows that silver plating causes a reduction in speed of about 12%, if the blank copper is taken as the 100% reference. The varnish layer decreases the speed by about 77%.

Observations in the experiments

Some observations made in the experiments might be worth mentioning. The scatter in the results for copper wires induced a number of variations in the performance of the experiments. It turned out that the best possible reproducibility was achieved by using the copper wires only once. Multiple use brings about uncontrollable deposits on the surface.

The main reason for the scatter of the experimental results was found in the appearance of bubbles along the wire. These bubbles were formed on the wire and were left behind in the ice while the wire moved downwards. About 15 minutes after their appearance, these bubbles vanished again in most cases. The occurence of bubbles was always connected with a decrease in wire speed.

In figure 5 these bubbles are shown in a block of ice. The decrease in speed is also recorded in this picture. If the wire remains longer in one place, more heat is conducted from the surroundings along the wire into the ice. This means that the incisions from both sides are deeper when bubbles appear. In the picture the deeper cuts indicate this bubble appearance.

In general the largest bubbles were observed on the thickest wire of 1 mm diameter, the most frequent appearance occurred on the 0.8 mm wire. Frequently used wires showed a greater tendency for bubble creation than new ones. Only little bubble formation was observed on silver-plated and varnished wires. So far no special investigation was performed on the nature and the causes of these bubbles.

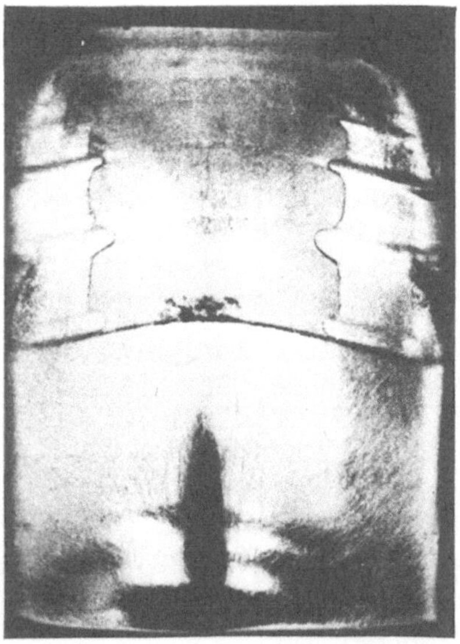

Fig.5 Bubbles on the wire

Theoretical considerations

The strong depencence of the penetration speed on the thermal conductivity of the wire led to the consideration that heat conduction is the significant parameter.

Two different models, for high and low thermal conductivity specimen were assumed and are presented in figure 6. Only steady state conditions will be considered, as far as heat conduction is concerned. In either case, the specimen is assumed to be surrounded by a very thin layer of liquid water.

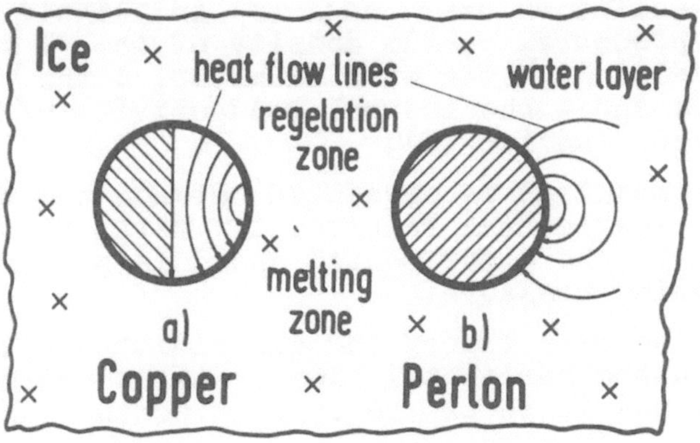

Fig.6 Heat conduction model

This originates from the melting of the ice below the wire, is pressed around the wire and freezes again above it. In freezing, the latent heat of melting is set free above the wire, while below the wire it is needed to melt the ice.

So it is assumed that heat will flow from the upper freezing section to the lower melting section. The paths which it takes depend on the thermal resistance or the thermal conductivity of the layers passed.

With copper of a high thermal conductivity and consequently a low resistance, the flow lines are assumed to go through the water layers and the wire as shown in a) of figure 6.

Perlon on the contrary has a low thermal conductivity, so the string represents a high resistance and the flow lines are assumed to go around the wire through the ice, which has a thermal conductivity greater by an order of magnitude than that of Perlon.

For these model cases, the Laplace equation was solved analytically. Using Fourier's equation and the Clausius-Clapeyron relation a heat balance was established from which the penetration speed could be determined as given in the following equations.*

For model fig.6a:

$$\underline{w}_a = \pi T (\underline{v}_s - \underline{v}_l) P / 4 \rho_I l^2 (d/\lambda_{Cu} + 2\delta/\lambda_{H2O}) \quad (1)$$

and for model fig.6b:

$$\underline{w}_b = \pi T (\underline{v}_s - \underline{v}_l) P \lambda_I / 4 d \rho_I l^2 \quad (2)$$

where $\underline{w}$ is the penetration speed, T the melting temperature, $\underline{v}$ the specific volume of solid (s) and liquid (l), p the bearing pressure, ρ_I the density of ice, l the latent heat of melting, d the wire diameter, δ the water layer thickness and λ the thermal conductivity of copper (Cu), water (H_2O) and ice (I).

In figure 7, comparison is presented between the theoretical and the experimental results. The ratio of the theoretical speed to the experimental speed is plotted against the bearing pressure.

Full agreement between theory and experiment would result in data on the horizontal line 1. As an explanation for the deviations it should be considered that the

*The detailed calculation procedure will be published later.

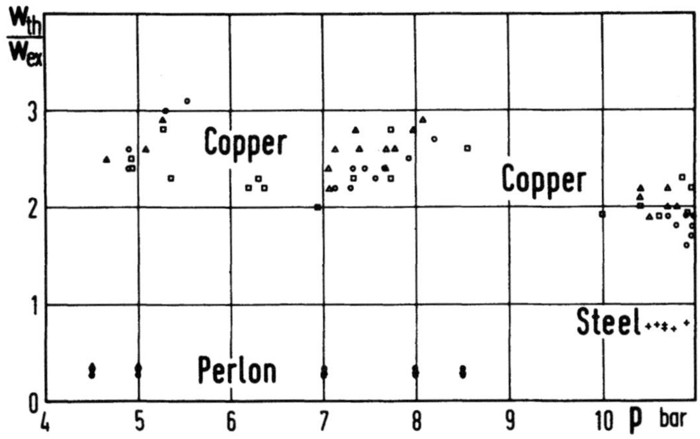

Fig.7 Comparison between theory and experiment

theoretical speed in the copper model requires an assumption for the waterlayer thickness. For the sake of simplicity a value of 1 μm was chosen. An increase to 2.3 μm, however, would bring the data points well around the full agreement line.

For the Perlon model it has to be realised that the ratio of the thermal conductivities of ice and Perlon only amounts to 7 while this ratio for copper and ice is about 170.

This means that the model a) for copper seems appropriate while the model b) for Perlon needs corrections for that part of the latent heat which passes through the Perlon string. Such a correction based on the thermal conductivities and the heat path lengths brings the data also closer to the full agreement line. The correction, however, again depends on assumptions. For the spring steel data also it has to be considered that heat will flow through the wire and around it, so that none of the mentioned models applies exclusively.

The theoretical consideration of the silver-plated wires predicts a decrease in speed of about 1% due to the additional silver layer. The experiments, however, give a decrease of 12% for copper and 18% for steel. Since the heat path lines should not be changed very much by the silver layer, the model a) should still be valid and the extra decrease in speed might indicate an additional effect.

The small difference in penetration speed with the varnished wires might be explained by the drastic change, from the model a) to the model b) heat flow pattern, where the wire material is of no influence.

Concluding it might be stated that heat conduction seems to play the dominant part in regelation.
The deviations of the silver-plated wires from theory indicate, that additional effects such as surface interactions might be present.

Acknowledgement

The authors gratefully acknowledge the assistance of R. Habeck and J. Hoiss in the execution of the experiments.

References

/1/ Weyl, W.A., J.Coll.Sci 6, 389-405 (1951)

/2/* Nye, J.F., Phil.Mag. Ser.8, 16, 1249-1266 (1967)

/3/* Frank, F.C., Phil.Mag.Ser. 8, 16, 1267-1274 (1967)

/4/* Townsend, D.W., R.P.Vickery, Phil.Mag.Ser. 8, 16 1275-1280, (1967)

/5/* Nunn, K.R., D.M. Rowell, Phil.Mag.Ser.8, 16 1281-1283 (1967)

* After submitting this paper, the authors learned that work on regelation had been carried out by Dr.J.F.Nye, Professor F.C.Frank and others at H.H.Wills Physics Laboratory, University of Bristol.

CALORIMETRIC STUDY OF GLASS TRANSITION OF THE AMORPHOUS ICE AND OF THE PHASE TRANSFORMATION BETWEEN THE CUBIC AND THE HEXAGONAL ICES

Masayasu Sugisaki, Hiroshi Suga and Syûzô Seki

Department of Chemistry, Faculty of Science

Osaka University, Toyonaka, Osaka, Japan

Abstract

By making use of the vapor-condensation type calorimeter, heat capacity measurements of amorphous, cubic and hexagonal ices were made between 20°K and 250°K. The glass transition phenomenon was found near 135°K with the sudden change of the heat capacity amounting to 35 J/(mole °K). The drastic crystallization with the exothermic effect which amounts to 1.64 kJ/mole was then followed at this temperature. The transformation of the cubic crystal to the hexagonal one was confirmed to occur in the temperature region from 160° to 210°K and proceeds in two steps with the activation energies of 21.3 and 44.7 kJ/mole, respectively. The magnitude of the exothermic effect accompanying this transformation was found to be about 160 J/mole.

1. Introduction

Since the work by Burton and Oliver /1/ in 1935, it has been reported by many workers /2-14/ that so-called amorphous state of water may be obtained when the water vapor is condensed onto the chilled substrate below about 140°K and that the amorphous ice changes spontaneously into the cubic ice upon heating and ultimately transforms into the hexagonal ice very sluggishly. The historical survey of the earlier works is tabulated in table 1, though it may be not exhaustive. As to the glass transition and to the temperature interval of the cubic ice, however, there exist confusion and contradiction as shown in this table.

In 1952 Pryde and Jones /3/ have first tried to find out the glass transition phenomenon of the amorphous state deposited on a chilled substrate, and such trials have been followed by many workers. They are, for example, Ghormley (1956) /4/, de Nordwall and Staveley (1956) /5/, and McMillan and Los (1965) /11/. Nevertheless, there had been so successful work before the study with a differential thermal analysis method by McMillan and Los who have found an endothermic effect at 139°K just below a crystallization temperature for an amorphous ice. Ghormley (1967, 1968) /12/ has recently reinvestigated thermal analysis by means of warming curves. He suspected the interpretation of DTA curve by McMillan and Los and denied strongly the existence of the glass transition phenomenon. More recently, Yannas (1968) /13/ pointed out the elusive nature of the glass transition point and derived this temperature indirectly.

In order to clarify the contradiction existing in the literature, and further to study the quantitative thermodynamic properties of various states of aggregation of water, we have constructed a novel type of calorimeter and really succeeded to observe the glass transition phenomenon for amorphous ice. Also we have measured the heat of crystallization of the glassy ice and the heat of transformation from the cubic to the hexagonal ices.

2. Experimental

Calorimeter. The calorimeter consists of a filling tube made of monel and soft copper, a copper cell with a platinum resistance thermometer, inner and outer adiabatic shields, a copper block as thermal station, and a vacuum jacket and others. The filling tube, two adiabatic shields and the copper block are all wound closely with constantan heaters.

The total view of the calorimeter and the sketch of the sample cell are given in figs. 1 and 2. As is shown in fig.2, a part of the filling tube enters into the sample cell. During the procedure of sampling, this part of the filling tube is warmed by the heat conduction from the upper part of the filling tube which is wound closely with the constantan heater. The part of the filling tube inside the cell can be kept at about 15°C, when the temperature of the wall of the sample cell is adjusted

Table 1. Summary of Previous Experimental Data.

Experimental Method	Temperature Region (°C) -180 -160 -140 -120 -100 -80 -60	Worker
X-ray diffraction	amorphous / semi-crystalline / hexagonal	Burton & Oliver (1935)
thermal analysis	amorphous / crystalline	Staronka (1939)
X-ray diffraction	small crystals / intermediate range not investigated / hexagonal	Vegard & Hillesund (1942)
electron diffraction	small crystals / cubic / hexagonal	Konig (1942)
thermal analysis	amorphous / crystalline	Pryde & Jones (1952)
electron diffraction	crystal growth poor / cubic / hexagonal	Honjo et al. (1956)
thermal analysis	amorphous / crystalline	Ghormley (1956)
thermal analysis	amorphous / crystalline	de Nordwall & Staveley (1956)
electron diffraction	amorphous or small crystals / cubic / hexagonal	Blackman & Linsgarten (1957)
X-ray diffraction	amorphous / cubic / hexagonal	Shallcross & Carpenter (1957)
X-ray diffraction	amorphous / cubic / hexagonal	Dowell & Rinfret (1960)
electron diffraction	halo pattern / cubic / hexagonal	Shimaoka (1960)
X-ray diffraction	amorphous / cubic / hexagonal	Beaumont et al. (1961)
differential thermal analysis	glass / cubic / hexagonal	McMillan & Los (1965)
	supercooled liquid?	
differential thermal analysis	glass	Angell et al. (1967)
thermal expansion	glass	Yannas (1968)
thermal analysis	amorphous / cubic / hexagonal	Ghormley (1968)

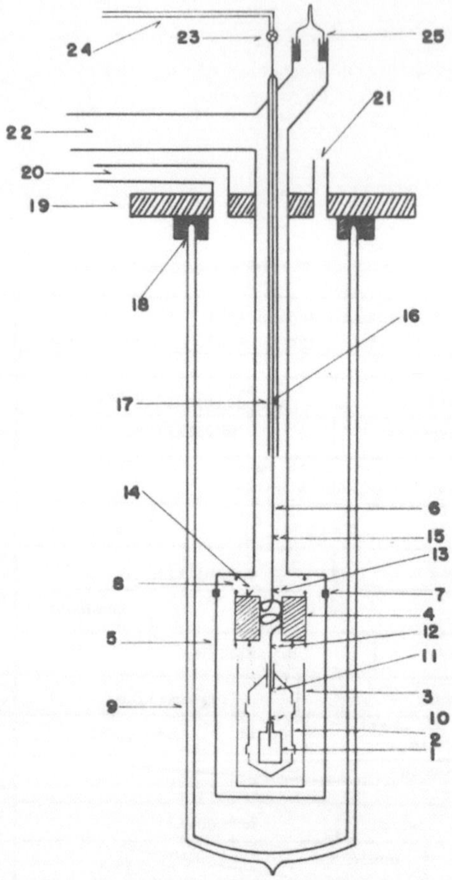

Fig. 1 Description of the vapor-condensed calorimeter: 1) sample container, 2) inner shield, 3) outer shield, 4) copper block, 5) vacuum jacket, 6) filling tube, 7) wood alloy joint, 8) nylon thread, 9) glass dewar, 10) - 16) thermo-couples, 17) sheath of the filling tube, 18) picein joint, 19) brass plate for supporting the apparatus, 20) outlet tube, 21) tube for introducing liquid hydrogen or nitrogen, 22) to vacuum pumps, 23) needle valve for sampling, 24) glass tube for introducing the vapor sample, 25) picein joint for taking out of lead wires

to be about $80^{o}K$ and the outer part of the filling tube is warmed around $30^{o}C$ under high vacuum condition of about 10^{-6} Torr inside the cell. In this case the heat conduction between the filling tube and the sample cell

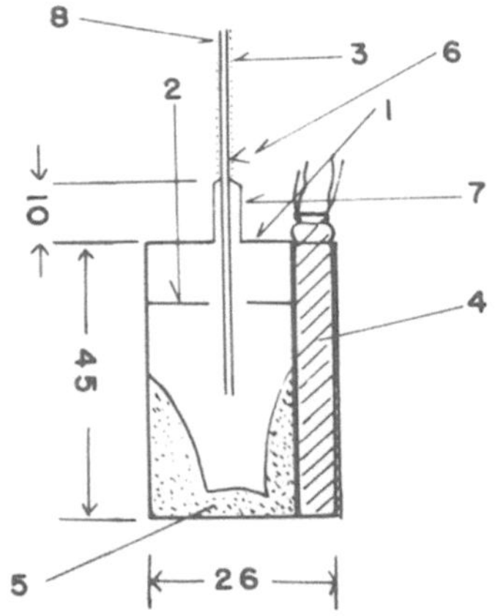

Fig.2 Sectional view of the sample cell:

1) wall of the cell (made of copper), 2) radiation shield, 3) filling tube, 4) platinum resistance thermometer provided with heater for energy-input, 5) sample, 6) thermocouple, 7) stainless tubing, 8) heater for sampling

occurs spontaneously through the outer short stainless steel tubing put on the top of the cell. The amount of this heat conduction is estimated as 1.3 (J/s), when the above temperature difference is maintained. Therefore, the effect of the thermal switch is necessitated whose power of the heat exchange is at least larger than 1.3 (J/s). For this purpose the conduction gas method by using helium gas is sufficiently adequate when the apparatus is operated above the boiling point of liquid nitrogen.

Preparation of the vitreous water. The vacuum jacket is immersed into the liquid nitrogen, and helium gas at one atmospheric pressure is introduced as a conduction gas into the vacuum jacket, the filling tube and the copper block being simultaneously warmed up at the desired temperature. Their temperatures are continuously registered by several thermocouples. The temperature of the sample cell is watched by recording the resistance of the platinum thermometer on the "Speedomax" recorder (type G, Leeds & Northrup).

When the appropriate temperature distribution is established within the calorimeter, the water vapor is slowly condensed into the sample cell by opening the needle valve situated at the top of the calorimeter. The rate

Table 2. Vitrification condition and some experimental data on glassy water.

Number of sample	Rate of condensation (g/hr)	Total amount of sample (g)	Heat of crystallization (kJ/mole)	ΔC_p at Tg (J/mole°K)	Amorphous part of sample (%)
1	0.032	0.6054	1.64 ± 0.05	35	100(assumed)
2 ·	0.033	0.5897	1.46 ± 0.05	29	89.0
3 ··	0.024	0.6959	1.06 ± 0.05	19	64.6
4 ···	0.030	0.7745	0.85 ± 0.05	11	49.0

· annealed at 128°K for 4 hours.
·· annealed at 128°K for 16 hours.
··· annealed at 128°K for 13 hours.

of condensation must be regulated to be very slow, since the surface of the amorphous sample produced in the cell may be warmed up above a devitrification temperature with the accumulation of the heat of condensation. The rate of condensation, the total amount of the sample actually employed and the annealing condition prior to the measurement are listed in table 2. During the vitrification, the temperature of the sample cell was kept between 103° and 106°K for sample (1), and 120° and 123°K for sample (2), (3) and (4).

Heat capacity measurement. After the sampling is over, the heaters of the filling-tube and of the copper block are all switched off, and the conduction gas in the vacuum jacket is evacuated down to 10^{-6} Torr for the realization of the adiabatic condition. Then helium gas (20 Torr at 80°K) is introduced in the calorimeter cell in order to facilitate the attainment of the thermal equilibrium.

Heat capacity measurements were done in the temperature range from 20° to 250°K. The calorimetry and thermometry circuits were already published /15/. When the measurement is finished, the calorimeter cell is heated up and the sample in the cell is gathered into the glass tube equipped outside the calorimeter for measuring the weight.

3. Experimental results*

The experimental data of heat capacity of amorphous ice are drawn in figs. 3 and 4, together with those of hexagonal ice. As is seen in these figures, the amorphous ice obtained by vapor condensation method exhibits an anomalous heat capacity jump characteristic of the glass transition phenomenon at about 135°K. On further heating, the drastic exothermic effect due to its crystallization is found to occur just above the glass transition point T_g. This jump of heat capacity at T_g and the total amount of the exothermic effect accompanying the crystallization are summarized in table 2.

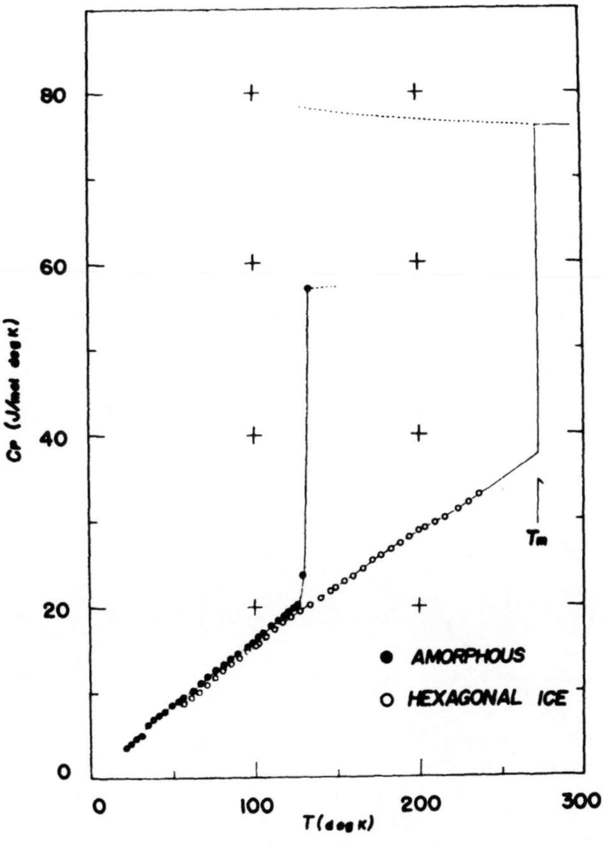

Fig.3 Heat capacity curve of water (sample(1))

* The detailed numerical data of the heat capacities will be published in Bull.Chem.Soc. Japan 41, No.11 (1968)

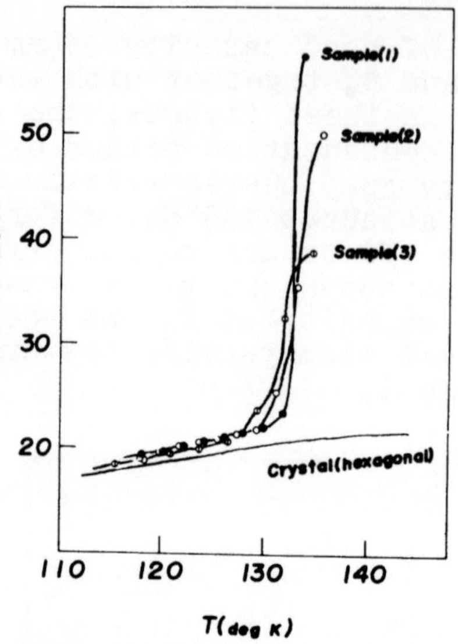

Fig.4 Heat capacity curve of water near the glass transition temperature

Results of heat capacity of cubic ice are given in figs. 5 and 6 in comparision with those of hexagonal ice.

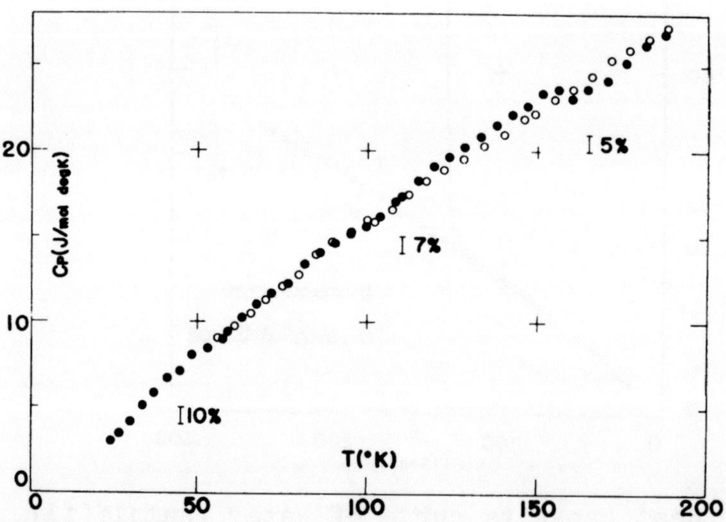

Fig.5 Heat capacity of ice: The filled circle represents the cubic ice and the hollow circle the hexagonal one

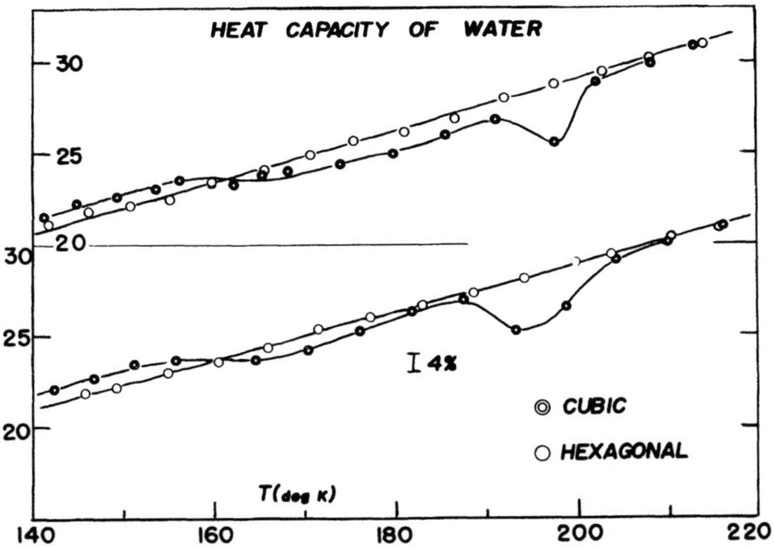

Fig.6 Heat capacity of ice near the transformation temperature.
The lower figure is result for the sample (3) and the upper for the sample (4)

Heat capacity of cubic ice shows anomalous region between 160° and 210°K. The total exothermic effect was estimated for each sample as is given in table 3.

Table 3. Total amount of exothermic effect accompanying irreversible transformation of the cubic ice to the hexagonal one.

Number of sample	$-\Delta H$ (J/mole)
2	158.9 ± 15
3	164.4 ± 15
4	159.5 ± 15

4. Discussion

Realization of a glassy state. From the results of our heat-capacity measurement, it is quite clear that the amorphous water condensed from the vapor onto the substrate chilled below 135°K shows a glass transition phenomenon. This fact shows that the condensed amorphous water exists in the glassy state below 135°K. This new finding of the glass transition phenomenon gives a definite resolution to the question whether the non-crystalline solid obtained by the vapor-condensation method is really a glass or not.

It is to be noted that the jump height of heat capacity at the glass transition point, and the heat of crystallization have different four values from sample to sample (see table 2). This fact suggests that each of samples contains a few amount of a crystalline part. This crystalline part of the sample is considered to appear during the process of annealing. This is evidenced by the fact that the exothermic effect was observed even after the annealing at 128°K for 16 hr, and that the heat of crystallization gets smaller as the time of duration for annealing is longer.

Sample (1) exhibits the hightest jump of heat capacity ΔC_p at T_g among all samples. The heat capacity just above T_g of this sample is, however, appreciably smaller than that of the supercooled liquid at the same temperature assumed by the simple extrapolation of the heat capacity of the liquid above the melting point. This fact is, at the first glance, considered to be due to the existence of crystalline part even in this sample. This interpretation seems, however, to be inadequate because of some thermodynamical contradiction. If C_p values of the supercooled liquid are assumed to be extended like a dotted line in fig.3, the corresponding enthalpy curve of the supercooled liquid comes out to be drawn in fig.7. The extrapolated curve of the supercooled liquid intersects that of the crystal far above 135°K ($=T_g$). This circumstance is evidently incompatible with the fact that the devitrification is necessarily accompanied by the exothermic effect. Consequently, it turns out to be inappropriate to assume that the C_p values of the supercooled liquid are equal to those obtained with extrapolating the C_p values of the liquid above the melting point.

Although at the present time we cannot investigate the details of temperature dependence of the heat capacity of the supercooled water by the usual experiment, we can expect that this phenomenon belongs undoubtedly to the same category that the physical properties of water are often abnormal compared with normal liquids.

<u>Appearance of the cubic ice and its transformation into the hexagonal ice</u>. Up to the present, the existence of

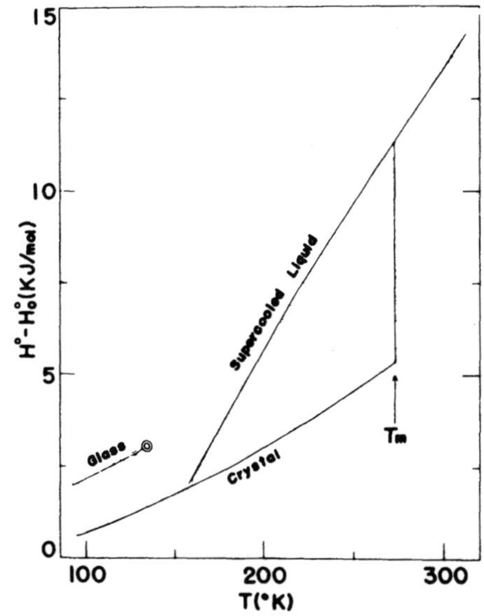

Fig.7 Enthalpy relation of various states of water

sluggish and irreversible transformation of the cubic ice to the hexagonal one has been observed by several workers. There has been, however, no quantitative experiment on the sample at the thermodynamical equilibrium state, so there remains still a confusion about the temperature interval of the appearance of the cubic ice, and also a more important problem whether the cubic ice is more stable than the hexagonal ice or not in the lower temperature region.

According to our present heat capacity measurements, the cubic ice transforms sluggishly and irreversibly into the hexagonal ice with an appreciable exothermic effect between 160°K and 210°K. The average total amount of the exothermic enthalpy change, 160 J/mole, is close to the value, 150 J/mole, previously reported by McMillan and Los.[+] This enthalpy change is considered to be due to the difference of the lattice energy between these two structures. Such an exothermic phenomenon seems to be concordant with the result calculated by Bjerrum /16/ i.e., the hexagonal ice is more stable than the cubic ice with regard to its lattice energy.

[+] This agreement is just fortuitous, since their value is based on the Ghormley's (1956) old value (1.26 kJ/mole) instead of his revised one (1968) (1.89 kJ/mole).

Now, the kinetics of this transformation has been studied by several previous workers who have interpreted this phenomenon in terms of the activation energy /11, 17/. McMillan and Los have reported that this irreversible transformation may be characterized by the activation energy of about 7.4 kcal/mole, which is calculated based on the X-ray data by Dowell and Rinfret /8/. On the other hand Nozik and Kaplan /17/, by use of the Mössbauer resonance method for the ice doped with ferrous ion, have reported that the transformation occurs at least in two stages with a certain intermediate state, and that the activation energy for the process of the cubic to intermediate comes out to be about 29 kcal/mole.

We have also estimated the activation energy for this irreversible phenomenon from the results of heat-capacity measurement. When the cubic ice transforms into the hexagonal one with the exothermic effect, the temperature of the sample rises spontaneously if an adiabatic condition is well maintained. Concerning this phenomenon, the following equations were assumed to hold;

$$\dot{T} = (\Delta H/C_p)\cdot\dot{x} = 1/\tau(\Delta H/C_p)\cdot x$$

and

$$\dot{x} = (1/\tau)\cdot x \ , \tau \sim \exp(E/RT) \ ,$$

here $\dot{T}$ is the time-derivative of the temperature of the sample, C_p the heat capacity of the sample including the sample-container, ΔH enthalpy change accompanying the cubic-hexagonal phase transition , $\dot{x}$ the time-derivative of the mole number of the cubic ice, τ the relaxation time characteristic of this phenomenon, x the mole number of the cubic ice, and E the activation energy for this rate process. The values of $\dot{T}$, C_p, ΔH, and x are all obtained as a function of temperature from the results of heat-capacity measurement, so τ is easily calculated for all temperatures. The obtained relation between $\log\tau$ and 1/T is drawn in fig.8. From this figure we see that this rate process may be separated into two processes, one of which is characterized with the activation energy of about 5 kcal/mole, the other with the activation energy of about 11 kcal/ mole. The fact that this phenomenon occurs in two stages in our study is quite concordant with the results by Nozik and Kaplan, but the values of the activation energy are different from each other. Taking into consideration the difference that our experiments are based on the pure ice, whereas Nozik and Kaplan's are based on the frozen solution, it is not quite clear to what extent such a comparison is plausible.

It is to be noted here that the activation energy of about 11 kcal/mole concerning the higher-temperature process is concordant with the result of Dowell and Rinfret. As their experiment was carried out with the X-ray analysis method, it may be possible that they have observed mainly the higher-temperature process for which the activation energy is pretty high.

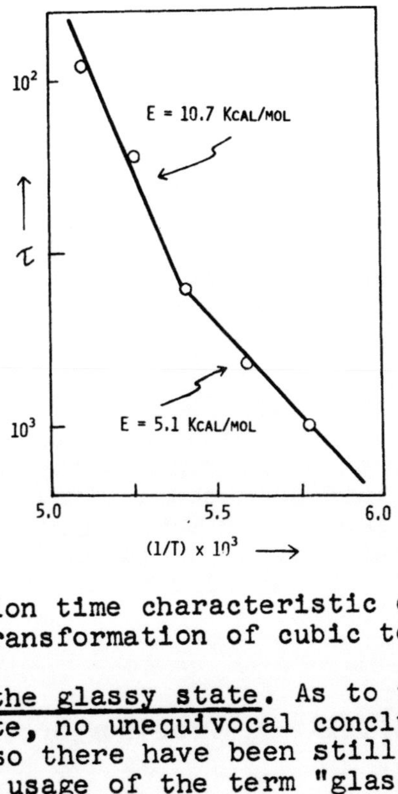

Fig.8 Relaxation time characteristic of the irreversible transformation of cubic to hexagonal ice

Definition of the glassy state. As to the definition of the glassy state, no unequivocal conclusion has been given hitherto, so there have been still a few confusions concerning the usage of the term "glassy state" /18/. This seems to be due to the fact that the past definition of the glassy state are almost concerned with the chemical composition of the glass or with the methods of obtaining the glassy state.

Commonly, a glassy state has been considered to be equivalent to a non-crystalline or amorphous solid obtained by supercooling the liquid. However, the noncrystalline solid can be obtained also by using other rather uncommon methods, such as vapor condensation, neutron bombardment, and others/18/. Such a non-crystalline solid has been observed to crystallize on heating. For this kind of solid, the glass transition phenomenon has not ever been verified to exist. As far as the glass transition phenomena are not observed, we may feel a certain hesitation to call this non-crystalline solid as a glass. However, our previous finding of the glass transition phenomena of methanol /19/ and the same result for water prepared by the vapor condensation method in the present study conclude that the non-crystalline solids of these materials prepared by this uncommon method must be called a glass. On the basis of our new observation we may propose more general properties of the glassy state as follows:

(a) The glassy state must be distinguished from other states only by the fact that the thermodynamically equilibrium state is never attained.
(b) The glassy state must show the glass transition phenomenon which characterizes the boundary between the glassy and the supercooled liquid states.
(c) The glassy state is established irrespective of either the method of preparation or the chemical composition of the material.

5. Conclusion

The confusion or contradiction concerning the long-unsettled problems, i.e. the existence of the glassy state as well as the temperature interval of the appearance of the cubic crystal, has been conclusively clarified by our present experiment under thermodynamically equilibrium conditions. The intricate status in the past of many investigations may be attributed to the dynamical nature of their experiments.

References

/1/ Burton, E.F., W.F.Oliver, Proc.Roy.Soc. A 153, 166 (1935)

/2/ Staronka, L., Roczniki Chem. 19, 201 (1939)

/3/ Pryde, J.A., G.O.Jones, Nature 170. 685 (1952)

/4/ Ghormley, J.A., J.Chem.Phys. 25, 599 (1956)

/5/ de Nordwall, H.J., L.A.K.Staveley, Trans. Faraday Soc. 52, 1061 (1956)

/6/ Blackman, M., N.D.Lisgarten, Proc.Roy.Soc. A 239, 93 (1957)

Lisgarten, N.D., M.Blackman, Nature 178, 39 (1956)

/7/ Shallcross, F.V., G.B.Carpenter, J.Chem.Phys. 26, 782 (1957)

/8/ Dowell, J.G., A.P.Linfret, Nature 188, 1144 (1960)

/9/ Shimaoka, K., J.Phys.Soc.Japan 15, 106 (1960)

/10/ Beaumont, R.H., H.Chihara, J.A.Morrison, J.Chem.Phys. 34, 1456 (1961).

/11/ McMillan, J.A., S.C.Los, Nature 206, 806 (1965)

/12/ Ghormley., J.A., J.Chem.Phys. 46, 1321 (1967); 48, 503 (1968)

/13/ Yannas, I., Science 160, 298 (1968)

/14/ König, H., Nachr.Akad.Wiss. Göttingen, No.1, 1 (1942);
Z.Krist. 105, 279 (1943)

/15/ Suga , H., S.Seki, Bull.Chem.Soc.Japan 38, 1000 (1965)

/16/ Bjerrum, N., Kgl.Danskab.Selskab. Math-fys.Medd. 27, 1 (1951)
Science 115, 385 (1952)

/17/ Nozik, A.J., M.Kaplan, Chem.Phys.Letters 1, 391 (1967)

/18/ Secrist, D.R., J.D.Mackenzie, "Modern Aspect of the Vitreous State" ed. by J.D.Mackenzie (Butter Worths Scientific Publication, Ltd., London,) vol.3, chapter 6 (1964)

/19/ Sugisaki, M., H.Suga and S.Seki, Bull.Chem.Soc. Japan 40, 2984 (1967)

THE SPECIFIC HEAT OF ICE Ih

M.A. Pick

Physik-Department

Technische Hochschule München, Germany

Abstract

The specific heats of pure and doped ice Ih single crystals were measured from 80°K to 230°K with a differential heat-flux calorimeter. In the region of ≈ 120°K the specific heat exhibits a positive anomaly.

Introduction

The specific heat of pure, polycrystalline ice Ih was measured very precisely by Giauque and Stout /1/ in 1936. With the help of these measurements it was discovered that ice has a zero-point entropy equal to (0.82±0.05) cal/°K mole. It is now generally agreed that this zero-point entropy is due to the completely frozen-in configurational disorder of the hydrogen lattice at low temperatures. A number of authors have calculated the theoretical value for this zero-point entropy /2,3,4,5,6/. Nagle /5/ solved the problem for three dimensional ice to a high degree of accuracy and his value, S_o = (0.8145±0.0002)cal/°K mole, agrees very well with experiment.

The results of various other experiments on ice have been interpreted as being due to an ordering of the proton orientations in the temperature range around 100°K /7,8,9/. An ordering of the proton orientations is connected with a change of entropy of the system and this, in turn, will cause an anomaly in the specific heat. Giauque and Stout /1/ found a small but definite anomaly between 85 and 100°K. An appreciable amount of order at low temperatures would, of course, contradict the experimentally found zero-point entropy.

It was of interest, therefore, to reinvestigate the specific heat of pure and doped ice as a function of temperature in the temperature range in question.

Experimental

The specific heat of pure and doped, mono- and poly-crystalline ice was measured by means of a heat-flux calorimeter /10,11/. This type of calorimeter allows the continual measurement of the heat capacity during heating. The ice sample was usually cooled at a rate of about 1°K/min, then kept at the boiling-point of nitrogen for 2 h. It was then heated at a rate of approximately 1°K/min. The particular calorimeter in use was characterized by a very good relative but a less satisfactory absolute accuracy. It was possible to measure changes in the specific heat amounting to less than 2×10^{-4} cal/°K gm. Variations in the absolute value of up to 10% were measured in different measurements. In view of the fact that the thermal conductivity of ice is rather poor, very small samples were used. The monocrystalline samples consisted of four or five cylinders of ice, each measuring approximately 3 mm in diameter and 30 mm in length. Together, they amounted to about 0.5 g. These samples were cut from a large single crystal with a lathe.

The cylinders were introduced into a thin gold-plated copper tube and consecutive measurements were made with and without the ice and subsequently substracted from one another. The polycrystalline samples were made by simply freezing the highly distilled, and if required, doped water in the copper tube itself.

Results

A very small but well reproducible effect was observed in every measurement made on pure ice. The gradient of the specific heat versus the temperature changed abruptly at about 100°K and then again at about 120°K, fig.1. No difference in the shape or size of the anomaly could be found between mono- and polycrystalline ice. In the measurements made on HF doped ice (concentration range 1×10^{-4} mole/ℓ ..3×10^{-3} mole/ℓ), the effect increased in size but the shape was not very well reproducible for different samples. The nature of the anomaly found in pure ice is similar to that found by Giauque and Stout /1/ between 85 and 100°K.

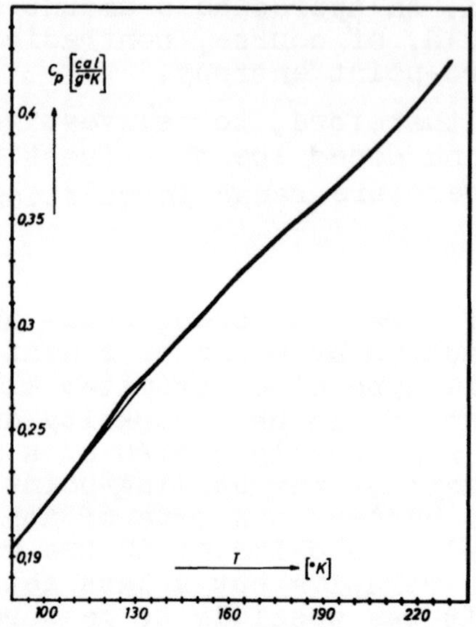

Fig.1 Specific heat of pure single crystal of ice

Assuming the anomaly to be the result of an order-disorder transition it is not possible to deduce from the shape of the anomaly whether the ordered phase is above or below the transition temperature. Kahane /9/ believes in an increasingly ordered ice at higher temperatures. Assuming, nevertheless, that the more ordered ice is stable at lower temperatures, as should be the case if the 3rd law of thermodynamics is to apply, approximate values for the total excess heat Q and the transition entropy S can be calculated. Values for the total excess heat range from

$Q = 1.4$ cal/mole to 4.1 cal/mole ($\pm$ 25%)

for the purest crystals to the highly doped crystals respectively. The corresponding values for the transition entropy are:

$S = 0.013$ cal/^{o}K mole to 0.036 cal/^{o}K mole.

No appreciable difference in the result was observed when the rates of either heating or cooling or both were changed by a factor of two.

Discussion

As the temperature is lowered, a point T_c is reached at which ice would, if it could, get ordered. A change of the dipole orientation of an H_2O molecule in ice can only be brought about by the passing of some defect: OH^-, H_3O^+ or Bjerrum defects. The concentration and the mobility (if it is not the result of a tunnel effect) of the defects decrease exponentially with temperature. It can thus be qualitatively deduced that the time required for an ordering of the ice increases exponentially as the temperature decreases. The temperature T_c, which could be called a hypothetical Curie temperature, must lie below 100°K because otherwise the observed ordering effects ought to be larger /12/. But, on the other hand, it is possible that this temperature is changed into a rather broad temperature range by the presence of lattice disturbances and defects in ice. This range could well extend above 100°K. It is then possible that regions of short-range-order are formed well above the Curie temperature.

The experimentally found transition entropy ΔS amounts to 1-3% of the zero-point entropy. This means that the ordered regions occupy only a very small part of the crystal volume. This agrees very well with the result of the measurement of the dielectric constant at 100°K /7/.

The anomaly in the specific heat can be explained by the additional energy required to destroy the order formed in regions around inhomogenities during cooling.

References

/1/ Giauque, W.F., J.W.Stout, J.Am.Chem.Soc. 58, 1144 (1936)

/2/ Pauling, L., J.Am.Chem.Soc. 57, 2680 (1935)

/3/ Takahasi, H., Proc.Phys.Math.Soc.Jap. 23, 1069 (1941)

/4/ DiMarzio, E.A., F.H.Stillinger, J.Chem.Phys. 40, 1577 (1964)

/5/ Nagle, J.F., J.Math.Phys. 7, 1484 (1966)

/6/ Lieb , E.H., Phys.Rev. 162, 162 (1967)

/7/ Eckener, U., Thesis, TH.München (1964)

/8/ Helmreich, D., Physics of Ice, Proc.Int.Symp., Munich, Plenum Press Corp. (1969), New York

/9/ Kahane, A., Thesis, University of Paris (1962)

/10/ Calvet, E., H.Prat: "Récents Progrès en Microcalorimétrie", Paris (1958)

/11/ Süssmann, W., Thesis, TH. München (1967)

/12/ Pitzer,K.S., J.Polisar, J.Phys.Chem. 60,1140 (1956)

THERMOELECTRIC EFFECT IN ICE

C.Jaccard

Institut de Physique, Université de

Neuchâtel, Switzerland

Abstract

The experimental results are reviewed and the following values are retained: -2.3±0.3 mV/°C for pure ice; -1.1±0.1 and -0.3±0.1 mV/°C for low and high HF doping; +2.2±0.2 and +0.4±0.2 mV/°C for low and high ammonia doping. The comparison with the theoretical values obtained from the dielectric data reveals a discrepancy of about 50% (showing up also in the electrochemical potential). This cannot be accounted for by a deviation from Einstein's relation for the ratio diffusity/mobility, as the electrochemical effect is a non-dissipative equilibrium process expressible by thermostatic relations for dilute solutions.

The origin of the discrepancy is assumed to be in the metal-ice potentials and their variation with temperature and impurity concentration, which have always been measured together with the other effects. A simple model of semiconductor/metal contact, with the impurity ions as donors or acceptors, explains part of the experimental results.

A method is proposed to eliminate in the measurements the influence of the ice-metal potentials.

1. Experiments and theories

Experimental evidence for charge separation effects in ice can be traced back to a century ago when Faraday /1/ discovered positive charges on ice hit by a stream of water. Many related observations have been reported since then and this phenomenon has gained a broad interest in the last two decades in its relation with the genesis of thunderstorms. Among many measurements, some indicate a definite connection between a charge transfer (or building up of a potential difference) and a thermal gradient in the crystal. On the other hand, the experiments performed by Eigen and De Maeyer /2/ on the ionic dissociation in ice suggest a difference in the mobilities of the positive and negative ions.

This led Latham and Mason /3/ to measure in 1960 the homogeneous thermoelectric coefficient, the value of which was found to be about 2 mV/^{o}C in pure ice and depending on the impurity concentration. Beside the calculation of the transient charge transfer between two pieces of ice having different temperatures and being brought into contact, Latham and Mason gave a theoretical explanation based on the larger diffusivity of the positive ions, which accounted for the correct sign and magnitude of the effect. However, they disregarded the influence of the L and D valence defects, which was taken into account only in the case of the transient process by means of the static permittivity (it cancels out in the formula for the static effect).

The combined action of the four defect species was considered by Jaccard /4/ in 1963, the basic equations used in the theory of the dielectric constant /5/ for the defect currents being extended to the case of a non-vanishing concentration gradient. They lead to formulae for the homogeneous thermoelectric potential and for the electrochemical potential in which each defect species contributes according to its energies of activation for dissociation and transfer. However, a formal asymmetry remains between ionic and valence defects (factors of 27 and 32, respectively), resulting probably from an error in the statistical evaluations for the currents.

The complete formal symmetry has been restored in a second theory /6/, based on the general relations for non-equilibrium processes, yielding the following relations for the homogeneous thermoelectric effect and the electrochemical potential:

$$T \left(\frac{\partial U_t}{\partial T}\right) = - \sum \left(\frac{t_k}{e_k}\right) (q_k^* - h_k) ,$$

$$U_c = - \sum \left(\frac{t_k}{e_k}\right) g_k .$$

Here t_k is the transfer number for dc conductivity of species k, e_k the defect charge, q_k^* the transfer heat, h_k the enthalpy and g_k the free enthalpy. These relations should remain valid in the case of doping; the impurities, assumed to be unable to move, influence all the parameters but the charges.

In the following years, several measurements were reported (table 1). In 1965, Bryant and Fletcher /7/ extended the observations to impure ice, doped with hydrofluoric acid or with ammonia, thereby increasing at low concentration the number of the positive (resp.negative) ions, and at high concentration the number of the L-(resp. D-) defects. The first measurements performed on monocrystals are those by Takahashi /8/ in 1966, differing from the previous results. To avoid spurious effects due to the electrodes which are frozen on the ice samples, Brownscombe and Mason /9/ devised in 1966 a new method based on the electrostatic induction produced by turning an ice bar possessing a thermal or a concentration gradient between two metal electrodes. The thermoelectric coefficient measured this way is somewhat higher (2.3 mV/^{o}C) than the original one, and the electrochemical potential is found to be a linear function of the natural logarithm of the HF concentration with a slope of -17±5 mV (for low concentrations). In 1967, Bryant /10/ repeated his measurements on doped ice, but this time with monocrystals; he obtained for HF in the low and in the high concentration range -1.1 and -0.3 mV/^{o}C respectively, and for NH_3 the corresponding values of +2.2 and +0.4 mV/^{o}C.

These experimental results agree at least qualitatively with the theory with respect to the order of magnitude and the overall dependence of the impurity concentration, as they can be deduced from the dielectric results.

A small HF concentration does not modify the D-L equilibrium, but increases strongly the number of the positive ions, which are then endowed with a smaller dissociation enthalpy from the acid than from the ice, whereas a large concentration enhances the concentration of the L-defects. This is translated with increasing concentration by the algebraic increase of the thermoelectric coefficient in two steps, the corresponding effect for the negative ions and the D-defects occuring with the ammonia. At very low concentration, a sharp negative peak indicates the predominance of the positive ions, with a large dissociation enthalpy. A first quantitative discrepancy shows up in the critical concentrations defining the steps; they are too large by a factor of about 5 with respect to the dielectric results, which however are not very suitable for a precise determination of the critical concentrations. Therefore, the consistency within an order of magnitude can be considered as satisfying.

A more serious difference occurs in the magnitude of the effect: the measured coefficient is by 50% larger than the theoretical one and this is also the case for the dependence of the electrochemical potential from the logarithm of the acid concentration. The authors cited above attribute this to an error in the factor G (occuring in the first theory), which together with kT/e multiplies the whole expression; its value should range between 1/27 and 1/32, whereas the experiments indicate rather something between 1/20 (Brownscombe and Mason /9/) and 1/23 (Bryant /10/). In the following, we show that this cannot be the case and that the discrepancy has to be looked for rather in the metal-ice contact potential.

2. The electrochemical potential and Einstein's relation

In the first theory the electrochemical potential is obtained from a balance defined by a vanishing total electric current; this balance occurs mainly between diffusive movements and the movements resulting from the electric field, so that the ratio "diffusivity/mobility" appears as a multiplier in the expression for the potential. A deviation from Einstein's law

$$D/\mu = kT/e$$

could be invoked to explain the experimental results, but this would not give the right answer for the following reason.

The diffusivity and the mobility characterize typical dissipative phenomena, whereas the electrochemical potential is a reversible effect, in a state of thermostatic equilibrium (stationary state of zeroth order); only in appearence it is bound to irreversible processes through the transfer numbers t_k. These must obey the relations /5/

$$t_+ + t_- = e_\pm / e \qquad t_D + t_L = e_{DL}/e$$

and on the other hand the dissociation reactions of the pure ice require for the free enthalpies:

$$g_+ + g_- = 0 \qquad g_D + g_L = 0$$

so that the electrochemical potential is in fact expressed only by means of thermostatic parameters:

$$U_c = -(g_+ + g_D)/e \qquad (\text{as } e_\pm + e_{DL} = e) .$$

Table 1. Thermoelectric effect in ice
Experimental values near $-20^{o}C$ in $mV/^{o}C$

Doping ranges: (I) $10^{20} - 10^{22}\ m^{-3}$
(II) $> 10^{23}\ m^{-3}$

Authors	Pure ice	Hf doping (I)	Hf doping (II)
1961 J.Latham B.J. Mason /3/	-2.0±0.2		-3 ±0.3
1965 G.W. Bryant N.H.Fletcher /7/	-3.5±0.5	-1.6±0.2	+0.2±0.3
1966 T.Takahashi /8/	-1.5±0.5	-1 ±0.5	+0.5±0.2
1966 J.L.Brownscombe B.J.Mason /9/	-2.3±0.3	†	
1967 G.W. Bryant /10/		-1.1±0.1	-0.3±0.1

† for the electrochemical potential:
$dU_c/d \ln c(HF) = -17 \pm 5$ mV

The question is then: how does the free enthalpy depend on the concentration? The very small concentrations encountered here (10^{-6} for the valence defects, 10^{-10} for the ions) preclude any significant interaction between the defects and this justifies the use of the formula for ideal dilute solutions:

$$g(p,T,c) = g^{o}(p,T) + kT \ln c$$

which is valid also for quantum systems with vanishing interaction Hamiltonian. Therefore, assuming a square root dependence for the ionic concentration (mass action law), the coefficient relating the logarithm of the HF concentration with the potential must have the value $1/2\ kT/e = 11$ mV.

The factor θ /4/, introduced in the theory of the dielectric properties and multiplying the ration D/μ, should not differ from unity, neither for the electro-

(concentration of D and L unaffected)

(D or L majority)

NH_3 doping (I)	NH_3 doping (II)	Electrodes (c) contact (i) induction	Crystal
		Cu (c)	?
$+2.0 \pm 0.3$	$+0.0 \pm 0.3$	Brass (c) Pd (c)	poly.
		Pd (c)	mono
		Brass (i)	mono
$+2.2 \pm 0.$	$+0.4 \pm 0.2$	Pd (c)	mono

chemical potential, nor for the thermoelectric coefficient, because it then loses its meaning. This can be seen from the following argument.

Bernard and Callen /11/ consider a general quantum system characterized by extensivé operator parameters Q_i, and under the influence of applied forces F_i which are conjugated to the Q_i by the Hamiltonian

$$H = H^{(o)} + \sum F_i(t)\, Q_i \; .$$

The response of the system is the average $\langle Q_i(t) \rangle$,"currents" are defined $\langle \dot{Q}_i(t) \rangle = J_i(t)$. Introducing the Fourier spectra $\hat{F}_i(\omega)$ of the forces and $\hat{J}_i(\omega)$ of the currents, the admittance Y is defined by the relation

$$\hat{J}_j(\omega) = \sum_i \hat{F}_i(\omega)\, Y_{ij}(\omega)$$

and it characterizes the response of the system of the applied forces. It is connected to the second moment of the fluctuations of the Q_i by the formula (fluctuation dissipation theorem)

$$1/2 \, \langle Q_i Q_j(t) + Q_j(t) Q_i \rangle^{(o)} =$$

$$= - \frac{2}{\pi\beta} \int_0^\infty d\omega \cos \omega t \cdot \mathrm{Re} Y_{ij}(\omega)/\omega^2 \theta(\hbar\omega\beta/2)$$

where the average $\langle \ \rangle^{(o)}$ denotes the equilibrium value in the absence of applied forces and

$$\beta = 1/kT \ , \ \theta(x) = \tanh x \ / \ x \ .$$

This formula is valid in the absence of a magnetic field provided $\langle Q_i \rangle^{(o)} = 0$.

This can be applied to a particle (or to one of our defects), $Q_i = Q$ being its displacement parallel to a coordinate axis. To satisfy the condition above (and to prevent the integral from diverging for ω going towards zero) we assume that the particle is in a very flat parabolic potential, the curvature of which will be made arbitrarily small and with it the corresponding oscillation frequency ω_o. If the particle charge is e and if the only applied force is an homogeneous electric field, the admittance is related to the particle mobility $\mu(\omega)$:

$$Y(\omega) = -\mu(\omega)/e.$$

We have then:

$$\langle Q(0)Q(t) \rangle = \frac{kT}{e} \frac{2}{\pi} \int_0^\infty d\omega \cos \omega t \ \mathrm{Re} \ \mu(\omega)/\omega^2 \theta(\hbar\omega\beta/2).$$

The diffusion coefficient D of the particle is given by

$$D = \langle [Q(t) - Q(0)]^2 \rangle \ / \ 2t.$$

Let us now consider only a time t which has a physical meaning, that is:
1) it should be much smaller than the oscillation period $1/\omega_o$. This makes no difficulty as ω_o is arbitrarily small.
2) It should be much larger than the correlation time.
Under these conditions the diffusion process is not influenced by the flat potential and we may expand the second member of the last equation to the first degree only

$$D = \langle \dot{Q}(t)Q(t) \rangle \ - \ \langle \dot{Q}(t)Q(0) \rangle$$

Assuming no correlation between position and velocity of the particle, the first average is zero and taking the second from the derivative of the integral above, we obtain

$$D = \frac{kT}{e} \cdot \frac{2}{\pi} \int_0^\infty d\omega \ (\sin \omega t/\omega) \ \mathrm{Re}\ \mu(\omega)/\theta(\hbar\omega\beta/2) \ .$$

It is no longer necessary to bother about the limit $\omega = 0$ as it introduces no divergence, provided $\mu(0)$ is finite (existence of a dissipative mechanism). The function $\mathrm{Re}\ \mu(\omega)/\theta\ (\hbar\omega\beta/2)$ is certainly complicated but it must have at the origin a plateau of height $\mu(0)$ and width of the order of $kT/\hbar$. Consequently, for $t \gg \hbar/kT \sim 10^{-13}$ s only this plateau contributes significantly to the integral; the remaining part of $\mathrm{Re}\ \mu/\theta$, multiplied by the "tail" of $\sin \omega t/\omega$ is irrelevant. We get the classical result:

$$D = \frac{kT}{e}\ \mu(0) \ . \ \frac{2}{\pi} \int_0^\infty d\omega \sin \omega t/\omega = (kT/e)\mu(0) \ .$$

A deviation from this formula is possible only for times of the order of $\hbar/kT$; as even the relaxation time $\mu(0)m/e$ of the particle velocity is of the order of 10^{-10} s (with $\mu(0) \sim 10^{-5}$ m^2/Vs, m/e of the proton) the formula should keep its validity for transport processes, at least for temperatures of the order of $100^\circ K$. In any case, if T goes towards zero the formula cannot be "corrected" by putting the function θ in front of the integral, but a detailed knowledge of the quantum mechanical properties of the system is required, and the expressions for the particle currents, established for a classical model of the thermoelectric and of the electrochemical effect, have also to be revised.

Therefore, Einstein's formula applies to the experiments already performed and the origin of the discrepancies between them and the theory has to be looked for somewhere else. In the next section, certain possibilities are discussed.

3. Interfering effects

a) A change of the dissociation enthalpies resulting from the electrostatic interaction of the defects with the charge clouds around them is much too small to explain the experiments. The interaction potential, using the formulae of the second theory /6/ is of the order of

$$U_{es} = (n_D + n_L)^{1/2} \ . \ 3.10^{-14} \ V \quad (n_{DL} \text{ in } m^{-3})$$

In the high concentration ranges, this yields at most 10 mV, a quantity below the experimental accuracy.

b) The whole theory rests on the assumption that only the ions and the valence defects contribute to the currents. A fifth mobile species, with a transfer number t_5, and not modifying the proton configuration, would change one of the relations above into

$$t_+ + t_- = (e_\pm /e)\ (1 - t_5).$$

Such a species could be a negative fluorine ion, which is known to have a diffusivity larger than the other impurity ions by five orders of magnitude. However, it is still much less mobile than the positive ions H_3O^+, which are in equal number so that the transfer number t_5 remains smaller than 10^{-4} and therefore quite negligible.

c) Electrons in the conduction band, even in small concentration, would show up easier because of their large mobility, but the electrolysis experiments /12/ performed at -10°C indicate an upper limit of some percent for t_5 which should not change much if the temperature decreases by 10°C. Therefore, no dynamical effect may be attributed to the free electrons (it is no longer the case for static effects, as we show below).

d) A source of error lies certainly in the surface of the ice samples. Latham and Mason /3/ indicate an influence of the surface-to-volume ratio on the thermoelectric coefficient above -8°C and it is known from conductivity measurements /13/ that the surface layers may show abnormal properties. It has therefore to be kept in mind that the results obtained so far might be biased by the surface, but no suitably designed experiment allows until now to infer the magnitude of the effect.

e) It remains here to look at another source of error, which has always been more or less tacitely disregarded, but which is systematically included in all the measurements, even with Brownscombe and Mason's induction method: the contact potential between ice and metal and its variations with temperature and concentration. In the following section, a simple model is presented, which shows that corrections of the correct magnitude can be expected.

4. The ice-metal potential

The ice-metal contact can be considered from the side of the electrons, for which the ice is a semiconductor; the band gap amounts to 7.4 eV, as it can be deduced from the ultraviolet absorption edge /14/. Introducing the work functions W_M and W_I in metal and in ice respectively, the energies E_V and E_C of the upper edge of the valence band and of the lower edge of the conduction band and the Fermi energy E_F, the simple model of the metal-semiconductor contact (e.g. as described by Kittel /15/) gives for the metal-ice potential:

$$U_{IM} = U_I - U_M = (1/e)(W_M - W_I - E_C + E_F).$$

We assume now that only the Fermi energy is influenced by changes in impurity concentration or temperature. If ΔX stands for the change of any quantity X going from the left to the right side of an ice bar, then the potential difference

$$\Delta U = - \Delta E_F/e$$

adds in measurements to the thermoelectric or to the electrochemical potential.

We assume further a simplified band scheme with only one donor and one acceptor impurity level at E_d and E_a; the concentrations N_d and N_a are much higher than the intrinsic free electron concentration, which is in any case very low because of the large band gap.

The Fermi energy is then

with donors only: $E_F=1/2\ (E_C+E_d)+1/2\ kT \ln\ (N_d/n_-^o)$,

with acceptors only: $E_F=1/2\ (E_V+E_a)-1/2\ kT \ln\ (N_a/n_+^o)$,

where $n_\pm^o = 2\ (2\pi m_{h,e} kT/h^2)^{3/2} = 2\times10^{25}(m_{h,e}/m)^{3/2}\ m^{-3}$

(the effective masses for holes and electrons have been introduced in the last equation).

If an impurity concentration gradient alone is present in the sample, the only variation of E_F is due to the change in N_d or N_a. To explain the case of low HF doping the donor concentration is required to vary with the square root of the acid concentration; for example, a fraction of the negative fluorine ions sitting at suitable places in the lattice could be acting as donors. The change in the Fermi energy is then

$$\Delta E_F = 1/4\ kT \Delta \ln\ n_{HF}$$

so that the measured potential difference is the sum of the electrochemical potential $-(kT/2e)\Delta \ln\ n_{HF}$ and of the contact potential $-(kT/4e)\Delta \ln\ n_{HF}$, i.e. the coefficient is $-(3kT/4e) = -16.5$ mV, which agrees with Brownscombe and Mason's result - 17 ± 5 mV.

The argument put forward in Sect. 2 about reversibility of the electrochemical potential and vanishing interaction between the defects appears to be strong enough to require a (kT/4e)-correction from boundary effects, provided the experimental results are correct. Therefore, their duplication by another group, perhaps with another method, is necessary to give to the contact potential hypothesis a firmer standing.

If a thermal gradient is present, one has to consider

$$\frac{\partial E_F}{\partial T} = \pm 1/2 \frac{\partial}{\partial T} kT \ln (N_{d,a}/n^o_\pm) .$$

As the concentrations $n^o_\pm$ are of the order of 10^{24} m^{-3}, we can safely assume that N_d and N_a are lower than these, so that the logarithm is negative. Furthermore, we shall start from the traditional hypothesis that the positive ions have a much larger mobility than the negative ones, and that both valence defects have about the same one (in the notation of the first theory /4/: $\phi_\pm << 1$, $\phi_{DL} \approx 1$).

In pure ice, the thermoelectric potential is then produced only by the positive ions; as the dissociation enthalpy for an ion pair amounts to 0.96 eV /16/, and as the tunnelling does not require a transfer energy, the thermoelectric potential is -0.96 V/(2×250°C) = -1.9 mV/°C. The experimental value of -2.3 mV/°C indicates a contact contribution of -0.4 mV/°C, which could be accounted for by a concentration of acceptors in pure ice amounting to 10^{20} m^{-3}.

The possible presence of unknown acceptors or donators in an unknown concentration renders quite illusory an estimation of the absolute value of the effect, but in HF doped ice, the difference of the thermoelectric coefficient between the two concentration ranges should reflect the change in the impurity concentration. In the low range, where only the positive ions contribute, but with the dissociation enthalpy of 0.32 eV /5/, the coefficient amounts to -1.3 mV/°C. In the high range, one has to add the contribution of the L-defects, lacking a dissociation enthalpy but with a transfer energy of about 0.23 eV, to obtain -0.4 mV/°C. The temperature dependence of the Fermi energy is given by

$$\frac{\partial E_F}{\partial T} = \frac{k}{2} \left\{ \ln \left(\sqrt{n_{HF} k_{Fo}} \;/n^o_-\right) - 3/2 \right\}$$

where k_{F0} is the preexponential factor of the dissociation constant if we assume as before that the donor concentration varies as $\sqrt{n_{HF}}$. Between the two ranges, the HF concentration increases by a factor of about 3000 and this gives a contribution of about -0.15 mV/^{o}C. Thus we obtain for the total difference $1.3 - 0.4 - 0.15 = 0.75$mV/^{o}C; this agrees with the experimental value of $1.1 - 0.3 = 0.8 \pm 0.2$ mV/^{o}C.

The formulae used so far are valid under the assumption

$$4(N_d/n^o_-) \exp\left[(E_C - E_d)/kT\right] >> 1 .$$

As N_d is of the order of the positive ion concentration (10^{18}-10^{19} m^{-3}) the inequality requires that the donor level is at least 0.3 eV under the conduction band; this condition can be easily satisfied with an energy gap of 7.4 eV.

In the case of NH_3 doped ice and if the acceptors predominate, the complementary model applies with a square root dependency on the ammonia concentration, as if part of the ammonium ions were acting as acceptors. The change in concentration between the low and the high range accounts then for 0.15 mV/^{o}C. The difference due only to the homogeneous thermoelectric effect is then $2.2 - 0.4 + 0.15 = 1.95$ mV/^{o}C, which corresponds to an energy of 0.5 eV.

But this difference has to be attributed to the D-defects; according to Levi and Lubart /17/ they seem to be liberated from the ammonia molecules with a dissociation enthalpy of about 0.3 eV; this implies that their mobility requires a transfer energy of about 0.25 eV, a value corresponding to another experimental estimation by these authors.

However, the agreement between the experimental and the calculated values should be considered as largely fortuitous because of the simplicity of the model which cannot account for the absolute value of the contact thermoelectric potentials; as we have seen for pure ice, it seems that a certain acceptor concentration must exist even in pure ice and the system is probably more complicated with a background of possible states near the surface, on dislocation lines or at point defects of the vacancy or interstitial type. However, the considerations above show that systematic experimental errors of the correct magnitude can be expected from the ice-metal potential, which should no longer be disregarded in the interpretation of the measurements.

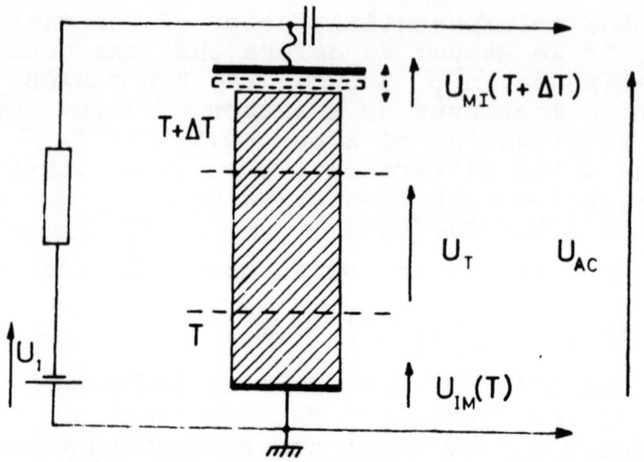

Fig.1 Principle of the independent measurements of the thermoelectric effect and of the ice-metal contact potential (see text)

5. Experimental method eliminating the contact potential

We are then left the problem of finding a method eliminating the influence of the ice-metal contact. The answer lies in the property of a vibrating capacitor to give a signal proportional only to the static potential between the two vibrating plates.

In the circuit of fig.1 we have for the static part

$$U_1 = U_{IM}(T) + U_t + U_{MI}(T + \Delta T).$$

But only the last term of the right member gives rise to the alternating potential U_{AC}. Therefore, if U_1 is a calibrated tension adjusted to let U_{AC} disappear, the measurement yields:

$$U_t = U_1 - U_{IM}(T) .$$

But a complementary measurement can also be performed without a thermal gradient ($U_t = 0$) and yields $U_{IM}(T)$, so that the thermoelectric potential U_t can be measured unequivocally, with the contact potential as a byproduct. If the capacitor is symmetrical, as in fig.2, the contact potential is eliminated from the beginning and the thermoelectric coefficient is just equal to the applied battery potential when the ac signal vanishes. The same

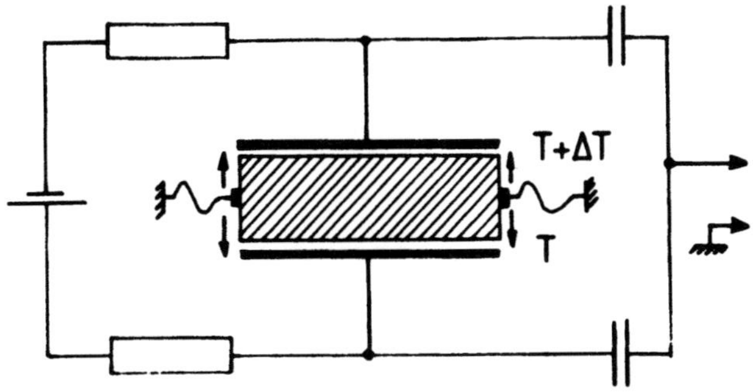

Fig.2 Measurement of the thermoelectric effect unbiased by the ice-metal contact potential

consideration applies to the electrochemical potential. It is essential for the vibrating capacity to be between the ice surface and a metal electrode: the configuration used by Brownscombe and Mason for example does not eliminate the temperature dependence of the ice-metal potential in the gap between the electrode and the end of the turning bar. The method described above is certainly sensitive to the state of the surface of the ice sample as well as of the counterelectrode, but this is also the case for electrodes frozen on the ice (then the conditions may be more difficult to control).

6. Conclusion

The differences observed between theoretical and experimental values of thermoelectric and electrochemical effects in pure and doped ice cannot be explained by the anomalous mobility of the ionic defects, by electrostatic interaction or by an electronic contribution. The origin of the discrepancy resides more probably in the variation with temperature and impurity concentration of the ice-metal potential, for which a simple contact model yields values agreeing with the observed concentration dependence. A suitable experimental method, with a vibrating capacity between the ice and the metal electrode, should give unbiased results.

Acknowledgement

This work was carried out under the support of the Swiss National Foundation for Scientific Research.

References

/1/ Faraday, M., Poggendorfs Ann. 60, 342 (1845)

/2/ Eigen, M., L.De Maeyer, Proc.Roy.Soc. A, 247, 505 (1958)

/3/ Latham, J., B.J.Mason, Proc.Roy.Soc. A, 260, 523 (1961)

/4/ Jaccard, C., Phys.Kondens.Materie 1, 143 (1963)

/5/ Jaccard, C., Helv.Phys.Acta 32, 89 (1959)

/6/ Jaccard, C., Phys.Kondens.Materie 3, 99 (1964)

/7/ Bryant, G.W., N.H.Fletcher, Phil.Mag. 12, 165 (1965)

/8/ Takahashi, T., J.Atmos.Sci. 23, 74 (1966)

/9/ Brownscombe, J.L., B.J.Mason, Phil.Mag. 14, 1037 (1966)

/10/ Bryant, G.W., Phil.Mag. 16, 495 (1967)

/11/ Bernard, W., H.B.Callen, Phys.Rev. 31, 1017 (1959)

/12/ Decroly, J.C., H.Gränicher, C.Jaccard, Helv.Phys. Acta 30, 465 (1957)

/13/ Bullemer, B., N.Riehl, Solid State Comm. 4, 447 (1966)
Jaccard, C., Physics of Snow and Ice, vol. I, 179 (1967) (Ed. by H. Oura, Sapporo, Japan)

/14/ Onaka, R., T.Takahashi, J.Phys.Soc. Japan 24, 548 (1968)

/15/ Kittel, C., Introduction to Solid State Physics, 2nd ed.,p. 358 (J.Wiley, New York) (1956)

/16/ Eigen, M., L.De Maeyer, H.C.Spatz, Ber.Bunsenges. 68, 19 (1964)

/17/ Levi, L., L.Lubart, J.Chim.Phys. 58, 836 (1961)

PROTONIC SEMICONDUCTORS

Lars Onsager

Yale University, New Haven, Connecticut 06520

Ice is the most important member of a class of solids in which the electric current is carried almost entirely by mobile protons. Known protonic semiconductors include all the disordered forms of solid water, several salt hydrates as well as representative examples of the gas hydrates, potassium hydrogen fosfate and its analogs, and one modification of formic acid, studied before the polymorphism of that substance was discovered (Johnson and Cole 1951).

A common structural feature is a network of hydrogen bonds, disordered in such a manner that a systematic change of orientations can be produced by an electric field; one observes dielectric relaxation as well as electrolytic conduction.

The ordinary mechanisms of electrolytic conduction in solids involve interstitial ions or ion vacancies - or both. Similarly, in ice and other protonic semiconductors we infer the presence of excess protons (H^3O^+) and proton vacancies (^-OH) on a fraction of the molecules. From the high mobilities of these intrinsic ions in liquid water we are at least prepared to find frequent intermolecular transfers of excess or defect protons in the solid as well.

Bjerrum (1951) found an analogous explanation for the dielectric relaxation in terms of bonding defects, where misfits of the orientations of neighboring neutral molecules produce either vacant bonds (L defects) or situations where two protons compete for the same bond (D defects). Both kinds of defects can move by successive rotations of the participating molecules, and a systematic drift produces a polarization current.

Bjerrum recognized that ionic as well as bridging defects can alter the state of polarization; but his estimates of the energies involved led him to expect that the bridging defects would predominate. Onsager and Dupuis (1962) arrived at the same conclusion by comparing the observed entropies of activation. This inference is strongly supported by more recent studies (Chan et al 1965) of dielectric relaxation at high pressures; in every disordered form of ice the relaxation times lengthen with increasing pressure, as one should expect for defects which tend to loosen the structure.

By contrast, Takagi (1948) ascribed the very fast dielectric relaxation in KH_2PO_4 (10^{-11}sec) to the intermolecular shuttling of protons, and this view was verified when Bjorkstam and Uehling (1959) studied the quadrupole fine strucure of the deuteron magnetic resonance. To complete the picture, Schmidt and Uehling (1962) measured the transfer frequencies between neighboring bonds by the spin-echo technique and obtained just 4 sec^{-1}at 25°C, with an activation energy of 0.58 e.v. Moreover, these transfer frequencies correspond to the measured electrical conductivities over a wide range of temperatures.

Steinemann (1957) showed that the dominance of the bridging defects in ice may be reversed by doping, and recognized that the motions of the various defects are coupled by way of the ambient polarization. Onsager and Dupuis (1960, 1962) pointed out that the interactions of the defects with electric fields and the polarization entail interactions among the defects, with predictable consequences for the recombination kinetics. In their formulation the polarization becomes essentially a traffic phenomenon; dominance depends on which kind of defects (bridging or ionic) generates more random traffic. Forward orientation of the molecules obviously entails backward polarization of the bridges, and vice versa. Space charges as well as time lags tend to spoil the classical relations between field and polarization. The ratios of currents to drift velocities determine effective charges; a certain fraction e_B/e of an elementary charge is carried by a bridging defect, the remaining $(e - e_B)/e$ by a mobile ionic defect. Strictly speaking, e_B is a tensor, and a modest anisotropy is reflected in the dielectric tensor of ice (Jaccard 1965). A large anisotropy of e_B would tend to promote ordering.

The interactions between the defects and the polarization are mainly statistical; the resultant motion tends to randomize the orientations. The defects are the only internal sources of either the electrostatic or the polarization field, according to the following table:

	D	L	H^+	HO^-
$(\epsilon_\infty/4\pi)\ (\nabla\cdot\underline{E})$	$-e_B$	$+e_B$	$-e+e_B$	$+e-e_B$
$(\nabla\cdot\underline{P})$	$-e_B$	$+e_B$	$+e_B$	$-e_B$

The values for H_2O agree well enough with earlier measurements. Diffusion-limited kinetics is inferred from the variation of the saturation current with the field intensity; the observed rate of increase indicates that the dominant mechanism of dissociation produces a pair of mobile ions directly and does not involve bridging defects as intermediates. Assuming that this is the only mechanism, and $e_B = 2 \times 10^{-10}$ esu., one can estimate that ice at -10° contains about 8×10^{10} ion pairs per cm^3, and that the sum of the mobilities equals 0.075 cm^2/ volt sec. On other grounds we have cause to believe that the positive ion is faster than the negative by one or two orders of magnitude.

Since the inferred mobility of the positive ion corresponds to transfer frequencies of the order 10^{13} $sec.^{-1}$, no activation energy is expected, and indeed the activation energy for the conduction is just half that of the dissociation. Further confirmation for this remarkable result was obtained occasionally when efforts to produce very pure ice proved less than adequate, and the d.c. conductivity remained almost independent of the temperature as long as the h.f. conductivity was substantially greater. Moreover, Steinemann (1957) combined the observed capacitive impedances at the electrodes with the measured resistivities of ice lightly doped with HF to infer a value of about 0.08 cm^2/ volt sec for the proton mobility. All this adds up to significant if indirect confirmation of the results obtained by Eigen et al.

Nevertheless, recent observations of surface conductivity have thrown doubt on all previous measurements of the bulk conductivity. For the time being, we can only hope that good experiments and adequate analysis will soon resolve the apparent contradictions.

Studies of doped ice are beset by many difficulties. Concentrations of impurity too small for easy assay produce large effects, and uniform distribution is not easily attained. Under the circumstances almost any set of data leaves some room for debate about the interpretation. Even so, the consistent features of the results strongly confirm our general theory; but it is not easy to extract the important parameters.

The implications of the theory may be summarized briefly. We recognize the carriers D, L, H^+ and HO^-. Each may be characterized by its coefficient of diffusion D; and by its charge ej; every carrier contributes to the high frequency conductivity according to the Einstein relation

$$\lambda_j = e_j^2 D_j / kT \tag{3}$$

so that

$$\sigma_\infty = \sigma_B + \sigma_I = \Sigma c_j \lambda_j \tag{4}$$

Both fields possess potentials. The proportionality constant for P is fairly small and varies with the temperature; in case of a purely statistical interaction direct proportionality is expected. Interpreting the observed dielectric constant of ice at -10° on that basis, Onsager and Dupuis estimated

$$e_B = 1.85 \times 10^{-10} \text{ esu}$$

for ice; but they allowed that the actual temperature coefficients reported by Auty and Cole (1952) might indicate a value as high as 2.3×10^{-10} esu.

In the case of KH_2PO_4 the spontaneous polarization below 125°K (215° for the deuterated compound) is a measure for the net ion charge; one obtains

$$e - e_B = 10^{-10} \text{ esu.}$$

(for the direction of the tetragonal axis). The small activation energies (0.08 e.v.) for ionization and deuteron shuttling (Schmidt and Uehling 1962) fit well into the picture.

The dielectric relaxation of pure ice exhibits a simple Debye dispersion

$$\epsilon - \epsilon_\infty = (\epsilon_s - \epsilon_\infty)/\ (1 + i\omega\tau) \qquad (1)$$

where ϵ_s exhibits a moderate anisotropy, while ϵ_∞ and τ seem to be practically isotropic. A constant activation energy of 0.575 e.v. describes the variation of τ over a wide range of temperatures; ϵ_∞ is practically constant, and the increment $(\epsilon_s - \epsilon_\infty)$ varies about inversely with the temperature. As yet, it would seem that the charge partition (e_B/e) is best estimated from ϵ_s although there is some room for conjecture. Jaccard (1965) prefers estimates slightly different from that of Onsager and Dupuis, and he attributes the anisotropy of ϵ_s to that of e_B. Along with the dielectric properties, the high frequency conductivity σ_∞ is well known according to the standard relation

$$4\pi\ (\sigma_\infty - \sigma_o) = (\epsilon_s - \epsilon_\infty)/\ \tau \qquad (2)$$

and from direct measurements too; for pure ice σ_∞ is several orders of magnitude greater than the d.c. conductivity σ_o. Eigen, de Maeyer and Spatz (1964) studied d.c. conduction and saturation phenomena in ordinary and heavy ice. Their results are summarized in the following table.

	H_2O	D_2O
$10^{-9}\ \sigma_o$ (-10°C)	1.o ± 0.15	0.36 ± 0.11
Activation energy (kcal)	11 ± 1.5	13 ± 1.5
$10^{10}\ K_{diss}$ (sec^{-1})	32 ± 3	0.27 ± 0.05
Activation energy (kcal)	22.5 ± 1.5	25 ± 3

We distinguish the rotational conductivity

$$\sigma_B = \sigma_D + \sigma_L \tag{5}$$

and the ion conductivity

$$\sigma_I = \sigma_H + \sigma_{OH} \tag{6}$$

Then from the condition that the number currents must agree, we obtain for the d.c. conductivity σ_o

$$\frac{e^2}{\sigma_o} = \frac{eB^2}{\sigma_B} + \frac{(e-eB)^2}{\sigma_I} = \frac{eB^2}{\sigma_B} + \frac{eI^2}{\sigma_I} \tag{7}$$

The relaxation time τ varied inversely as the total random traffic; if c_o denotes the number of molecules per unit volume and d the distance between neighbors, then in the geometry of ice

$$c_o \frac{d^2}{3\tau} = \Sigma\, c_j\, D_j = kT\left(\frac{\sigma_B}{e_B^2} + \frac{\sigma_I}{e_I^2}\right) \tag{8}$$

The dielectric dispersion can be computed from the above relations together with equations (1) and (2) . The formula

$$\sigma_\infty - \sigma_o = (e_I\, \sigma_B - e_B\, \sigma_I)^2 / (e_I^2\, \sigma_B + e_B^2\, \sigma_I) \tag{9}$$

is useful and instructive in this context.

Some interesting results were based on the observation that D and L faults can be sequestered by neutral moelcules (HF and NH_3 , respectively). Thus Jaccard (1959) was able to determine the mobility of L faults by measuring the high frequency conductivity of samples heavily doped with measureable concentrations of HF. More recently Bryant (1967) studied thermoelectric effects in HF doped as well as NH_3 doped ice over a range of concentrations, and interpreted the results according to Jaccard's theory (1964) in terms of varying transference numbers. Even though his electrodes were not as well defined as one might wish, he obtained consistent results indicating that the transference number for L faults

$$\tau_L = \sigma_L / (\sigma_L + \sigma_D)$$

in pure ice is about 0.7; L. faults are a little more than twice as mobile as the D faults. For the energy of formation of D + L he obtained 1.3 kcal/mole, in good agreement with Jaccard; but the absolute concentrations of Bjerrum defects inferred by these two authors differ by a factor of 5. Again, we can only hope that discrepancies of this kind may soon be resolved.

References

Auty, R. P. and Cole, R. H. 1952, J. Chem. Phys. 20, 1309.

Bjerrum, N. 1951, K. Danske Vid. Selsk. 27, 1.

Bjorkstam, J. L. and Uehling, E. A. 1959, Phys. Rev. 114, 961.

Bryant, G. W., 1967, Phil. Mag. (8) 16, 495.

Chan, R. K., Davidson, D. W. and Whalley, E. 1965, J. Chem. Phys. 43, 2376

Eigen, M., de Maeyer, L. and Spatz, H. C. 1964, Ber. Bunsenges. 68, 19.

Jaccard, C. 1959, Helv. Phys. Acta 32, 89.

Jaccard, C. 1964, Phys. Kond. Mat. 3, 99.

Jaccard, C. 1965, Ann. N.Y. Acad. Sci. 125, 2, 390.

Johnson, J. F. and Cole, R. H. 1951, J. Am. Chem. Soc. 73, 4536.

Onsager, L. and Dupuis, M. 1960, Rendiconti S.I.F. X Corso 294.

Onsager, L. and Dupuis, M. 1962, in Electrolytes, p. 27, Pergamon Press, Oxford.

Schmidt, V. H. and Uehling, E. A. 1962, Phys. Rev. 126, 447.

Steinemann, A. 1957, Helv. Phys. Acta. 30, 581.

Takagi, Y. 1948, J. Phys. Soc. Japan 3, 271, 273.

THEORY OF THE MOBILITY OF STRUCTURAL DEFECTS IN ICE

S.F. Fischer, G.L. Hofacker+

Department of Chemistry, Northwestern

University, Evanstone, Illinois / USA

1. Construction of a particle picture for the defects

The hypothesis of four conductivity-related defects in ice as it was developed in its various stages of sophistication by Bjerrum, Onsager, Gränicher, Eigen, and others /1-7/ seems to provide a satisfactory basis for the interpretation of the crucial experiments of protonic transport in ice. Even though we know very little about the detailed structure of the defects, their principal nature follows in a straightforward way from the peculiar molecular arrangement in an ice crystal. This at least enables us to deal with the problem of defect mobility in terms of a model approach. If we found that a defect model can be constructed which complies with experimental data on one hand and characteristic microscopic parameters (structural as well as dynamic ones) for ice on the other, we had added a good piece of evidence to Onsager's model of protonic transport. Furthermore, one may think of possibilities how to gain more experimental insight into the nature of the defects in ice and how to recognize similar situations in other ionic transport systems.

The principal difficulty of the theory of defect mobility is immediately evident: although there is transport of charge and mass, neither in the case of Bjerrum nor in the ionic defects is there an individual particle moving

---

+ Present address: Lehrstuhl für Theoretische Chemie, Technische Hochschule, München, Germany.

through the crystal. Both mass and charge propagation are due to local displacements of protons between or within hydrogen bonds along the chain of hydrogen bonds via which the defect moves. A defect in an ice crystal may simply be characterized by the asymptotic polarization of hydrogen bonds far from it, e.g., in a one-dimensional model, like in fig. 1.

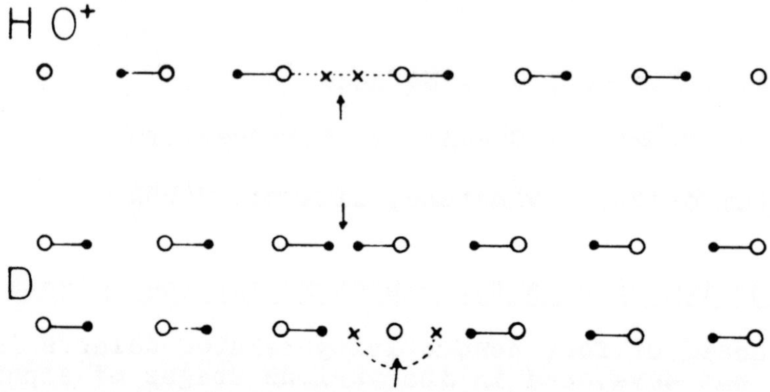

Fig.1 Linear models of defects in ice. Arrows indicate symmetry centers of metastable configurations. x symbolizes a proton distributed over two potential minima

Saying the defect be localized at a particular lattice site involves a statement about all protons along the infinite chain of hydrogen bonds and the Hamiltonian of the defect structure must comprise all the protons, not just the ones close to the defect site. Since the motion of all protons is correlated, we can speak of the potential acting on a single proton only in terms of a one-proton Hartree approximation. The repulsive proton-proton interactions are not exactly of the two-body force type but we shall approximate it in that way for the sake of simplicity. Then the proton in each hydrogen bond is subject to an inherent potential and interactions with surrounding protons, prevalently its neighbors. The interactions lead to asymmetric Hartree potentials for all protons, except in the case of the middle proton in the metastable configuration b) of fig. 1, as shown in fig.2.

There is no principal difference in the theoretical treatment of all four defects in ice. We have a protonic Hamiltonian of the form

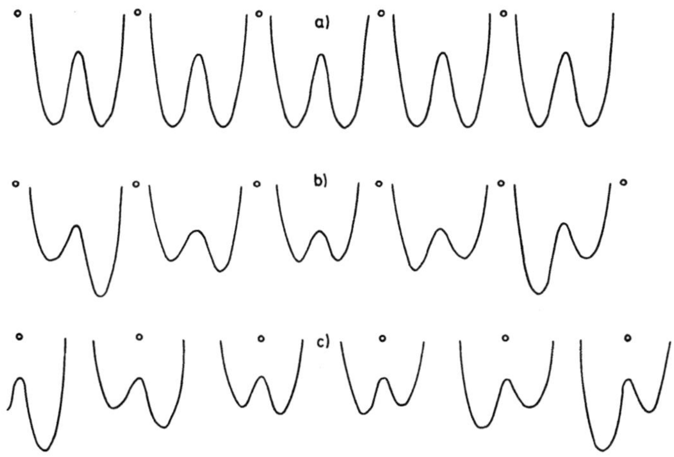

Fig.2 a) Effective potentials for noninteracting protons.
b,c) Hartree potentials for two metastable protonic configurations

$$H^P = H^o + H^{Tun} + H^{Int} + H^{Tra} \quad ;$$

$$H^o = \sum_{\substack{i=-N \\ \alpha}}^{N} \varepsilon_\alpha (r_i^{\alpha+} r_i^{\alpha} + l_i^{\alpha+} l_i^{\alpha}) \quad ;$$

$$H^{Tun} = \sum_{\substack{i=-N \\ \alpha}}^{N} T_\alpha (r_i^{\alpha+} l_i^{\alpha} + l_i^{\alpha+} r_i^{\alpha}) \quad ; \qquad (1)$$

$$H^{Int} = \sum_{\substack{i=-N \\ \alpha}}^{N} V_\alpha \, r_i^{\alpha+} r_i^{\alpha} l_{i+1}^{\alpha+} l_{i+1}^{\alpha} \quad ;$$

$$H^{Tra} = \sum_{\substack{i=-N \\ \alpha\neq\beta}}^{N} W_{\alpha\beta} (r_i^{\alpha+} r_i^{\beta} + r_i^{\beta+} r_i^{\alpha})(l_{i+1}^{\alpha} l_{i+1}^{\beta+} + l_{i+1}^{\alpha+} l_{i+1}^{\beta}) \quad ;$$

$\alpha, \beta = 0,1,2,\ldots$ number of local states.

For the sake of simplicity, in (1) we took only interactions into account between protons located in neighboring potential minima. Further on we are going to limit ourselves to two states in each potential well, i.e. $\alpha,\beta=0,1$. The local creation and destruction operators, $r_i^{\alpha+}$, r_i^{α}, $l_i^{\alpha+}$, l_i^{α}, operate within the Fock space of 2N-protons occupying 4N one-particle oscillators, two for each hydrogen bond. ε_α denotes the eigenstates in the unperturbed oscillator potentials, T_α the tunneling frequency in the state α, V_α is the matrix element of direct proton-proton interaction and $W_{\alpha\beta}$ the matrix element of band transition between two adjacent bands.

The Hilbert space of eigenstates of the Hamiltonian (1) comprises the total dynamics of 2N protons whereas we are interested only in the peculiar mechanism, involving just a few protons, which causes defect propagation. Therefore we project out a subspace of the above Hilbert space comprising only states which describe a localized defect with the proper asymptotic arrangement of protons.

This can be done by defining the following operators:

$$a_n^o = \prod_{i=-N}^{n} r_i^o \prod_{i=n+1}^{N} l_i^o \quad ;$$

$$a_n^{o+} = \prod_{i=-N}^{n} r_i^{o+} \prod_{i=n+1}^{N} l_i^{o+} \quad ;$$

$$a_n^{\alpha'} = \frac{1}{\sqrt{2}} \prod_{i=-N}^{n-1} r_i^o \,(r_n^\alpha \pm l_n^\alpha) \prod_{i=n+1}^{N} l_i^o \quad ; \qquad (2)$$

$$a_n^{\alpha'+} = \frac{1}{\sqrt{2}} \prod_{i=-N}^{n-1} r_i^{o+} \,(r_n^{\alpha+} \pm l_n^{\alpha+}) \prod_{i=n+1}^{N} l_i^{o+} \quad ;$$

$\alpha' = 2\alpha$ for + sign

$\alpha' = 2\alpha+1$ for − sign

and $\alpha \neq 0$.

These are formally one-particle operators which create a subspace of the total Hilbert space

$$\{a_n^{\alpha+} \mid \Omega > \; = \; \mid n^\alpha > \} \quad \alpha = 0,1,\ldots \qquad (3)$$

and the states in (3) can be considered one-particle states. It should be noted here that states with more than one hydrogen bond excited need not be taken into account because of energy reasons. The Hamiltonian operating within the subspace of defect states can then be expressed in terms of the operators $a_n^{\alpha+}$, a_n^{α}, we call it the defect Hamiltonian:

$$H^{Def} = \sum_{n^\alpha n^\beta} \langle n^\alpha \mid H^P \mid n^\beta \rangle a_n^{\alpha+} a_n^\beta$$

$$= \sum_{n,\alpha} E'_\alpha a_n^{\alpha+} a_n^\alpha + \sum_n T_o (a_n^{o+} a_{n+1}^o + a_{n+1}^{o+} a_n^o)$$

$$+ \sum_{\substack{n \\ \alpha' \neq 0}} W_\alpha (a_{n+1}^{\alpha'+} a_n^{\alpha'} + a_n^{\alpha'+} a_{n+1}^{\alpha'}) \quad ; \qquad (4)$$

$$E'_o \equiv \varepsilon_o + V_o \quad ;$$

$$E'_1 = \varepsilon_1 + V_1 - T_1 \quad ;$$

$$E'_2 = \varepsilon_2 + V_1 + T_1 \quad .$$

The first four terms in equation (4) describe the defect-propagation by means of interacting ground and first excited states, whereas the last term permits the exciton transfer of vibrational excitations between different hydrogen bonds. W will be found small later on for the H_3O^+ defect. Concentrating on this case we pay no attention to the W term when we diagonalize this Hamiltonian.

H^{Def} can be diagonalized by a Fourier transformation in the usual way. We introduce the non-local particle operators

$$a_n^\alpha = \frac{1}{\sqrt{2N}} \sum_k e^{iknd} a_k^\alpha \qquad (5)$$

and obtain

$$H^{Def} = \sum_{k,\alpha} E_\alpha (k) a_k^{\alpha+} a_k^\alpha \quad ;$$

$$E_o = \varepsilon_o + V + T_o \cos kd \quad ;$$

$$E_1 = \varepsilon_1 + V + T_1 \quad ;$$

$$E_2 = \varepsilon_2 + V - T_1 \quad .$$

We have simplified the dispersion relations slightly, setting $V_0=V_1=V_2=V$ and allowing a band only for the ground states in view of the fact that later on we shall discover $W \ll T_1$. The one lower band which we obtain has the width T_0.

2. Defect-phonon interaction

Structural defects in ice are coupled to the phonons in various ways with each phonon branch contributing in a peculiar way to the conduction mechanism. The defect Hamiltonian, H^{Def}, comprises only degrees of freedom of the protons along the hydrogen bond axis. Therefore, the phonon part of the total Hamiltonian must arise from all the rest of the bound motions of the atoms in the ice crystal, i.e. the vibrations of the oxygen sublattice and the vibrations of the protons perpendicular to the bond. Significant coupling of the defects will occur in the ground band to the phonons in the acoustical branch along the chain, the optical branches originating from the O....O vibration (220 cm^{-1}) and wagging vibrations (780 cm^{-1}). Furthermore, coupling to the upper protonic states can be caused by all phonons, mainly the acoustical ones. The latter scattering mechanism was previously discussed by P. Gosar and we refer to it by his name.

The pure phonon Hamiltonian, H^{Ph}, has the form

$$H^{Ph} = \sum_{q} \omega(q)\, b_q^{+} b_q \tag{7}$$

and we consider only the simplified phonon spectrum

$$\omega_q = \begin{cases} < |q| & \text{for } |q| < q_D\,; \\ \omega_o + \omega_1 \cos qd & ; \\ \omega_o' + \omega_1' \cos qd & ; \end{cases} \tag{8}$$

where

$$\omega_o \approx 780\ cm^{-1} \quad \text{(wagging vibrations)},$$
$$\omega_o' \approx 220\ cm^{-1} \quad \text{(O...O vibrations)}.$$

Then there is the coupling to the diagonal part of the defect operators,

$$H_d^I = \sum_{\substack{nq\\ \alpha}} \omega_q\, g_q^{\alpha}\, e^{iqR_n^o}\, a_n^{\alpha+}\, a_n^{\alpha}\, (b_q^{+} + b_{-q}) \tag{9a}$$

which expresses the changes in the double well potentials if the lattice vibration of wave vector q displaces the atoms. The coupling to the non-diagonal elements in nearest neighbor approximation is given by

$$H_{nd}^I = \sum_{nq} \omega_q\, \Lambda_q^o\, e^{iqR_n^o}(a_n^{o+} a_{n+1}^{o} + a_{n+1}^{o+} a_{n+1}^{o})(b_q^{+} + b_{-q}) \tag{9b}$$

and represents the perturbation in the interaction between protons in neighboring hydrogen bonds due to the lattice wave. Finally, the coupling between the ground band and higher excited states can be expressed as

$$H^I_{int} = \sum_{nq} \omega_q \hat{g}_q e^{iqR^o_n}(a^{o+}_n a^1_n + a^{1+}_n a^o_n + a^{o+}_n a^2_n + a^{2+}_n a^1_n)(b^+_q + b_{-q}). \quad (9c)$$

Altogether, we adopt the interaction Hamiltonian

$$H^I = H^I_d + H^I_{nd} + H^I_{int} \quad . \quad (10)$$

We are going to deal with the problem of the coupling constants later on. At this point we should like to complete the formalism under the premises that coupling to the diagonal part of the local defect operators is strong whereas the other couplings can be dealt with in a weak coupling scheme. A verification of these assumptions on the basis of a model calculation will be given later on for the case of the excess proton.

The strong interaction with the diagonal part of the local defect Hamiltonian has to be treated in infinite order of the coupling constant. This can be achieved best by using a canonical transformation technique which was developed in the theory of electronic conductivity for the case of strong electron-phonon coupling (Holstein, Sewell, Firsow, and others /9/). We concern ourselves only with the ground band here and define a unitary operator

$$U = e^Q$$

with

$$Q = \sum_{nq\alpha} g^\alpha_q e^{iqR_n} a^{\alpha+}_n a^\alpha_n (b^+_q - b_{-q}). \quad (11)$$

This can be rewritten as

$$U = \sum_{n\alpha} B^\alpha_n a^{\alpha+}_n a^\alpha_n \quad (12)$$

with

$$B^\alpha_n = \exp \{\sum_q g^\alpha_q e^{iqR_n} (b^+_q - b_{-q})\}$$

and the exponential operators B^α_n obey the following commutation relations:

$$[B_n^\alpha, B_{n'}^\beta] = [B_n^{\alpha+}, B_{n'}^{\beta+}] = 0 \quad ;$$

$$[B_n^\alpha, b_q\ ] = [B_n^\alpha, b_q^+] = g_q^\alpha\, e^{iqR_n}\, B_n^\alpha \quad ; \tag{13}$$

$$[B_n^{\alpha+}, b_q\ ] = [B_n^{\alpha+}, b_q^+\ ] = -g_q^\alpha\, e^{-iqR_n}\, B_n^{\alpha+} \quad .$$

The canonical transformation (11), applied to the total Hamiltonian,

$$H^{Tot} = H^{Def} + H^{Ph} + H^I \quad , \tag{14}$$

produces a transformed Hamiltonian, $\tilde{H}^{Tot}$, which we split into 3 different parts, $\tilde{H}^0$, $\tilde{H}^1$ and $\tilde{H}^2$:

$$\tilde{H}^{Tot} = U^{-1}\ H\ U = \tilde{H}^o + \tilde{H}^1 + \tilde{H}^2 \tag{15}$$

with

$$\tilde{H}^0 = \sum_{n\alpha} (E_\alpha - \eta_\alpha) a_n^{\alpha+} a_n^\alpha \quad , \qquad \eta_\alpha = \sum_q \omega_q g_q^{\alpha^2} ; \tag{16a}$$

$$\tilde{H}^1 = T_o \sum_{nq} (a_n^{o+} a_{n+1}^o B_n B_{n+1}^+ + a_{n+1}^{o+} a_n^o B_{n+1} B_n^+)(b_q^+ + b_{-q}) ; \tag{16b}$$

$$\tilde{H}^2 = \sum_{nq} \omega_q \hat{g}_q (a_n^{o+} a_{n+1}^1 + a_n^{1+} a_n^o + a_n^{o+} a_n^2 + a_n^{2+} a_n^o)(b_q^+ + b_{-q}) . \tag{16c}$$

In equs. (16a) E_0 is the center of the lowest band, $E_\alpha - \eta_\alpha$ denotes the shifted local states due to the strong coupling to the phonons, thus H^o may be considered the unperturbed Hamiltonian of a new particle with different local eigenstates. This quasi-particle is known as the small polaron in electronic transport theory. It should be noted that for the new particle the requirements of weak coupling are fulfilled $-\tilde{H}^1$ and $\tilde{H}^2$ contain only weak coupling terms.

The band width for the new particle in the ground band can be obtained by averaging over the phonons:

$$\langle \tilde{H}^{Tot} \rangle_{phonons} = (E_o - \eta_o) \sum_n a_n^{o+} a_n^o + C_o \sum_n (a_n^{o+} a_{n+1}^o + a_{n+1}^{o+} a_n^o) \tag{17}$$

with

$$C_o = (T_o + \sum_q \omega_q \Lambda_q^o (1 + q^2 \coth\frac{\beta\omega_q}{2})) \cdot \exp\{-2 \sum_q g_q^{o^2} \coth\frac{\beta\omega_q}{2} \sin^2 \frac{qd}{2}\} .$$

In place of the tunneling frequency for the free defect, a more complicated term arises for the polaron which, at high temperatures, shrinks the band width and at low temperatures may even broaden the band. We can also see that strong coupling to low frequency vibrations leads to total collapse of the band. The strongly coupled defects in ice have $\beta\omega_q/2>1$ and, as will be seen later, $\Lambda_q^o < 1$, whereas $g_q^{o2} > 1$. Consequently, we are dealing with cases where the defect-phonon-interaction always tends to shrink the band, in other words, trap the defect at a certain lattice site.

3. Defect propagation through the ground band

The conductivity of a defect species can be calculated by way of the Kubo formula /8/

$$\sigma = \lim_{\varepsilon \to 0} i \int_0^\infty dt\, e^{-\varepsilon t} \int_0^{-i\beta} d\lambda < j\ (\lambda-t)\ j > , \tag{18}$$

which can be simplified in the absence of a magnetic field by partial integration:

$$\sigma = \beta \lim_{\varepsilon \to 0} \int_0^\infty dt\, e^{-\varepsilon t} < j\ j(t + \tfrac{i}{2}\beta) > . \tag{19}$$

The flux operator can again be obtained from the many-proton flux operator by quantizing with respect to the defect states:

$$J = id_1\ e^* \sum_{n,m} (n-m)\ T_{o(n-m)} a_n^{o+} a_m^o \quad , \tag{20}$$

where $T_{o(n-m)}=0$ if $n-m\neq 1$ in the nearest neighbor approximation which we adopted, d_1 is the distance between the minima in the double well potential. This distance appears since the dipole operator is $(d_1/2)\sum_i(r_i^+r_i-l_i^+l_i)$. e^* is the effective charge of the defect.

If we carry out the canonical transformation (11) the flux operator reads

$$\tilde{J} = id_1\ e^* \sum_{m,n} (n-m)\ T_{o(n-m)}\ a_n^{o+} a_m^o\ B_n B_{-m} \quad , \tag{21}$$

substituting back into equ. (19) we obtain

$$\sigma = -e^{*2}\beta d_1^2 \sum_{mnm'n'} T_{o(n-m)} T_{o(n'-m')} (n-m)(n'-m') \cdot$$
$$\cdot \lim_{\varepsilon\to o} \int_o^\infty dt\, e^{-\varepsilon t} \langle a_n^{o+} a_m^o B_{-m} B_n \cdot a_{n'}^{o+} a_{m'}^o B_{-n'} B_{m'} (t+\tfrac{1}{2}\beta)\rangle. \quad (22)$$

The correlation function in (22) can be evaluated if the interaction term in the Heisenberg operators is neglected:

$$\langle a_n^{o+} a_m^o B_{-m} B_n \cdot a_n^{o+} a_{m'}^o B_{-n'} B_{m'} (t+\tfrac{1}{2}\beta)\rangle =$$
$$= \langle a_n^{o+} a_m^o a_{n'}^{o+} a_{m'}^o \rangle_o \langle B_{-m} B_n B_{-n'} B_{m'} (t+\tfrac{1}{2}\beta)\rangle_o = \quad (23)$$
$$= \delta_{nm'}\delta_{mn'} \exp\{-2\sum_q g_q^{o2} \sin^2\frac{qd}{2}(n-m)(\coth\frac{\beta\omega_q}{2} - \frac{\cos(t-\frac{1}{2}\beta)\omega_q}{\sinh\frac{\beta\omega_q}{2}})\}.$$

This function has a sharp strong peak near $t=0$ and, possibly after some oscillations, an exponential long time behavior.

The strong peak for short times reflects the strong correlation between the lattice and the defect while the defect undergoes correlated jumps. The jump-time, τ_h, is the half-width of this peak. If we calculate the contribution of the jump-mechanism to the conductivity by a saddle point approximation (expanding $\cos \omega_q t$) we obtain

$$\sigma_h = \frac{\pi}{2}\beta e^{*2} d_1^2 \rho N T_o^2 \tau_h \cdot \exp\{-2\sum_q g_q^{o2} \sin^2\frac{qd}{2} \tanh\frac{\beta\omega_q}{4}\} \quad (24)$$

with the jump time

$$\tau_h = (\sum_q g_q^{o2} \omega_q^2 \frac{\sin^2 q d}{\sinh\frac{\beta\omega_q}{2}})^{1/2} . \quad (25)$$

At high temperatures expansion of the tanh $\frac{\beta\omega_q}{4}$ in eq. (24) gives rise to a positive activation energy:

$$E_a = \frac{1}{2}\sum_q g_q^{o2} \omega_q \sin^2\frac{qd}{2} \quad (26)$$

and we have come to the description of a purely diffusive process. Notable deviations from the simple activation mechanism and random jumps from site to site will be found in ice. First because the strong coupling frequencies are large in comparison to kT, secondly because the long time correlations are likely to contribute, particularly in the case of the excess proton.

The integral over the long time part of the correlation function diverges. This stems from neglecting the interaction term in the Heisenberg operators which give rise to a damping. To remedy this deficiency we introduce a relaxation time and resume a damping ansatz for the energy-conserving diagonal processes:

$$a_n(t) = e^{-t/\tau_b}\, a_n \quad ; \quad a_n^+(t) = e^{-t/\tau_b}\, a_n^+ \, . \qquad (27)$$

The problem is then to find a microscopic expression for τ_b. Expressions for τ_b, considering only the part $\tilde{H}^1$, equ. (16b), of the perturbation have been worked out by Schnakenberg, Lang and Firsow, but their formulas do not hold very well in our particular case. We have therefore calculated the second order processes for scattering in the lower band, finding

$$\tau_{b_1}^{-1} = 8\pi\, T_o^2 \sum_{qq'} g_q^{o2}\, g_{q'}^{o2} \sin^2 \frac{qd}{2}\, N_q\, (N_q+1)\, \delta\, (\omega_q-\omega_{q'}) . \qquad (28)$$

The interaction term $\tilde{H}^2$, Gosar's coupling equ. (16c), contributes to the scattering time in the band as well. Following Gosar we obtain for this scattering mechanism

$$\tau_{b_2} = 8\pi \sum_{qq'} \frac{\hat{g}_q^2\, \hat{g}_{q'}^2}{(\Delta E)^2}\, \omega_q^4 \sin^2 \frac{qd}{2}\, N_q\, (N_q+1)\, \delta\, (\omega_q-\omega_{q'}), \qquad (29)$$

his estimate of $\tau_{b_2}^{-1}$ is $3\cdot 10^{-12}$ sec for the H_3O^+ defect. The contribution of band conduction to the conductivity is then

$$\sigma_b = \tau_b\, e^{*2}\, d_1^2\, T_o^2\, \beta\, \rho \cdot \exp\{-2\sum_q g_q^2 \sin^2 \frac{qd}{2} \coth \frac{\beta\omega_q}{2}\} \qquad (30)$$

with

$$\tau_b^{-1} = \tau_{b_1}^{-1} + \tau_{b_2}^{-1}$$

and the ratio of conductivities comes out as

$$\frac{\sigma_b}{\sigma_h} = \frac{\tau_b}{\tau_h} \cdot \exp\{ -2 \sum_q g_q^2 \sin^2 \frac{qd}{2} (\coth \frac{\beta\omega_q}{2} - \tanh \frac{\beta\omega_q}{4})\} . \qquad (31)$$

At high temperatures σ_h prevails, at low temperatures σ_b.

The considerations, outlined before, were made primarily in view of the H_3O^+ defect which, of all four defects, has the best chance to propagate through the ground band. We tend to assume this mechanism since the energy of activation for the H_3O^+ mobility is found experimentally to be very low. On the other hand, an exciton-like mechanism of propagation is likely to hold for the D defects. Analogous arguments, as in the H_3O^+ case, can be made but are of lesser interest for lack of experimental information on the nature of the Bjerrum defects. We can only express our intuitive belief that for D defects the width of the exciton band will by far exceed the width of the lower bands, however, we are unable to predict the splitting of the two upper states, from which the exciton band is formed, in relation to the width of the exciton band.

4. Coupling constants

An estimate of the coupling constants g_q^o and $\hat{g}_q$ for a phonon branch steming from the local O....O vibrations in the 220 cm^{-1} region was given by Fischer, Hofacker, and Sabin /10/ on the grounds of a simple model calculation. A proton-oxygen coupling term is assumed of the form

$$H_o^I = \sum \left(V(R_n - x_n) - V(R_n - x_{n-1})\right) \qquad (32)$$

where R_n are the oxygen- x_n the proton-coordinates. Expansion around equilibrium positions R_n^o yields /11,12/

$$H_o^I = \sum_n \left(\frac{\partial V}{\partial R_n}(R_n^o - x_{n-1}) + \frac{\partial V}{\partial R_n}(R_n^o - x_n)\right)\cdot(R_n - R_n^o) \qquad (33)$$

with

$$\frac{\partial V}{\partial R_n}(R_n^o - x_n) = \sum_{\alpha\beta} < \phi_n^\alpha \mid \frac{\partial V}{\partial R_n}(R_n^o - x_n)|\phi_n^\beta> a_n^{\alpha+} a_n^\beta \,, \qquad (34)$$

where ϕ_n^α are the protonic wavefunctions for band n, we obtain for the lowest band, $\alpha=\beta=0$,

$$H_o^I = \frac{1}{\sqrt{2N}} \sum_{nq} g_{nq}^o \; h\omega_q \; e^{iqR_n^o} \, a_n^{o+} a_n^o \, (b_q^+ + b_{-q}) \qquad (35)$$

and the coupling constant g_{nq}^o is given by (cf. /16/)

$$g_{nq} = \{<\phi_n^o|\frac{\partial V}{\partial R_n}(R_n^o-x_n)|\phi_n^o>+<\phi_{n-1}^o|\frac{\partial V}{\partial R_n}(R_n^o-x_{n-1})|\phi_{n-1}^o>\}\sqrt{\frac{\hbar}{2M\omega_q^3}}\cdot e^{iqR_n^o} . \quad (36)$$

In contrast to the intraband coupling case, the ground band-upper states interaction must be derived by a Born Oppenheimer scheme (since energy gap $\Delta E >>$ phonon frequency and band width /13,14/). Thus

$$H_1^I = -\frac{\hbar^2}{M}\sum_{\substack{n\\ \alpha\beta}} <\phi_n^\alpha(R_n,R_{n+1},x_n)|\frac{\partial}{\partial R_n}|\phi_n^\beta(R_n,R_{n+1},x_n)>\frac{\partial}{\partial R_n} a_n^{\alpha +} a_n^\beta$$

$$= \frac{1}{\sqrt{2N}}\sum_{\substack{nq\\ \alpha}} g_{nq}^{\alpha\beta}\,\hbar\,\omega_q\, a_n^{\alpha +} a_n^\beta\,(b_q^+ + b_{-q})\, e^{iqR_n^o} \quad (37)$$

and

$$g_{nq}^{\alpha\beta} = \sqrt{\frac{\hbar}{2M\omega_q}}\cdot\{<\phi_n^\alpha|\frac{\partial}{\partial R_n}|\phi_n^\beta>+<\ _{n-1}^\alpha|\frac{\partial}{\partial R_n}|\phi_{n-1}^\beta>\}. \quad (38)$$

In (9c) the notation

$$\hat{g}_q \equiv g_{nq}^{o1}$$

is used. With the approximation

$$\phi_n^\alpha(R_n, R_{n+1}, x_n) = \phi_n^\alpha(\kappa R_n - x_n ; \kappa R_{n+1} - x_n), \quad (39)$$

where $\kappa = 1$ corresponds to a model which was considered previously by Gosar /15/, we arrive at

$$<\phi_n^o|\frac{\partial}{\partial R_n}|\phi_n^\alpha> = -\frac{\kappa}{2}\sqrt{\frac{m\Delta E}{\hbar^2}} , \quad \alpha = 1,2 \quad (40)$$

and

$$\hat{g}_q = -\frac{\kappa}{2}\sqrt{\frac{m\,\omega_o}{2M\omega_q}} , \quad (41)$$

$$\omega_o \equiv \frac{\Delta E}{\hbar} .$$

A LCAO-SCF calculation with a minimal basis set on the linear 7- center system

O....H-O....H....O-H....O

leads to

$$g_q^o \approx 3.4 \;, \qquad \hat{g}_q \approx 0.5 \;,$$

indicating strong coupling to the longitudinal 220 cm^{-1} phonon band. These numbers should be taken as an indication of the order of magnitude of the coupling strength. A potential, as it is deemed reasonable, for the middle proton and its change under displacement of a neighboring oxygen are given in fig. 3 to 5.

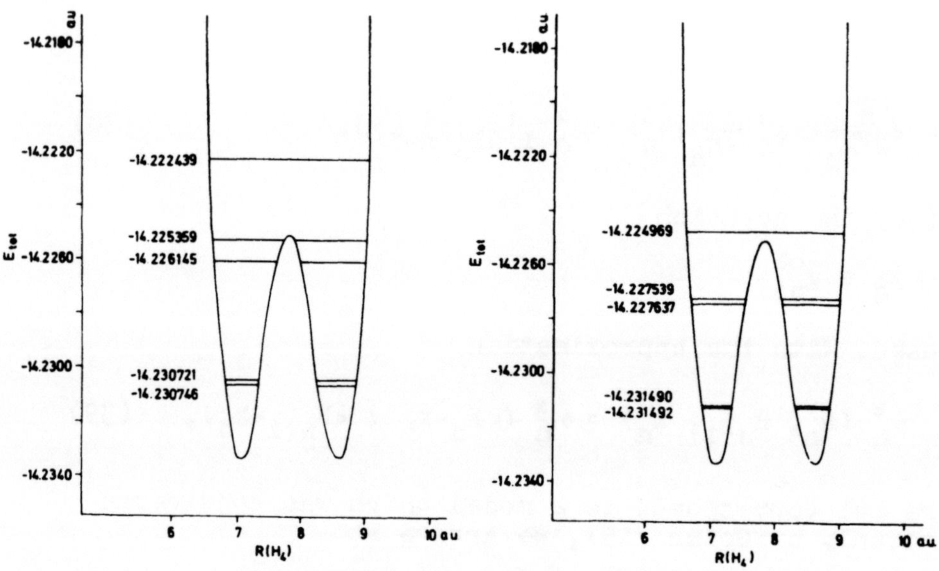

Fig.3 and 4 Lowest protonic/deuteronic energy levels (symmetric oxygen arrangement)

The coupling of the diagonal part of the defect operators to the wagging vibrations can easily be estimated on the basis of any reasonable effective-charge model (charges on protons between 0.3 and 0.5 e). One finds readily that this coupling will be of about the same order as the longitudinal one, though temperature effects will diminish its overall importance.

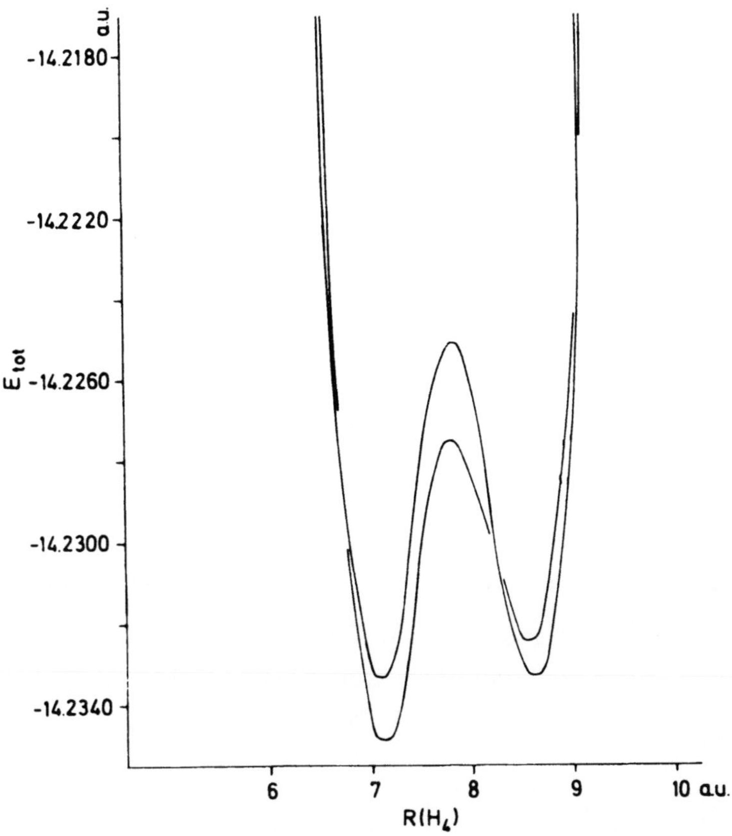

Fig.5 Potential for the middle proton after simultaneous displacement of the oxygen and hydrogen atoms to the left of it

References

/1/ Bjerrum, N., Kgl.Dan.Videnskab.Selskab. Math.-Phys. Medd. 27, 1 (1951)

/2/ Onsager, L., M. Dupuis, S.I.F. Rendiconti, X Corso, 294 (1960)

/3/ Gränicher, H., C. Jaccard, P. Scherrer, A. Steinemann, Disc.Faraday Soc. 23, 50 (1957)

/4/ Gränicher, H., Phys.kondens.Materie 1, 1 (1963)

/5/ Eigen, M.,L. De Maeyer, Z.Elektrochem.Ber.Bunsenges.physik.Chemie 60, 1037 (1956)

/6/ Eigen, M.,L. De Maeyer, H.-Ch. Spatz, Z.Elektrochem.Ber.Bunsenges.physik.Chem. 68, 19 (1964)

/7/ Jaccard, C., Helv.Phys.Acta 32, 89 (1958)

/8/ Kubo R., J.Phys.Soc. Japan 12, 570 (1957)

/9/ cf. e.g. Holstein T., Ann.Phys. 8, 343 (1959); Sewell G., Phys.Rev. 129, 597 (1963); Polarons and Excitons, ed. Kuper, C., and G. Whitfield, Plenum, New York (1963); Fröhlich, H., Arch.Sci. (Geneva) 10, 5 (1957); Schnakenberg, J., Z.Phys. 190, 209 (1966); Lang, I., Firsow, Yu., Soviet Physics J.E.T.P. 16, 1301 (1963); Kudinov, E.K., Firsow, Yu., Soviet Physics J.E.T.P. 20, 400 (1965); 22, 603 (1966)

/10/ Sabin, J.R., S.F. Fischer, G.L. Hofacker, Phys. kondens.Materie, in press

/11/ Bloch, F., Z.Phys. 52, 555 (1928)

/12/ Fröhlich, H., Phys.Rev. 79, 845 (1950)

/13/ Ziman, I.M., Proc.Cambr.Phil.Soc. 51, 707 (1955)

/14/ Haug,A.,Z.Phys. 146, 75 (1956)

/15/ Gosar, P., Nuovo Cimento 30, 931 (1963)

/16/ Sussmann, J.A., Physik kondens.Materie 2, 146 (1964)

SPECTRAL BEHAVIOR OF DEFECTS IN ICE - QUASIPARTICLE MODEL *

S.F. Fischer, G.L. Hofacker**, M.A. Ratner

Department of Chemistry, Northwestern University, Evanston, Illinois / USA

In connection with one of Professor Gränicher's list of interesting problems in the chemistry and physics of ice /1/, I should like to describe work we have done on the nature of the infrared absorption spectrum for ice defects in the region above 800 cm^{-1}. The analysis involves a model for one-dimensional linear ice. The four defects can be seen as boundary-and-charge conditions, and the problem can be reduced to its essential nature, without the complications introduced by the normal lattice, by projection onto the subspace of defect states. The resulting defect Hamiltonian will be taken as the starting point for this talk. Its formal derivation can be found in the lecture by Professor Hofacker /2/.

The actual defect concentration in ice, even in doped ice, remains very low /1/; and the infrared spectrum of ice, as described by Dr. Whalley /3/, is very rich, so that the spectrum predicted by this work will be hard to observe. The predicted behavior might be seen more easily in some other hydrogen-bonded substances, such as linear imidazole polymer or simpler solids such as $HCrO_2$; but it can be important in ice in several possible experiments and mechanisms, to be discussed below.

---

* Lecture delivered by MAR. An N.I.H. predoctoral fellowship to MAR is herewith gratefully acknowledged. This work was supported by NIH.

** Present address: Lehrstuhl für Theoretische Chemie, Technische Hochschule, München, Germany

1. The model

We treat the case of the defect interacting strongly with the lattice vibrations (phonons) of the host ice crystal. This assumption of strong interaction is the basis of our further work, and was definitely implied by a calculation on a model of the longitudinal optical vibrations interacting with the excess proton (H_3O^+) /4/. Strong coupling with the "wagging" /5/ vibrations was first suggested to us by Onsager /6/, and a model calculation yields a coupling constant here of the order 2 for the wagging-excess proton interaction. Strong interaction seems indicated also by the careful observations of Ibers, Delaplane, and coworkers /7/ on the chromous and cobaltous acid systems. These contain O-H...O hydrogen bonds, and show great variation of the observed spectral behavior upon substitution of deuterium for hydrogen. Low order perturbation theory of the O-H and O-O vibrations seems incapable of accounting for the observed behavior, so indicating a strong coupling problem /7a/. Finally, the large mass of the proton makes the Born-Oppenheimer expansion parameter, $(M_{proton}/M_{ion})^{1/4}$, a number of order 1/2, so indicating slow convergence of the Born-Oppenheimer perturbation series in this case, and suggesting a strong-coupling situation /8/.

The other assumption made for the present work is the existence of a local symmetric double-minimum potential for the defect. This is not a necessary assumption, as the treatment could also be carried out in the case of a highly asymmetric single-minimum system, such as probably exists for many hydrogen-bonded systems, or for a symmetric double minimum with appreciable tunnelling in the upper state. The symmetric double minimum model seems the correct one for the ice defects, however, and will be adopted here.

We will outline the derivation for the general case of a defect in one-dimensional ice. We include the tunnelling in the upper and lower states (our double-well is chosen to have only two local energy levels, or four split levels, beneath the barrier), the direct interaction of two protons near the same oxygen (this force, and all interactions, we simplify to two-body type), and the exciton-site energy-transfer term, which describes the deexcitation of a proton at lattice site n and the excitation, simultaneously, of a proton at site n+1. This Hamiltonian, which will then characterize all the defects and their lattice interaction, can be written in the notation of Fischer and Hofacker as:

$$H = \sum_{\alpha=0}^{2} \sum_{n=-N}^{N} E'_\alpha \, a_n^{\alpha+} a_n^\alpha + \sum_n T_o (a_n^{o+} a_{n+1}^o + a_{n+1}^{o+} a_n^o) + \sum_q \omega_q b_q^+ b_q$$

$$+ \sum_n \sum_{\alpha \neq 0} W_\alpha \, (a_{n+1}^{\alpha+} a_n^\alpha + a_n^{\alpha+} a_{n+1}^\alpha)$$

$$+ \frac{1}{2N} \sum_n \sum_\alpha \sum_q \omega_q \, g_q^\alpha \, e^{iqRn} \, a_n^{\alpha+} a_n^\alpha \, (b_q^+ + b_{-q}) \; .$$

We have defined

$$E'_o \equiv \varepsilon_o + V_o \quad ,$$

$$E'_1 \equiv \varepsilon_1 + V_1 - T_1 \; ,$$

$$E'_2 \equiv \varepsilon_1 + V_1 + T_1 \; .$$

Here b_q^+ creates an optical phonon of energy ω_q with wave vector q; $a_n^{+\alpha}$ creates a defect in energy state E'_α at lattice site n; and T_o describes the motion of the defect from lattice site n to lattice site n+1, corresponding to the tunnelling of the local proton. (In the case of the ionic defects, this corresponds to a tunnelling of the local proton at lattice site n; in the case of the Bjerrum faults, it corresponds to a rotation around the oxygen at site n). We choose to write:

$$\sum_q \omega_q g_q^\alpha \, e^{iqRn} \, (b_q^+ + b_{-q})$$

as

$$\sum_q \omega_q \, (u_{\alpha nq} \, b_q^+ + u_{\alpha nq}^* \, b_q) \; ,$$

where

$$u_{\alpha nq} \equiv g_q^\alpha \, e^{iqRn} \quad .$$

Here R_n is the coordinate of oxygen n and g_n^α is the coupling constant for interaction of the phonon of wave-vector q with the defect. We assume that this coupling is strong: $\frac{1}{2N} \sum_q \omega_q |g_q^\alpha|^2 >> 1$. We have neglected the nondiagonal coupling of Gosar /17/ and any coupling nondiagonal in the site number. These couplings are important for scattering processes, lifetime, and linewidths, but they

---

*We use atomic units throughout $\hbar = m_e \equiv a_o = 1$.

should be less important than the diagonal coupling for the qualitative spectral behavior. We also allow the coupling constants to differ in the different energy levels; that is, the g_q^α can differ for differing α; the upper state g_q should be much larger than g_q^o. We have also neglected the tunnelling in the lower state of the excess proton which would appear as splitting of the E_o energy level. This tunnelling is quite small, and its neglect is justified.

Since the interaction is strong, its effects cannot be taken into account solely by low order perturbation theory; we require knowledge of the multiphonon processes. We therefore take the strong interaction into account in infinite order by a canonical transformation:

$$\tilde{H} \equiv e^Q H e^{-Q} \text{ where } Q = \frac{1}{2N} \sum_q \sum_\alpha a_n^{\alpha +} a_n^\alpha \left(u_{\alpha nq} b_q^+ - u_{\alpha nq}^* b_q\right).$$

We have included details of this transformation, and of all other mathematical details of this talk, in a paper on strong coupling in hydrogen-bonded systems, to be published in J.Chem.Phys. /10/. This transformation yields a new Hamiltonian for our defect problem

$$\tilde{H} = \sum_q \omega_q b_q^+ b_q + \sum_n T_o (a_{on}^+ a_{on+1} B_n^o B_{n+1}^{o+} + a_{on+1}^+ a_{on} B_{n+1}^o B_n^{o+})$$

$$+ \sum_n \sum_{\alpha=0} \tilde{\varepsilon}_\alpha a_n^{\alpha +} a_n^\alpha + \sum_n \sum_{\alpha \neq 0} W_\alpha \left(a_{n+1}^{\alpha +} a_n^\alpha B_{n+1}^\alpha B_n^{\alpha +} + a_n^{\alpha +} a_{n+1}^\alpha B_n^\alpha B_{n+1}^{\alpha +}\right).$$

Here we have defined

$$B_n^\alpha \equiv e^{\sum_q \left(u_{\alpha nq} b_q^+ - u_{\alpha nq}^* b_q\right)},$$

$$\tilde{\varepsilon}_\alpha \equiv E_\alpha - \Delta E_\alpha ,$$

$$\Delta E_\alpha \equiv \sum_q \omega_q g_q^{\alpha 2} .$$

The physical interpretation of this Hamiltonian involves a quasiparticle description of the system. We now say that the $a_n^{+\alpha}$ operator creates a quasiparticle in the state α at site n. This quasiparticle is analogous to the small polaron and consists of the defect dressed in a cloud of virtual phonons, which describe the relaxation effect exerted on the lattice by the defect. The lattice relaxes about the defect position, so reducing the energy of the defect quasiparticle in state α by an energy ΔE_α (often called the polaron shift energy). The B_n operators appear in the interaction term because the motion of a quasiparticle in the lattice now consists in both the motion of the defect and the readjustment of the phonon system about the new defect state. In the derivation of $\tilde{H}$, we have assumed a low density of defects, and therefore neglected any defect-defect interactions.

If this Hamiltonian is Fourier transformed, we arrive at a band picture: the lower state of the local defect is broadened into a narrow band, whose width is proportional to T_o /8/. The upper states are broadened also, by the exciton transfer (this effect, while dominant in the case of the D-defect, should be very small for the excess proton) /9,10/. The defects are able to move through the lattice by means of these narrow bands.

2. The method - calculation of K(ν)

We are interested in calculating the absorbtion coefficient $K(\nu)$, where ν is the frequency of the radiation shined on the sample.

$$K(\nu) = \frac{\nu}{c} \cdot \sqrt{\frac{1}{2}\left[-\varepsilon_T' + \left((\varepsilon_T')^2 + (\varepsilon_T'')^2\right)^{\frac{1}{2}}\right]},$$

where $\varepsilon_T(\nu) = \varepsilon_T'(\nu) + i\varepsilon_T''(\nu)$ is the complex dielectric constant. If we assume a slowly-varying background absorption due to processes unrelated to the quasiparticle absorption, and a weak variation in the refractive index $\bar{n}$, then

$$K(\nu) = \frac{\nu}{c}\frac{\varepsilon''}{2\bar{n}}$$ where $\varepsilon(\nu) = \varepsilon'(\nu) + i\varepsilon''(\nu)$ is the complex dielectric constant for the quasiparticle processes. We actually calculate a complex conductivity, which can be related to the dielectric constant by

$$\varepsilon(\nu) = \frac{4\pi i\,\sigma(\nu)}{\nu} \quad /11,12/ \quad .$$

The expression for the conductivity can be derived by linear response theory /13,14/. We have

$$\sigma(\nu) = \frac{-1}{\nu}\int_0^\infty dt e^{-i\nu t}\int_0^\beta d\lambda < \tilde{j}(-i\lambda)\tilde{j}(t) >$$

where $\beta = (k_B T)^{-1}$ and $\tilde{j}$, the quasiparticle current, is given by the Heisenberg equation of motion as $\tilde{j} = i[\tilde{H},\tilde{X}]$ and where

$$\tilde{j}(t) \equiv e^{i\tilde{H}t}\,\tilde{j}e^{-i\tilde{H}t} \; .$$

The actual calculation of $\sigma(\nu)$ and $\tilde{j}(t)$ is rather involved, and will be published elsewhere /10/. The exact time dependence of the state is not easy to find. We use a damping approximation within the correlator

$$a^{\alpha}(t) = \exp\,(i\tilde{\varepsilon}_{\alpha}t)\exp(-\omega_{\alpha}|t|)a^{\alpha}$$

This type of damping behavior is very familiar for correlation functions in general /15/, and agrees with the behavior found for similar systems (polarons) by Schnakenberg /16/ and by Kudinov /11/ using Green's function and diagram techniques. The damping coefficients w_{α} are obviously crucial parameters of the theory. They depend upon the non-diagonal coupling processes and on the longitudinal phonons, as well as on the diagonal coupling included in $\tilde{H}$. Gosar /17/ has estimated $(w_0+w_1)^{-1}$ as $3\cdot 10^{-12}$ sec, considering only a coupling non-diagonal in the defect states; this seems to be of the right order of magnitude /2,9,10/. We define $w_{\alpha\beta} \equiv w_{\alpha}+w_{\beta}$.

We can use the Kubo identity /13/ to transform the expression for $\sigma(\nu)$ into one containing only single integrals:

$$\sigma(\nu) = \frac{-1}{\nu}\int_0^{\infty} dt e^{-i\nu t}\langle\tilde{j}(t)\tilde{j}(o)-\tilde{j}(o)\tilde{j}(t)\rangle - \frac{i}{\nu}\int_0^{\infty} d\lambda \langle\tilde{j}(-i\lambda)\tilde{j}(o)\rangle .$$

The major contribution to the spectrum will come from the first integral. In addition to the longtime exponential decay that comes from the damping, a term like $\langle\tilde{j}(t)\tilde{j}(o)\rangle$ will contain B operators, coming from the canonical transformation, which lead to an oscillation in the correlator. This oscillation comes from the effect of the virtual phonon cloud around the defect. $\langle\tilde{j}(-t)\tilde{j}(o)\rangle$ might typically look like the curve in fig. 3. The integrations cannot be simply performed, and a steepest-ascents method is used in the high temperature range /18/.

3. Results

A schematic absorption curve for the IR range 1000-4000 cm^{-1} is plotted in figure 2. It can be understood in conjunction with the correlator of figure 3 and the energy-level diagram of figure 1.[x] The results quoted are for the high-temperature limit; the low temperature behavior is qualitatively the same, and will be published subsequently.

The spectrum consists of four major features. The large peaks 1 and 2 are roughly gaussian in shape. They come about, mathematically, from the short-time behavior of the correlation function (that is - from the area under the first peak in figure 3).[xx] They correspond to an excitation from an initial stable state of energy $\tilde{\varepsilon}_0$, in which the lattice has relaxed about the defect, to an excited state in which the lattice has not yet fully relaxed, since the transition time for this excitation is shorter than a typical phonon relaxation time. Since this unrelaxed upper state is unstable with respect to the totally relaxed state, the absorption line to it will be quite wide. This behavior is analogous to the Stepanov mechanism /19/ for "predissociation" in the spectra of hydrogen-bonded compounds in which the line is broadened by upper-state interaction with the local O-O mode. There are two such peaks, corresponding to transitions to the two split upper levels. The width of these peaks is proportional to the square root of the sum of the polaron shift energies in the upper and lower states. In the case of high temperature, these peaks could be as wide as 800 cm^{-1}. See also the appendix.

---

[x] This is greatly oversimplified for the purposes of the present discussion. In fact, the excited states cannot be drawn in the same potential as the ground states, because of the strong coupling. The actual equilibrium potentials are different in the two states, and Franck-Condon curves should be drawn, with different equilibrium positions, for the two relaxed states. This is an important point, and is fully discussed in ref. /10/.

[xx] Again, this is an oversimplification, as the steepest-ascents expansion is made about a complex point, not around t=0; but the peak is a short time one. Also, the peak is not actually Gaussian. See ref. /10/, and the appendix.

The two triplet peaks (345 and 678) come from the long-time tail in the correlator $\langle\tilde{j}(-t)j(o)\rangle$. They correspond to a slow excitation of the defect to a lower, stable state of energy $\tilde{\varepsilon}_1$ or $\tilde{\varepsilon}_2$. This transition can be considered "adiabatic" in a sense converse to that of the usual Born-Oppenheimer sense: the lattice follows the motion of the defect particle almost adiabatically, since the transition time is of the order of, or longer than the phonon relaxation time (this situation occurs also in very-strong-coupling limit polaron theory /20/). Since the upper state for the transitions is a stable one, these lines are narrow. The largest ones (4 and 7) have widths w_{o1} and w_{o2} respectively, and are approximately Lorentzian in shape.

The sidepeaks (3,5,6,8) correspond to this same transition, with the additional absorption (3,6) or emission (5,8) of a lattice phonon of frequency ω_o (we have assumed an optical phonon dispersion relation of the form: $\omega_q=\omega_o+\omega_1 \cos dq$, where d is a lattice spacing). These sidepeaks will be considerably broadened by anharmonic terms in the lattice displacements, which lead to a finite lifetime for the optical phonons, and may be very difficult to observe experimentally.[x] The actual appearance of the spectrum may differ considerably from figure 2. In particular, the values of E_α' and T_o come from the shape of the original symmetric double-minimum potential chosen for the defect. The two large peaks (1 and 2) are separated by $E_2^1-E_1^1$, while the Lorentzian peaks (4 and 7) are separated by $\tilde{\varepsilon}_2-\tilde{\varepsilon}_1$. They may occlude each other, or change their relative positions on the wavelength axis. The actual spectrum of figure 2 is based on a barrier height of 5000 cm^{-1} /23/. The spectrum shown corresponds to strong interaction with only one optical branch. Similar interaction with a second branch would introduce a second pair of triplets into the spectrum.

One of the advantages of using linear-response theory, which is a density matrix method, is that we obtain automatically the temperature dependence of the spectrum. In this case, we see that the positions of the peaks are roughly independent of T, although the widths and heights are temperature-dependent.

The appearance of a doublet in the absorption spectrum, corresponding to the g→u and u→g transitions, has often

---

[x] The sidepeaks occur in this treatment due to an assumed form of the q-dependence of the coupling constant g_q^α: $g_q^\alpha \propto q^{-2}$ /11/. They would not occur, if this dependence were of some other functional form.

been considered an inescapable result of a symmetric double-minimum potential, and its absence has occasionally been adduced as proof of the absence of a potential of this sort. The present treatment shows that even if T_1 is large, the doublet (peaks 1 and 2 of figure 2) can merge into a single absorption if the polaron shifts $\Delta E_2, \Delta E_1$, and ΔE_0 are large enough. This should happen in the case of strong coupling, and thus the appearance of the doublet is therefore not diagnostic for the symmetric double-minimum model.

The mass dependence of our model is quite complicated, since the parameters g_q^{α} and w_{α} depend on the defect mass. Thus it is improbable that a simple product rule will hold for our system, while the observed strong dependence upon mass /7/ may be expected to come from numerical calculations based on this formalism.

4. Specific application to ice

Consider first the H_3O^+ defect in common ice Ih. Its actual IR spectrum would be quite difficult to observe, due to the richness of the background absorption (in the language of section 2, $\varepsilon_T^{\,\prime\prime} \gg \varepsilon''$) /3/. Even the maximum amount of acid doping would, probably, not increase the H_3O^+ concentration enough to make the expected defect absorption identifiable. The D defect represents the complementary case to the excess proton, in that here we might expect the exciton-like energy-transfer process, in which a de-excitation of the defect from energy level $\tilde{\varepsilon}_2$ or $\tilde{\varepsilon}_1$ at lattice site n leads to an excitation of the defect from level $\tilde{\varepsilon}_0$ to $\tilde{\varepsilon}_1$ or $\tilde{\varepsilon}_2$ at site n+1, so moving the defect along the chain, to be dominant. In this case, in Fourier space, the two upper states are broadened into bands, with bandwidths proportional to W_1 and W_2. The lower state band, of width T_0, is much narrower. In the case of the absorption spectrum, this will simply result in a greater width for the absorption lines.

The most interesting application of the present results to the case of a real ice crystal would probably lie in photoconductivity observations. Since the exciton energy transfer, in the case of the D-defect, results in a band in the upper states, whose width far greater than the width of the lower band, we would expect the mobility of the D-defect to increase markedly after photoexcitation to energy levels $\tilde{\varepsilon}_1$ or $\tilde{\varepsilon}_2$ /9,10,21/. Thus an ac conductivity measurement on ice subjected to IR radiation of frequency $\cong \tilde{\varepsilon}_2 - \tilde{\varepsilon}_0$ should yield a considerably larger value of the conductivity than in the non-irradiated sample. Also, theories of mobility or conductivity which involve Bjerrum-defect motion as the slow, rate-determining step /24/ should be subject to test by the photoconductivity experiments: if the process is not enhanced by irradiation then the proposed mechanism would be incorrect.

The mobility of the H_3O^+ defect might be expected to decrease upon irradiation, due to the removal of defects from the narrow lower band to the self-trapped upper states. This would be quite difficult to see experimentally, due to heating of the crystal and inefficiency of the excitation process, but might be searched for using a pulsed laser as the source of IR radiation in a dc photoconductivity measurement.

Finally, it should be pointed out that we have nowhere restricted our one-dimensional model to ice Ih. Any of the other phases of ice /3,22/ which contain the defects in a double-minimum potential and fulfill the strong coupling criterion should also be suitable systems in which to test the present theory experimentally.

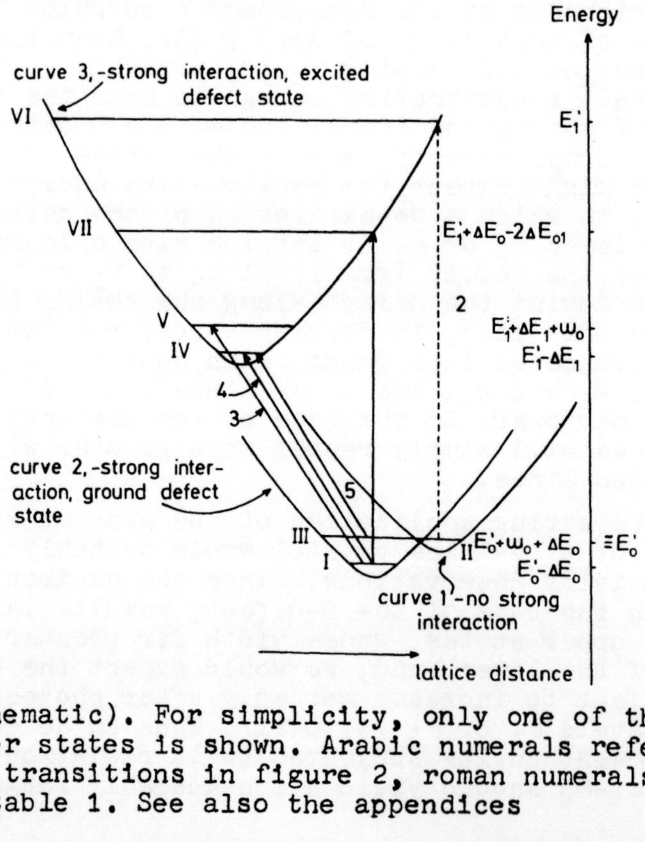

Fig.1 (schematic). For simplicity, only one of the split upper states is shown. Arabic numerals refer to the transitions in figure 2, roman numerals refer to table 1. See also the appendices

Table 1 Legend for figure 1

State	Energy	Remarks
I	$E_o' - \Delta E_o$	Lower level for optical transitions
II	E_o'	Ground state of uncoupled defect
III	$E_o' + \omega_o - \Delta E_o$	Lattice phonon + (I)
IV	$E_1' - \Delta E_1$	Upper level for adiabatic transition
V	$E_1' + \omega_o - \Delta E_1$	Lattice phonon + (IV)
VI	E_1'	Upper level for uncoupled defect
VII	$E_1' + \Delta E_o - 2\Delta E_{o1}$	Upper level for vertical transition

Appendix 1

In this appendix, a brief description is given of the broad curve labelled 1 in figure 2. It occurs at a position $\nu \cong E_2'-E_1'+2\Delta E_o-2\Delta E_{o1}$. The position is shifted by $2\Delta E_o-2\Delta E_{o1}$; of this ΔE_o, comes from a lowering of the initial state energy due to the lattice relaxation, while the rest comes from the shifted position of the upper potential curve caused by the lattice relaxation. The peak is not exactly Gaussian in shape - the proper expression includes, besides the Gaussian factor which has a maximum $\nu = E_2'-E_o'+2\Delta E_o-2\Delta E_{o2}$ and gives the height and width listed in table 2, two additional factors. One of these is an activation-like factor, such as is familiar in polaron theory /18/. The other is a denominator which is symmetric about its minimum at $\nu = E_2'-E_o'+\Delta E_o-\Delta E_2$. It causes asymmetry in the curve, so that the figures for the height and width given in table 2, which come from the Gaussian part only, must be considered approximate.

The peak labelled 2 is the analogous feature for the absorption to the lower of the two split upper levels. This shift has not been previously described - it is more fully investigated in ref. /10/.

Appendix 2

The Franck-Condon curves shown in figure 1 correspond to different energy states of the defect. They are similar to Franck-Condon curves for diatomic molecules, except that the defect here plays the role of the electron in the diatomic case. The adiabatic coordinate (abscissa) which is the internuclear distance in the diatomic case, can be seen here as a lattice distance, expressed as a linear combination of the various phonon modes, chosen to change slowly with protonic state. The effect of the strong coupling is manifested in the downward energy shift (polaron shift) of curves 2 and 3, and in their movement to the left, corresponding to a relaxation of the lattice distance.

The ratio of the distance shift of curve 3 to that of curve 2 is $\cong \sum_q g_q^1/\sum_q g_q^o \cong 2.65$. We can see this bond-shortening effect in the measured distances in actual crystals of ice Ih - the lattice distance in normal ice is 2.76 Å, while the lattice distance of the excess proton is only 2.55 Å.

The energy levels drawn include a zero-point vibrational energy of 1/2 ω_o. The dotted transition is the one that would be observed if there were no coupling in the ground state (ΔE_o=0). Due to the coupling, we observe instead the vertical transition labelled (2). This is the transition which results from the steepest-ascents integration. It corresponds to an excitation to an energy state which contains many phonons-it is thus similar to the vertical transition in the diatomic case, in which the upper state contains many vibrational quanta. It has a linewidth which depends on the phonons.

The absorption (4) is the adiabatic peak. It comes from singularities in the undamped correlator $\langle \tilde{J}(-t)\tilde{J}(0)\rangle$, and has the lattice-phonon sidepeaks (3) and (5). (5) starts from a lower state which contains a lattice phonon. We should expect its intensity to be smaller by a Boltzmann factor exp $(-\omega_o/kt)$; this does indeed come out of the calculation /10/.

Figure 1 was drawn assuming:

$$E_1' - E_o' = 2600 \text{ cm}^{-1}$$
$$\omega_o = 150 \text{ cm}^{-1}$$
$$\Delta E_o = 150 \text{ cm}^{-1}$$
$$\Delta E_{o1} = 450 \text{ cm}^{-1}$$
$$\Delta E_1 = 1450 \text{ cm}^{-1}$$

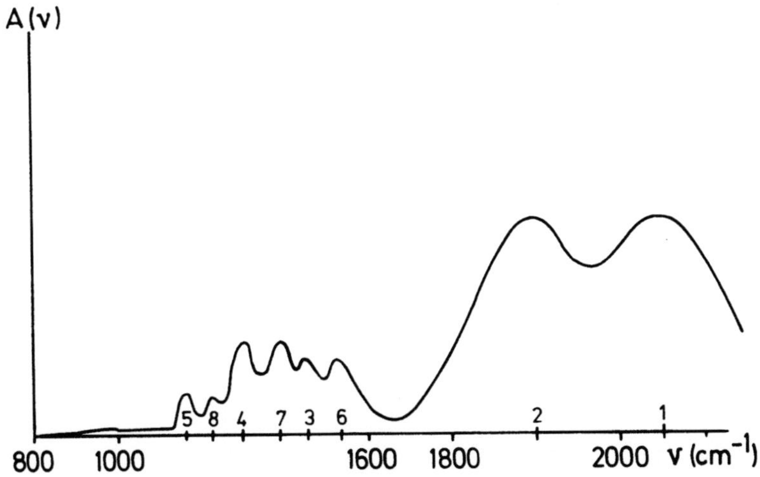

Fig.2 (schematic)

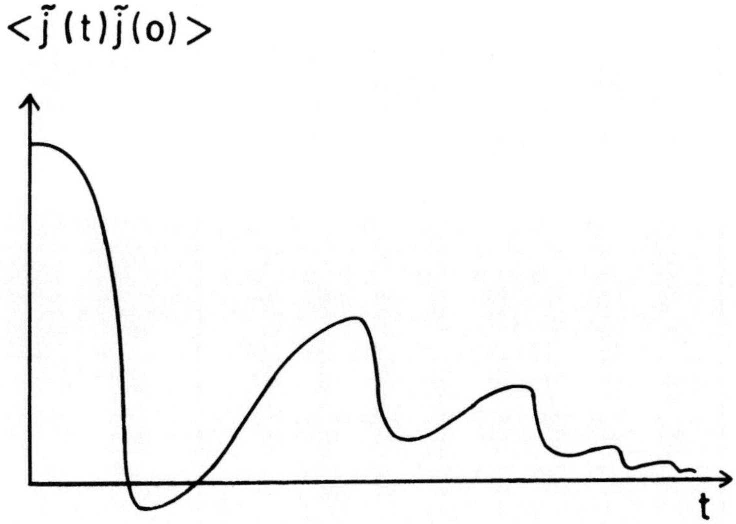

Fig.3 (schematic)

Table 2 [×]

Peak number	Position	Width	Height
1	$E_2'-E_o+2\Delta E_o-2\Delta E_{o2}$	$\left(kT(\Delta E_2+\Delta E_o-2\Delta E_{o2})\right)^{1/2}$	$\propto\frac{1}{\nu}\left(kT(\Delta E_2+\Delta E_o-2\Delta E_{o2})\right)^{-1/2}$
2	$E_1'-E_o'+2\Delta E_o-2\Delta E_{o1}$	$\left(kT(\Delta E_1+\Delta E_o-2\Delta E_{o1})\right)^{1/2}$	$\propto\frac{1}{\nu}\left(kT(\Delta E_1+\Delta E_o-2\Delta E_{o1})\right)^{-1/2}$
3	$E_1'-E_o'-\Delta E_1+\Delta E_o+\omega_o$	See text	
4	$E_1'-E_o'-\Delta E_1+\Delta E_o$	$\propto w_{o1}$	$\propto\frac{1}{w_{o1}}$
5	$E_1'-E_o'-\Delta E_1+\Delta E_o-\omega_o$	See text	
6	$E_2'-E_o'-\Delta E_2+\Delta E_o+\omega_o$	See text	
7	$E_2'-E_o'-\Delta E_2+\Delta E_o$	$\propto w_{o2}$	$\propto\frac{1}{w_{o2}}$
8	$E_2'-E_o'-\Delta E_2+\Delta E_o-\omega_o$	See text	

[×] $\Delta E_{\alpha\beta} \equiv \sum_q \omega_q\, g_q^\alpha\, g_q^\beta$

$\Delta E_\alpha \equiv \sum_q \omega_q\, g_q^\alpha\, g_q^\alpha$

References

/1/ Gränicher, H., Physics of Ice, Proc.Int.Symp., Munich, Plenum Press, New York (1969)

/2/ Hofacker, G.L., Physics of Ice, Proc.Int.Symp., Munich, Plenum Press, New York (1969)

/3/ Whalley, E., Physics of Ice, Proc.Int.Symp., Munich, Plenum Press, New York (1969)

/4/ Sabin, J.R., S.F. Fischer, G.L. Hofacker, Phys. kondens.Materie, in press

/5/ cf. e.g. N. Ockman, Advances in Physics 1, 199 (1958)

/6/ Lars Onsager, private communication

/7/ Hamilton, W.C., J.A. Ibers, Acta Cryst. 16, 1209 (1963); J.A. Ibers, J.Chem.Phys. 41, 25 (1964); 48, 539 (1968); R. Delaplane, J.A. Ibers, J. Rush, J. Ferraro, J.Chem.Phys., in press

/7a/ Delaplane, R., J.A. Ibers, private communication

/8/ Born, M., K. Huang, The Dynamical Theory of Crystal Lattices, Oxford, Appendix 7

/9/ Fischer, S.F., G.L. Hofacker, J.Chem.Phys., to be published

/10/ Fischer, S.F., G.L. Hofacker, M.A. Ratner, J.Chem. Phys.,to be published

/11/ Kudinov,E.K.,Yu. Firsov, J.E.T.P. 20, 400 (1965)

/12/ Smyth, C.P., Dielectric Behavior and Structure, McGraw-Hill, 1955, p.72

/13/ Kubo, R., J.Phys.Soc. Jap. 12, 570 (1957)

/14/ Zwanzig, R., Ann.Rev.Phys.Chem. 16, 67 (1965)

/15/ Ferrell, R.A.,Bat-sheva Summer School in Quantum Fluids, Haifa, 1968 (to be published by Gordon and Breach)

/16/ Schnakenberg , J.,Z.Phys. 190, 209 (1966)

/17/ Gosar, P., Nuovo Cimento 30, 931 (1963)

/18/ Holstein, T., Ann.Phys. 8,343 (1959)

/19/ Stepanov, B.I., Zh.Fiz.Khim. 19, 507 (1945)

/20/ cf. e.g. G.R. Allcock, in Polarons and Excitons, ed. C. Kuper and G. Whitfield, Plenum, New York (1963)

/21/ Lang, I., Yu. Firsov, J.E.T.P. 16, 1301 (1963)

/22/ Kamb, B., Physics of Ice, Proc.Int.Symp., Munich, Plenum Press, New York (1969)

/23/ Somorjai, R., D. Hornig, J.Chem.Phys. 36, 1980 (1962)

/24/ Engelhardt, H., B. Bullemer, N. Riehl in Physics of Ice, Proc.Int.Symp., Munich, Plenum Press, New York (1969)

PROTON — PROTON AND PROTON — LATTICE INTERACTIONS IN ICE

P. Gosar

Institute "J. Stefan" and University of

Ljubljana, Ljubljana, Yugoslavia

Abstract

The migration of Bjerrum L-defects caused by thermally activated rotations of the adjacent water molecules is investigated. It is shown that the activation energy for the rotation can be transferred from one molecule to another due to the proton-proton interactions. Several models for the rotational collisions are presented. A short discussion of the dissipation of the activation energy to the lattice is given too. Further, the effect of the multiple rotational collisions on the dielectric relaxation is studied.

1. Introduction

In this paper we report about some new ideas related to the problem of the rotation of water molecules in ice crystals. The rotation of molecules is responsible for many physical properties of ice, for instance the dielectric and mechanical relaxation.

Ice crystals show the Debye dielectric relaxation at low frequencies and the corresponding relaxation time is given by

$$\tau = \tau_o \exp(E/kT) \qquad (1)$$

with $\tau_o=5.3\cdot10^{-16}$ s and E=0.575 eV = 13.2 kcal $mole^{-1}$ /1-4/. In order to explain the electrical properties of ice two kinds of defects, ionic and orientational, are usually postulated /5/. For the dielectric relaxation the most important are orientational or Bjerrum defects /6/. Namely, the rotation of the water dipoles or molecules is possible only in the neighborhood of the orientational defects. Such rotations produce the motion of these defects. The theory of the dynamic dielectric constant has been worked out by Steinemann and Gränicher /3/, and in a more general form by Jaccard /7/. Both theories are based on the assumption that the rotations of water molecules are uncorrelated. Using the experimental data on the Debye relaxation, one obtains from these theories the information on the transition probability for the rotational jumps of water dipoles. It seems that in pure ice only the motion of the vacant bond or L-orientational defect is important. Then, according to Jaccard

$$\frac{1}{\tau} = \frac{3}{4}\left(\frac{n_L}{n_H}\right) p_L \quad , \tag{2}$$

where n_L is the concentration of L-defects and n_H the number of hydrogen bonds per unit volume. p_L is the transition probability for the rotational jump of the water molecule adjoining the defect. The activation energy for the relaxation time (eq.(1)) is equal to the sum of one half of the formation energy $E_f=(0.68\pm0.04)$eV for Bjerrum defects and the activation energy $E_r=(0.235\pm0.01)$ eV for the rotation /7/. Approximately,

$$n_L = n_H \exp\left(-E_f/2kT\right) \; . \tag{3}$$

It follows from eq.(1) and (2) that

$$p_L = 2.5 \cdot 10^{15} \exp(-E_r/kT) \; s^{-1} \quad . \tag{4}$$

At -10 °C the transition probability is of the order 10^{11} s^{-1}.

What is remarkable with the expression (4) is the very high pre-exponential factor. According to the classical absolute-rate theory for the thermally activated processes /8/ this factor should be equal to

$$\frac{kT}{h} \exp\left(\Delta S/k\right) \quad , \tag{5}$$

where h is the Planck's constant and ΔS the increase of the entropy at the transition of the molecule from the ground to the activated state. Now $kT/h=5.5\cdot10^{12}$ s^{-1} at -10 °C. An appreciable change ΔS of the entropy is unbelievable. The change of the stretching or bending frequencies of protons of the water molecule in the activated state has no effect on ΔS because the involved vibrational quanta are much greater than kT. The only contribution to ΔS could come from the translational degrees of freedom. In the activated state the crystal binding is weakened because of the breakage of one hydrogen bond. One could expect an increase of the amplitude of lattice vibrations. But, since the proton mass is small, the rotation of the water molecule is fast and the transition takes place before the lattice could adjust to the new situation in the transition state. Therefore we must exclude also this possibility for an increase of ΔS.

The formula (5) is correct only if the vibration, which leads to the jumping process when it is strongly excited, has greater frequency than kT/h. In the opposite case kT/h should be replaced by this vibrational frequency /9/. In ice the librational modes are responsible for the rotational jumps. Their frequency is near 800 cm^{-1} /10,11/ and hence greater than kT/h.

It seems that no simple explanation exists for the high pre-exponential factor in eq. (4). Therefore we investigated if perhaps something is wrong with the basic assumptions of the theory. One could better explain the experimental data if the assumption of the uncorrelated rotations is replaced by the idea of the concerted rotations of many water molecules. We have in mind the possibility that the rotation of one molecule induces the rotation of the neighboring one and so on. This is illustrated with the aid of fig. 1, which shows, as an example, one L-defect

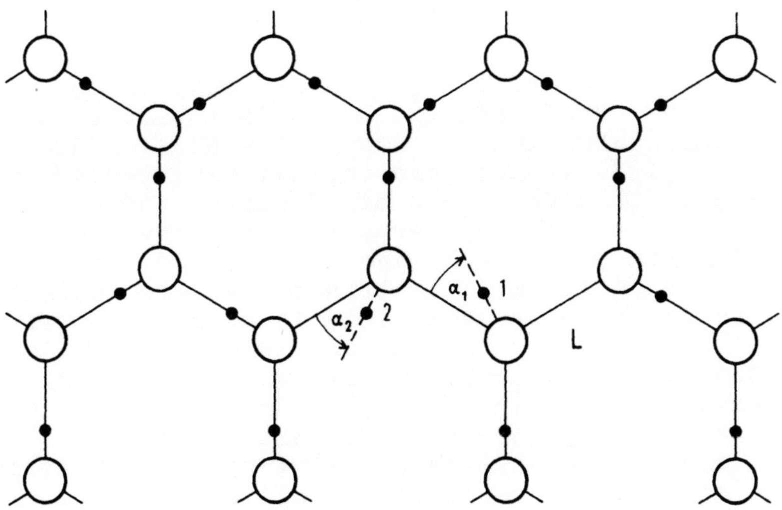

Fig.1 L-defect in the crystal structure of ice

perpendicular to the c-axis of the ice crystal. The open circles are oxygen and the dots hydrogen atoms. The protons are shown in their equilibrium positions except for two in the neighborhood of the defect, which are labeled with 1 and 2, respectively. The motion of the L-defect takes place e.g., by the thermal excitation of proton 1 and its subsequent jump in the direction of the arrow towards the unoccupied bond of the defect. This type of motion has been taken into account in previous theories of the dielectric relaxation. We show in this paper that it is very likely that another process of motion also takes place. Namely, if proton 2 is thermally excited, it is able to transfer the activation energy to proton 1 during its motion in the direction of the arrow and push this proton in the direction of the vacant bond. The transfer of the activation energy is possible because of the strong proton-proton repulsion forces. The final result of this single thermal activation is that proton 2 occupies the bond previously occupied by proton 1 and proton 1 fills the vacant bond of the L-defect. We deal here with the concerted rotation of two water molecules. As a consequence of the single stochastic thermal process two water molecules changed their orientation and the L-defect made a two step displacement. The extension of the idea of the concerted rotations to the case of more than two molecules is straight-forward and this needs not to be discussed in further detail at this point. Note that the basic process in the proposed model consists in single or multiple successive collisions between protons. Actually, it is more appropriate to talk about rotational collisions between the water molecules.

The multiple collision problem, which has many common features with the case discussed here, is a problem of replacement collisions in the theory of radiation damage in crystals /12,13/. It has been intensively studied in the last decade. Consider a row of identical atoms. If one atom is given a kinetic energy from the incident radiation with its momentum in the direction of the row, then the kinetic energy will be transmitted from atom to atom down the row without loss. This is a familiar result from the collision theory for two particles of equal mass if one of the colliding particles is initially at rest. If the transmitted kinetic energy is high enough, it may happen that the atoms, when come at rest after the collisions, occupy the lattice sites of the struck atoms. This is the case of the replacement collisions. Here the kinetic energy and also a series of replacements propagates down the row of atoms. This is exactly what we expect to happen in ice.

The problem of rotational collisions of water molecules is mathematically extremely difficult. Also the potentials for the hindered rotation of the water molecule and the proton-proton interaction are not well known. We will be able to get only some rough estimates, by using different models and approximations. This is done in the next Section. Further, a separate Section is devoted to the question of the dissipation of the rotational energy to the lattice vibrations. The dissipation is important for the estimates of the number of molecules which can rotate in a coherent way. In the last Section the problem of the dielectric relaxation is discussed in view of the new mechanism.

2. Proton-proton interactions and rotational collisions

An extremely rough estimate for the interaction between two water molecules or protons in ice can be obtained from the dipole-dipole interaction of water dipoles. Let us consider, for example, the interaction of protons 1 and 2 in fig. 1. The axes of rotation of the corresponding water molecules have the direction of the crystal c-axis. The dipole approximation gives the following result for the interaction energy

$$H_{d.-d.} = \frac{\mu^2}{6\pi\,\varepsilon_o r^3}\, f(\alpha_1,\alpha_2) \quad , \qquad (6)$$

with

$$f(\alpha_1,\alpha_2) = -\frac{1}{4}\cos(\alpha_1+\alpha_2) - \frac{3}{4}\cos(\alpha_1-\alpha_2) + \frac{1}{4}\sqrt{3}\sin(\alpha_1+\alpha_2) - \frac{3}{4}\sqrt{3}\sin(\alpha_1-\alpha_2) \quad , \qquad (7)$$

where μ=1.84 Debye is the dipole moment of the water molecule /14/ and r=2.76 Å the intermolecular distance. Therefore, $\mu^2/(6\pi\varepsilon_o r^3)$=0.067 eV = 1.6 kcal/mole. For example, the variation of the interaction energy with the rotational angle α_2 if proton 1 is in the equilibrium position $\alpha_1=0^o$ is given by

$$f(0,\alpha_2) = -2\cos(60^o + \alpha_2) \quad . \qquad (8)$$

The change in the interaction energy between the original $\alpha_2=0^o$ and the collinear O-H...H-O situation $\alpha_2=120^o$, where the interaction is strongest, is 4.7 kcal/mole. This is of the order of the hydrogen bond formation energy. The interaction energy can be considerably reduced if we assume the correlated motion of both protons. In the case $\alpha_1=\alpha_2=\alpha$ we obtain

$$f(\alpha,\alpha) = -\frac{1}{2}\cos(60^\circ + 2\alpha) - \frac{3}{4} \qquad (9)$$

The variation of the interaction energy between $\alpha=0^\circ$ and the maximum at $\alpha=60^\circ$ is here only 1/4 of the previous value.

We must emphasize that the dipole-dipole approximation actually fails completely for the situations like collinear state where the real interaction between protons is much stronger. Better estimates for the proton-proton interaction are found in the paper of Dunitz /15/ concerning the nature of orientational defects. For the region of interest here, the following estimates for the interaction of two non-bonded hydrogen atoms can be used:

$$E(r) = 10\,000 \exp(-4.6\, r) - 49.2/r^6 \quad /16/ \quad (10)$$

$$E(r) = 6\,600 \exp(-4.08\, r) - 49.2/r^6 \quad /17/ \quad (11)$$

where E is in kcal/mole and the interprotonic distance in Å. According to these formulas the interaction between protons, in the equilibrium position, of two neighboring hydrogen bonds is negligible. For the cases $\alpha_2=60^\circ$ and $\alpha_1=0^\circ$, we obtain for E the value 6.7 and 11.5 kcal/mole from eq. (10) and (11), respectively. In the collinear situation the interaction energies are about 60 kcal/mole. We note a great increase of the interaction at small distances with respect to the dipole-dipole approximation.

In addition to the interactions between neighboring protons, we have in ice also the long range dipole-dipole interactions between distant protons.

We proceed to the study of the rotational collisions. Because of the mathematical complexity of the problem, we shall first try to get some information by investigating the collisions between the freely rotating water molecules. The case of the hindered rotation, which corresponds to the real situation in the ice crystal, will be discussed afterwards.

Let us consider the collision between the molecules or protons 1 and 2 in fig. 1. One important question is: at what angle α_2 starts the collision of the rotating molecule 2 with the molecule 1 at rest ($\alpha_1=0$)? The final positions after the collision are also of interest. The multiple replacement collision process is only possible without the appreciable loss of the excitation energy if the individual collisions are mutually well separated processes without any interference. This means that in the free rotating model the starting angle and the final orientation should be 60° and 120°, respectively. The requirement for the starting angle in the neighborhood of 60° seems to be compatible with the range of the interaction forces if we exclude the long range part which can be better described by the concept of the polarization fields. We shall discuss this problem of internal electric fields later in the concluding Section.

The equations of motion for molecules 1 and 2 are

$$I\ddot{\alpha}_1 = T(\alpha_1,\alpha_2) ,\qquad I\ddot{\alpha}_2 = -T(\alpha_1,\alpha_2) , \tag{12}$$

where I is the moment of inertia of the water molecule for the rotation about the c-axis and $T(\alpha_1,\alpha_2)$ the torque due to the proton-proton interaction. Let us assume the most simple expression for $T(\alpha_1,\alpha_2)$, namely

$$T(\alpha_1,\alpha_2) = \eta(\alpha_2-\alpha_1-\pi/3) \tag{13}$$

for the region of interaction $(\alpha_2-\alpha_1)>60^{o}$. Here η is the coupling constant. The solution of equations (12) is now easy and straightforward. We obtain

$$\alpha_1 = \frac{\omega_2}{2}\left(t - \frac{\sin\omega t}{\omega}\right) ,\qquad \alpha_2 = \frac{\pi}{3} + \frac{\omega_2}{2}\left(t + \frac{\sin\omega t}{\omega}\right) , \tag{14}$$

where ω_2 is the initial angular velocity of molecule 2 and $\omega=\sqrt{2}\,\eta\sqrt{I}$. The time t is measured from the beginning of the collision. The duration of the collision is π/ω and the final positions of the molecules, just after the collision, are

$$\alpha_1 = \frac{\pi}{2}\frac{\omega_2}{\omega} , \tag{15}$$

and

$$\alpha_2 = \frac{\pi}{3} + \frac{\pi}{2}\frac{\omega_2}{\omega} , \tag{16}$$

respectively. In order to get from eq. (16) $\alpha_2=120^{o}$ the ratio ω_2/ω must be 2/3. This equality is reached when the rotational enery $1/2\ I\omega_2{}^2$ is equal to $4\ \eta/9$. The energy $4\ \eta/9$ is the proton-proton interaction energy at the angle of approach of $\alpha_2-\alpha_1=104^{o}$ in our model. If we assume for the rotational energy the energy of the order of the activated state energy or the hydrogen bond formation energy, we see that the required proton-proton coupling $4\ \eta/9$ is relatively weak as compared with some of the estimates given at the beginning of this Section. This disagreement looses its importance if one takes into account the fact that the real low kinetic energy motion of the water molecule is a hindered rotation. The crystal potential in which the molecule rotates forces it to assume the correct orientation after the end of the collision.

The problem of the collision between molecules which rotate in the field of the crystal potential is mathematically difficult. If the kinetic energy is low, the rotation of the individual molecule is strongly hindered and one gets the liberational oscillations. The collision between two molecules takes place only if one of the molecules has high enough energy to pass from one to another well of the crystal potential. The corresponding saddle point energy is considered to be determined by the activation energy for the migration of L-defects. We can assume that the most important stage of the collision starts when the molecule is just over the saddle point. During the collision both molecules move in the identical potential $V(\alpha)$, which can be roughly approximated by a simple harmonic potential as shown in fig. 2. By supposing the same interaction (eq. (13)) between protons, as in the previous case, we obtain the following equations of motion

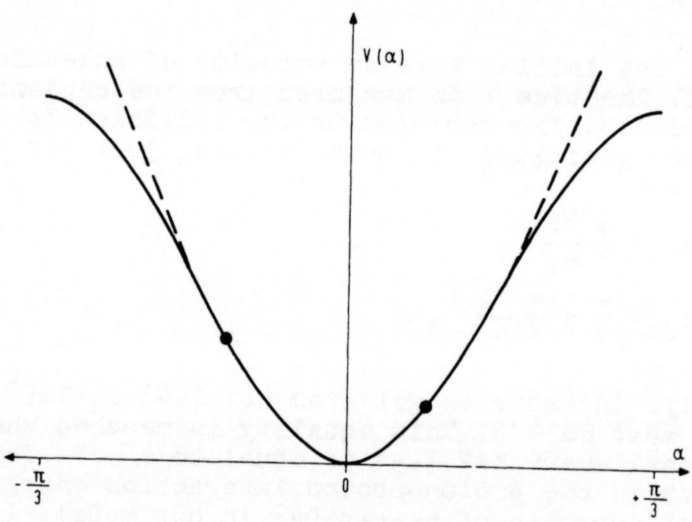

Fig.2 Potential $V(\alpha)$ for the hindered rotation (full line) and its harmonic approximation (dashed line)

$$I\,\ddot{\alpha}_1 = -K\,\alpha_1 + \eta\left(\frac{\pi}{3} + \alpha_2 - \alpha_1\right) ,$$
$$I\,\ddot{\alpha}_2 = -K\,\alpha_2 - \eta\left(\frac{\pi}{3} + \alpha_2 - \alpha_1\right) , \qquad (17)$$

where K is the force constant of the potential for the hindered rotation. The zero for the angle α_2 is here the orientation of the hydrogen bond 1. The initial state of protons at t=0, just before the collision, can be taken as $\alpha_1=0$, $\dot{\alpha}_1=0$, and $\alpha_2=-60^o$, $\dot{\alpha}_2=0$. One can assume $\dot{\alpha}_2=0$, because the angular velocity at the saddle point is determined by the kinetic energy of the order kT. This is a small energy as compared with the saddle point energy which determines the angular velocities at zero or small angles. The solutions of equations (17) which correspond to the given initial conditions are

$$\alpha_1 = -\frac{\pi}{6}(\cos\omega t - \cos\omega_1 t) + \frac{\pi}{3}\frac{\eta}{K+2\eta}(1-\cos\omega_1 t) ,$$
$$\alpha_2 = -\frac{\pi}{6}(\cos\omega t + \cos\omega_1 t) - \frac{\pi}{3}\frac{\eta}{K+2\eta}(1-\cos\omega_1 t) , \qquad (18)$$

where $\omega=\sqrt{K/I}$ is the angular librational frequency of the proton in the potential well of the hydrogen bond and $\omega_1=\sqrt{(K+2\eta)/I}$. The oscillating solutions describe the transfer of the energy from one proton to the other as the result of the interaction. The angular distance between protons during the collision is given, according to eq. (18), by

$$\alpha_1 - \alpha_2 = \frac{\pi}{3}\frac{2\eta}{K+2\eta} + \frac{\pi}{3}\left(1-\frac{2\eta}{K+2\eta}\right)\cos\omega_1 t . \qquad (19)$$

In case of strong interactions, $\eta \gg K$, the oscillating term in eq. (19) is small and the distance between protons remains nearly constant. On the basis of the estimates for the interaction energies, we can conclude with certainty that in ice $\eta \gg K$. Therefore, we have approximately

$$\alpha_1 = +\frac{\pi}{6}(1-\cos\omega t) ,$$
$$\alpha_2 = -\frac{\pi}{6}(1+\cos\omega t) . \qquad (20)$$

During the collision the energy is exchanged in time π/ω.

So far we treated the problem of collisions completely classically. It would be interesting to look in the quantum aspects of the collision too. The wave functions $\psi(\alpha)$ for the energy eigen states of the hindered rotator must satisfy the Schrödinger equation

$$- \frac{\hbar^2}{2I} \frac{d^2\psi(\alpha)}{d\alpha^2} + V(\alpha)\ \psi(\alpha) = E\ \psi(\alpha) \quad , \qquad (21)$$

where E is the energy.

The solutions of eq. (21) in terms of the known functions can be obtained if we assume for $V(\alpha)$ the form

$$V(\alpha) = \frac{V}{2}\ (1-\cos 3\alpha) \quad , \qquad (22)$$

where V is the saddle point energy. This approximation for $V(\alpha)$ is quite satisfactory. By introducing a new variable $x=3\alpha/2+\pi/2$ we get the Mathieu's differential equation

$$\frac{d^2\psi(x)}{dx^2} + (b-s\ \cos^2 x)\ \psi(x) = 0 \ , \qquad (23)$$

with $b=8IE/9\hbar^2$ and $s=8IV/9\hbar^2$. The value of the parameter s is obtained from the requirement that the energy difference between the saddle point energy and the ground state energy of the rotator is equal to the rotational activation energy 0.235 eV. We obtain s=50. The tables /18/ of the characteristic values of Mathieu's differential equation reveal the existence of 4 eigen states or narrow energy bands below the top of the potential barrier, as shown in fig. 3. The energy difference between two adjacent states or bands is greater than kT. Therefore, the quantum effects at the collision process of two protons

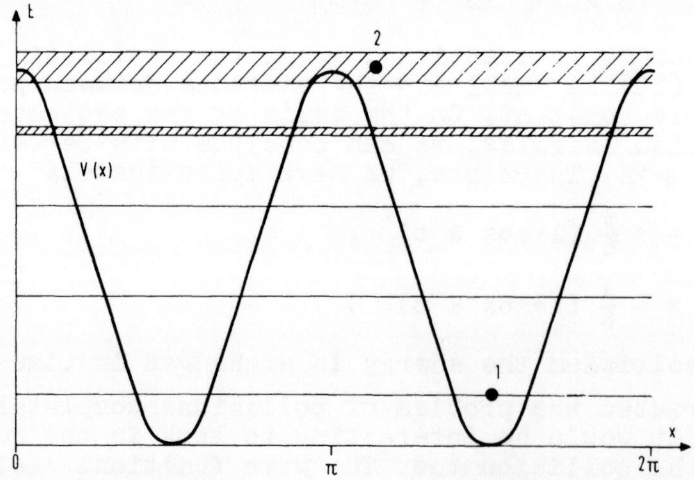

Fig.3 Eigen-state energies of the hindered rotator

could be important. The mechanism of the collision in the quantum picture is very complicated because of too many intermediate states between the band at the top of the potential barrier, which is occupied by the excited proton 2, and the ground state, where proton 1 lies. Since the proton-proton interaction is strong, the possibility of the exchange of the energy and the occupation of states between both protons exists. Proton 2 makes a transition from the excited state to the ground state and proton 1 the transition in the reverse direction. The transition takes place in the complicated fashion by steps through the intermediate states and the corresponding probability depends on the values of the transition matrix elements of type

$$\langle \psi_f(x_1,x_2) \mid V(x_1,x_2) \mid \psi_i(x_1,x_2) \rangle \quad , \tag{24}$$

where $\psi(x_1,x_2)$ is the product wave function of protons and the labels i and f denote the states involved in the transition. $V(x_1,x_2)$ is the interaction potential. The matrix element (24) is great especially for transitions between the neighboring states in the level diagram in fig. 3, because of the strong overlap of the corresponding wave functions $\psi_i(x_1,x_2)$ and $\psi_f(x_1,x_2)$. Unfortunately we do not have reliable estimates for eq. (24). On the basis of some plausible arguments we expect that the greatest matrix elements are at least of the order of few kcal/mole. In the intermediate states the energy is not conserved. But the initial and final state have the same energy and are therefore the resonant states. The duration of the collision could be derived from detailed study of the transition process. The magnitude of the transition matrix elements indicates that this time should be of the order of the inverse of the librational frequency.

We conclude that the quantum treatment also confirms the essential results of the classical approach to the collision problem. At the end of this discussion we would like to point out, that the transfer of the activation energy from one proton to another is very similar to the transfer of the excitation energy in the case of the Frenkel excitations.

3. Proton-lattice interactions

The main source of the dissipation of the rotational energy to the lattice vibrations seems to be the collision process itself. During the collision the repulsive force between protons acts also on the oxygens of the interacting water molecules. In order to estimate the amount of the transferred rotational energy to the oxygens we shall use the impulse approximation, since the duration of the collision is much shorter than the shortest period of the lattice translational modes.

The forces between the colliding protons can be decomposed into the torques and the forces acting on the corresponding water molecules. The torques give rise to rotational collisions, the problem which has been extensively studied in the previous Section. The direction of the force on the particular oxygen does not change very much in the course of the collision. The impulse received by the oxygen is $\Delta p = Ft$, where F is the average force of the interaction and t the duration of the collision. The amount of the transferred kinetic energy to the lattice for each collision is therefore

$$\frac{(\Delta p)^2}{M} , \qquad (25)$$

where M is the water molecule mass. Note that the standard factor 2 in the denominator of eq. (25) is missing because two molecules are involved in the collision.

A reliable calculation of Δp is at present impossible. But, one can get the order of magnitude for Δp on the basis of the following argument. Consider the case of freely rotating water molecules. The impulse exchanged between protons during the rotational collision of two molecules is just Δp. One of the protons is initially at rest and the other has kinetic energy of the order E_r. After the collision the velocities of protons are exchanged. Therefore,

$$E_r \approx \frac{(\Delta p)^2}{2m} \cos^2\beta \quad , \qquad (26)$$

where m is the proton mass and β an average angle between the direction of the force and the tangent on the path of the rotating proton. The angle β is about 80°. We conclude from eq. (26) and (25) that the energy transferred to the lattice per collision is of the order $(2m/M)\, E_r$, i.e., approximately 1/10 of the activation energy is lost. This order of magnitude estimate should be basically correct also if one considers a more realistic case of the hindered rotation of water molecules.

4. Dielectric relaxation

In this concluding Section some further aspects of the concerted rotation of water molecules will be discussed.

The multiple rotational collisions with the activation energy of only E_r are not possible in the defect free part of the crystal. The result of such a collision would be a formation of the pair of Bjerrum defects. The formation energy E_f of the stable pair, this means L- and D-defect which are well separated in the crystal, is much higher, i.e., E_f=0.68 eV.

The situation is different if we consider the region in the neighborhood of the L-defect. In the study of the multiple collisions we must, in fact, take into account also the long range forces between protons. In Section 2 we considered individual collisions as mutually separated processes. In ice this is not completely correct. The rotations of molecules are part of the usual process for the formation of the Bjerrum defects. At the formation the long range attractive Coulomb forces between the L- and D-defect "in statu nascendi" must be overcome too. On the contrary, if the rotations take place in the neighborhood of the L-defect, its electric field compensates the above attractive force. The effective charge of the L-defect is 0.6 e /7/, where e is the electronic charge. For instance the interaction energy of two charges of 0.6 e at distance 5 Å is 0.32 eV if we take for the dielectric constant of the medium the high frequency value 3.2 of ice. The strong electric field in the neighborhood of the L-defects seems to allow, in this region, the multiple collisions and the corresponding motion of the L-defect with the activation energy E_r. At present we are not able to calculate the average number z of the molecules which rotate in a single multiple collision process. z depends on the electric field, activation energy, and the dissipation of the rotational energy to the lattice.

The dependency of the Debye relaxation time τ on the number z can be obtained from the well known formula, which expresses τ in terms of the high frequency conductivity and the static dielectric constant. In case of ice this formula reads /7/:

$$\frac{1}{\tau} = \frac{\sigma_L}{\varepsilon_o(\varepsilon_s-\varepsilon_\infty)} , \qquad (27)$$

where σ_L is the high frequency conductivity due to the motion of L-defects. ε_s and ε_∞ are the static and high frequency dielectric constant, respectively. The conductivity σ_L, mobility, and the diffusion constant of L-defects are proportional to the square of the jump distance of the defect in the thermally activated elementary process of motion. Since the jump distance linearly increases with z, we conclude that the new expression for $1/\tau$ must have the form

$$\frac{1}{\tau} = c\, z^2 \left(\frac{n_L}{n_H}\right) p_L \quad , \tag{28}$$

where the exact value of the numerical constant c is still an open question, which a more detailed theory should answer. The value of c probably is not much different from 3/4 in the formula (2). In order to get a reasonable pre-exponential factor for p_L the number z should be around 20, what seems to us more than we would expect if the rotational energy of the water molecule is only E_r.

Further theoretical and experimental investigations are needed, in order to get a more clear picture about the elementary events which cause the dielectric relaxation.

References

/1/ Auty, R.P., R.H. Cole, J.Chem.Phys. 20, 1309 (1952)

/2/ Humbel, F., F. Jona, P. Scherrer, Helv.Phys.Acta 26, 17 (1953)

/3/ Steinemann, A., H. Gränicher, Helv.Phys.Acta 30, 553 (1957)

/4/ Dengel, O., N. Riehl, A. Schleippmann, Phys.kondens. Materie 5, 83 (1966)

/5/ Gränicher, H., Phys.kondens.Materie 1, 1 (1963) (review article)

/6/ Bjerrum, N., Dan.Mat.Phys.Medd. 27, 56 (1951)

/7/ Jaccard, C., Helv.Phys.Acta 32, 89 (1959)

/8/ Glasstone, S., K.J. Laidler, H. Eyring: The Theory of Rate Processes (McGraw-Hill Book Company, New York 1941)

/9/ Wert, C.A., C. Zener, Phys.Rev. 76, 1169 (1949)

/10/ Giguère, P.A., K.B. Harvey, Can.J.Chem. 34, 798 (1956)

/11/ Giguère, P.A., K.B. Harvey, J.Mol.Spectroscopy 3, 36 (1959)

/12/ Gibson, J.B., A.N. Goland, M.Milgram, G.H. Vineyard, Phys.Rev. 120, 1229 (1960)

/13/ Leibfried, G.: Bestrahlungseffekte in Festkörpern (B.G. Teubner Verlagsgesellschaft, Stuttgart 1965)

/14/ Sänger, R., O. Steiger, K. Gächter, Helv.Phys.Acta 5, 200 (1932)

/15/ Dunitz, J.D., Nature 197, 860 (1963)

/16/ Hendrickson, J.B., J.Amer.Chem.Soc. 83, 4537 (1961)

/17/ Bartell, C.S., J.Chem.Phys. 32, 827 (1960)

/18/ Tables Relating to Mathieu Functions (Columbia University Press, New York, 1951)

PROTONIC CONDUCTION OF ICE +)

PART I: HIGH TEMPERATURE REGION

B. Bullemer, H. Engelhardt, N. Riehl

Physik-Department

Technische Hochschule München / Germany

Abstract

To separate and measure bulk and surface dc conductivity a guarded potential probe method was used. A number of electrode materials and combinations was studied. Surface currents exceed bulk values above -30°C and have activation energies between 20 and 30 kcal/mole. The intrinsic bulk conductivity was found to be $\sigma = (1.1 \pm 0.5)\cdot 10^{-10}$ $(\Omega cm)^{-1}$ at -10°C with a corresponding activation energy of 8.0 ± 0.5 kcal/mole. These values cover all measurements except those with proton injecting Pd-black electrodes. Protonic mobilities were determined from bulk saturation currents in 0.05 to 0.4mm monocrystaline ice sheets. Activation energy for the dissociation rate is 17.6 ± 1.3 kcal/mole. Drift mobility (deduced from saturation current and intrinsic conductivity) increases with falling temperatures and amounts to $(2.4 \pm 1.6)\cdot 10^{-3}$ $cm^2/(Vsec)$ at -10°C. Hall-effect measurements are rediscussed and commented. New evidence for the protonic character of conduction in ice is provided by comparing the injecting properties of proton saturated and deuterium saturated Pd-black electrodes.

1. Introduction

DC conduction of ice is one of the more controversial

+) work performed in collaboration with: U.Eckener, W.Egle, I.Eisele, A.Gattinger, C.Schröder-Etzdorf, P.Seige topics in ice physics. Among the numerous measurements which have been published, the discordance seems rather complete. One might, however, observe something like a falling tendency over the years in the values of both, conductivity and activation energy. To illustrate the situation, we take two prominent examples: i) the measurements of Jaccard /1/ covering a wide temperature range down to -46°C and ii) those by Eigen et al./2/ based on careful potential probe measurements but restricted in temperature range to above -19°C by the water electrodes used. In both cases no current-time dependence was observed and the conclusion imposed itself that electrode polarization could be neglected. In the second case a linear drop of potential seemed to provide additional evidence. But in spite of these similarities the results disagree.

There are two major sources of error: i) more or less pronounced electrode polarization, always present, may evolve sufficiently fast to pass unnoticed by the dc worker; ii) surface conduction sufficiently important at higher temperatures, may obscure bulk conductivity completely. To take care of both difficulties we have developed experimental techniques that lead to coherent results even for different electrode systems used.

Four-probe measurements are widely used to assign currents to their bulk fields correctly. But in the case of ice, unless proper care is taken, it is very likely that only surface potentials are detected. We may regard the system ice-electrodes as a circuit having two potential dividers in parallel, the first one corresponding to a chain of surface resistances R_{si} (i = 1,2,....) the other one to a chain of bulk resistances R_{bi} of ice (fig.1, A). It is now generally accepted that R_s is small compared to R_b at least above -30°C /3,4,5,6/. Therefore, a conventional probe, coupling to both chains R_{si} and R_{bi}, is bound to "see" only surface potentials. The obvious remedy in this case is to protect the probes by guard rings from surface potentials. If one uses unity gain amplifiers to drive the guard rings, it even becomes possible to observe bulk potentials attaining equilibrium values. It goes without saying that bulk currents are to be measured in guarded circuits too (fig.1, B).

2. DC conductivity by guarded potential probe method

The water used for growing ice single crystals was deionized, distilled with $KMnO_4$ added in alkaline solution

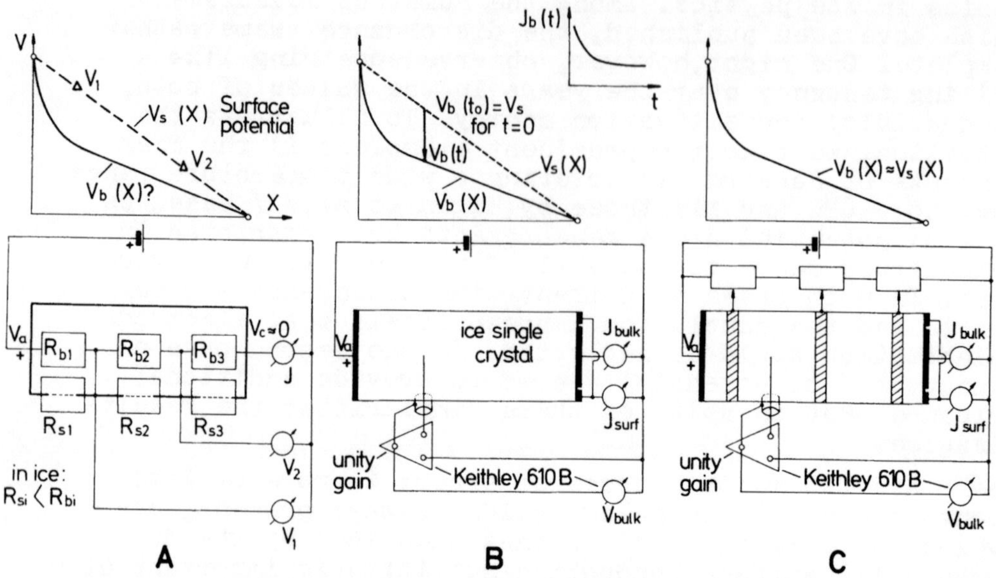

Fig.1 Guarded potential probe method: experimental setup and schematic potential profiles.
A: general considerations (see text).
B: fundamental type of experiment.
C: modified experiment

and consequently treated in a 5-stage distilling apparatus under highly purified nitrogen atmosphere. The distilling apparatus was similar to the one used in ref./2/ described in ref./7/. Relative concentration of ionic impurities should be less than 10^{-8}. The purity of the ice crystals may be even higher due to 'zone refining' in the growing process. Ice single crystals of 8.5 cm in diameter and more than 20 cm in length were grown upwards from a seed crystal in vacuum-tight plexiglass containers. The growing rate never exceeds 1.5 cm/day. The c-axis is parallel to the direction of growth. Crystallographic perfection as tested by partial melting under IR-irradiation /8/ is high. Samples were cut to shape on a lathe or on a microtome. Crystals used in the potential probe measurements had cylindrical shape, 4 or 6.5 cm in diameter and 10 to 15 cm in length, with c-axis parallel to the cylinder axis. Guarded palladium cathodes were used in combination with various anodes (water, ion exchange membranes, gold or palladium). Potential probes of 0.5 or 1 mm diameter Pd-wire were introduced radially so as to reach the cylinder axis. In most cases, probes were frozen into holes slightly larger in diameter using a small amount of additional, very pure, water. Tight fitting was also used. Guard rings for potential probes were Pd-plated or palladium tubes 10 or 5 mm in diameter frozen about 1 mm deep into the surface.

On surface conduction, our results can be summarized as follows:
i) The drop of surface potential along the sample is linear regardless of the electrodes used (water, ion exchange membranes, Au, Pd), provided the experiments are made in an inert atmosphere like nitrogen gas. Current-voltage characteristics are also linear.
ii) The activation energy seems independent of the type of electrode used. But it shows marked variation with surface purity. If every care is taken to avoid contamination of the surface, values as low as 21 kcal/mole can be reached. Usually, values for the activation energy of surface conduction scatter around 30 kcal/mole.
iii) The magnitudes of surface currents depend markedly on electrode materials. They exceed bulk values by 100 for water electrodes and by a factor of 10 otherwise at temperatures near -10°C.

We interpret these observations on the basis of a 'water-like surface' of ice /9,10/. The semi-liquid state is stable at temperatures not too far from the melting point. The width of this layer, decreasing with falling temperature, should account for the activation energies observed if we admit an additional thermal activation for the conduction process itself. Impurities which depress the melting point should result in an increased width of the layer thereby accentuating the effect. Fletcher points out in his most recent theory that carrier concentrations in the surface layer exceed bulk values considerably /10/.

As we will see the troubles caused by surface effects have not yet been overcome completely. To determine bulk conductivity we measure bulk potentials and bulk currents as functions of the applied voltage and plot bulk currents versus bulk fields on doubly logarithmic scale. At temperatures above -20°C where surface conduction is high we observed ohmic behaviour only for very low bulk fields (< 4 V/cm) if water-, ion-exchange membrane- and Au-anodes were used. At higher fields bulk currents increase more like a square root law. This effect may be due to

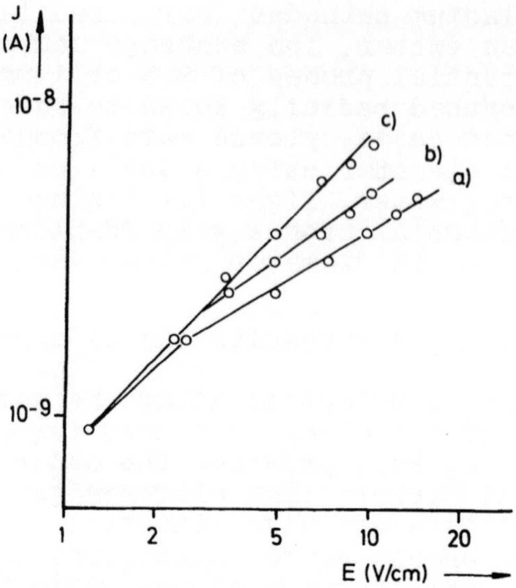

Fig.2 Bulk current a) vs. applied field, b) vs. bulk field, c) vs. bulk field with surface potential equal to bulk potential (at -14.5 °C)

differences of bulk and surface potentials which cause a radial field component in the bulk of the crystal. This field component constricts the bulk-current lines to a smaller area than they would fill normally, i.e., if bulk and surface potentials were equal at equal distances from the electrodes. At the x-coordinate where these differences are maximum (fig.1), current lines are constricted to resemble a bottle-neck; the effective cross section of the sample decreases as applied voltage increases. The result is a less than linear rise of the bulk current vs. field characteristic (fig.2).

An experiment has been designed to test this interpretation (fig.1, C). We attach a number of additional electrodes in radial belts around the sample and force them to the potential measured in the bulk of the crystal at the corresponding site. Now the radial component is suppressed, current lines become parallel and, as a result, we get a linear current-field dependence for the whole range covered. Fig.2 demonstrates the effect for a sample contacted with ion exchange membranes (anion permeable at the anode, cation permeable backed with Pd-black at the cathode).

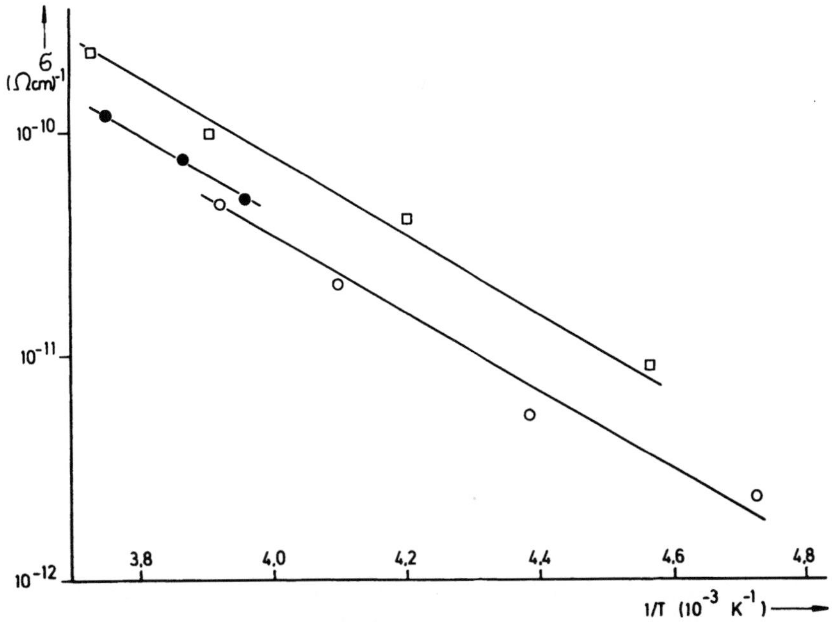

Fig.3 Temperature variation of bulk conductivity for Au (+), Pd (-) electrodes (three different samples). Activation energy 8.0 ± 0.5 kcal/mole

From ohmic current vs. field plots (like fig.2, curve C) one can readily deduce bulk conductivity values. For the gold (+) palladium (-) system we find $\sigma = (1.1 \pm 0.5) \cdot 10^{-10}\ (\Omega cm)^{-1}$ at -10°C and a corresponding activation energy of 8.0 ± 0.5 kcal/mole for temperatures above -60°C, as shown in fig.3. Results for water (+)-Pd(-) /11/ and for the combination of ion exchange membranes agree within the limits of error. The same is true for values of activation energy obtained independently by Kahane et al. /12/. It has to be emphasized that the systems Au (+), water (+) and anion permeable membrane (+) all exhibit a large potential drop at the anode (as shown in fig.4 for Au (+))and a correspondingly falling current-time dependence. A classification of these anodes as 'poorly injecting' seems justified. We may conclude that only thermally generated carriers effect the charge transfer. Therefore, the values given are the intrinsic bulk conductivity of pure ice.

By contrast, hydrogen saturated palladium black anodes yield conductivities one order of magnitude higher and

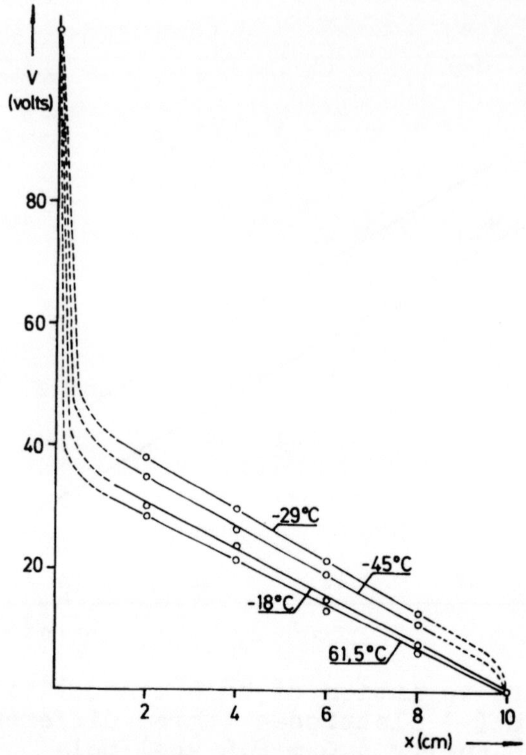

Fig.4 Variation of bulk potential along the axis of cylindrical sample. Au (+), Pd (-) electrodes

activation energies 10 to 25% lower than those above. The scatter in activation energies is also larger. Only with electrodes used for a long time or low in hydrogen content, activation energies up to 8 kcal/mole are obtained. These resemble more closely poorly injecting electrodes. It seems that these palladium anodes inject charge carriers greatly outnumbering intrinsic ones. Activation energies in this case are determined by temperature dependent injection properties. Part II will have to resume this topic in more detail.

3. Drift mobility from saturation currents

Having learnt about the product $n(\mu^+ + \mu^-) = \sigma/e$ from conductivity measurements, we continue along the lines traced by Eigen and his group. We resumed their saturation current measurements /2/ to determine, among other interesting quantities, the drift mobility μ^+ of protonic charge carriers in ice. We restrict ourselves to a brief outline of the evaluation process; you may find it worked out in detail in the papers by Eigen et al. /13,2/. Suppose, we have no injection from the electrodes, all carriers are generated by thermal dissociation and, moreover, carriers reaching the electrodes are discharged or absorbed immediately. Then, fields high enough to draw out all carriers before they are able to recombine yield saturation currents given by $I_s = e\ A d k_D\ n$. (e elementary charge, A electrode area, d electrode distance, n concentration of neutral molecules). Obviously, saturation currents have to stand both tests: i) are they field independent, ii) are they increasing proportional to volume $V = A\ d$ of the sample? Or, to cite professor Onsager 'You can't cheat with saturation currents'. This simple formula for I_s can at best describe a very limited range of fieldstrength. We have to consider a transition region before reaching true saturation where part of the carriers is lost by recombination. The loss depends on the product of both positive and negative carrier mobility. On the other hand saturation current slightly increases with the field applied(dissociation field effect). From this we get information on the effective dielectric constant.

Samples were cut on a microtome from the parent crystal to thicknesses varying between 0.4 and 0.05 mm, diameter of the bulk electrode was 8 mm. Gold anodes were used as they seem as near to the blocking case as one can get to, while palladium cathodes should provide for the proton absorption assumed to be ideal in theory. Rectangular voltage pulses were applied to the sample and stationary currents observed after decay of displacement currents. Again, guard rings had to be used. There seems to be a saturation also of surface currents but it is very hard to observe. All numbers given in this chapter refer to bulk values. The measurements plotted in fig.5 clearly show the different regimes: ohmic, transition, saturation and field effect. At the highest fields used (above 20 kV/cm) the field effect seems exaggerated, due to the onset of weak injection possible even from a gold anode. On changing polarity, saturation gets lost completely as the former cathode (Pd) now is bound to inject protons. Temperature variation of saturation current is shown in fig.6, the parameter being sample volume in terms of electrode distance. The results are: dissociation rate $k_D = (1.1 \pm 0.2)10^{-9}\ sec^{-1}$ at $-10^{\circ}C$ with an activation energy of 17.6 ± 1.3 kcal/mole. Combined with the conductivity of $(1.1 \pm 0.5)\cdot 10^{-10}\ (\Omega cm)^{-1}$ already given, we have for protonic mobility $\mu^+ = (2.4 \pm 1.6)10^{-3}\ cm^2/Vsec$. Comparing activation energies for k_D and for σ suggests that μ^+ varies like a weak positive exponential with 1/T,

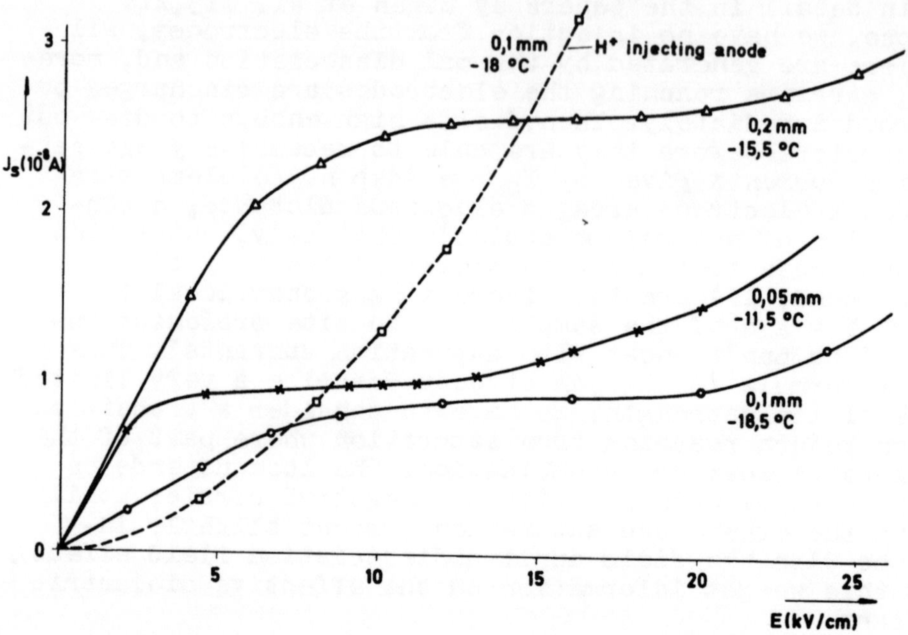

Fig.5 Bulk current vs. applied field. Saturation behaviour for Au (+) Pd (-) electrodes. Injecting behaviour for Pd (+), Au (-) electrodes dotted line

i.e., mobility increases with falling temperatures. Experimental errors, however, are still too large to insist on numbers in this case. From the recombination term we get $\mu^+ \approx 50\ \mu^-$. From dissociation field effect we obtain an effective dielectric constant between 7 and 11 in good agreement with the value 9.5 from Onsager's theory /14/.

4. Hall effect and Hall mobility

A different method to determine carrier mobility, very common in semiconductor research, are Hall effect measurements. The interpretation of Hall data, however, becomes a problem of its own if we cannot assume quasi-free charge carriers. This is exactly what we are not allowed to do in the case of ice. Experimental difficulties common to all high resistance Hall probes are complicated further by asymmetries in the electrode polarization and, again, by surface conductivity. We

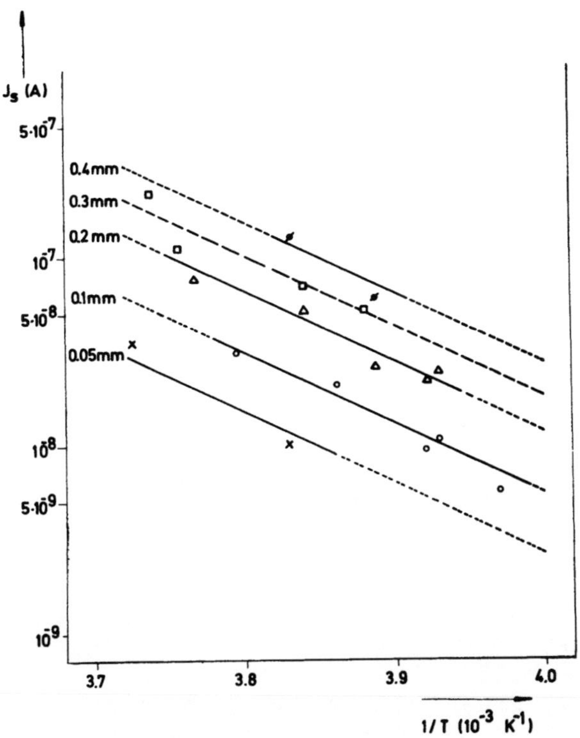

Fig. 6 Temperature variation of saturation current Parameter is volume of sample in terms of thickness d. Activation energy 17.6 ± 1.3 kcal/mole

were able to observe the Hall effect in ice /15,16/ using the experimental arrangement shown in fig.7. By splitting the driving current into two separate, ground-free circuits, polarization problems could be met and reasonable zero stability was obtained. Considering the sample as a Hall voltage source, we realize that surface conduction between the two Hall electrodes will short out the Hall voltage. Or better: it will prevent Hall voltage from developing . Guarding of Hall electrodes, therefore, is a must. Guarding is less of a problem with currents. Therefore, we contented ourselves to determining Hall currents. Experience gained in our guarded potential probe measurements should prove useful in future Hall voltage experiments.

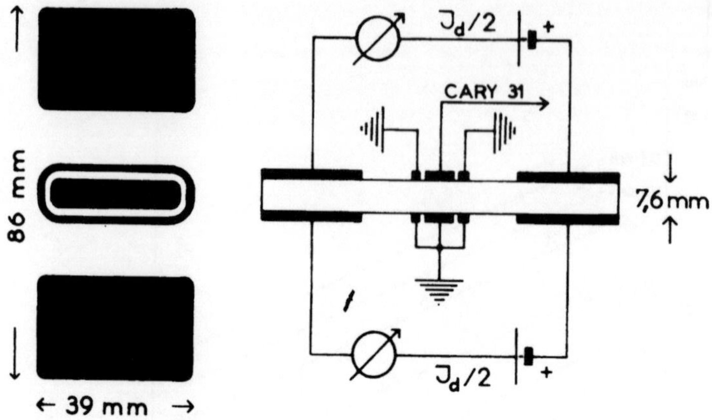

Fig.7 Experimental setup and electrode system for Hall effect measurements

The Hall constant $R = 1/en_H$ is a measure of the number of charge carriers actually in motion. Only these carriers will be deflected by a Lorentz force. Their velocity is described by a Hall mobility defined as $\mu_H = R\sigma = en\mu/en_H$. From this relation we get the ratio of carriers in motion n_H to the total number n of carriers, included those waiting for the next jump: $n_H/n = \mu/\mu_H$. Inserting our observed mobilities ($\mu = 2.4 \cdot 10^{-3}$ cm^2/Vsec, $\mu_H = 1.4$ cm^2/Vsec near -10°C) we are led to the conclusion that only $1.7 \cdot 10^{-3}$ times the total number of carriers can be considered to be quasi-free. The Hall mobility μ_H may seem very high. But actually it is related more closely than the drift mobility μ to the elementary step of charge transfer. If we accept $\approx 10^{-12}$ sec as the characteristic time constant determining the elementary step /15,17/, a microscopic mobility even higher than the Hall mobility will be required to move an H_3O^+ ion more than a single O-O distance per step. The concept of correlated proton motion is basic to any modern theory of protonic conduction in ice /17,18/. Without correlation the Hall effect in ice would be hard to understand. You may have wondered why the numerical value of Hall mobility is not affected by the new intrinsic conductivity data presented. But it must be remembered that Pd-black electrodes were used and that bulk conductivity value with injected carriers nearly equals the older data /2/. This coincidence, however, is merely accidental.

5. Comparison of proton and deuteron injection

Even it looks somewhat like adding water to the sea, we should reexamine the question: what are the charge carriers in ice? The amount of hydrogen evolved at the cathode in the coulometric work of Gränicher et al./19/ corresponds within 3% to the charge passed through the sample which confirms earlier results of Workman et al. /20/. Admittedly, we have a protonic conductor at the temperature of -10°C where these measurements were taken. But since surface conduction may greatly exceed the bulk values as we have seen, we are left with the question: is the ice surface or is the bulk of the ice crystal or are both of them protonic conductors? Independent evidence is desirable to settle the question. From saturation current behaviour we know that positive carriers are more mobile than negative ones by a factor of 50. The observed positive Hall-effect points in the same direction. Still, one might object, that the carriers could possibly be electron holes.

We therefore studied and compared injecting properties of proton-saturated and of deuteron-saturated palladium black electrodes /21/. These electrodes did not contact the ice crystal directly but only via an intermediate layer of pure palladium, some 100 Å in thickness which had been evaporated onto the sample. If the saturated palladium black was allowed to touch the virgin palladium layer only a short time before positive voltage is applied, we observed a slow but steady increase in current amounting to three orders of magnitude over some 20 hours in the case of H^+. The current from a deuterated electrode attains its highest value within less than 10 hours. But D^+ currents are lower by two orders of magnitude than H^+ currents (fig.8). No effect of that kind was ever observed when an electrode was used for a second, similar, run. If we assume that the hydrogen isotopes diffusing into the palladium from the Pd-black layer change it in such a way as to increase electron hole injection we can not understand why H^+ and D^+ currents saturate at extremely different levels. On the other hand, interpretation is straightforward if we admit injection of protons and deuterons into the bulk of the ice crystals once they have crossed the palladium layer by diffusion. The proton is propagated readily by correlated tunneling while the deuteron is not. Thus, the protonic character of conduction in ice can be established independently from coulometric evidence.

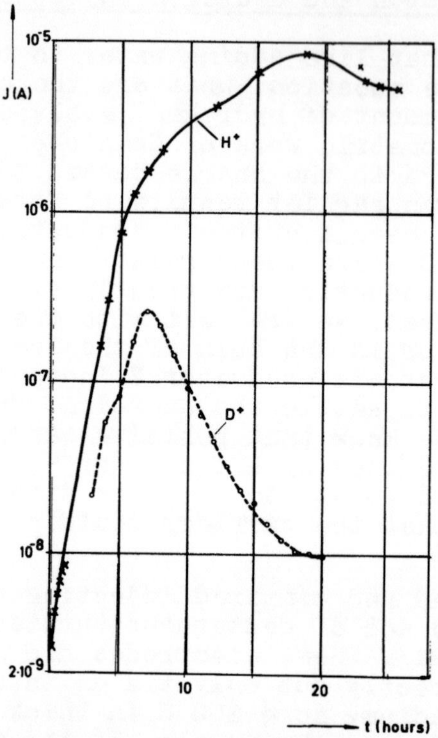

Fig.8 Time dependence of currents from delayed injection of protons and deutrons

References

/1/ Jaccard, C., Helv.Phys.Acta 32, 89 (1959)

/2/ Eigen, M., L.De Maeyer, H.Ch.Spatz, Ber.Bunsenges. Phys.Chemie 68, 19 (1964)

/3/ Bullemer, B., N.Riehl, Solid State Comm. 4, 447 (1966)

/4/ Jaccard, C., "Physics of Snow and Ice; Sapporo (1966)" H.Ōura (ed.), Inst.Low Temp.Sciences, Hokkaido University, Sapporo 1967, p.173

/5/ Camp, P.R., W.Kiszenick and D.A.Arnold, CRREL Res.Rep. 198, Hanover, N.H. (1967)

/6/ Gross, G.W., Advances in Chem. 73, 27-97 (1968); see page 84-86

/7/ Eigen, M., L.De Maeyer, Z.Elektrochem., Ber. Bunsenges.Phys.Chemie 59, 986 (1955)

/8/ Müller-Krumbhaar, H., Physics of Ice, Proc.Intern. Symp., Munich, Plenum Press, New York (1969)

/9/ Jellinek, H.H.G., J.Colloid Interface Sci 25, 192 (1967);
Weyl, W.A., J.Colloid Sci. 6, 389 (1951)

/10/ Fletcher, N.H., Phil.Mag. 18, 1287 (1968)

/11/ Bullemer, B., I.Eisele, H.Engelhardt, N.Riehl, P.Seige, Solid State Commun. 6, 663 (1968)

/12/ Durand, M., M.Deleplanque, A.Kahane, Solid State Commun. 5, 759 (1967);
Kahane, A., Physics of Ice, Proc.Intern.Symp., Munich, Plenum Press, New York (1969)

/13/ Eigen, M., L.De Meayer, Z.Elektrochem., Ber. Bunsenges.Phys.Chem., 60, 1037 (1956)

/14/ Onsager, L., M.Dupuis, in "Electrolytes", p.27-46, Pergamon Press (1962)

/15/ Bullemer, B., N.Riehl, Phys.Letters 22, 411 (1966)

/16/ Bullemer, B., N.Riehl, Phys.kondens.Materie 7, 248 (1968)

/17/ Hofacker, G.L., Physics of Ice, Proc.Intern.Symp., Munich, Plenum Press, New York (1969)

/18/ Gosar, P., Physics of Ice, Proc.Intern.Symp., Munich, Plenum Press, New York (1969)

/19/ Decroly, J.C., H.Gränicher, C.Jaccard, Helv.Phys. Acta 30, 465 (1957)

/20/ Workman, E.J., F.K.Truby, W.Drost-Hansen, Phys.Rev. 94, 1073 (1954)

/21/ Bullemer, B., H.Engelhardt, L.Knoblauch, N.Riehl, C.Schröder-Etzdorf, Solid State Commun. 6, 545 (1968)

PROTONIC CONDUCTION OF ICE [x]

PART II: LOW TEMPERATURE REGION

H. Engelhardt, B. Bullemer, N. Riehl

Physik-Department, Technische Hochschule

München / Germany

Abstract

Hydrogenated Pd-layer electrodes readily inject protons into the ice throughout the entire temperature range down to liquid helium temperature. With these electrodes we could measure space charge limited currents and determine the mobility of proton and of vacant bond migration. The results are interpreted by a correlated motion of protons at H_3O^+ sites along chains of hydrogen bonds. The reorientation of these chains is established by the migration of vacant bonds thermally activated or field injected from the surface.

1. Introduction

In the experiments described in part I, we have been concerned mainly with the conductivity of ice caused by charge carriers which were generated by thermal dissociation. Therefore, an evaporated gold electrode was taken as anode at the price of putting up with electrode polarization which has been overcome by the guarded potential probe method. Gold is generally accepted as a blocking electrode. As cathode, we used an evaporated palladium electrode which should be able to accept protons readily from the ice. The proton saturation current experiment was carried out with the same electrode

[x] Work performed in collaboration with: U. Eckener, W. Egle, I. Eisele, A. Gattinger, C. Schröder-Etzdorf, P. Seige.

arrangement. No saturation currents have been measured with palladium electrodes as anode.

2. Mobility

The quantity of interest is the mobility of the charge carriers. The conductivity and saturation current experiments together supply the mobility only indirectly. The mobility evaluated from these results shows no apparent activation energy and seems to increase with decreasing temperature. These two facts are decisive for any interpretation of the underlying mechanism, whether by a hopping process or a quantum mechanical tunnelling mechanism with phonon interaction.

In this paper, we shall concentrate on carrier mobility. From the experimental point of view, conductivity measurements can now be well controlled. In contrast, the experimental investigation of the mobility is still in its initial stages. Precise direct measurements of the proton mobility are very difficult. A comparison of values obtained by different methods would be useful to strengthen confidence in the results.

Two techniques, however, will not be appropriate for the solution of the problem. There is first the measurement of the diffusion coefficient which yields a value for the mobility using the Nernst-Einstein formula $\mu = eD/kT$. Taking the value of D from diffusion experiments /1,2/ we get $\mu = 10^{-9}$ cm^2/Vsec at -10 °C. This low value shows that the mechanism responsible for diffusion is completely different from the electrical charge transfer mechanism. Secondly, Heinmets /3/ measured the proton mobility directly by boundary movement in a frozen solution of hydrochloric acid and sodium hydroxide with proton indicator molecules. The mobilities found ($\mu = 10^{-4}$ to 10^{-5} cm^2/Vsec) decrease with decreasing temperature. This method is restricted to strongly protonated ice and therefore the results are hardly applicable to pure ice.

Besides the indirect determination of the proton mobility from conductivity and saturation current experiments, from Hall-effect measurements discussed above by Bullemer /4/ and by the thermoelectric power method described by Jaccard /5/, there are two methods available which are more direct. The most direct method to determine the mobility would be to measure the transit time of protons injected from protonic electrodes. But the experimental difficulties resulting from the low conductivity and the high dielectric constant of ice are appreciable. Therefore, reliable results have not yet been obtained. The experimental technique can probably be refined sufficiently for performing this interesting experiment. The measurements of space-charge limited currents, on the other hand, have been carried out repeatedly /6,7,8/.

Child's law in its simplest form reads:

$$j = \frac{9}{8} \mu \varepsilon \varepsilon_o \frac{U^2}{L^3}$$

under conditions of good charge carrier injection from the electrode. Here the carrier concentration does no longer appear and the mobility can be evaluated directly. The high frequency value of ε for ice seems to be justified here, because the long dielectric relaxation time does not affect very much the fast moving proton state.

3. Proton injecting electrodes

Since the injecting properties of the electrodes are of decisive importance, these properties have been carefully tested. The electrodes consist of different layers (fig. 1). A layer of palladium metal, 200-300 Å thick, is evaporated directly onto the ice surface. The evaporation takes place under high vacuum and therefore the poisoning of the electrode-ice interface by adsorbed gas can

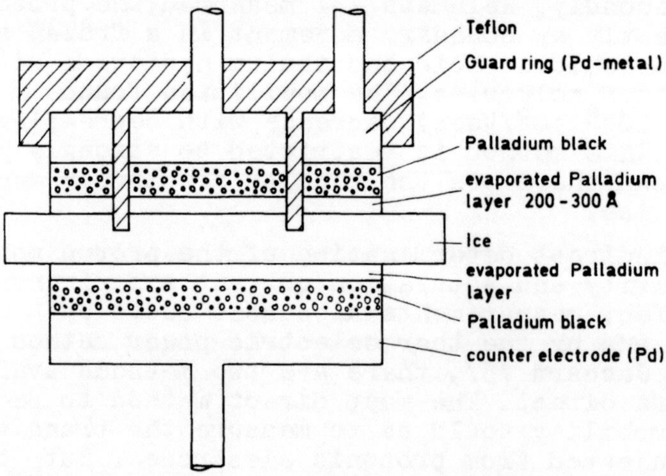

Fig.1 Ice electrode arrangement

obviously be avoided. Then follows a layer of palladium-black of spongy consistence saturated electrolytically with hydrogen. On top there is compact metal, either palladium or gold-plated brass. The evaporated layer reduces electrode noise and makes much better contact than a palladium-black electrode alone. Palladium is used because of its well-known properties of absorption and desorption of hydrogen even at low temperatures /9/. Palladium-black can be loaded with hydrogen to the extent of about one proton per palladium atom. Protons diffuse readily with the low energy of activation of 1.3-1.9 kcal/mole below -60 °C.

The observations regarding the properties of these electrodes are the following:
1. The absolute values of conductivity have been compared with those determined with gold electrodes (fig. 2) by using an evaporated gold electrode and a palladium-layer

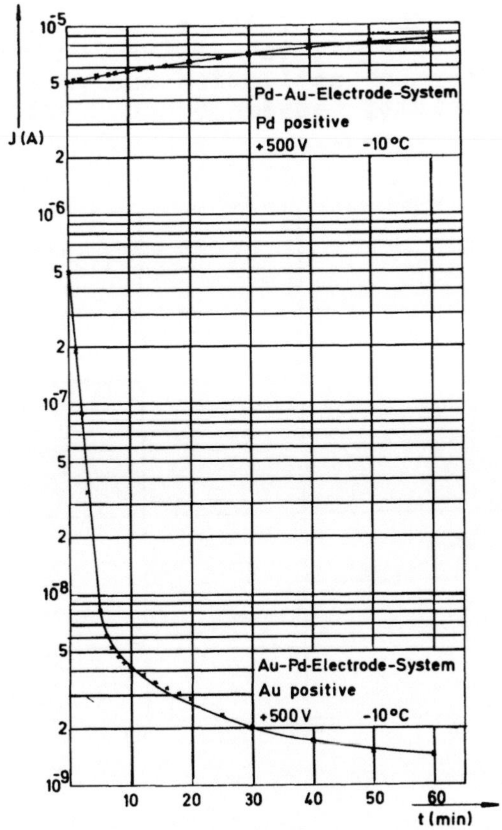

Fig.2 Time dependence of the current. Sample thickness 0.3 cm; electrode area 3.8 cm^2

counterelectrode on the ice specimens. The current observed with positive palladium electrodes is one to four orders of magnitude greater (depending on the applied voltage) than the current with positive gold electrodes. The time dependence is also remarkable. The injecting palladium electrode exhibits a nearly time independent current - even a slight increase is observed - while the blocking gold electrode shows a strongly time-dependent current decrease caused by polarization.

ii. The temperature dependence of the conductivity with injecting palladium electrodes has been studied with different applied voltages (fig. 3). The activation energy is found to depend on the applied voltage. At high temperatures, there is a competition between thermally dissociated and injected charge carriers. With increasing voltage, injection predominates more and more. The apparent activation energy for the conductivity vanishes or becomes equal to the activation energy of the diffusion of protons in palladium-black. This happens when the supply of protons at the palladium surface approaches exhaustion. At low voltages, the activation energy tends towards the limit for the activation energy of the intrinsic bulk conductivity between 8 and 9 kcal/mole.

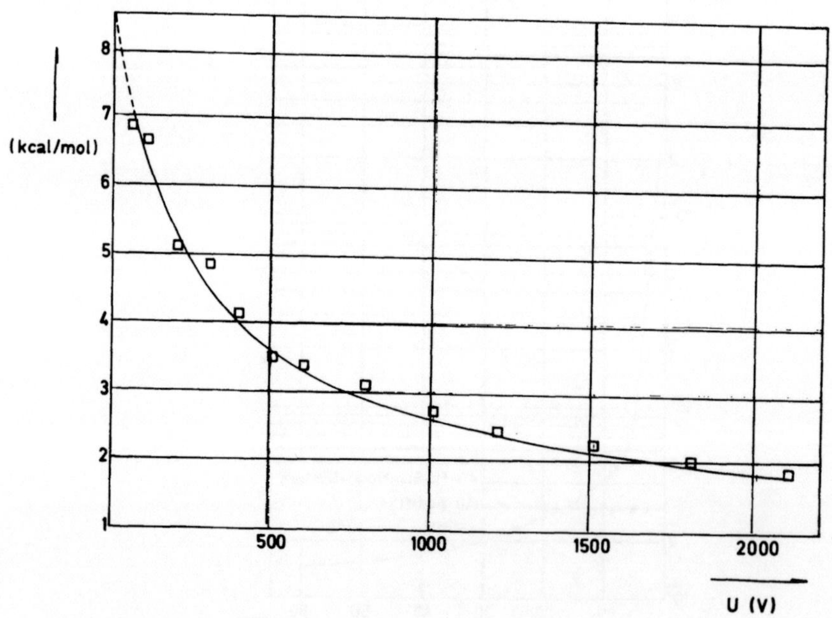

Fig.3 Apparent activation energy of the dc conductivity with injecting Pd(H)-electrodes depending on the applied voltage. Sample thickness 0.3 cm

iii. The potential distribution within the ice is shown in fig. 4. The upper curve represents the first measurement with a freshly hydrogenated palladium-black electrode. This electrode floods the ice with protons and causes a region of positive space-charge indicated by the flat portion of the curve. The lower curve shows the situation at the end of a series of measurements. Exhaustion of the proton supply prevents space charge from being built up as in the first experiment. The shape of the potential distribution curve also indicates preferential injection of carriers, protons, from the anode. Protons arriving at the cathode are readily released and discharged, while obviously no negative charge carriers are injected. We can therefore hope that the functional relation of the space-charge limited current will obey a simple law like that given above.

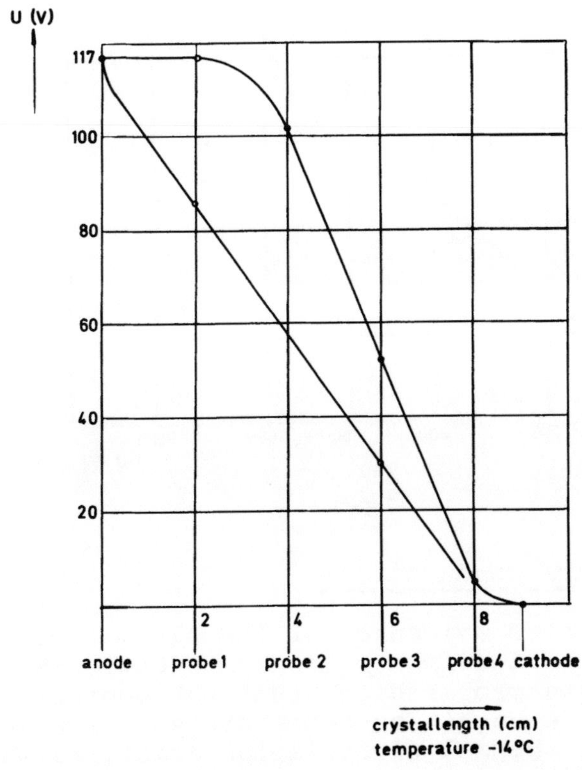

Fig.4 Potential distribution in the bulk

iiii. The injecting properties of the electrodes are reflected by the current-voltage characteristic (fig. 5). At low voltages, an ohmic current is observed; at high voltage, we get a space-charge limited current almost exactly proportional to U^2.

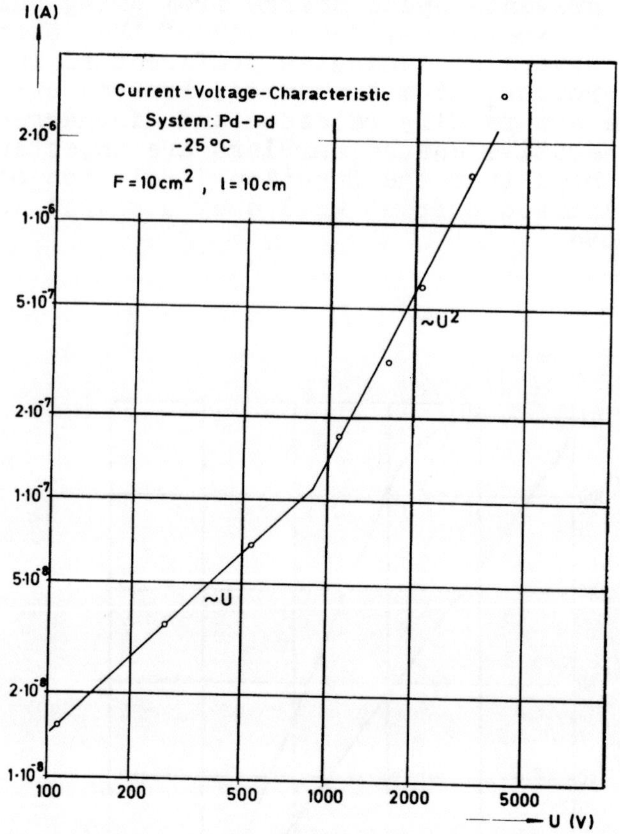

Fig.5

4. Low temperature conductivity

Having now presented evidence for the claim that proton injection into the ice from palladium electrodes works quite well, we are prepared to study the conductivity of an ice sample even at low temperatures where no protons produced by thermal dissociation exist. In this section, we shall discuss the temperature dependence of the current. In fig. 6 the current is plotted versus 1/T. Comparatively low voltages are applied so that low injection rates preserve the capability of injection from

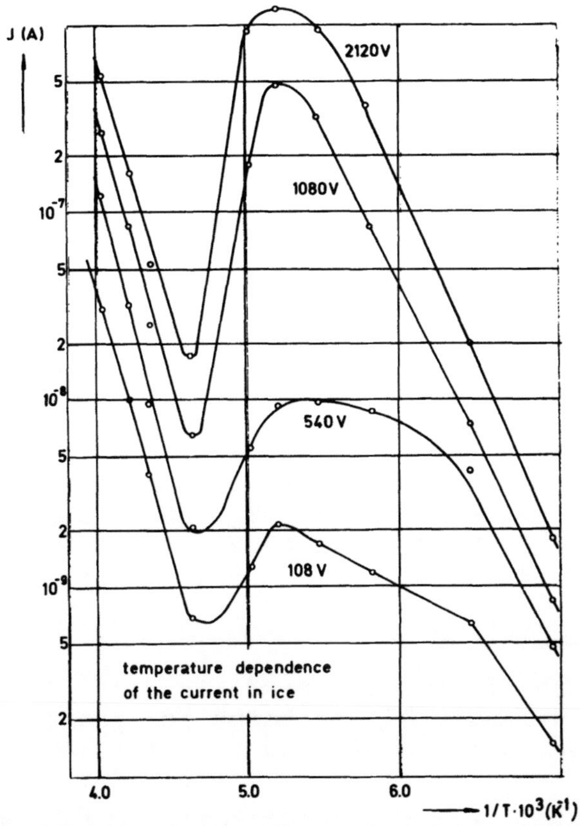

Fig.6 Temperature dependence of current. F=35 cm^2, L=10 cm

the palladium electrodes. Several temperature regions are separated fairly well from each other by two distinct changes in the conduction behaviour of the ice monocrystals at about -60 °C and between -80 and -100 °C.

Above -60 °C there is a competition between injected and thermally activated charge carriers (protons). But with decreasing temperature, the thermal carriers gradually die out while the injected carriers remain. Below -60 °C, the number of thermal carriers becomes negligible. Yet the current increases remarkably with decreasing temperature. Since the injection is temperature independent, or even drops slightly with falling temperature, this upturn must be due to an increase in the mobility of the injected protons. From the space charge limited current plot we get μ = 5 cm^2/Vsec for the mobility at -60 °C. Compared with the mobility obtained from conductivity and saturation current measurements, this value seems to be very high.

But there we are measuring the drift mobility of thermally dissociated charge carriers which undergo recombination and which have only a very short lifetime. In the case of the space-charge limited current experiment, no recombination with counterions occurs. Therefore, the mobility obtained is closer to the microscopic mobility for the molecular transfer of the H_3O^+ state. These findings are thus comparable to the mobility value from Hall-effect measurements. At -8 °C the Hall-mobility was about 1.4 cm^2/Vsec. The value of 5 cm^2/Vsec at -60 °C then represents a rather drastic increase of mobility with decreasing temperature. The proton transfer is to be interpreted as correlated motion of the protons along hydrogen bonds by quantum mechanical tunnelling in low double-minimum potentials. For a detailed theoretical explanation of this phenomenon refer to Gosar and Hofacker /10,11/. In order to explain the broad maximum between -80 and -100 °C where we get currents as high as at high temperatures, the conductivity model should first be discussed in more detail.

5. Vacant bond injection

A symbolic picture of hydrogen bonds is drawn in fig. 7a and 7b. They are suitably oriented for passage between the electrodes of an H_3O^+ state and a vacant bond respectively.

The proton injected from the anode forms an H_3O^+ state at the crystal surface. This ion state moves through the crystal along a chain of hydrogen bonds by a mechanism of cooperative tunnelling. When the ion state reaches the other electrode, the chain is oriented in such a way that no further H_3O^+ ion can pass the same chain. One should expect a current decrease with time. The experiments, however, yield time-independent currents, even at this low temperature. Therefore, a reorientation mechanism for the molecules oriented by the passage of an H_3O^+ has to be postulated. It might be possible that the discharge

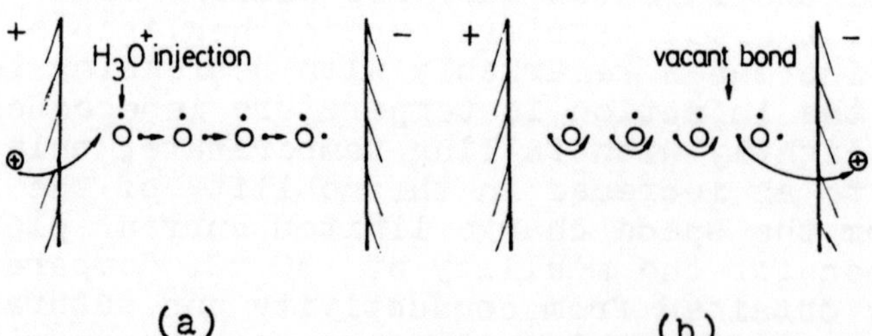

Fig.7

of an ion at the cathode is correlated with the formation of a vacant bond defect (fig. 7b) /15/. This "V-defect" travels by rotational motion of the molecules from right to left , thus reorienting the chain. At the end the initial configuration is reestablished. This vacant bond migration acts in the same way as do the well-known Bjerrum L-defects. But the latter are formed in a completely different manner. At high temperatures, enough vacant bonds are formed by thermal activation to provide prompt reorientation. The rate determining step for the conduction is the ion-state migration mechanism. The two mechanisms will reach a point of balance with decreasing temperature. Exactly that happens between -80 and -100 °C. Here, the two mechanisms, ion-state migration and vacant-bond reorientation, must wait for each other. This is also well illustrated by our space-charge limited currents at these temperatures (fig. 8).

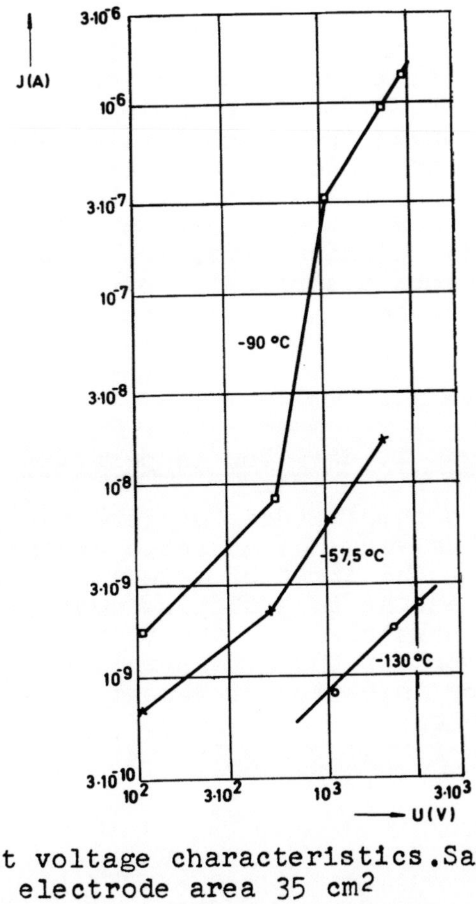

Fig.8 Current voltage characteristics.Sample thickness 10 cm, electrode area 35 cm^2

At -90 °C we observe a very steep increase of the current with voltage. This behaviour indicates that the current is controlled by double injection: protons from the anode and vacant bonds from the cathode. At the end of the transition region the rate determining step of the conduction has changed from ion-state to vacant-bond transfer. From the slope of the curve between -100 and -150 °C we can deduce an activation energy of about 10 kcal/mole. The values of the mobility obtained from the space-charge limited current experiments in this temperature region are much lower than above -80 °C, about 10^{-5} cm^2/Vsec at -130 °C. They decrease with decreasing temperature. The activation energy therefore is to be attributed to injection and migration of the vacant bonds.

6. Proton traps

At very high fields, one additional effect occurs at these low temperatures. The current-voltage characteristic shows a very steep rise at one well defined voltage (fig. 9). To explain this behaviour, we postulate the existence of traps for the charge carriers. One simple possibility would be that vacant bonds form traps for protons and vice versa. By their trapping, a space-charge is built up which changes the conductivity behaviour of the ice. This effect is certainly connected with charge storage processes in ice which occur in this temperature range and which are being studied extensively by the electrical glow curve technique /12,13/. If we assume for simplicity a homogeneous distribution of this trapped space-charge, the well-known trap-filled limit law /14/, $U_{tfl} = N_t L^2 e/2\varepsilon\varepsilon_0$ (L = electrode distance), yields a trap density N_t of about $5 \cdot 10^{12}/cm^3$ taking the dynamical value of 3.2 for ε.

7. Polarisation current at very low temperature

Below liquid nitrogen temperature, no time-independent currents could be observed. The currents dropped with time below $5 \cdot 10^{-15}$ A, the limit of detection. Protons are still injected leading to a displacement current which polarizes the crystal. But the reorientation mechanism does not work at these temperatures: The activation energy for injection of vacant bond defects is no longer available.

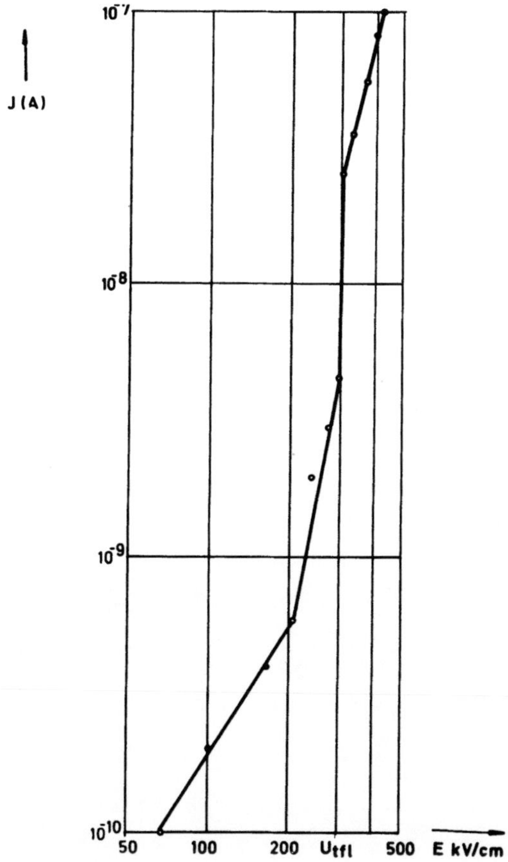

Fig.9 Current voltage characteristic at 77°K

References

/1/ Dengel, O., N. Riehl, Phys.kondens. Materie 1, 191 (1963)

/2/ Blicks, H., O. Dengel, N. Riehl, Phys.kondens. Materie 4, 375 (1966)

/3/ Heinmets, F., Nature 4754, 925 (1960)

/4/ Bullemer, B., H. Engelhardt, N. Riehl, Physics of Ice, Proc.Intern.Symp., Plenum Press, New York (1969)

/5/ Jaccard, C., Physics of Ice, Proc.Intern.Symp., Plenum Press, New York (1969)

/6/ Magnan, D., A. Kahane, Comptes Rendus 256, 5539 (1963)

/7/ Engelhardt, H., N. Riehl, Phys.Letters 14, 20 (1965)

/8/ Engelhardt, H., N. Riehl, Phys.kondens.Materie 5, 73 (1966)

/9/ Engelhard Industries, Technical Bulletin Vol.II, Nos. 1/2 (1966)

/10/ Gosar, P., Physics of Ice, Proc.Intern.Symp., Plenum Press, New York (1969)

/11/ Hofacker, G.L., Physics of Ice, Proc.Intern.Symp., Plenum Press, New York (1969)

/12/ Bishop, P.G., J.W. Glen, Physics of Ice, Proc. Intern.Symp., Plenum Press, New York (1969)

/13/ Glockmann, H.P., Physics of Ice, Proc.Intern.Symp., Plenum Press, New York (1969)

/14/ Lampert, M.A., Reports on Progress in Physics, 27, 329 (1964)

/15/ Hippel v., A., Laboratory for Insulation Research MIT, Cambridge Mass., Technical Report 3 (1968)

EXPERIMENTAL AND THEORETICAL STUDIES ON THE DC CONDUCTIVITY OF ICE

A. Kahane

Faculté des Sciences de Grenoble

St.-Martin-d'Hères / France

Abstract

The dc conductivity of ice single crystals has been studied with the aid of ion exchange membranes. With this method very stable currents were obtained after several hours so that the activation energy could be determined with good precision. A value of 9 kcal/mole was found between -20 and -70°C while the value of 14 kcal/mole was obtained for the temperature range between -70 and -100°C. The existence of two different activation energies has been interpreted on the basis of a theory of protonic conduction in ice. According to this theory the transport of protons is governed by the concentration of OH^- and H_3O^+ ions and by the mechanism of molecular rotation. The two energies of activation are related to the energies of activation for ion formation and for dielectric relaxation.

1. Introduction

In this colloquium, several studies of dc conductivity of ice are reported. In comparison with these studies the peculiarities of our work /1/ are the following:
1) Apparatus: using ion-exchange membranes close to the electrodes enables us to obtain steady and reproducible currents.
2) Results: the activation energy, close to 9 kcal/mole, measured between -20°C and -70°C is in conformity with the data of Bullemer and Riehl /2/, but between -70°C

and -100°C the activation energy is higher, about 13 kcal/mole.
3) Interpretation: The present results may be accounted for by the existence of a double mechanism of proton transfer, by means of ionic defects and by dipolar relaxation.

2. Apparatus

Cylindric single crystals (diameter: 20 mm, length: 7.5mm) are used.
The applied electric field, parallel to the c-axis, varies between 500 V/cm and 2000 V/cm. The measurements are made below -20°C in order to avoid the effect of surface conductivity and above -100°C. The temperature of the crystal is kept constant until the current becomes stable (about 12 hours). The current is measured with a galvanometer for the range 10^{-4} to 10^{-7} A, and with a dc amplifier for 10^{-7} to 10^{-11} A.

The crystal is maintained between two ion-exchange membranes A and C. C being permeable to cations and A permeable to anions /3/. Their functions are shown in fig.1: the positive impurities are taken away from the crystal through C and the negative impurities through A. It may be seen that, in principle, the passing of the current corresponds to a loss of matter; it seems that the presence of gaseous hydrogen in the sample-bearing cell (fig.2) is necessary to regenerate the crystal. We notice that the current is very unsteady at the beginning of each experiment with a new crystal. After 48 hours,

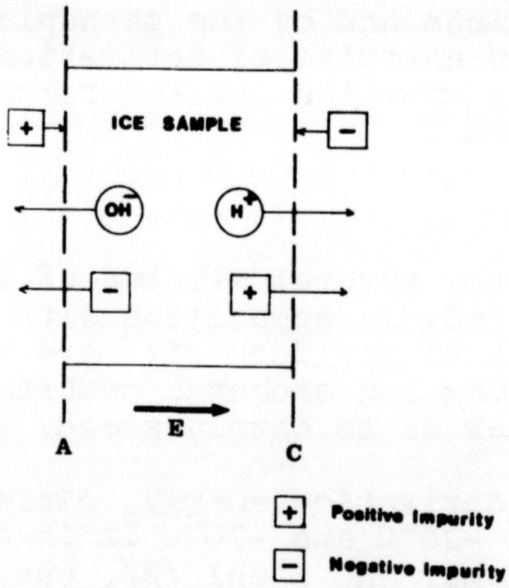

Fig.1 Ice electrode arrangement. C = membrane permeable to cations. A = membrane permeable to anions

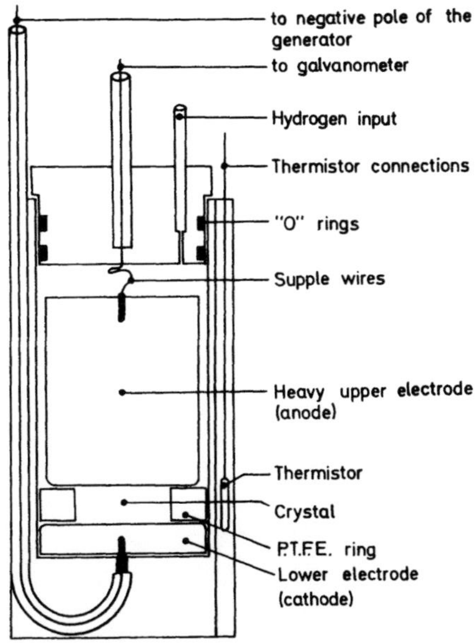

Fig.2 Design of the measure cell

crystal cleaning seems to be realized and the registered currents are very steady.

3. Results

The results already obtained /4/ do not enable us to determine the absolute value of the conductivity of ice since the contact surface between the sample and the electrodes is undefined, but they allow us to determine, with a good reproducibility, an activation energy of 9 kcal/mole between -20°C and -70°C and an activation energy close to 14 kcal/mole between -70°C and -100°C, where the measurements are less precise. Figure 3 represents the data which were obtained from experiments done on different samples, A and B.

4. Interpretation

First hypothesis:
We postulate the existence of ionic defects OH^- and H_3O^+, the concentration of which, ρ, follows the law:

$$\rho = \rho_0 \exp\left(-\frac{W}{RT}\right). \qquad (1)$$

Taking into account the work of Bullemer and Riehl /2/, we may assign a value of 8.5 kcal/mole to W. We relate the transport of ion-states to ν (jump frequency of a

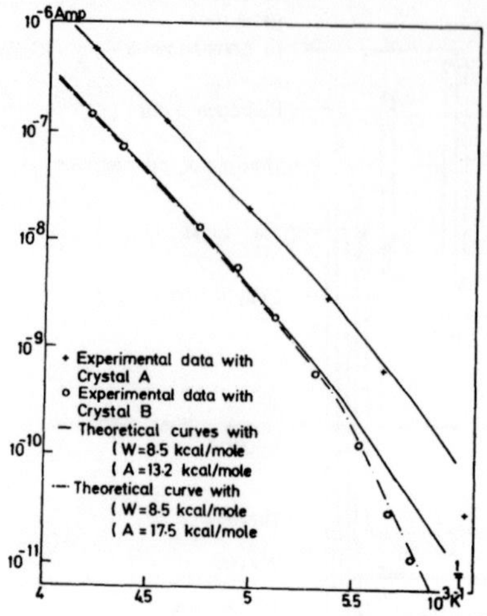

Figure 3

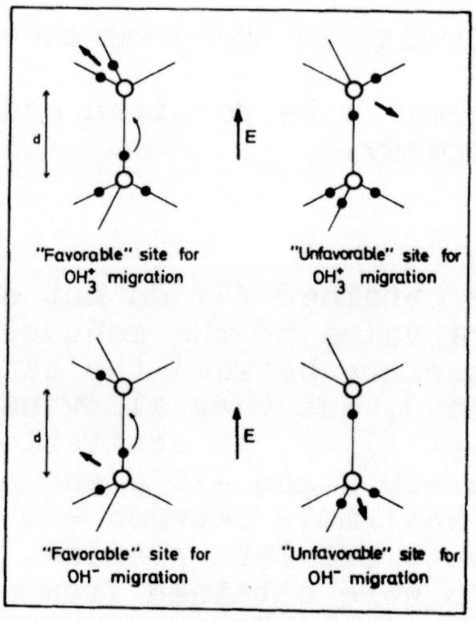

Figure 4

proton along the O-O bond).

Second hypothesis:
We introduce the idea of "favourable" and "unfavourable" sites (fig.4). A water molecule lies in a favourable site if it allows proton motion along the bond parallel to the optical axis, in the same direction as the applied field. The proton motion results in an ionic migration corresponding to the displacement of an effective charge

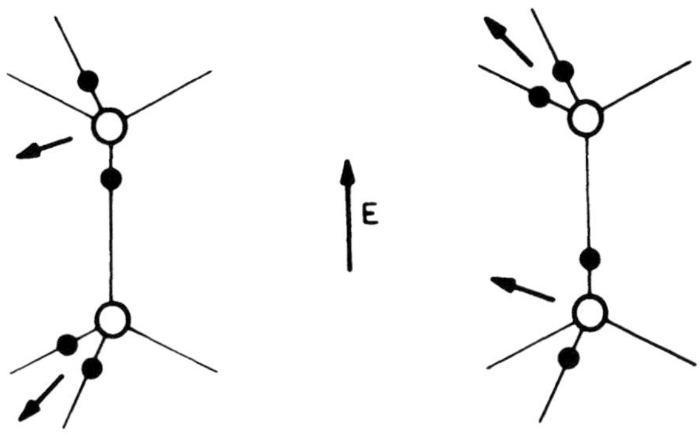

Fig.5 Regeneration of a "favourable" site from an "unfavourable" site characterised by a change in direction of the local dipole moment

q along a length d. A water molecule lies in an unfavourable site when the contrary is the case. The motion of a proton in the same direction as the applied field converts a favourable site into an unfavourable site. In order to compensate this depletion of the favourable sites, we admit a relaxation mechanism with a dipolar character as shown in figure 5. Hence we admit that the relaxation mechanism has an activation energy of A = 13.2 kcal/mole, this being the value obtained by Auty and Cole /5/ during dielectric relaxation measurements giving a Debye dipolar relaxation time:

$$\tau = \tau_o \exp\left(-\frac{A}{RT}\right) . \qquad (2)$$

Starting from hypothesis 1 and 2, we study the concentration of favourable sites N^+ and unfavourable sites N^- when the field is applied, taking into account the proton transfer and relaxation mechanism ($N^+ + N^- = 2 N_o$ = total number of molecules in the crystal). We relate N^+ and N^- to j^+ and j^-, the current densities which correspond to proton transfer in the same direction as the applied field or opposed to it, ν^+ and ν^- are the proton jumps in the same direction as the applied field E or opposed to it.

We may write the following equations:

$$j^+ = N^+ \rho \nu^+ q d , \qquad (3)$$

$$j^- = N^- \rho \nu^- q d , \qquad (4)$$

$$dN^+ = (-N^+\nu^+\rho + N^-\nu^-\rho) - \frac{N^+ - N_o^+}{\tau} dt , \qquad (5)$$

$$dN^- = (-N^-\nu^-\rho + N^+\nu^+\rho) - \frac{N^+ - N_o^-}{\tau} dt . \qquad (6)$$

We assume that ν^+ and ν^-, N^+ and N^- depend linearly on the applied field:

$$\nu^+ = \nu(1+\alpha E) \; ; \; \nu^- = \nu\,(1-\alpha E) \; ; \qquad (7)$$

$$N_o^{\,+} = N_o\,(1+\beta E) \; ; \; N_o^{\,-} = N_o(1-\beta E) \; . \qquad (8)$$

All these equations allow us to calculate the whole current density and to set up the equation of the dc current when the applied field remains constant and parallel to the c-axis:

$$\log I = C_1 + \log \frac{\exp(-W/RT)}{1+C_2 \exp \frac{A-W}{RT}} \; . \qquad (9)$$

C_1 and C_2 can be determined from this theoretical equation by using two pairs of experimental values. Figure 3 shows two sets of experimental data, each being compared with the theoretical curve (C_1 and C_2 were adjusted to give the best fit).

5. Conclusion

In this model, first we are using the idea of change of proton jump frequency rather than the idea of ionic mobility in the applied field: this seems more convenient in order to connect a phenomenological model with a quantum mechanical theory; secondly we don't use the D and L defects of Bjerrum but we suppose that the rotation of the molecules is a cooperative phenomenon perhaps connected with point defects, clusters, dislocations,and grain boundaries.

On the other hand, the existence of two basic mechanisms which are dominant at high and at low temperature respectively seems to be connected with the question of ordered and disordered structure of ice /6/:

- at high temperatures, the dominant mechanism corresponds to a partial rotation of the molecules and the number of free ions limits the direct current; this dominant mechanism of a cooperative type is compatible with a partial ordering of the protons in the crystal;

- at low temperatures, the dominant mechanism is that of ionic motion and dipolar relaxation will limit the direct current; this dominant mechanism may destroy an ordered protonic structure in the crystal lattice of ice. This is in conformity with Pauling's theory of the existence of a zero-point entropy.

References

/1/ Durand, M., M.Deleplanque, A.Kahane, Sol.St.Comm. 5, 759, (1967)

/2/ Bullemer, B., N.Riehl, Sol.St.Comm. 4. 447, (1966)

/3/ Barret, S., G.Briere, Journ.Chimie Physique, 970, (1965)

/4/ Durand, M., Diplôme d'Etudes Supérieures, Grenoble, (1967)

/5/ Auty, R.P., R.H.Cole, J.Chem.Phys. 20, 1309 (1952)

/6/ Kahane, A., Thèse, Paris, (1962)

ELECTRICAL CONDUCTION IN ICE

P.R. Camp*, W. Kiszenick**, D. Arnold ***

US Army Cold Regions Research and Engineering

Laboratory, Hanover, New Hampshire / USA

Abstract

In an attempt to resolve the conflict existing in the literature as to the dc electrical conductivity of pure ice, an extensive series of measurements has been made of both the dc conductance and audio-frequency reactance of ice. Evidence was found for significant surface conductivity when slight surface contamination was present. In order to explain these results quantitatively, it appears necessary to postulate a surface conduction region whose thickness varies with temperature. Extrinsic bulk conductivity due to trace impurities was also found to play an important role. These two processes together probably account for much of the disagreement found in the literature. Special efforts were made to develop techniques which would insure the highest sample purity. For what we believe to have been our purest samples, both ac and dc conductivity measurements show that, for a fresh sample, the dc conductivity is nearly independent of the temperature down to temperatures at which the ac and dc conductivities are roughly equal. This temperature appears to vary with sample purity. Below this temperature, the dc conductivity drops rapidly with decreasing temperature and the high frequency ac conductivity tends to level off. The results suggest that the high frequency conductivity is limited by two processes in parallel and that the dc conductivity is limited by the same two processes in series. Plateau values as low as 2×10^{-11} $(\Omega\ \mathrm{cm})^{-1}$ have been found.

*, **, *** see last page of this paper

This plateau disappears after the crystals have been annealed for a day at -5°C. The disappearance of the plateau with time may result from the diffusion of impurities into the sample from the walls of the container. In this case, the concentration of ions in pure ice, and hence the dissociation constant, must be much lower than has been commonly believed. On the other hand, the plateau may be associated with traps or internal boundaries which are modified by annealing.

1. Introduction

For the past three years, the demands of other activities have kept me out of the laboratory. Only this past summer have I had a chance to think in terms of renewing my own research activity. Thus this symposium is a most happily timed event for me and I'm here more to listen than I am to speak. However, the invitation which resulted in my leaving the Cold Regions Research and Engineering laboratory came suddenly and prevented me from reporting adequately on a series of experiments on electrical processes in pure ice which were conducted there in 1963 and 1964 /1/. Although part of what we did has been supplanted by work done by others in the intervening years, there are several features of it that may still be of interest. Today I will discuss two such features, one having to do with conduction at the ice surface and the second having to to with conduction in the bulk.

At the time these studies were undertaken there was a great deal of conflict, at least at the experimental level, as to the true direct current electrical conductivity of ice /2-9/.

There isn't space here to review this work but two conclusions seem to be warranted from it; first, because responsible experimenters in different laboratories found such widely different activation energies for conduction, they really must have been measuring different processes. Second, the fact that their data all seemed to converge to values of the order of 10^{-9} ohm-cm^{-1} at -10 °C indicates that at this temperature dc conductivity is an insensitive test for differentiating between different models of conduction.

The activation energies for conduction as determined by experiment could be roughly grouped into 3 categories: low values of the order of 7 kilocalories per mole, intermediate values of the order of 13 kilocalories per mole, approximately equal to the well established values for dielectric relaxation (high frequency dielectric conductivity), and values of 20 kilocalories per mole and higher. It having been well established /6,7,10,11/ that the effect of impurities such as HF and NH_4F is to markedly increase the conductivity through a decrease in the activation energy for conduction, it was natural to attribute the lowest values to contamination. It occurred to us that one of the two remaining groups might be due to surface phenomena and consequently we set out initially to study surface conduction. Ultimately we were led to a rather long study of both surface and bulk properties including both ac and dc measurements. Because we had been handicapped by the lack of experimental detail in the work cited, we have given a fairly complete account of our major experiments together with the occasional pitfalls and frustrations we've met along the way in Reference 1, copies of which are available as indicated. Time will not allow a full discussion of these experiments here and I will merely summarize those which bear on the two phenomena mentioned at the start.

2. Surface Experiments

In order to distinguish between surface and bulk phenomena we constructed the apparatus shown in fig.1. It is essentially a guarded electrode structure in which the area of the guard ring is just equal to the area of the central disk. If the gap between the ring and the disk

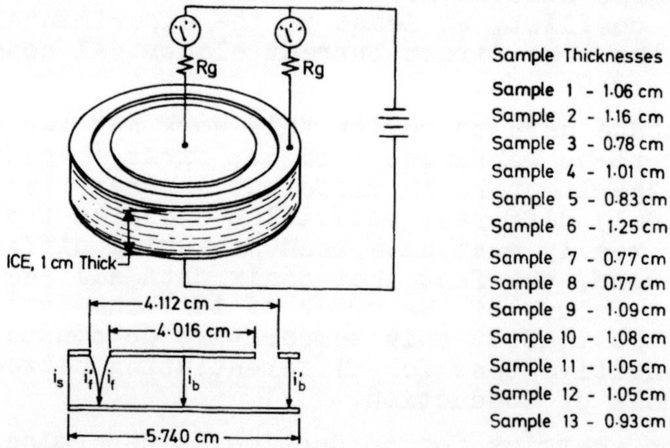

Fig.1 Guarded ring apparatus for separating bulk and surface effects

is small, the fringing effect introduced by it should be very nearly equal for both the guard current and the disk current so that the only significant difference between the two currents should be that resulting from the outside surface. This experimental arrangement has some drawbacks (for example, it does not allow for a different electrode polarization behavior), nonetheless it seemed a useful experiment. Our measurements with this apparatus indicated a very much larger ring current than disk current, a phenomenon which we interpreted at the time as indicating large surface currents. Later experiments indicate that what we were really seeing was the fact that the concentration of impurities in the outer zone was greater than in the inner zone and that there was substantial impurity "surface" conduction.

Having established the existence of large surface currents, we sought to overcome some of the objections to the guard ring experiment by four probe measurements on slender rods. By making the rod diameter small, it is possible to emphasize the surface effects in comparison with bulk effects. Thus experiments were performed on polycrystalline ice cylinders 1/8 inch in diameter prepared by freezing distilled demineralized water, conductivity 5×10^{-7} (ohm cm)$^{-1}$, in clean pyrex tubes and then extruding the ice onto a frame of platinum electrodes. The data shown in fig.2 are typical of the measurements made on these samples. By virtue of the geometry, the calculated surface conductivity in ohms^{-1} is numerically equal to the conductance measured. A total of 11 different temperature runs on two different samples with this apparatus gave an activation energy of $19.25 \pm .25$ kcal/mole. One sample was repeatedly measured over a period of 10 days and the extremes of this data are shown in fig.2. The slow decrease in conductance with time suggested that there was a gradual cleaning of the surface (or poisoning depending on the point of view one took as to the nature of conductance). To explore this matter further, the atmosphere of one experiment was changed from air to CO_2 and no measurable change in conductance or activation energy resulted. In another experiment the sample was immersed in kerosene, again with no apparent change. Thus our interpretation is that the surface was already well contaminated before the first experiments were begun. Nearly unavoidable surface contamination of this kind is consistent with the hypothesis that the high activation energies reported by some observers were consequences of surface conductance.

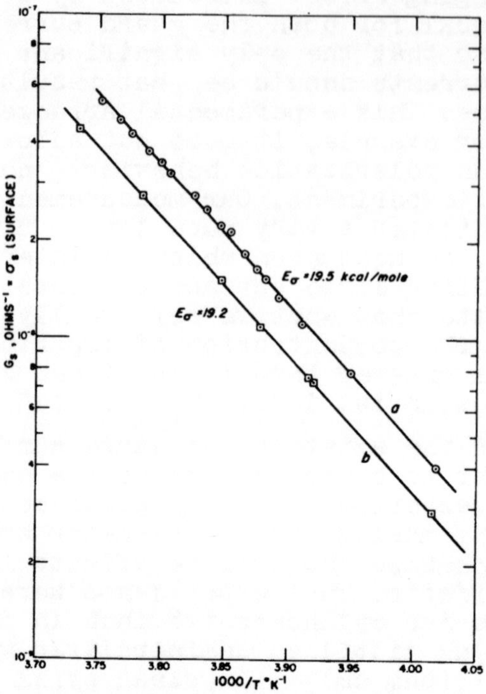

Fig.2 Conductance (ohms)$^{-1}$ vs. 1000/T $^{o}K^{-1}$ for slim rod experiments in air. Data of line b were taken 10 days after those of line a

As a further test of the surface current hypothesis, 2 experiments were done with larger rods (diameter 2.8 centimeters with voltage probes 2.2 centimeters apart). The purpose of these experiments was to show the transition from predominently surface current at high temperature to predominently bulk current at low temperatures which should come about as a consequence of the difference in activation energies. The first sample which was not perfectly free from bubbles gave erratic results and was discarded. A second sample which was perfectly clear polycrystalline ice gave the results shown in fig.3. With the exception of the very low temperature branch of this curve, which we attribute to systematic errors in the temperature measurement, the sample showed the behavior expected. However, further experiments with this sample also gave erratic results and we concluded that random surface contamination was so difficult to avoid with this arrangement that another approach was needed.

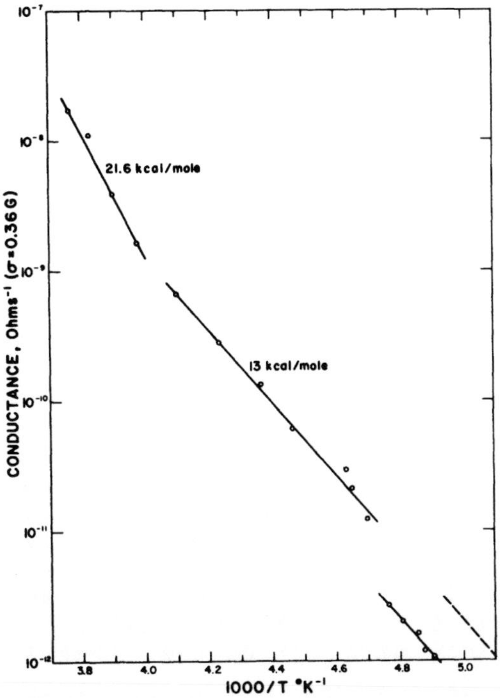

Fig.3 Conductance (ohms)$^{-1}$ vs. 1000/T $^{O}K^{-1}$ of 2.8 cm diameter four probe sample in air. Voltage probes 2.2 cm apart

Our next step was to see what we could learn about the properties of the surface from ac measurements where electrode polarization effects would be minimized and the guard ring technique could be used (for which sample preparation and handling were easier). This type of measurement has been enormously simplified by the introduction of the toroidal transformer bridge illustrated in fig.4. Our bridge was a general radio type 1615A

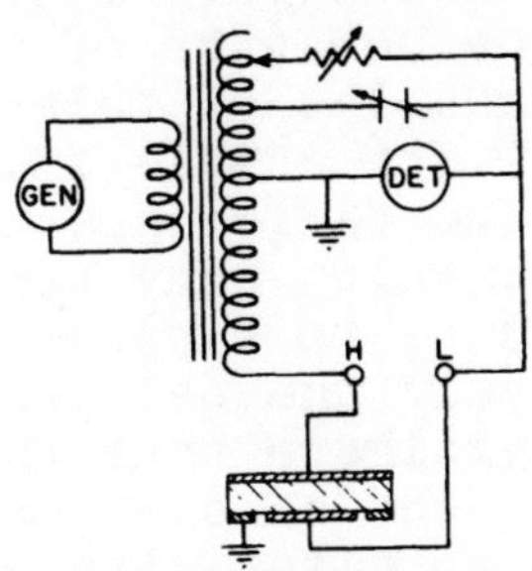

Fig.4 Simplified schematic diagram of the bridge circuit for ac measurements

capacitance bridge with a general radio type 1232A tuned null detector. First, the standard guarded circuit arrangement was used to measure bulk properties. Then the guard and disk connections were interchanged and the outer ring (bulk plus surface) properties were measured. The second arrangement does not prevent a slight fringing flux in the air surrounding the sample introducing a small nearly constant shunt capacitance of the order of one picofarad.

In all, 9 physically separate samples were prepared from demineralized distilled water having an initial resistivity of 2 x 10^6 ohm·cm. They were grown at a rate of 3 mm/hour in lucite tubes having an inside diameter of 5.74 centimeters. Both polycrystalline and single crystal samples were prepared. For the latter, the technique was improved by evacuating the sample tube to remove gases dissolved in the water. The behavior of these samples was orderly and has been discussed elsewhere /1/. We shall consider here specifically only sample 13, a section of single crystal, cut without extruding it from the lucite tube in which it was grown, which was then mounted in a platinum plated sample holder with the c-axis parallel to the electric field.

Equivalent parallel conductance and capacitance for both bulk and bulk plus surface were measured (for all samples) as a function of frequency in the range 20 to 20.000 cps for a number of different temperatures between -5 °C and -30 °C. Data for single crystal sample No.13 (platinum plated electrodes) at -12.6 °C are shown in fig.5. For contrast, the -11.6 °C data for one of our poorer polycrystalline samples (sample 10) are included in fig.6. All samples were well behaved in their dielectric relaxation and semi-log plots of the relaxation frequency versus 1000/T for samples 11, 12, and 13 yield an average activation energy of 13.4 ± .2 kcal/mole as shown in fig. 7.

A principle purpose of these experiments was to study the difference between the behavior of the outer zone and that of the inner. Examples of this difference data are shown in fig. 8 and 9 for sample 13. The similarity between the shapes of these difference curves and those of the parent curves from which they were derived is suggestive. It would lead us to suppose that they are a consequence of effectively unequal areas for the ring and disk electrodes. We estimate that the misalignment of our plates relative to the sample and the possible misfit of the sample due to eliptіcity of the lucite tube in which it was grown can only account for a

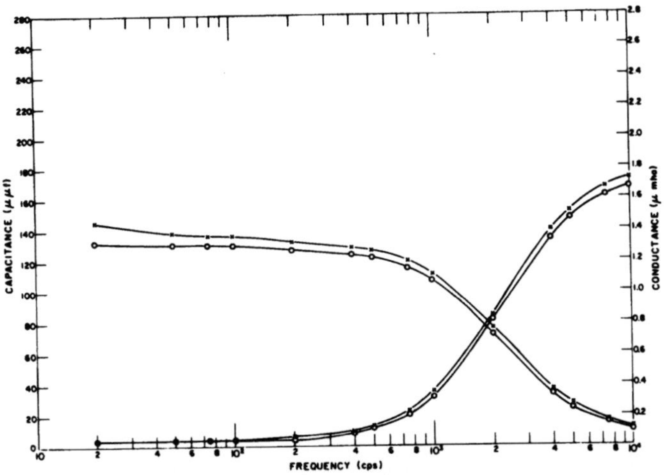

Fig.5 Equivalent parallel conductance and capacitance for single crystal sample 13 at -12.6 °C. Platinum plated electrodes. Crosses indicate bulk plus surface (ring electrode) and circles indicate bulk (center electrode)

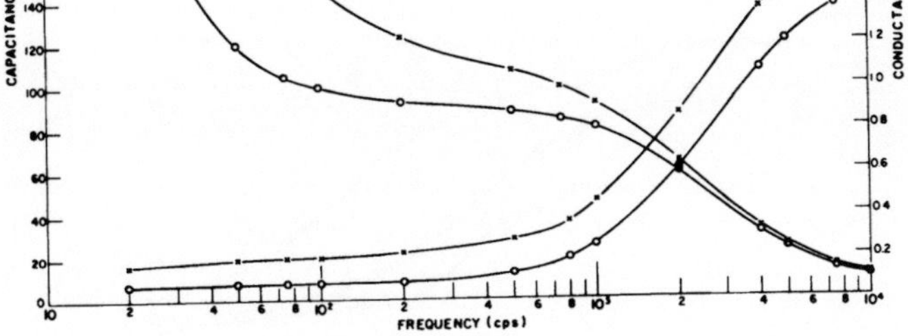

Fig.6 Equivalent parallel conductance and capacitance for polycrystalline sample 10 at -11.6 °C. Brass electrodes. Crosses indicate bulk plus surface (ring electrode) and circles indicate bulk (center electrode)

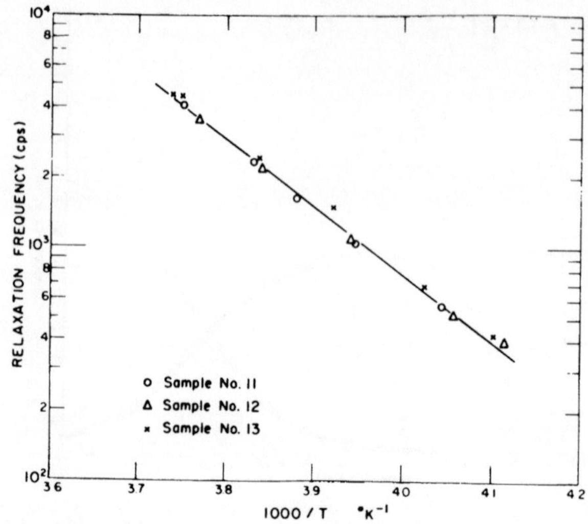

Fig.7 Semi-log plot of relaxation frequency vs. 1000/T ($^{o}K^{-1}$) for samples 11, 12 and 13. (Single crystal)

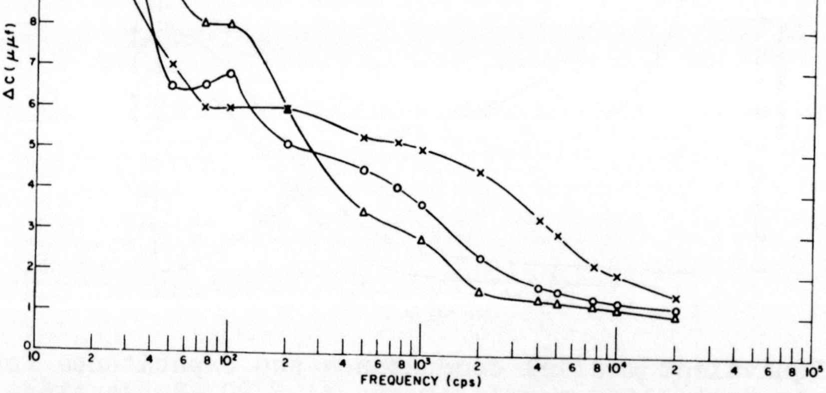

Fig.8 Difference in capacitance (bulk plus surface, minus bulk) for sample 13 at indicated temperatures

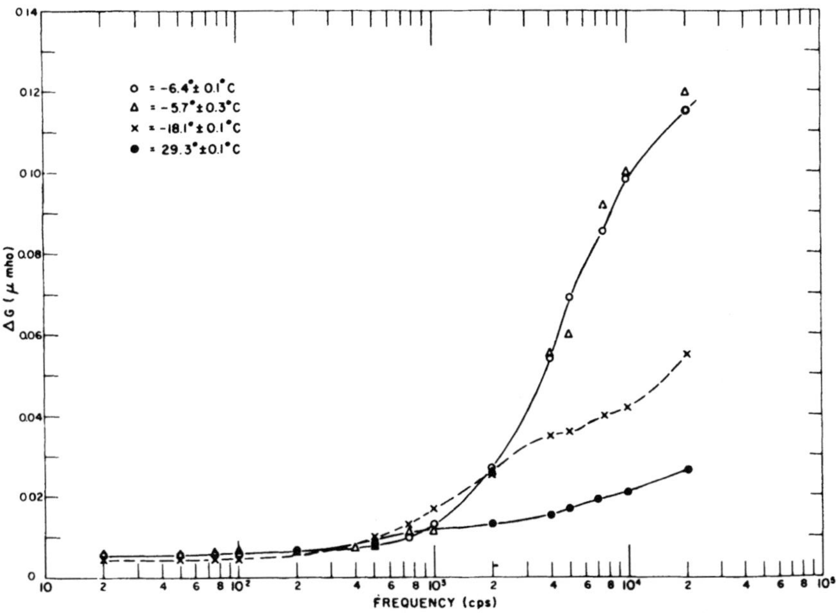

Fig.9 Difference in conductance (bulk plus surface, minus bulk) for sample 13 at indicated temperatures

difference of a few percent at most. Even for the single crystal, the differences are larger than that and for the less-pure, polycrystalline samples the differences are substantially larger. In addition, a more careful look shows that the frequency dependence is not really in accord with this explanation.

Our experience with growing single crystals has led us to the realization that when impurities are present in a melt they usually do not distribute themselves uniformly throughout the crystal /12/. This is largely the result of differences in the circulation velocity in the immediate vicinity of different parts of the liquid solid interface. Quantitative measurements of this behavior have been reported by Blicks et al./13/. We believe that the difference, ΔG, is a result of such a non-uniform distribution of impurities and consequently an indication of the presence of impurities. The difference, ΔC, appears to be an electrode effect quite possibly also associated with a non-uniform distribution of impurities.

These results led us to seek conditions of highest purity and then to repeat our slim rod experiment. Eventually we were able to obtain water having a room temperature conductivity of about 7×10^{-8} (ohm cm)$^{-1}$ and, using a well leached Teflon cell of 1/8 inch inside diameter containing 6 platinum electrodes, we obtained the data of fig. 10A. More will be said about this curve shortly. In order to measure an equivalent section in air, the platinum leads were then removed and a section adjacent to the one already measured was extruded onto a set of four platinum wire electrodes. This sample was measured with the result shown in fig. 10B. The fact that this section was found to contain some tiny bubbles may have contributed to the difference between A and B. The significant matter here is the change in conductivity when it was then deliberately contaminated by running a finger along its length with the results shown in fig. 10C. This affirms our previous conclusions that

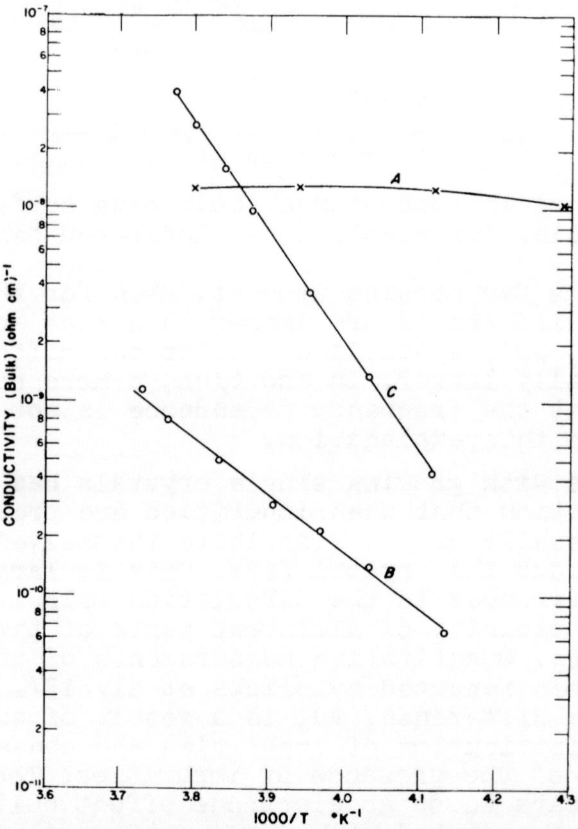

Fig.10 Direct current conductivity vs. 1000/T ($^{o}K^{-1}$) for 4-probe slim rod experiments using high purity water. A. 1/8 inch rod encased in Teflon; B. Sample adjacent to A after extrusion from Teflon tube; C. Sample B after deliberate contamination of the surface

the conductivity in the slim rod experiments was due to surface contamination. However a quantitative interpretation in terms of surface conductivity comes to grief. When we consider the actual numbers involved, we find that for the contaminated case, the concept of simple surface conductivity is not adequate. The reason is that the pre-exponential term in the Arrhenius expression cannot be arbitrarily large. This means that (when the activation energy, E, is much larger than RT) the larger the activation energy, the smaller the actual conductivity must be. Large conductivities and large activation energies cannot go hand in hand in this simple way.

The general framework of the argument is as follows: Current is carried by mobile charges q which may be present as a result of thermal ionization of water or impurity molecules, the activation energy for ionization being ε_i. There are m such mobile ions per cubic centimeter (or m_s per square centimeter of surface). The transport of charge takes place by jumps between adjacent lattice sites separated by a distance λ. Each jump requires the traverse of a potential energy barrier ε_μ.

By the law of mass action, m is temperature dependent according to $m = \alpha n e^{\varepsilon_i/2kT}$, where, n, is the number of ionizable molecules per unit volume and α is a numerical factor which we will approximate as 1*. The mobility, $\mu = \lambda\nu/E$ where E is the electric field and ν is a jump frequency which is an exponential function of temperature and a barrier energy ε_μ. Because $\sigma = mq\mu$, the small signal approximation leads to an equation of the form

$$\sigma = 10^7 \frac{\lambda^2 q^2}{kT} n \nu_o e^{-(\varepsilon_i/2+\varepsilon_\mu)/kT} \text{ (ohm cm)}^{-1} .$$

Since q cannot reasonably be much larger than the electronic charge, λ cannot be larger than the intermolecular distance, n cannot be larger than the number of water molecules per cubic centimeter and ν_o cannot exceed the lattice frequency, the maximum value which sigma can take on will be very strongly controlled by the exponential factor. The argument for surface conduction is the same. With these simple assumptions we are led to the conclusion that bulk and surface conductivity cannot exceed the values given in Table I below for the activation energies listed. This is in basic conflict

*This would be the case for $m = 10^9/cm^3$ at -10 °C (in the range deduced by Eigen /8/ and $N_o\varepsilon_i$ = 28 kcal/mole (Jaccard /7/).

with our experimental evidence. Even if we attribute the conductivity of our thin rod samples to impurities distributed throughout the bulk (this is the extreme case of a very thick surface layer), our slim rod values for σ at -10 °C are then in excess of 10^{-7} (ohm·cm)$^{-1}$. Table I shows that the maximum bulk conductivity we can account for in this way having an activation energy of 19 kcal/mole is 10^{-12} (ohm cm)$^{-1}$. We are forced to conclude that the temperature dependence we have measured does not give the activation energy for the conduction process alone.

Table I*

Maximum values possible for bulk and surface conductivity at -10 C (α=1)

Activation energy (kcal/mole)	σ max (bulk) (ohm cm)$^{-1}$	σ_s max (surface) (ohm^{-1})
1	8×10^{2}	2.4×10^{-5}
5	4×10^{-1}	1.2×10^{-8}
10	3.0×10^{-5}	9×10^{-13}
13	1×10^{-7}	3.1×10^{-15}
19	1.1×10^{-12}	3.2×10^{-20}
24	8.4×10^{-17}	2.5×10^{-24}

*If, as we believe from our experiment, the actual value of d c conductivity of ice is much lower than that used by Eigen, m will be very different from his value and the numbers for σ_{max} and $\sigma_{s\ max}$ in Table I will all need to be changed. However the form of the argument will remain the same.

A way out of this dilemma can be found if we modify our model. We assume a disordered surface region whose thickness is an exponential function of temperature and which has an anomalously high conductivity. If the effective thickness of the surface layer is a strong function of temperature, we can then have a specific conductance in this region which has a low activation energy and hence a large numerical value. The temperature dependence of the measured conductance will still be large because the thickness of the current carrying layer is strongly temperature dependent. It is immaterial to this argument whether the heightened conductance is caused by the disorder itself or by the presence of impurities in this region. The idea of such a disordered region having mechanical properties quite different from those of ordinary ice is one which has received much support in recent years based on a variety of different experiments. Jellinek /14/ has reviewed this subject. Using values for the thickness of this region which are comparable with those proposed to explain the mechanical properties (e.g. 10 Å at -10 °C) one can satisfy both Table I and our slim rod data.

3. Bulk Experiments

The experiments relating to improvement in technique mentioned above were directed toward three major goals: improving the quality of the water from which the samples were grown, reducing the contamination from the sample tube and preventing the formation of bubbles in the samples grown in slim cylinders. Our search culminated in the use of a special mixed bed demineralizer, Barnstead model HN8902, which yields water with a conductivity as low as 5.5×10^{-8} (ohm cm)$^{-1}$ at room temperature /1/. It also resulted in the use of Teflon sample tubes with sealed in platinum electrodes and the development of a special technique for agitating the water in the small tubes during crystal growth. Non-polarizing, bubble-free samples exhibiting highly-linear voltage-current relations and normal ac behavior resulted.

As the purity of the samples was improved, an interesting phenomenon emerged. The dc conductivity of the samples did not at first decrease as the temperature was lowered from 0 °C but remained constant over a range of temperature which became larger as the sample purity was improved (fig. 10A). At sufficiently low temperatures the conductivity decreased. Meanwhile, the high frequency ac conductivity showed an equally anomalous behavior. It decreased with temperature in a "normal" fashion as long as the dc conductivity remained constant. Then as the latter began to fall off, the ac conductivity tended to level off. This behavior is illustrated in fig. 11 for 1.3 cm diameter polycrystalline samples grown from high purity water in a Teflon sample tube.

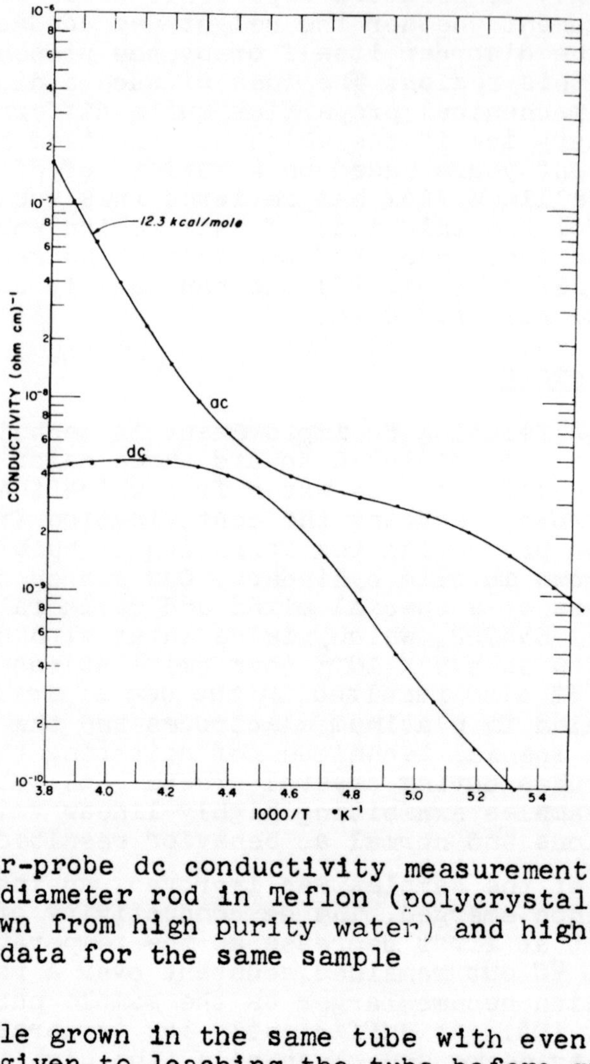

Fig.11 Four-probe dc conductivity measurements for a 1.3 cm diameter rod in Teflon (polycrystalline sample grown from high purity water) and high-frequency ac data for the same sample

A new sample grown in the same tube with even greater attention given to leaching the tube before growing the crystal gave the data of fig. 12a and b. Using a larger Teflon cell (diameter 3.2 cm) and putting some of the mixed bed demineralizer resin inside the tube itself, we obtained the dc data of fig. 12c. Still further purity was sought by constructing a new two-electrode apparatus which allowed the growing interface to be flushed continuously with freshly demineralized water at 0 °C. At this temperature, the conductivity of the water was 2.5×10^{-8} (ohm cm)$^{-1}$.

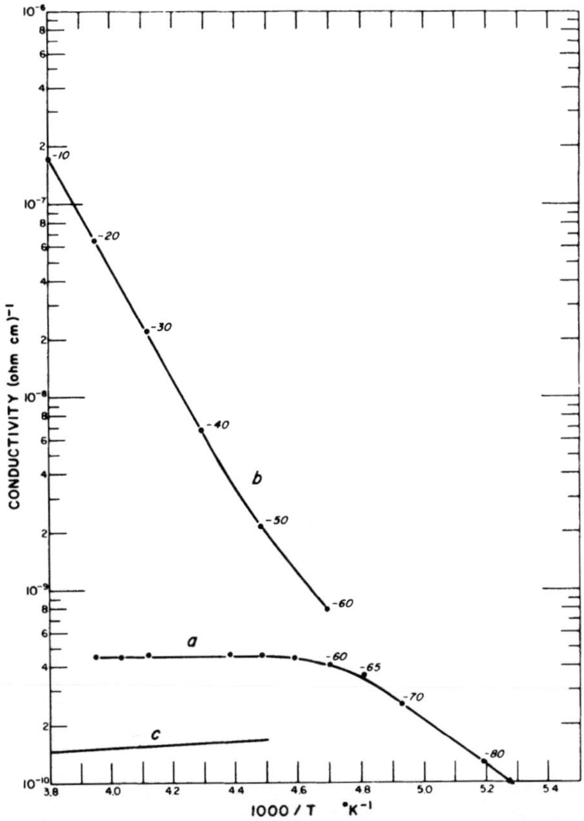

Fig.12 a. Four-probe dc conductivity measurements of a new sample similar to that of figure 11 but with the sample tube leached for a longer time.
b. High frequency ac measurements on same sample.
c. Four-probe dc conductivity data for sample grown in 3.2 cm diameter measuring cell containing some demineralizer resin

Measurements with this sample are shown in fig. 13c. The arrows show the direction of the temperature change during the run. The gradual rise of conductivity as temperature was increased followed by the nearly constant values as temperature was again decreased indicate a conductivity which changes with time and which does so more rapidly at higher temperatures than at lower temperatures. For convenience, the data already cited are also plotted on fig.13. Curve a is the data from fig. 12a plotted over the full range of temperatures used. This sample was remeasured after having been

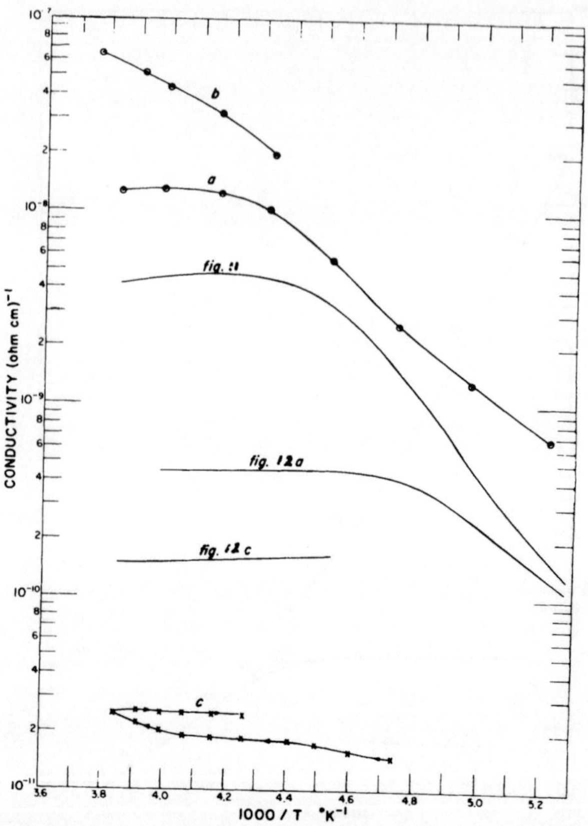

Fig.13 a. Four-probe dc conductivity data of figure 10A. b. Repeat of (a) after sample had been stored for 5 days at -20 C. c. Two-probe dc conductivity data for sample grown under conditions such that the water was continuously circulated through demineralizer. Arrows indicate sequence of measurements

stored for 5 days at -20 °C with the result shown in fig. 13b. This also indicates a substantial change in conductivity and gradual disappearance of the plateau with time.

In considering these results, two points should not be overlooked. One is that the dependence of conductivity on time, particularly as the temperature approaches 0 °C, means that the initial conductivities of these samples must have been lower than when they were measured. The second is that it is quite possible to prepare slightly impure samples which have a low dc conductivity. But in such cases we have found that they exhibit a large apparent activation energy. For these slightly impure samples, the high frequency ac conductivity is undisturbed down to the lowest temperatures used and give very good activation plots resulting in a value of about 13 kcal/mole*.

The symmetry of the ac and dc results suggests a situation in which there are two fundamental processes which contribute in parallel to the ac conductivity and in series to the dc conductivity. Possible electrical analogues are illustrated in fig.14.

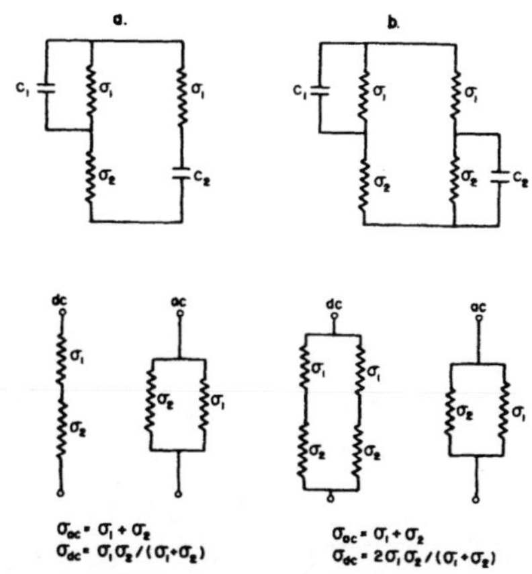

Fig.14 Possible electrical analogues of samples which exhibit a plateau in dc conductivity and a break in ac conductivity as in fig.11

This interpretation is strengthened if we plot log $\sigma_1 \times \sigma_2$ versus 1000/T ($^{\circ}K^{-1}$) as is done in fig.15 for the data of fig.13. The result is very nearly a straight line over the whole temperature range, minus 10 °C to minus 90 °C. This is just what we would expect if σ_2 were nearly independent of temperature and σ_1 obeyed the Arrhenius equation. It is very tempting to use the conduction model of Gränicher et al. and ascribe σ_1 to the rotation of dipoles and σ_2 to the translation of protons. This would imply that in our plateau producing samples a constant number of ions, independent of temperature, was available for conduction and that this concentration differed from sample to sample.

* We are speaking here of impurity concentrations well below those of doping experiments which do indeed alter the activation energies for hf ac conductivity.

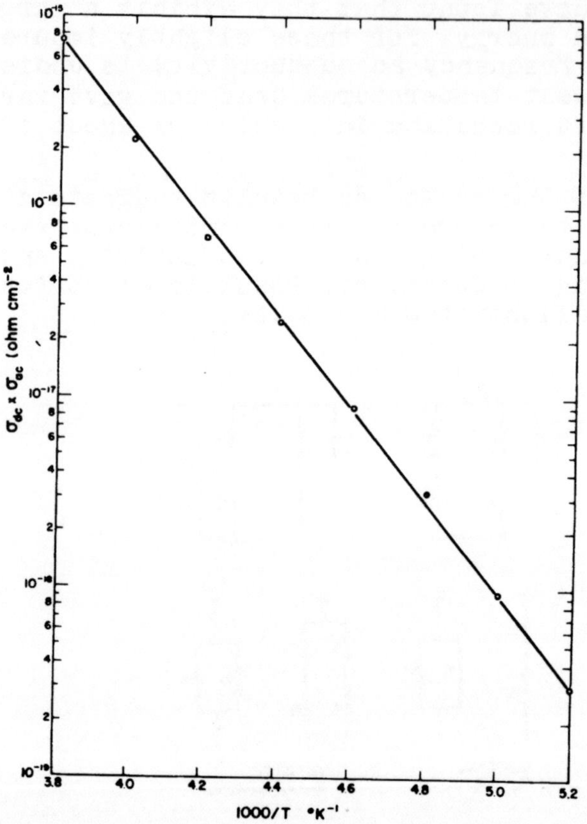

Fig.15 Semi-log plot of the product $\sigma_{ac} \times \sigma_{dc}$ as a function of 1000/T($^{o}K^{-1}$) for the data of fig.11

I came to this symposium without a really satisfactory explanation of how such a temperature independent concentration of ions could come about. This morning's papers by the München group seem to have provided a model. Proton injection from the platinum electrodes, the proton injecting ability of the electrodes varying with sample preparation and electrode construction as suggested by Prof. Riehl and his colleagues, would seem to account for our observations. If so, the true dc conductivity of ice at -10 ^{o}C must be lower than 2×10^{-11}(ohm cm)$^{-1}$, our lowest plateau value.

4. Conclusions

Many determinations of the dc electrical conductivity of ice have been made at different laboratories in various parts of the world. The results are most unsatisfactory for there is no unanimity as to its values. Activation energies for conduction of from 7 to 24 kcal/mole have been proposed. At the outset of this work it was suggested that this great disparity might arise from sample contamination on the one hand and surface conductivity on the other. That surface phenomena can play an important role seems to have been demonstrated by the slim rod experiments. However, the situation is more complicated than we had first supposed. In order to justify the numbers which we measured, large values of conductivity together with large apparent activation energies, we have been forced to postulate a surface conduction region whose effective thickness varies with temperature. This model agrees qualitatively with the picture of a disordered surface region of temperature dependent thickness which has been suggested by others in accounting for interfacial viscosity, regelation and other remarkable properties of ice. By a suitable choice of indetermined constants, the agreement can be made semiquantitative. As might be expected, the conducting properties of this region appear to be quite strongly modified by slight amounts of impurity.

The guard ring experiments showed that there was a large contribution to an apparent surface effect even in rather clean samples due to a non-uniform radial distribution of residual impurites. This emphasized the importance of extra efforts towards sample purity. They also put an upper limit on any surface contribution to the ac properties. After an intensive effort to improve the purity of our samples we found an anomalous behavior of the dc conductivity. Under what we believe to have been our conditions of highest purity we found that the dc conductivity of ice appeared to be nearly independent of temperature down to at least -65 °C when the sample had been freshly prepared. Samples of slightly lower purity exhibited a similar plateau but at a higher value of conductivity and over a smaller temperature range below which the conductivity dropped off with decreasing temperature. These samples exhibit a symmetry between the dc conductivity and the radio frequency conductivity such that their product is a straight line on a semi-log versus 1/T ($^{\circ}K^{-1}$) plot. This suggests that two processes contribute in series to the dc conductivity and that the same two processes contribute in parallel to the ac conductivity. The experiments reported this morning by Bullemer, Engelhardt and Riehl suggest that this plateau occurs because of proton injection from the platinum electrodes.

References

/1/ Camp*, P.R., W. Kiszenick** and D. Arnold***, "Electrical Conduction in Ice", Research Report 198, Sept. 1967, pp 52. Available on request from U.S. Army Terrestrial Science Center, P.O.Box 282, Hanover, N.H. 03755

/2/ Ayrton, W.E., J. Perry, Proc. Phys. Soc. (London) 2 (1875-8)

/3/ Johnstone, J.H.L., Proc.Trans.Nov.Sc.Inst. 13, 126-144 (1912)

/4/ Bradley, R.S., Trans.Faraday Soc. 53, 678-691 (1957)

/5/ Siksna, R., Arkiv för Fysik 43, 495-510 (1957)

/6/ Gränicher, H., C. Jaccard, P. Scherrer, A. Steinemann, Disc.Faraday Soc. Molecular Mechanism of Rate Processes in Solids, 50-63 (1957)

/7/ Jaccard, C.,Helv.Phys.Acta 32, 89-128 (1959)

/8/ Eigen, M., L. de Maeyer, Z.f.Elektrochem. 60, 1037-48 (1956)

/9/ Heinmets, F., R. Blum, Trans.Faraday Soc. 59, 1141-1146 (1963)

/10/ Brill, R., P.R. Camp, USA SIPRE Research Report 68, 75 p. (1961)

/11/ Brill, R., H. Ender, A. Feuersanger, Z.f.Elektrochem. 61, 1071-5 (1957)

/12/ Camp, P.R., U.S. Army Cold Regions Research and Engineering Laboratory Research Report 114,38 p. (1963)

/13/ Blicks, H., H. Egger, N. Riehl, Phys.kondens.Materie 2, 419-422 (1964)

/14/ Jellinek, H.H.G., J.Appl.Phys. 32, 1793 (1961)

* Department of Physics, University of Maine.

** Department of Physics, Polytechnic Inst.of Brooklyn

*** Department of Physics, Boston University.

IMPURITY STATISTICS IN ICE

R.G. Seidensticker

Westinghouse Research Laboratories

Pittsburgh, Pennsylvania 15235

and

R.L. Longini

Carnegie-Mellon University

Pittsburgh, Pennsylvania 15213

Abstract

The concept of ice as a protonic semiconductor has been extended by using a statistical treatment, analogous to that used for electronic semiconductors, to calculate the influence of impurities on the populations of electrically active defects. In addition to the states introduced by hydrofluoric acid and ammonia, the states occuring when an oxygen atom lacks the usual tetrahedral coordination are included. These "interface states" may have a pronounced effect on the Bjerrum defect population.

Finally, the results of the calculations are used to derive the solute partition coefficient for ice grown from HCl solutions, and good agreement is found between the theoretical predictions and experiment.

I. Introduction

Ice has frequently been called a protonic semiconductor, but little use has been made of the formalism developed for the analysis of localized impurity effects in more familiar electronic semiconductors. In analogy with the chemistry of aqueous solutions, defects in ice are usually treated by applying mass action equations to hypothesized chemical reactions in the solid. In this paper we present an alternative statistical treatment which closely parallels the analysis used for localized defects (donors, acceptors, etc.) in the electronic semiconductors /1/. The equations which finally result are in many cases the same as those derived from mass action considerations /2, 3/, as they should be. We feel, however, that our formalism clarifies the relationships between defects in the ice lattice and perhaps simplifies the development of suitable models for the description of observed phenomena.

II. Configuration States

The populations of hydrogen bond defects can be found by calculating the most probable distribution of protons among what we call "configuration states" in the ice lattice. These configuration states, somewhat analogous to chemical species, describe the number and position of the protons in the hydrogen bonds around a given lattice point. The configuration states, or more simply states, which we will consider in this paper are schematically shown in fig 1 and described in table 1. It should be

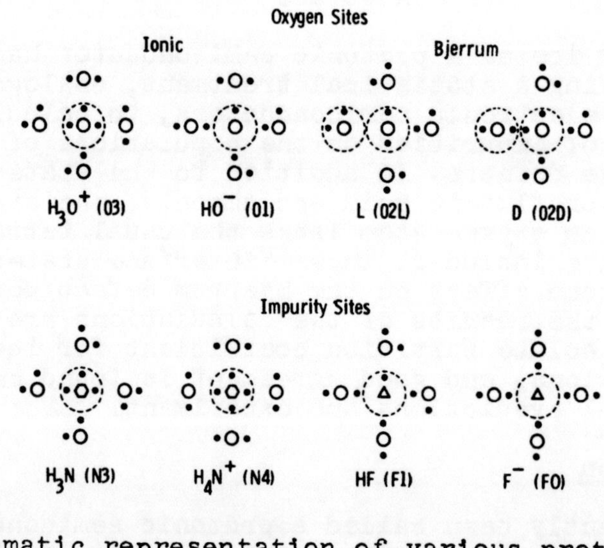

Fig.1 Schematic representation of various proton configuration states. Protons are represented by solid dots. The configuration states are circled, and the state designations are given in parentheses

Table I: Description of Configuration States

Name of State	Designation	Description of State
Oxygen one	O1	One proton near an oxygen site. No L or D defects terminating at the site. An hydroxyl ion state.
Oxygen two	O2	Two protons at an oxygen site in normal bonding.No L or D defects terminating at the site. The normal configuration for H_2O in ice.
Oxygen-two-D	O2D	Two protons at an oxygen site with one in a D defect. Two such states for each D defect.
Oxygen-two-L	O2L	Two protons at an oxygen site which is the terminus of an L defect. Two states for each L defect.
Interface Oxygen-two	IO2	Two protons at an interface oxygen site. One proton in a hydrogen bond,the other "dangling".
Oxygen-three	O3	Three protons at an oxygen site whith no L or D defects present. The hydronium ion state.
Fluorine-zero	FO	No protons at a substitutional fluorine site. A fluorine ion site.
Fluorine-one	F1	One proton at a fluorine site with no L or D defects present.
Nitrogen-three	N3	Three protons at a nitrogen site with no L or D defects present. Neutral ammonia.
Nitrogen-four	N4	Four protons at a nitrogen site with no L or D defects present. An ammonium ion.

emphasized that these descriptions are idealized representations of a complex reality; however, they provide a basis for writing the equations of constraint needed for the extremum problem.

For purposes of exposition in this paper , we shall consider ice doped with H_3N and HF (or HCl). It will be assumed that every nitrogen substituting for an oxygen in the lattice introduces a Bjerrum D defect and every fluorine an L defect. The appropriate configuration state is fairly obvious from corresponding chemical description. In addition to the obvious states, however, we also consider what we call "interface states" in order to illustrate the versatility of the approach.

At free surfaces and at some types of lattice defects such as vacancies and dislocations, there should be oxygen atoms lacking the usual tetrahedral coordination; in this paper these are called interface oxygen atoms. Some of the protons associated with these interface sites may not be involved in hydrogen bonds with neighboring sites. In such an instance, we shall consider this to be an IO2 configuration; if there are two protons at an interface oxygen and both are in hydrogen bonds, then we shall assume that we have a normal O2 state. Such a simplified model is justified only by lack of data to support a more complex picture. It would be naive to expect quantitative results from so simple a model; however, a qualitative description of the role played by these interface states may be developed.

In the present simple model of interface states, the most important interaction is with the Bjerrum defect population. Fig.2 illustrates the two configurations to be found at interface sites and also how Bjerrum defects can be generated or annihilated at the sites. In the bulk, Bjerrum defects can be generated only as complementary pairs; at the interface sites, single species can be produced.

III. Impurity Statistics

In this section, we will outline the development of the statistical treatment. Since ice is essentially incompressible, we minimize the Helmholtz free energy to determine the equilibrium populations in the system. We must write:

$$A = A_O + A_N + A_F + \Sigma_i A_i \qquad (1)$$

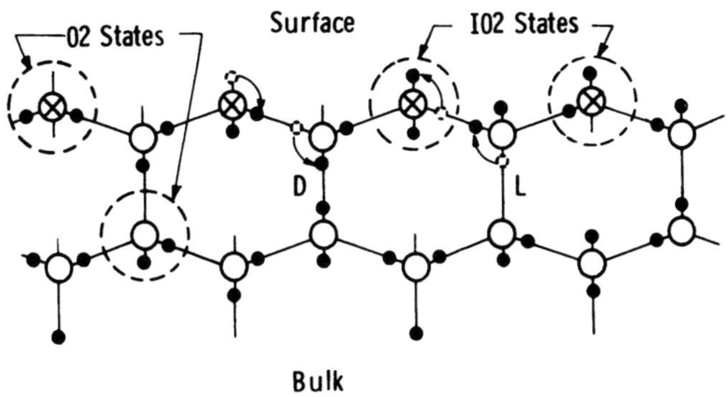

Fig.2 Schematic representation of interface states at a surface. Two I02 states are shown at the surface and 02 states are indicated both at the surface and in the bulk. The interface oxygen atoms are distinguished by a cross. Generation of an L and a D defect is shown; the broken dots represent prior positions of the protons and the arrows indicate the proton motion

where A_O, A_N, and A_F depend only on the total number of oxygen, nitrogen and fluorine atoms in the system; the entropy of mixing of the atoms over the lattice sites is the principle non-linear contribution to these terms. The A_i's in the summation are the contributions to the free energy from the ensembles of the various configuration states and may be written:

$$A_i = -kT \ln Z_i \tag{2}$$

where Z_i is the canonical partition function for the i-th kind of state. If the n_i members of the ensemble are only weakly interacting,

$$\ln Z_i = n_i \ln z_i + \ln W_i \tag{3}$$

where z_i is the partition function of a single state and W_i is the mixing permutation function of the states over the appropriate chemical sites, e.g. the mixing of all the configuration states involving fluorine among fluorine sites. The single state partition functions z_i are defined by

$$z_i = \Sigma_j \exp(-\varepsilon_{ij}/kT) \tag{4}$$

where the sum is over the internal (quantum mechanical) states of the configuration state and includes all the degeneracies of the state. The ε_{ij}'s are the appropriate energies.

If we now define a specific free energy a_i for the configuration state by $a_i = - kT \ln z_i$, then the system free energy may be written:

$$A = A_O + A_N + A_F + \Sigma_i n_i a_i - kT \Sigma_i \ln W_i . \quad (5)$$

Since the number of defect oxygen states is small compared to the total number of oxygen sites, the W_i's for the O1, O3 and IO2 states may be written as

$$W_i = \frac{N_O!}{n_i!(N_O - n_i)!} \quad (6a)$$

for the oxygen states and similarly for the IO2 states. The O2D and O2L states mix as pairs giving

$$W_i = \frac{(N_O/2)!\ 2^{n_i}}{(n_i/2)!\ (N_O - n_i)/2!} . \quad (6b)$$

The fluorine and nitrogen states mix simply as:

$$W_i = \frac{N!}{\pi_i(n_i!)} . \quad (6c)$$

Before the free energy can be minimized, several equations of constraint are needed. First, electrical neutrality is required which may be expressed by requiring the total number of protons to be the same as if the crystal was composed of H_2O, H_3N and HF molecules. This constraint is introduced by the Lagrange multiplier μ. The second constraint states that Bjerrum defects must be produced as pairs in the bulk, may be produced singly at interfaces and that one appropriate defect is introduced for each impurity atom. This constraint is associated with the Lagrange multiplier λ. Finally, the sum of the states associated with a given impurity must be the same as the total number of that impurity atom in the lattice. These conditions are expressed with the use of the Lagrange multipliers γ_N and γ_F.

The equation which is finally developed is

$$\begin{aligned} A = {} & A_0 + A_N + A_F + N_0 a_{02} - n_{01}(a_{02} - a_{01}) + n_{02D}(a_{02D} - a_{02}) \\ & + n_{02L}(a_{02L} - a_{02}) + n_{I02}(a_{I02} - a_{02}) + n_{03}(a_{03} - a_{02}) \\ & + n_{F0}a_{F0} + n_{F1}a_{F1} + n_{N3}a_{N3} + n_{N4}a_{N4} \\ & - kT\Sigma_i \ln W_i - \mu(n_{03} - n_{01} + n_{F1} + 3n_{N3} + 4n_{N4} - 3N_N - N_F) \\ & - \lambda(n_{02D} - n_{02L} + 2n_{I02} - 2N_N + 2N_F - N_{I0}) + \gamma_F(n_{F0} + n_{F1} - N_F) \\ & + \gamma_N(n_{N3} + n_{N4} - N_N) \end{aligned} \quad (7)$$

The populations of the various states are evaluated in the usual way by equating to zero the partial derivatives of A with respect to the n_i's. The resulting equations have a Fermi-like form, however in most cases, the populations are such that they may be written approximately as Boltzmann distributions. The four electrically active oxygen states may thus be written: ($\beta = 1/kT$)

$$n_{O1} = N_O \exp\{(a_{O2} - a_{O1} - \mu)\beta\} \qquad (8a)$$

$$n_{O3} = N_O \exp\{(a_{O2} - a_{O3} + \mu)\beta\} \qquad (8b)$$

$$n_{O2D} = 4N_O \exp\{-2(a_{O2D} - a_{O2} - \lambda)\beta\} \qquad (8c)$$

$$n_{O2L} = 4N_O \exp\{-2(a_{O2L} - a_{O2} + \lambda)\beta\}. \qquad (8d)$$

The population of interface states may not be small with respect to the number of interface oxygen atoms, so the Fermi-like form must be retained:

$$n_{IO2} = N_{IO} \{(1 + \exp(a_{IO2} - a_{O2} - 2\lambda)\beta\}^{-1}. \qquad (8e)$$

After evaluating the Lagrange multipliers γ_N and γ_F, the populations of the ionized impurity states is found to be:

$$N_{FO} = N_F \{1 + \exp(\mu - a_{FO} - a_{F1})\beta\}^{-1} \qquad (8f)$$

$$n_{N4} = N_N \{1 + \exp(a_{N4} - a_{N3} - \mu)\beta\}^{-1} \qquad (8g)$$

which may be put in Boltzmann form if the populations are small.

The Lagrange multiplier μ, which is, in fact, the proton chemical potential, may be evaluated from the distribution functions and the equations of constraint. The result is : $\exp \mu\beta =$

$$= \left[\frac{1 + \frac{N_F}{N_O} \exp\left[(a_{F1}-a_{FO}) - (a_{O2}-a_{O1})\right]\beta}{1 + \frac{N_N}{N_O} \exp\left[(a_{O3}-a_{O2}) - (a_{N4}-a_{N3})\right]\beta} \right]^{1/2} \exp \hat{\mu}\beta \qquad (9)$$

where $\hat{\mu} = \{(a_{03}-a_{02}) + (a_{02}-a_{01})\}/2$ is the value of μ when $N_N = N_F = 0$; $\hat{\mu}$ is the intrinsic value of μ in semiconductor parlance. Eq. (9) may be simplified when only one impuritiy is present or when the impurity content is large enough that unity may be neglected in the numerator and/or the denominator. If the simplified expressions for μ are explicitly used in the distribution functions, the resulting equations resemble the more familiar formulae derived from the mass action law. The present form, however, is particularly clear in exhibiting the concentration dependence under various regimes of N_N and N_F.

The Lagrange multiplier λ, which we shall call the bond defect parameter, plays a role for the Bjerrum defects analogous to that of μ for the ionic defects. Obtaining an analytical expression for λ in the general case is complicated since a cubic equation results. We shall therefore first consider the situation where interface effects are negligible; the resulting quadratic is easily solved. If we let $x = \exp(\lambda-\hat{\lambda})\beta$ where $\hat{\lambda}$ is the value of λ in pure ice, then

$$n_{02D} = \hat{n}_B\, x \tag{10a}$$

and

$$n_{02L} = \hat{n}_B / x \tag{10b}$$

where

$$n_{02D} = n_{02L} = \hat{n}_B \quad \text{when } N_N = N_F .$$

Using the distribution functions and the equations of constraint,

$$x = \frac{N_N - N_F}{\hat{n}_B} + \sqrt{1 + \left(\frac{N_N - N_F}{\hat{n}_B}\right)^2} \; . \tag{11}$$

In the limiting cases, this equation reduces to the more familiar form /2, 3/.

Our model for the interface states is so oversimplified, that the general solution of the cubic would be of dubious value. The qualitative behavior can be obtained by considering a special case. Consider the case where in pure ice $\hat{n}_{02L} = \hat{n}_{02D}$ which also implies $\hat{n}_{I02} = 1/2\, N_{I0}$. It might be noted that such conditions are unlikely since they require that $2(a_{02D}-a_{02L}) = a_{I02}-a_{02}$; we may, however, make the assumption for purposes of illustration. The appropriate solution for x is shown graphically in fig.3 in terms of a reduced impurity

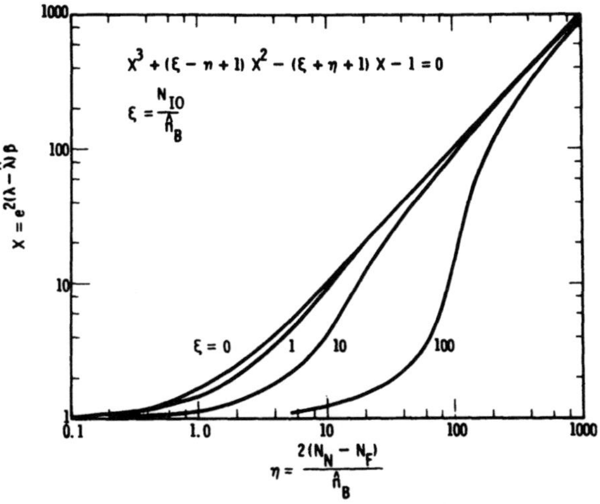

Fig.3 Curves for determining the influence of interface states

content, $(N_N - N_F)/\hat{n}_B$, and a reduced number of interface oxygen atoms, $N_{IO}/\hat{n}_B$. The curve $N_{IO} = 0$ corresponds to eq. (11); all the other curves lie below it and approach it at low and high concentrations. Physically, some of the Bjerrum defects introduced by the impurities are annihilated by the interface states and do not contribute to the population of mobile defects in the bulk. This annihilation of impurity induced Bjerrum defects is accompanied by a change in the net polarisation of the interface region. If there were no net polarization in pure ice, then the addition of fluorine would cause a negative polarization while the addition of nitrogen would cause a positive polarisation.

If only the interface states at outside surfaces were considered, then their effect would probably be negligible for sample sizes of the order of centimeters, a common case. The surface to volume ratio would be much increased if the sample contained many voids and the effect of these surface sites could become important. Finally, one must take into account the states which accompany dislocations and lattice vacancies. The latter may be quite numerous, perhaps 10^{17} cm^{-3} or more near the melting temperature. If the concentrations were indeed this high, then their contribution to the bulk Bjerrum defect population could be pronounced and one species of defect might be decidedly in the majority. The generation or annihilation of defects at a lattice vacancy is similar to the "invested vacancies" proposed by Kopp /4/; the present formalism provides a simple and unified method for analyzing the problem.

IV. Solute Partitioning

As an application of the formalism we have developed, we may calculate the influence of the hydrogen bonding defect populations on the partitioning of a solute between a solid and liquid solution in equilibrium. Following the established practice, we define the partition coefficient of a solute by

$$k = X_s / X_\ell \tag{13}$$

where X_s and X_ℓ are the atom fractions of the solute in the solid and liquid phases respectively. Continuity of chemical potential leads to the equation

$$k = \exp \beta \left\{ \frac{\partial G^{\bullet}(\ell)}{\partial N} - \frac{\partial G^{\bullet}(s)}{\partial N} \right\}. \tag{14}$$

Where $G^{\bullet}$ is the Gibbs free energy except for the entropy of mixing of the solute atoms, i.e., $G^{\bullet} = G + TS_{mix}$. This entropy of mixing refers to mixing of the solute atoms among the lattice sites but does not refer to the mixing of the configuration states considered earlier.

The contribution from the term involving the liquid phase is just f, the activity coefficient of the solute. The contribution from the solid may be calculated from eq. (7). If mechanical or electrostatic interactions between the configuration states are neglected, which seems reasonable, then the contribution from A_O, A_N and A_F to eq . (14) is just a constant factor. The use of the Lagrange multiplier γ_F and γ_N in the balance of eq.(7) allows us to consider the n_i's and N_i's to be independent variables, and the differentiation is straightforward. In the case where HF (or HCl) is the only solute, evaluation of eq. (14) gives

$$k = \text{const. } f \exp (2\lambda B) \tag{15a}$$

$$= f\, k_o\, x \tag{15b}$$

where f is the activity coefficient of the solute in the liquid, k_o is the limiting value of the partition coefficient at infinite dilution, and x is the function defined by eq.(11). Manipulation of eq.(15) gives the partition coefficient in terms of the solute concentration in the liquid:

$$k = \frac{f\ k_o}{\sqrt{1 + f\ k_o \left(\frac{N_F(\ell)}{\hat{n}_{DL}} \right)}} \tag{16}$$

where we have introduced $n_{DL} = \hat{n}_B/2$, the concentration of L or D defects in pure ice.

A comparison of the predictions of eq.(16) with experiment is shown in fig.4. The data represents some initial

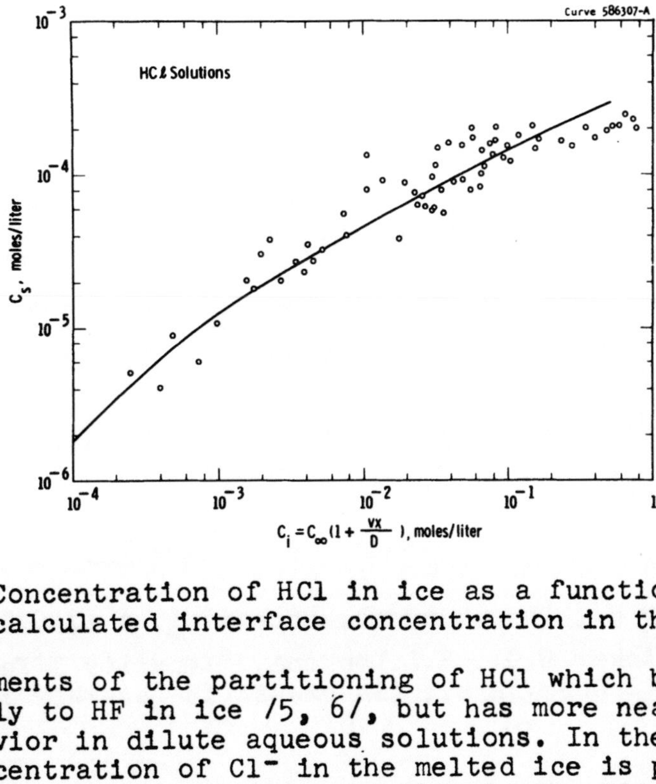

Fig.4 Concentration of HCl in ice as a function of the calculated interface concentration in the liquid

measurements of the partitioning of HCl which behaves similarly to HF in ice /5, 6/, but has more nearly ideal behavior in dilute aqueous solutions. In the figure, the concentration of Cl^- in the melted ice is plotted against the calculated concentration in the liquid at the interface. The solid curve represents the concentration in the solid calculated from the partition coefficient given by eq.(16). The parameters $k_o = 0.02$ and $\hat{n}_{DL} = 1.25 \times 10^{-5}$ M ($= 7.5 \times 10^{15}/cm^3$)were used. These values seem eminently reasonable on the basis of the partitioning of very dilute HF solutions/7/ and other estimates for the Bjerrum defect population in pure ice /8/.

V. Summary

We have briefly presented a statistical mechanical alternative to the mass action description of solute interactions with electrically active defects in ice; a more detailed analysis will appear elsewhere /9/. A simplified model of interface states has been treated and their interactions with the Bjerrum defect populations analysed. Finally, the model has been used to calculate the concentration dependence of solute partitioning during the freezing of a dilute solution.

Acknowledgements

One of us, R.G.S., was supported in this work by the Office of Saline Water, U.S. Department of the Interior, Contract 14-01-0001-1336.

References

/1/ Shockley W., J.T.Last, Phys.Rev. 107, 392 (1957)

/2/ Levi L., O.Milman, E.Suraski, Trans.Far.Soc. 59 2064 (1963)

/3/ Jaccard C., Phys.kondens.Materie 2, 143 (1963)

/4/ Kopp M., U.Pittsburgh, Department of Physics, Progress Report No. 19, May 1967

/5/ Gross G.W., Science 138, 520 (1962)

/6/ Young I.G., R.E. Salomon, J.Chem.Phys. 48, 1635 (1968)

/7/ Jaccard C., L.Levi, Z.Angew.Math.Phys. 12, 70 (1961)

/8/ Gränicher H., Phys.kondens. Materie 1, 1 (1963)

/9/ Seidensticker R.G., R.L.Longini, "Impurity Effects in Ice", J.Chem.Phys., (in press (15 December 1968)).

CHARGE AND POLARIZATION STORAGE IN ICE CRYSTALS *

S.Mascarenhas

Physics Dep., Escola Engenharia

S. Carlos, S. P., Brazil

Abstract

Charge storage in pure or doped ice crystals has been observed to occur under two conditions: a) during phase changes through the so called Costa Ribeiro effect /1/ in which a transient charge storage is observed and b) by the formation of thermo-electrets under the simultaneous action of electric and thermal treatments /2/. Polarization storage and release of charge has also been observed and is attributed to ferro-electricity /3/. In this communication we report on our results on the Costa Ribeiro effect studied under controlled conditions of phase-change and on the properties of the ice thermoelectret obtained with pure water and HF solutions. In the first case the charge storage is studied under dynamical conditions of phase change and in the second case by the thermally stimulated current technique. We also report on some recent results by Lôbo /4/ on a statistical model that predicts the thermoelectricity of hexagonal ice and some experimental results on HF-doped ice regarding to its ferroelectric behavior.

I. Introduction

Charge and polarization storage in ice is observed to occur for at least three types of phenomena: electret

* Work done under ONR contract and partial support of Nat. Res. Council of Brazil and Foundation for Research of S. Paulo

formation /1/, the Costa Ribeiro effect /2/ and ferroelectric behavior of ice /3/. We shall mention briefly the characteristics of the three effects and will then dedicate the rest of the paper to a description of our results on: the Costa Ribeiro effect in ice and electret formation and behavior of pure and HF doped ice. The Costa Ribeiro effect leads to the appearance of electric potentials during growth of ice from pure water and from solutions. The effect was studied by many investigators /5,6/ and it may be important to explain atmospheric electricity. As we will describe in more detail later, the current appearing during the phase change is proportional to the rate of phase change. Potentials of several hundred volts can be observed during the solidification of aqueous solutions. The same effect has been observed for other phase changes, including sublimation /6/ of other substances. With naphtalene, anisotropic behavior was also verified /7/. A basic difference however with ice is that the effect is not completely reversible in the sense that the current does not revert to the same values during melting. Physical reasons for this effect will be given in later sections of this paper.

Electret behavior of "pure" ice was first observed by Gelin /2/. The behaviour of pure and HF doped ice was investigated by Mascarenhas and Argüello /2/ and a more complete understanding of the nature of the heterocharge was possible. This will be described below. Essentially, electret behavior of ice may be analyzed by depolarization peaks as first used by B.Gross /8/ in investigating the classical carnauba electret. It was shown that the electret behavior of pure and HF doped ice is due to space-charge formation and not to dipole orientation. This is one of the reasons why we distinguish charge and polarization storage.

Finally, another way to storage polarization in ice is via the ferroelectric behavior first discovered by Riehl and collaborators /3/. In the final section of this paper we shall mention briefly some experiments conducted in our laboratory with HF doped samples to study its possible influence on the ferroelectric transition temperature. We also discuss a theoretical model of Lôbo and Hipólito in which a second order ferroelectric transition is predicted for hexagonal ice.

II. Investigations on the Costa Ribeiro effect in ice

The experimental system used was developed to provide a controlled rate of phase change and this may be useful

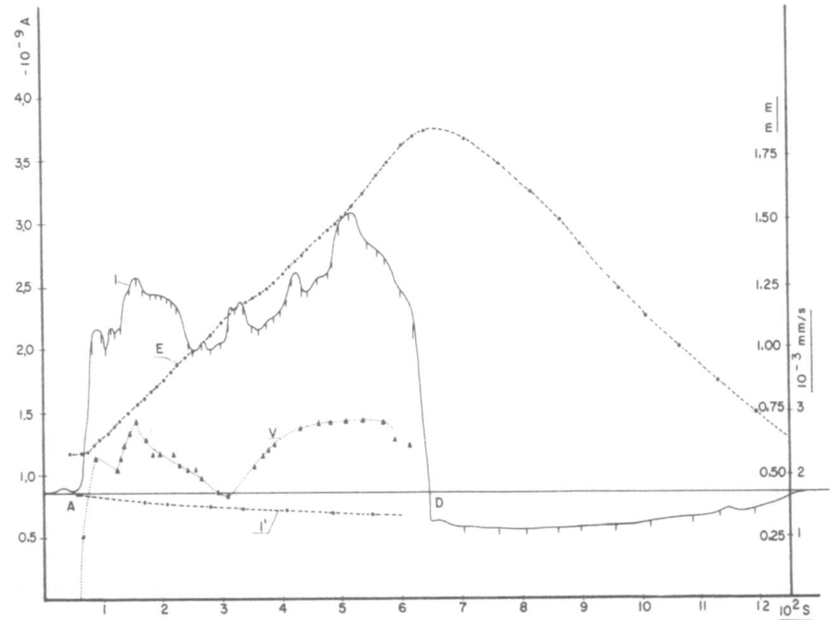

Fig 1 The electrical current during the phase change. I - total current as observed in the electrometer, with time marks superimposed; E - curve of thickness of solid deposit as the phase change proceeds; V - curve of rate of motion of solid-liquid interface (in mm per sec) showing that the structure of curve I can be interpreted as due to kinetics of phase change; I' - background current obtained with the interface still as a function of position

for other types of investigations with ice. It is described elsewhere /1/ in detail. It suffices to say that using that system, it was possible to measure simultaneously the current during phase change and the rate of phase change. Typical results are shown in fig.1. It is seen that the current is proportional to the rate of phase-change. Thus, there is no meaning in talking about freezing potentials during the phase change in the way other authors wanted to study. The potentials appearing are characteristic only of the kinetics of solidification. Another important result is that the effect is not reversible as it is for naphtalene and other good insulators, because during melting the current is very small and does not depend on the speed of melting. In the samples investigated ($3x10^{-6}(\Omega cm)^{-1}$) ammonia free, doubly distilled water), the ice phase was at a negative potential relative to the water phase during solidification. It is known, however, that impurities may change the sign and the value of the effect. Another important quantity that may be obtained from this type of data is the total charge Q which is given by

$$Q = \int I dt.$$

It is found that between 1 and 10 μCb of charge are associated with the change of 1 g of water from the liquid to the solid phase. This is a rather large amount of charge, capable in principle of explaining atmospheric phenomena, though a satisfactory mechanism has not yet been given.

The nature of the Costa Ribeiro effect seems to vary for different materials. For naphtalene it is most probably an electronic effect. However, for ice it is most probably related to selective trapping of ions in the solid phase, a suggestion first advanced by Workman and Reynolds /6/. B.Gross has developed a general model /9/ for the Costa Ribeiro effect which seems adequate to explain at least the general features of the phenomena. However, for the case of ice the Gross model cannot be applied because in the solution of his basic equations he assumes the solid to have infinite resistance which is obviously not the case for ice. A general picture may be offered for an ionic effect: during solidification the motion of the interface causes a continuous trapping of charge in the solid deposit. This charge does not dissipate immediately but has a certain relaxation time depending on the resistivity and other parameters of the solid phase. We will thus be led to a space charge distribution in the ice phase. Currents in the external circuit are then obtainable in principle by solving for the time-dependent field of this distribution. With this general picture in mind, we can now offer a simple explanation for the lack of reversibility in melting: Due to the small relaxation time of charges in ice these leak away and we do not observe the same charge release during melting. Evidence for space charge formation during the Costa Ribeiro effect is given by another class of experiments. We have measured the specific charge (charge per unit mass) as a function of interelectrode distance d. We did not find a uniform charge distribution as expected for a dipole-effect but rather a charge distribution that depends on the thickness of the sample (see /1/),

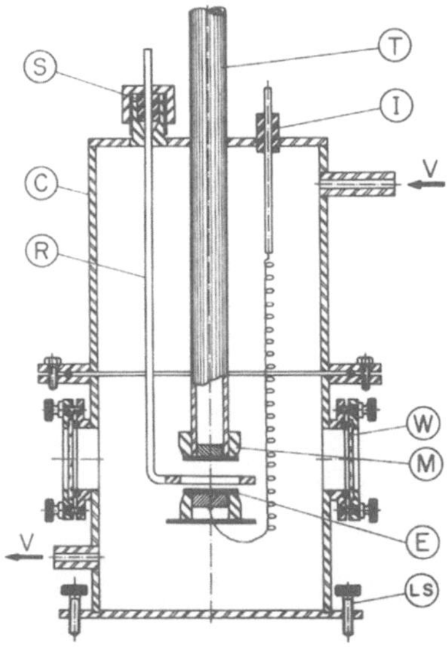

Fig.2 Apparatus for single-crystal growth and electret preparation and measurement. T, stainless steel tube; I, teflon insulator; S, vacuum seal and movable stopper; C, external metal cylinder; W, optical window; M, magnet; E, electrode (platinum); LS, levelling screws; V, vacuum outlets

III. Investigations on the pure and HF doped ice thermoelectret

The experimental set-up is that of fig.2. A single crystal is grown between the platinum electrodes E, held to each other by magnets M. Quality and single crystallinity of the sample can be observed through window W. Potentials are applied and currents measured with the help of leads I. Vacuum or argon atmosphere can be used during cooling and warming of samples. The stopper S keeps the electrodes separated by a chosen distance before solidification. It is then moved up for electrical measurements. The thermoelectret is prepared in the following way: A field is applied at a high polarization temperature T_p, usually $-10^{o}C$. The sample is then cooled in the field to a lower temperature where the polarization can be "frozen in". The sample is then warmed while no field is applied and the depolarization current is measured continuously as a function of temperature. A

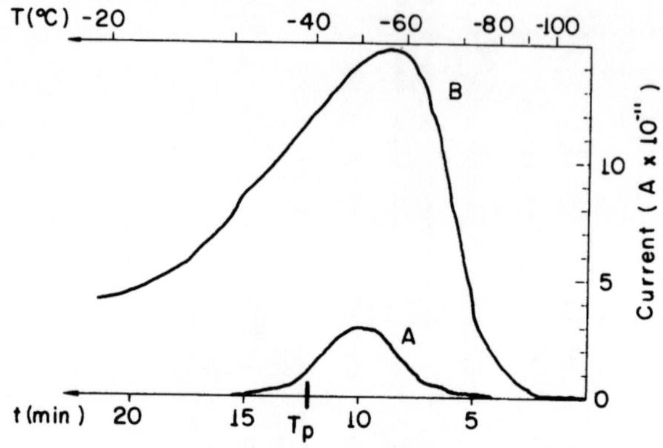

Fig.3 Electrical glow-peaks due to charge release in the ice thermoelectret. A, pure ice; B, HF - doped ice

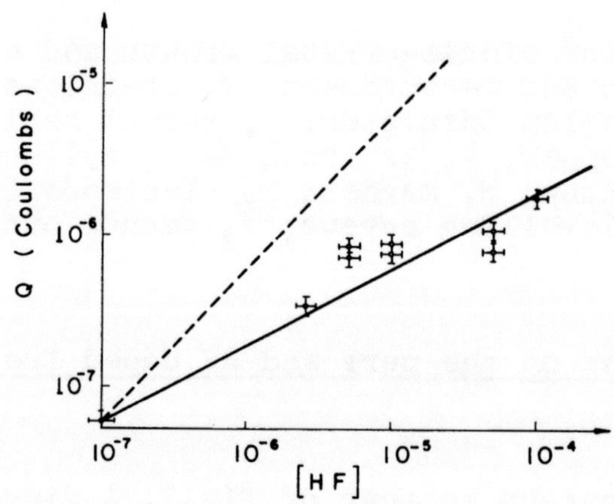

Fig.4 Dependence of released charge Q on HF doping. A square root-dependence is found. Dashed line indicates a linear function (see text for discussion)

very well defined "glow-peak was found for pure ice at around -55°C (fig.3, curve A). At first we were inclined to think of a dipole-orientation effect. In order to check this and other points, we investigated HF-doped ice. Doping resulted in an enormous increase in the effect, as shown by curve B. We thus investigated the dependence of Q, the total released charge, as a function of HF concentration (fig.4). A square root dependence was found. This points to a possible influence of ionic defects rather than that of Bjerrum defects. A frozen-in space-

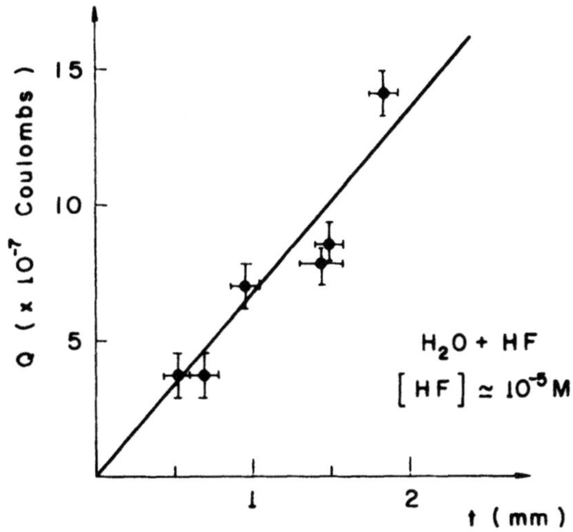

Fig.5 Dependence of released charge Q on thickness of sample, for constant field and concentration

-charge could then be responsible for the effect. In order to further investigate this point we studied the influence of the thickness of the samples on the released charge with equal applied fields. If dipoles were the cause of the polarization storage, a thickness independent effect should be expected; with space-charge, a linear dependence might be found. Changing the crystal thickness from 0.5 to 2 mm resulted in charge storage which was not constant but had a linear dependence on thickness as shown by fig.5. Finally, as is found in ionic thermal current /10/ or thermally stimulated current experiments /11/, a space-charge peak has its temperature T_m strongly dependent on polarization temperature. This is not the same for dipoles because a definite activation energy is involved in the process and this determines T_m. This was also found experimentally. Values of stored charge in ice thermoelectrets are enormous, of the order of 10^{-5} Cb/g for doped crystals.

We thus found satisfactory experimental evidence to be able to say that most probably the ice thermoelectret is not due to dipoles but to space-charge formation which is probably related to the high mobility of the protonic carrier as compared to the OH^- ion.

IV. Ferroelectricity of ice

The discovery of the Munich group of the ferroelectricity of ice was questioned as a still unsettled point /12/. Recently, R. Lôbo and O. Hipólito /13/ ouf our group proposed a simple statistical model in which they show that a second-order ferroelectric phase transition is expected for hexagonal ice. Starting from the existence of the molecular dipole and the knowledge of the interatomic structure vectors $\vec{r}_{ij}$ they calculate a tensor

$$\underline{k}_i = \sum_{i \neq j} \frac{(\underline{I} - 3\, \hat{r}_{ij}\, \hat{r}_{ij})}{r_{ij}^3}$$

related to the internal field. It is found that nearest neighbors of a dipole are not important. Next nearest neighbors assume an important role in the calculation of the internal field. Neglecting tunneling, the number $\bar{n}$ of molecules with dipole components oriented along the c-axis was calculated as a function of temperature. Lôbo and Hipólito can solve selfconsistently for a Curie temperature $T_c \cong 100^\circ K$ using direct experimental data and assuming that all water dipoles are involved. Defects or impurities should then only have a catalytic action as suggested by Onsager. Ferroelectricity then appears to be an intrinsic property of hexagonal ice. In order to check on this point, we repeated essentially Riehl's experiment and found no variation of the inversion current temperature range (associated by those authors to the ferroelectric transformation) by changing the concentration of HF in ice by a factor of 10.

We are presently investigating hysteresis and ferroelectric induced birefringence with a special photoelastic technique developed at S.Carlos /14/, to further investigate the effect. Other electro-optical and mechanical effects are also under study in order to clarify the situation.

Conclusions

The main result of all these investigations is that large amounts of charge (of the order of 10^{-5} coulombs or more) can be trapped and stored for a long time in ice if the temperature is low enough. The level of understanding of the various effects and main results are:

a) Costa Ribeiro effect - probably related to ionic carriers selective trapping during growth. B.Gross model with modifications could perhaps be applied.

b) Thermoelectret - was found by the results of the present author to be due to space-charge formation and to be linked to ionic defects.

c) Ferroelectricity - probably an intrinsic collective property, independent of defects or impurities that merely catalyze the transformation.

Acknowledgements

We thank Dr.R.Lôbo for stimulating discussions and O. Hipólito for help with measurements. Most of the work reported here was done in collaboration with Drs. C.Arguello and D.Pinatti.

References

/1/ Pinatti, D., S.Mascarenhas, J.Appl.Phys. 38, 2648 (1967)

/2/ Gellin, H., R.Stubbs, J.Chem.Phys. 42, 967 (1965); see also C.Arguello and S.Mascarenhas, Bull.Am. Phys.Soc. 12, 4, 558 (1967) and J.Electroch.Soc. 115, 386 (1968)

/3/ Dengel, O., U.Eckener, H.Plitz,N.Riehl, Phys.Letters 9, 921 (1964)

/4/ Lôbo, R., O.Hipólito (in publication)

/5/ Costa Ribeiro, S., thesis, Univ. of Brazil (1945): see also Ann.Brazil.Acad.Ci. 22, 325 (1950)

/6/ Workman, E., S.Reynolds, Phys.Rev. 78, 254 (1950): see also reference /1/ above

/7/ Mascarenhas, S., L.G.Freitas, J.Appl.Phys. 31, 1685 (1960)

/8/ Gross, B., S.Denard, Phys.Rev. 67. 253 (1945)

/9/ Gross, B., Phys.Rev. 94, 1545 (1954)

/10/ Jannuzzi, N., S.Mascarenhas, J.Electrochem.Soc. 115, 382 (1968)

/11/ Sinencio, F., S.Mascarenhas, B.Royce, Phys.Letters 26A, 70 (1967)

/12/ Onsager, L., Ferroelectricity of ice? in Ferroelectricity, E.Weller (editor), Elsevier Publ.Co., Amsterdam, 1967

/13/ Lôbo, R., O.Hipólito, On the Ferroelectricity of Ice (in publication)

/14/ Lüty, F., S.C.Ribeiro, S.Mascarenhas, V.Sverzut. Phys.Rev. 168, 1080 (1968).

ELECTRICAL POLARIZATION EFFECTS IN PURE AND DOPED ICE AT LOW TEMPERATURES

P.G. Bishop, J.W. Glen

Department of Physics

University of Birmingham / England

Abstract

If a single crystal of ice is cooled to about -190°C in the presence of an electric field of about 10 kV/cm, then, on subsequent warming up in the absence of an electric field, a current is observed which is related to the direction of the applied field during cooling. This current has maxima at reproducible temperatures, and in some cases the currents associated with the maxima can be reversed in sign relative to one another if the applied field had been reversed during cooling. This phenomenon is presumably due to thermally activated charge trapping centres. Crystals heavily doped with HF exhibit a large maximum at 107°K which is not observed in pure crystals. By applying activation theory to this HF maximum an activation energy has been obtained.

Single crystals of ice frozen from deionized water exhibit several thermostimulated current peaks within the temperature range from 120° to 200°K (fig. 1). These peaks were obtained by applying a high electric field (in the region of 10 kV/cm) to the crystal as it was cooled to liquid nitrogen temperature and then measuring the output current at zero applied field on a Keithley 610B electrometer as the crystal was allowed to warm at about 6 deg/min.

These thermostimulated peaks proved to be essentially independent of each other, since by selecting the temperature range over which the field was applied during cooling a single peak could be obtained on heating(fig2).

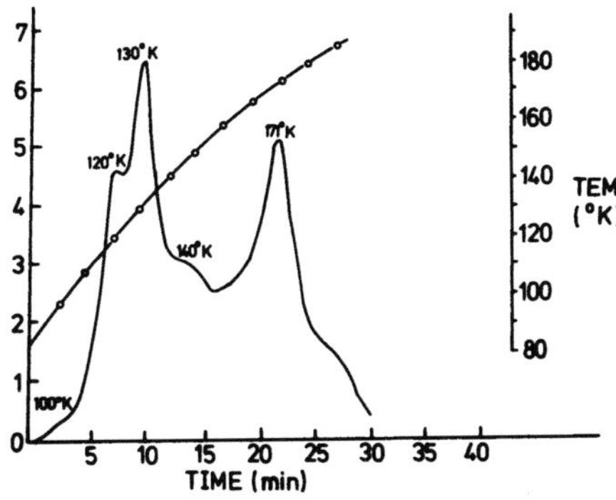

Fig.1 Thermostimulated current released on warming a "pure" ice specimen. The circles indicate temperature as a function of time, the peaked curve, transcribed from pen-recorder output, gives the current in units of 10^{-10}A. The electric field was applied throughout the time of cooling

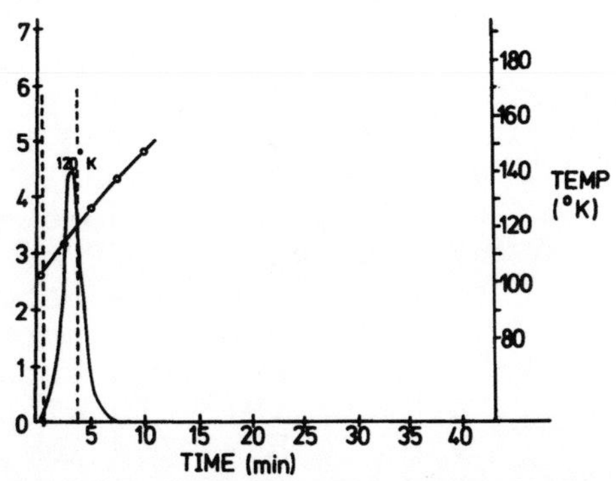

Fig.2 Thermostimulated current released on warming a "pure" ice specimen to which the electric field had been applied only when cooling between the temperatures indicated by the vertical dashed lines. The output current, in units of 10^{-10}A, is virtually restricted to the same temperature range

It was even possible (fig.3) to reverse the sign of adjacent peaks of current by reversing the electric field applied in the corresponding temperature range during cooling.

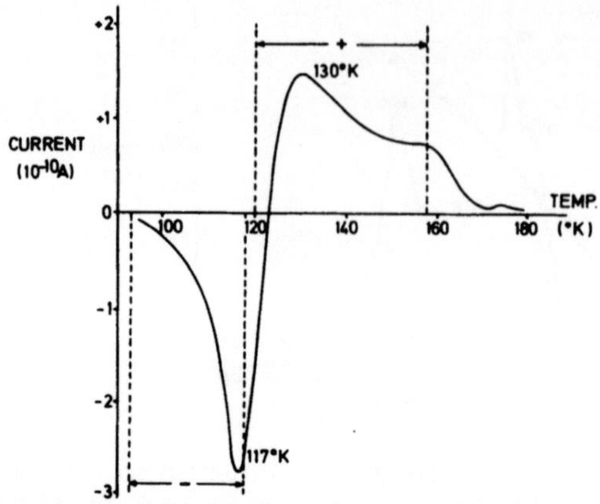

Fig.3 Thermostimulated current released on warming a "pure" ice specimen which was cooled with the field in one direction in the upper temperature range indicated, and with the field in the opposite direction in the lower temperature range. Note that the sign of the current released (with zero applied field) on warming changes sign at about the same temperature

However these peaks have proved difficult to study individually due to mutual interference, and they varied in magnitude for different samples, probably due to variations in residual impurity level. However it was discovered that by doping the ice with HF an additional peak was produced at about 107°K, a temperature lower than that of any of the peaks found for "pure" ice. Tests down to liquid helium temperatures showed that there were no major peaks below 107°K in either pure or HF-doped ice. The 107°K peak in HF-doped ice had a larger integrated current than any of the other peaks (possibly because more HF can be dissolved in ice than any other impurity), and measurements of this integrated current in a given crystal showed that the charge stored was proportional to applied voltage (fig.4). It was also found that for similarly sized specimens with different HF concentration the charge was approximately proportional to the 0.6 power of the HF concentration as determined colorimetrically. This result mightbe connected with the fact that, at higher temperatures, the number of positive ionic defects in HF-doped ice is proportional to the square root of the HF concentration (Gränicher /1/) and this type of defect is mobile at any temperature due to quantum mechanical tunnelling.

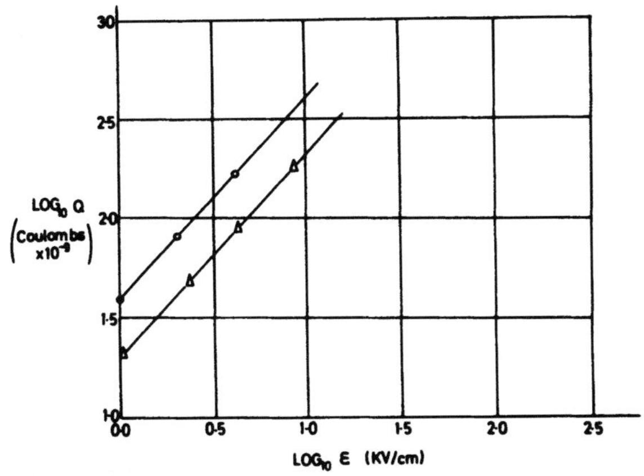

Fig.4 Logarithmic plot of total charge released during warming of HF-doped ice crystal against the electric field applied during cooling. Circles correspond to a specimen 7 mm thick, triangles to a specimen 3 mm thick cut from the same single crystal

Further experiments were performed on disc-shaped single ice crystals 3.3 mm thick in which one or both sides were doped with HF by applying a small amount of 1% HF solution which was then allowed to diffuse through the crystal at a temperature of -27°C. As can be seen from Figure 5, the magnitude of the HF peak increased as HF diffused through the crystal. This indicates that the polarization is a bulk property of the crystal unrelated to the surface, for when all the HF was initially

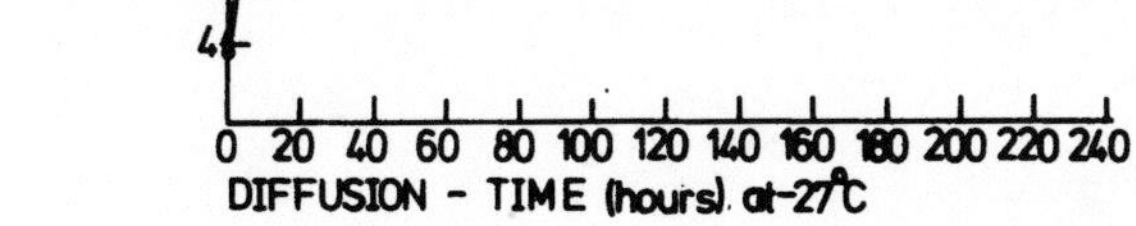

Fig.5 Total charge stored in ice crystals into which HF was diffused from one or both sides as a function of time of diffusion

near the surface only a relatively small peak was observed. Another interesting feature of the single-sided diffusion experiments was that the magnitude of the charge stored (integrated current) was independent of the direction of the field applied during cooling. Since the addition of HF creates charged point defects, one might have expected an asymmetry in the amount of charge stored or an asymmetry in peak temperature depending on the direction of the field, contrary to what was observed. It would therefore seem reasonable to regard the polarization effect as being localized to the HF-doped region.

If we postulate a dipole theory of polarization, this would account satisfactorily for the linear variation of polarization with applied voltage, and would also explain why the effect was localized in a particular region. However the absolute magnitude of the polarization effect is too large to be accounted for by the addition of impurity dipoles. This can be seen if we calculate the total polarization of the ice lattice. Using a value of 1.85×10^{-27} C·cm for the dipole moment of the water molecule and taking account of the geometrical constraints of the lattice, the maximum surface charge density can be shown to be 10^{-5} C·cm^{-2}.

In our experiments a field of 20 kV/cm typically produces a stored charge density in an HF-doped sample of 6×10^{-7} C·cm^{-2}, which is about 6% of the theoretical maximum and since the HF concentration is approximately 0.1% and the dipole moment of an HF molecule is essentially similar to that of a water molecule (1.9×10^{-27} Ccm), it can be seen that the polarization cannot be due to impurity dipoles alone.

From this discussion we may deduce that the polarization is due to the water molecules themselves. If we assume that free dipoles of dipole moment p are being oriented in a field E, and that they follow a Boltzmann distribution law, then the ratio of the number of dipoles n_1 oriented parallel to the field to the number n_2 oriented antiparallel will be given by

$$\frac{n_1}{n_2} = \exp\left(-\frac{2pE}{kT}\right).$$

Then if $pE \ll kT$,

$$\frac{n_1}{n_1 + n_2} \approx \frac{pE}{kT}$$

therefore the surface charge density σ will be given by

$$\sigma \approx \frac{pE\sigma_{max}}{kT}$$

where σ_{max} is the charge density for complete polarization. Substituting known values of σ/σ_{max}, T and p into this equation, the electric field is found to be E = 160 kV/cm which is eight times the field deduced from the applied potential and the crystal thickness. There are two reasons why this might not be the appropriate field to use; the local field at a molecule is not necessarily equal to the average applied field, and the potential drop across the specimen may not be uniform, but the calculation does show that the polarization frozen in is large, even compared with the dielectric polarization to be expected for free water molecules.

At temperatures well below the peak the logarithm of the current flowing from HF-doped ice cooled in a high field gives a straight line on an Arrhenius plot against 1/T (fig.6). The slope gives an activation energy, the average value of which from several samples is $U = 0.27 \pm 0.01$ eV. An attempt was made to fit the whole thermo-stimulated current curve using simple glow-curve theory (see e.g. Grossweiner /2/). If this is valid the rate of recovery of stored charge Q should be given by

$$\frac{dQ}{dt} = -A\, Q\, \omega \exp\left(-\frac{U}{kt}\right)$$

where $\omega \exp(-U/kT)$ is the probability of dipole activation per unit time.

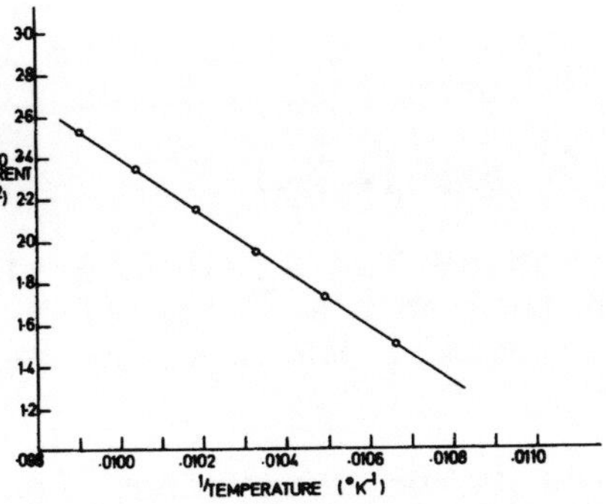

Fig.6 Arrhenius plot of logarithm of current flowing during low-temperature part of thermostimulated peak of HF-doped ice against reciprocal of absolute temperature

This model leads to a rather rapid fall-off in the high-temperature region when using the activation energy from the low-temperature region. By assuming a differential equation of the form

$$\frac{dQ}{dt} = -A\ Q^{\kappa}\ \omega \exp(-\frac{U}{kT})$$

a much better fit for the high-temperature end of the curve can be obtained. By plotting ln(dQ/dt)+ U/kT against ln Q , (fig.7) the best value of κ can be deduced from the gradient; the mean value from several curves is κ = 1.45 ± 0.05.

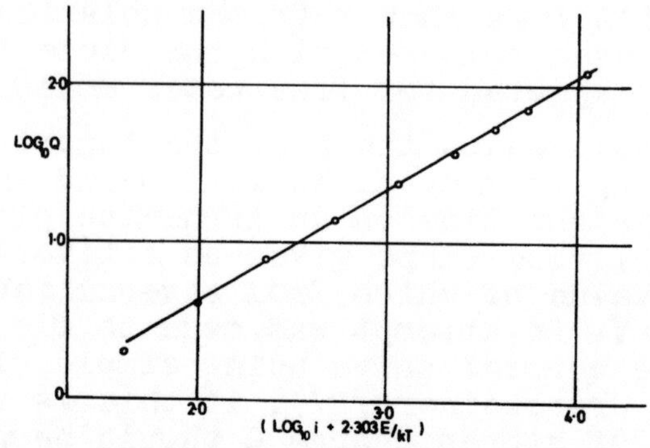

Fig.7 Plot of logarithm of residual charge against log i + 2.303 U/kT to determine parameter κ

The value of κ can be investigated by studying the relation between output current at a given temperature and voltage applied to the specimen during cooling. If it is assumed that the total charge stored Q_o is proportional to V (as has been demonstrated in fig.1) then the initial value of current i_o will be given by

$$i_o = -(\frac{dQ}{dt})_o \alpha\ V^x \exp\ (-\frac{U}{kT})$$

so a plot of log V versus log i_o should yield a straight line of slope x. In this way a value of x was determined for several specimens , the average being

$$x = 1.5 \pm 0.1$$

in agreement with the value determined from the glow curves. The isothermal current values at temperatures well below the peak can also be used to determine the activation energy and give a mean value of

$$E = 0.25 \pm 0.02\ \text{eV}.$$

These thermostimulated current peaks in ice can be compared with similar peaks in other materials such as the alkali halides (Bucci et al. /3/). These are attributed to impurity dipoles, and it is possible that some of the peaks in "pure" ice are of this nature, but as we have seen above, the peak in the HF-doped ice is too large for such an explanation and it seems more reasonable to suppose that the polarization arises from the orientation of water molecules brought about by the migration of electrical point defects (ionic defects or Bjerrum defects). The various peaks in the "pure" ice can then be attributed to the movement and trapping of various electrical point defects, though no specific attribution of the peaks to particular defects can be made at this stage. The fact that the peaks vary in magnitude from specimen to specimen suggests that we are dealing with uncontrolled variable concentrations of trapping centres.

In the case of HF-doped ice, the peak was reproducible in magnitude as well as temperature for a given HF concentration and, as has been mentioned above, the approximately square-root dependence on HF concentration suggests that it may be the positive ions released from HF that are responsible. The actual magnitude of the peak, which as seen above is larger than would be predicted from a simple Boltzmann distribution of free dipoles suggests that perhaps there are cooperative phenomena present, i.e. that the orientation of one water molecule lowers the energy for the orientation of its neighbour, as would be needed if a ferroelectric transition occured. The temperature of the peak is close to the reported ferroelectric transition of Dengel, Riehl et al. /4/, which was only observed in impure ice suggesting again that defects in excess of those present in thermal equilibrium in pure ice must be present to accomplish the reorientation of water molecules. It is also noteworthy that the polarization observed in these experiments (6×10^{-7} C/cm^2) produced by a field of 20 kV/cm corresponds to a permittivity of 340, a value very similar to that observed near the maximum permittivity by Dengel, Riehl et al. /4/. However as shown above the actual polarization observed is only about 6% of that for complete polarization of the sample. This could be due to the existence of ferroelectric domains oriented in opposition to the field, or it could be a result of a non-uniform field distribution through the sample; if the voltage drop is concentrated into about 6% of the specimen thickness then this portion could be completely polarized to give the observed results.

The activation energy observed for the thermostimulated current peak of about 0.25 to 0.27 eV is very close to the value of 0.235 eV quoted by Gränicher for the movement of Bjerrum defects; it does not correspond to any of the energies associated with positive ionic defects. This suggests that perhaps Bjerrum defects may be involved in the removal of the polarization. They might also be responsible for the formation of the polarization, but in this case another origin for the HF concentration dependence would have to be sought since the number of Bjerrum defects is believed to be directly proportional to the HF concentration. This result has however been postulated primarily at high temperatures. The number of HF molecules present is such that if each provides a Bjerrum defect, these defects would only have to travel about 1000 atomic distances to achieve complete orientation of the ice. Even allowing that at low temperatures there are many fewer ions released than Bjerrum defects, there are still adequate concentrations of either Bjerrum defects or ions to produce complete reorientation if the defects are free to move from the originating fluoride ion to the edge of the crystal. It does not seem necessary therefore to suppose that in the case of HF ice the reorientation is limited to trails of water molecules along the paths followed by individual point defects as suggested by Kahane.

The remaining possibility is that the bulk of the ice remains unoriented, but that positive ions move to one end of the specimen and L-defects to the other. If no net reorientation is to be produced, there must be equal numbers of the two kinds of defects moving, and so the number mobile would be controlled by the least numerous species, i.e. the positive ions, which would explain the HF-concentration dependence, whereas the actual movement would be controlled by the species which is hardest to move, i.e. the L-defect, thus explaining the activation energy found when the defects move back on thermostimulation.

The value of κ of 1.45 is not easy to explain. A simple first-order orientation kinetics would of course correspond to $\kappa = 1$, whereas defects moving in the field produced by a dipolar layer would have a velocity proportional to the charge on the end surfaces of the layer, so that with the number of defects proportional to the charge, we might expect a value of $\kappa = 2$. The observed value suggests that some intermediate model may be needed.

The present electrical experiments on HF-doped ice can be interpreted either in terms of ferroelectricity or in terms of defect migration. The evidence for ferroelectricity on other grounds has been discussed by Onsager /5/, and other papers in the present symposium give evidence for a phase change close to the temperature of the HF-doped peak (Helmreich /6/; Pick /7/), which are in favour of the ferroelectric interpretation. The oberservations on pure ice confirm that no comparable transition occurs in the times used in the present experiment (of the order of minutes).

References

/1/ Gränicher, H., Physik kondens. Materie 1, 1-12 (1963)

/2/ Grossweiner, L.I., J. Appl. Phys. 24, 1306-07 (1953)

/3/ Bucci, C., R. Cappelletti, R. Fieschi, G. Guidi and L. Pirola , Suppl. Nuovo Cimento 4, 607-29 (1966)

/4/ Dengel, O., U. Eckener, H. Plitz and N. Riehl, Physics Letters 9, 291-92 (1964)

/5/ Onsager, L., Ferroelectricity of ice? (In Weller, E. F., ed. Ferroelectricity. Proceedings of the Symposium on Ferroelectricity, General Motors Research Laboratories, Warren, Michigan, 1966. Amsterdam, etc. Elsevier Publishing Co., p. 16-19

/6/ Helmreich, D., Physics of Ice, Proc. Int. Symp., Munich, Plenum Press, New York (1969)

/7/ Pick, M., Physics of Ice, Proc. Int. Symp., Munich, Plenum Press, New York (1969)

CONDUCTION ANOMALIES AND POLARIZATION IN ICE AT LOW TEMPERATURES

H. Peter Glockmann

Physik-Department

Technische Hochschule München / Germany

Abstract

Pure ice single crystal specimens (70 mm dia., 1 mm thick) exhibit release of about 7 nC "stored" charge near 105°K, and a second peak near 155°K, on heating when temperature is cycled between 180 and 95°K at linear rates of up to 1 deg/min and if polarizing fields of up to 2 kV/cm are applied during cooling. The idea of thermally activated carrier trapping is made doubtful by the appearance of current peaks at corresponding temperatures during cooling. Short pulses (~10 msec) and their temperature dependence in the same region lend support to the idea that ice is a (partially) ordered ferroelectric above 105°K. One example of apparently spontaneous polarization is shown. Annealing near melting point does not seem to remove the order completely.

1. Introduction

This paper extends to the low temperature region between 180 and 95°K the systematic investigation of the conducting properties of pure ice single crystals which is being carried out, among other studies, by the Munich group under the direction of Professor N.Riehl. Various physical properties of ice are known to behave anomalously in this region, e.g., birefringerence studied by Kahane /1/, elastic constants measured by Helmreich /2/ or specific heat first studied by Giauque and Stout /3/ and recently re-examined by Pick /4/. "Charge storage" which appears to have been noted first by Engelhardt /5/ is being studied by Bishop and Glen /6/, by Mascarenhas /7/ and in this paper. Anomalies of the dielectric constant, first seen by Vaubel /8/, have been linked to ferroelectric behaviour by Eckener and others of the Munich group /9,10/. Onsager /11/, however, has questioned this interpretation. A ferroelectric transition in ice has been the subject of recent theoretical discussions by Lieb /12/ and others; it is also a subject of this paper.

2. The experimental situation

This experiment has been designed to study conduction phenomena together with evidence of collective polarization by providing two separate signal channels as illustrated schematically in fig.1 where pertinent performance data are given. All samples used in this study are circular discs of 70 mm diameter and 1 mm thickness prepared at -10°C with the aid of circular saw and lathe from the 85 mm diameter parent single crystals which are grown from highly purified water at this laboratory. The electrodes, made from gold plated brass sheet stock, have a diameter of 70 mm. The central 25 cm², 57 mm in diameter, are covered with proton-loaded palladium-black to provide good carrier injection.

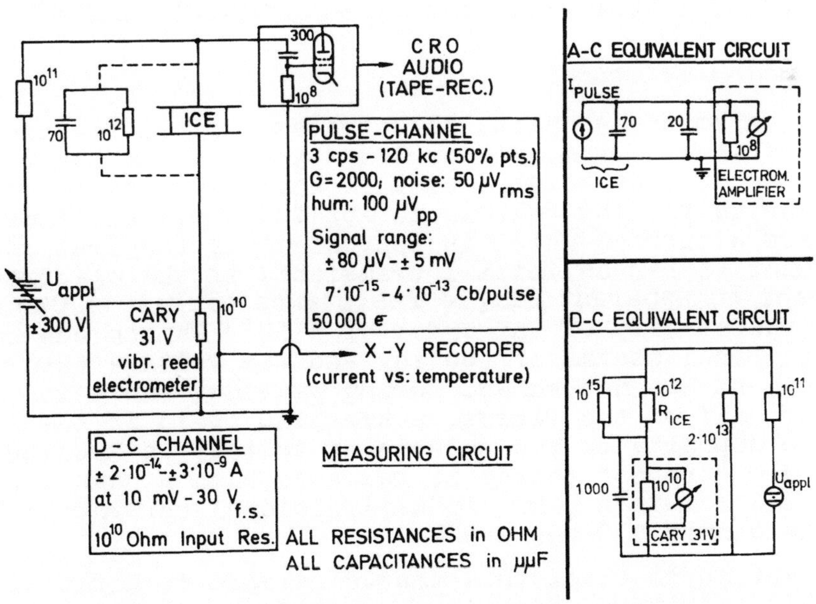

Fig.1 Measuring and equivalent circuit schematics including 2 leakage resistances in dc equivalent circuit

Under control of a motorized set-point programmer, linear heating and cooling rates of up to ± 1 deg/min can be maintained through the temperature range from the melting point to about 95°K. The temperature is averaged by a resistance wire thermometer encircling the bottom edge of the sample and measured independently at the bottom center. Observed temperature differences approach 1 deg. The temperature gradient which is measured between bottom center and a point on the upper edge of the sample can be appreciable and has considerable effect on the dc measurements.

The direct current through the ice is recorded continuously on an X-Y plotter as a function of temperature; replicas of such plots shown in this paper are either exact or smoothed by suppression of minor fluctuations. The pulse-channel output is monitored with cathode-ray oscilloscope, loudspeaker and, occasionally, tape-recorder.

All measurements reported here have been performed with well-injecting palladium-black electrodes at fixed or zero applied potentials and with temperatures varied linearly with time at rates from ± 0.3 to ± 1 deg/min. Studies at constant temperatures and with time-variable voltages as well as work with blocking electrodes and applied fields exceeding 2 kV/cm remain to be done. The results presented here are, therefore, somewhat limited.

3. Conduction anomalies

Figure 2, a smoothed plot, illustrates typical features of an experimental run. The cooling cycle in the upper half of the diagram starts from a storage temperature near 200°K. The polarizing voltage is applied somewhat later and causes a current which is approximately temperature-independent (after an initial transient) if the electrodes are fresh. An apparent sample resistance of $2 \cdot 10^{11}$ Ohm, resp. a resistivity of the order of $5 \cdot 10^{13}$ Ohm·cm can be assigned. Final stages of cooling and the voltage turn-off transient are not shown. During heating, shown in the lower half of the figure, a sweeping field in the direction opposite to the polarizing voltage is applied at 90°K and "stored" charge is released with a peak current of $2.1 \cdot 10^{-10}$A near 115°K. Additional current peaks are visible near 130 and 145°K.

It is tempting to treat this charge release in terms of a thermally activated trapping mechanism as done, e.g., by Bishop and Glen /6/. In this study, the position of the low temperature current peak varied in dependence on the heating rate in a manner consistent with such a

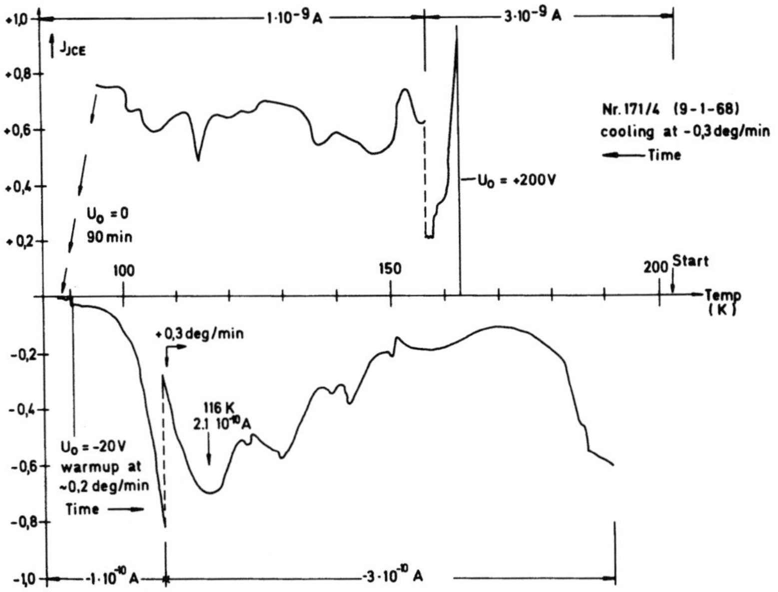

Fig.2 Typical dc vs. temperature plot, smoothed. U_0 is the applied voltage (polarizing on cooling, sweeping on heating). Full scale sensitivity is indicated by figures on horizontal bars at I = ± 1.0

mechanism. Scattering of the data, however, precluded a reliable determination of the activation energy.

Indeed, the concept of a thermally activated carrier mechanism becomes doubtful when fig.3 is considered. This is a smoothed cooling plot from a sample cut at 45° to the crystallographic c-axis. The obvious temperature-dependence of the current is probably due to reduced carrier injection caused by depletion of the proton supply on the electrode surfaces. The current increases on cooling near 155 and 115°K, marked a and e on the graph, are reproducible effects and correspond to current peaks on heating. Similar effects, current increases on cooling as well as release of "stored" charge on heating, are also observable on polycrystalline pure ice. They were even barely discernible when a piece of cardboard, soaked in highly purified water, was used in place of the single crystal sample in an early test.

Another reproducible effect is observable at marks b, c and d in fig.3. Here, the cooling rate has been reduced deliberately for the periods indicated on the diagram. This reduction causes a decrease of the temperature

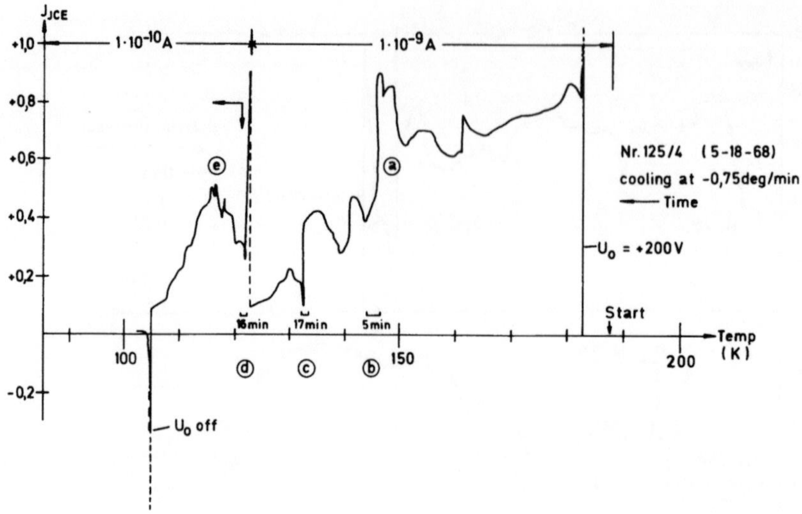

Fig.3 Dc vs. temperature cooling plot, smoothed. Current increases at a and e. Cooling rate reduction for indicated periods at b, c and d causes decrease in temperature gradient and correlated current change

gradient between bottom center and upper edge of the sample together with a correlated current change. While amounting to 0.5 to 2 deg at stationary temperatures, the gradient across the 1 mm sample thickness may reach 12 deg when the temperature is varied at a rate of ± 1 deg/min. The cause of this current variation has not been determined. It may be noted that the current is measured by the potential drop it causes across the 10^{10} Ohm CARY 31V input resistor. The voltage variations then approach 1 V, well within the range of thermo-electric and chemical potentials. An additional clue to the origin of these current variations may reside in the fact that persistent currents are observed in an external circuit of 10^{11} Ohm resistance when a sample is annealed without any applied voltage which had been polarized at low temperatures. These persistent currents show little change after several hours near melting point and disappear only after re-cooling the sample to -40 to -60°C

Figure 4 is the exact replica of a run with a sample cut at 45° to the c-axis. Cooling and heating proceed at 0.5 deg/min. The polarizing voltage is 200 V; no sweep field is applied during heating. The current increase on cooling near 165°K, marked a, and its heating cycle counter-

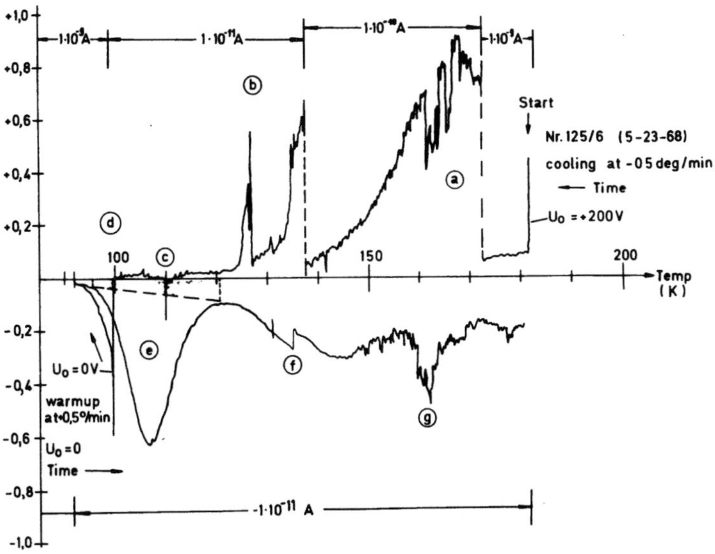

Fig.4 Exact dc vs. temperature plot of 45° cut sample. Current peaks on cooling at a and b; charge release peaks on heating at e and g. The spike at c is an artifact, that at d the voltage turn-off transient. At f, temperature held constant for 5 min. Run number 8

part, g on the graph, are accompanied by appreciable current fluctuations and by pulse discharges. The sharp current peak at 125°K, marked b, has a rise-time approaching the 10 sec dc channel input time constant and lasts only 4 min, corresponding to 2 deg. The release of "stored" charge on heating without sweep field peaks at 108°K (mark e). The charge released during this part of the experiment was found to be $6.95 \cdot 10^{-9}$C by graphical integration after subtraction of the shaded "background"; this corresponds to a charge density of $0.28 \cdot 10^{-9}$C/cm^2 on the 25 cm^2 electrode area. The discontinuity at f is the effect, this time negligibly small, of holding the sample temperature constant for 5 min.

This run, number 8 for that particular sample, taken after 220 hours of experimentation has a significant history. The preceeding run, number 7, was marked by continual current fluctuations with more than ten times the amplitudes of fig.4 which were independent of applied potentials. The sample was then annealed for several hours at 0.9 deg below the melting point before the data of fig.4 were

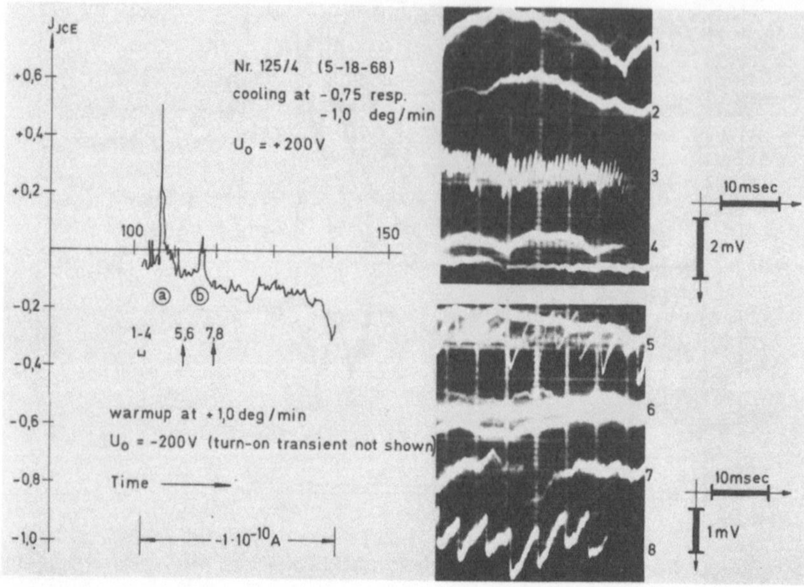

Fig.5 Exact dc vs. temperature heating plot of run number 6 on same specimen as shown in fig.4. Traces of pulse activity photographed at times of corresponding numbers

obtained. In the light of this experience, annealing between successive runs has been made standard practice.

4. Polarization phenomena

Figure 5 shows the low temperature heating plot of run number 6 on the same specimen. There is no "stored" charge released at 108°K. The current spike is in the direction of the polarizing voltage and opposed to the rather high sweeping field. At the same time, vigorous pulse activity is observed in the other signal channel. The numbered traces shown on the right of fig.5 were obtained at the times of the corresponding numbers on the graph. The exposures had to be taken at random; the most spectacular pulse events are not displayed. The saw-tooth shaped discharges of traces 3 and 8 are not observed very often. They may persist from 0.1 to 15 sec. The longer they last, the lower drops the amplitude and the higher rises the frequency. The observations made during runs 7 and 8 discussed above (which followed the run shown in fig.5) may be understood in terms of polarized regions created during this sixth run which persist and fluctuate during run 7 and are finally removed by annealing prior to run 8.

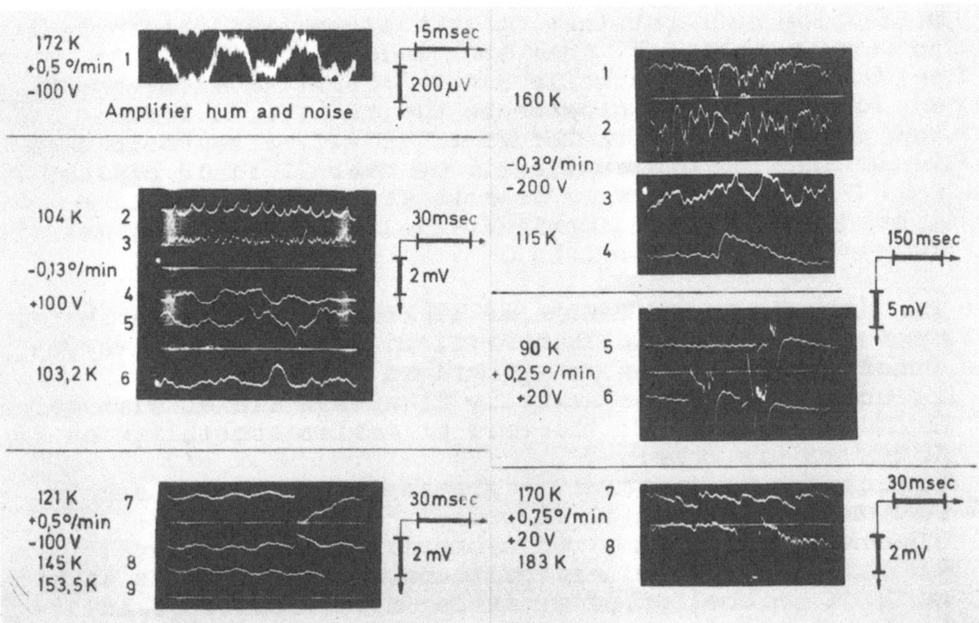

Fig.6 Barkhausen pulses, unretouched photographs. Temperature, rate of change and applied voltage listed to the left, time and amplitude scales to the right of the traces

Other pulse patterns which have been obtained with the aid of a storage-oscilloscope during several runs with a sample cut perpendicular to the c-axis are shown in fig.6. The comparatively slow, rounded type of activity (traces 4 through 6, left, and 1 through 3, right) is most often seen. It may persist with little change for as long as an hour. Individual pulses (traces 7 through 9, left, or 7 and 8, right) are mostly seen on warmup and may be widely separated in time. The time interval between traces 7 and 8, left, is 46 min. No intervening pulses could be recorded although the monitor had a 9% duty cycle.

I have become convinced that these pulses are evidence of collective polarization of larger or smaller regions within the sample; they are, in other words, evidence of formation, growth and switching of domains. The pulses then deserve the name of Barkhausen pulses just as in other ferroelectrics.

(In private communications, L. Onsager suggested motion of charged line dislocations, H. Gränicher considered local breakdown at small electrode regions as alternatives. The data obtained in this study are insufficient to reach a decision. This author finds it difficult, prima vista, to reconcile either objection with the observed temperature dependence of the effect. Induced surface charges of the order of 10^{-13}C/pulse would seem large for line dislocations. The overall field of less than 2 kV/cm would seem to make local breakdown improbable, even though much higher field gradients may exist at the electrode surfaces.)

The conditions for occurence of Barkhausen pulses in ice have not yet been fully clarified. In a qualitative way, the observations can be summarized as follows:
1) Appreciable pulse activity is always linked with large dc fluctucations; if the current varies smoothly, few or no pulses are observed.
2) Appreciable temperature gradients across the sample seem to be detrimental to pulses.
3) Annealing before a run promotes pulse occurence.
4) Few ,isolated pulses are found at temperatures as high as -40°C on cooling after annealing a previously polarized sample.
5) On <u>cooling</u>, vigorous pulse activity starts near 160°K and reaches a first maximum about 150°K. Below this temperature, activity decreases but does not stop. A moderately strong maximum occurs near 125°K on most runs, while the largest amount of pulse activity is observed near 115°K. No further pulses were observable to the lowest attainable temperature around 90°K.
6) On <u>heating</u>, pulse activity starts at or below 100°K. A less pronounced maximum near 125°K is observed only if there has been no smooth charge release ("glow-curve") near 106°K. At least one extended burst of pulses is observed near 155°K on almost every run. The highest observed temperature of a strong pulse burst on heating has been 183°K (= -89°C! cf. fig.6, trace 8, right).
7) In one 45° cut sample, pulses were observed only if the polarizing field had exceeded a threshold of 1.5 kV/cm.

Evidence of <u>spontaneous polarization</u> seems to have been obtained in the latest experiment. The sample, cut perpendicular to the c-axis, was stored at -10°C for two days after preparation and annealed at -1°C for several hours immediately before the first temperature cycle. The low temperature portion of this "virgin curve" is shown in fig.7. The external circuit is closed with 10^{11} Ohm; no voltage had ever been applied to this crystal. On cooling (top), there is a sharp current maximum of $4 \cdot 10^{-12}$A amplitude at 105°K. It has a rise-time close to

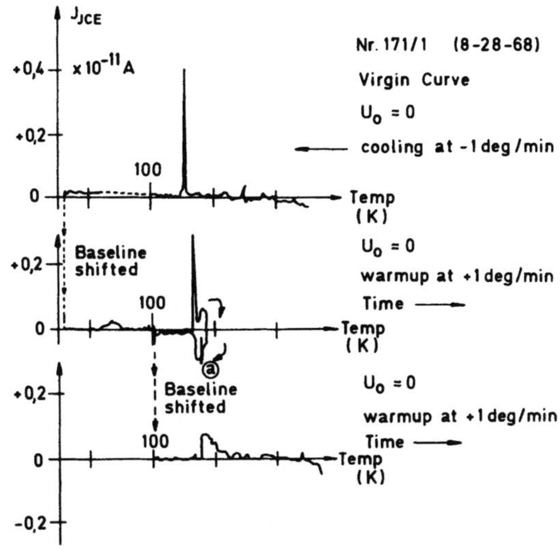

Fig.7 Exact dc vs. temperature plot; 0° cut sample. Virgin curve exhibiting spontaneous polarization(?). Spike at a may be artifact. Baseline displaced twice

10 sec and lasts 30 sec, corresponding to 0.5 deg. The heating curve (center) exhibits a similar maximum of $3 \cdot 10^{-12}$A amplitude at 106.5°K followed by a small one. The sample was immediately cooled again to 100°K; the small spike, marked a, may be an artifact. When heated once more, the small maximum is reproduced. No pulses occurred during this part of the experiment. This sequence of events can be understood in terms of spontaneous polarization which has almost reached whatever its final state may be at the time of the second heating branch.

Discussion

Although it is too early to attempt final conclusions on the base of these limited data, support seems to be given to the idea that ice is ferroelectric. This study suggests a primary transition at a temperature near 105°K and a polarizability of the order of $0.3 \cdot 10^{-9}$C/cm^2. There may be a further ferroelectric transition in the 150 to 160°K range. When combined with the findings of Pick /4/ and Helmreich /2/, one may conclude that the state of higher, but incomplete, order exists above the transition temperature of 105°K. Some partial ordering seems to persist even after annealing a previously polarized sample close to the melting point. No evidence of changes in the hexagonal oxygen sublattice has been reported. Any ordering should, therefore, concern the proton arrangement only. This picture is compatible with ideas proposed by Kahane /1/: Ice is a locally ordered ferroelectric at elevated temperatures and a disordered system at low temperature.

A very tentative model, suggested but certainly not proved by this study may be an aid to further investigation: The transition at 105°K may involve the protons on those hydrogen bonds which are parallel to the c-axis. The effect should be pronounced because only two possible arrangements exist. The transition around 155°K could involve the oblique bonds where six naturally preferred directions exist. If domains are formed which are comparable in size to the microstructures of Workman /13/ and Truby /14/, then local ordering along any one of these directions could still leave the extended sample in an average state of no or very little ordering. The pulse activity between these transition regions might then indicate gradual nucleation, growth and switching of such ordered domains.

References

/1/ Kahane, A., Thèse, Université de Paris, Série A no. 3996 (1962)

/2/ Helmreich, D., Physics of Ice. Proc.Int.Symp., Munich, Plenum Press, New York (1969)

/3/ Giauque, W.F., J.W.Stout, J.Am.Chem.Soc. 58, 1144 (1938)

/4/ Pick, M., Physics of Ice, Proc.Int.Symp., Munich, Plenum Press, New York (1969)

/5/ Engelhardt, H., N.Riehl, Physics Letters 14, 20 (1965); Phys.kondens.Materie 5, 73 (1966)

/6/ Bishop, P.G., J.W.Glen, Physics of Ice, Proc.Int. Symp., Munich, Plenum Press, New York (1969)

/7/ Mascarenhas, S., Physics of Ice, Proc.Int.Symp., Munich, Plenum Press, New York (1969)

/8/ Vaubel, G., Thesis, Lab.Technische Physik, Technische Hochschule, München (1963)

/9/ Dengel, O., U.Eckener, H.Plitz, N.Riehl, Physics Letters 9, 291 (1964)

/10/ Otte, E., Thesis, Physik-Department, Technische Hochschule München (1965);
Hecker, S., Thesis, Physik-Department, Technische Hochschule München (1965)

/11/ Onsager, L., "Ferroelectricity of Ice?" in Weller, E.F. (ed.): "Ferroelectricity", Proceedings of the International Symposium on Ferroelectricity, General Motors Research Laboratories, Warren, Mich., 1966, pg. 16, Elsevier Publishing Co. Amsterdam (1967)

/12/ Lieb, E.H., Phys.Rev.Letters 18, 692 (1967);
ibid. 18, 1046 (1967);
Sutherland, B., ibid 19, 103 (1967);
Yang, C.P., ibid. 19, 587 (1967);
Sutherland, B., C.N.Yang, C.P. Yang, ibid 19, 588 (1967)

/13/ Workman, E.J., Phys.Rev. 92, 544 (1953)

/14/ Truby, F.K., J.Appl.Phys. 26, 1416 (1955)

DIFFUSION AND RELAXATION PHENOMENA IN ICE *

L. K. Runnels **

Coates Chemical Laboratories

Louisiana State University, Baton Rouge, La.

Abstract

The several types of lattice imperfections proposed for ice are discussed in terms of their ability to account for the observed rate processes. We feel that the orientational defects as originally suggested by Bjerrum still represent the most probable explanation for the dielectric and elastic relaxation phenomena. They are not, however, able to explain the motional effects seen by magnetic resonance experiments. The spin-lattice relaxation process, as well as the linewidth behavior, appears to be more closely associated with the self-diffusion of water molecules. Although the evidence is somewhat contradictory, the imperfection deemed the most likely explanation of the resonance and diffusion experiments is the interstitial molecule. Suggestions are offered for further experiments and calculations.

---

* Based largely on a dissertation submitted by L.K.Runnels to the Graduate School of Yale University, 1963, in partial fulfillment of the requirements for the degree of Doctor of Philosophy. Supported in part by a National Science Foundation Predoctoral Fellowship and by National Science Foundation Grant No. GP-6822. A summary report is given by L. Onsager and L. K. Runnels, Proc.Nat.Acad.Sci. U.S. 50, 208 (1963).
** Alfred P. Sloan Foundation Fellow

1. Introduction

Apart from the electrical conductivity of ice, by itself a widely studied property, the activation energies of four rate processes in ice have been reported - all yielding values in the vicinity of 14 kcal/mole, or 0.6 eV. Presented in table 1, the processes are those of dielectric /1,2/ and elastic /3/ relaxation, proton spin-lattice relaxation /4,5/ and diffusion /6-10/ of hydrogen and oxygen. The intriguing question raised is whether the similarity of these activation energies is an accident or whether it reflects microscopic mechanisms held in common.

Table 1 Activation energies in ice

	kcal/mole	eV	
Dielectric relaxation	13.2	0.572	/1/
	13.27±.05	0.575±.002	/2/
Elastic relaxation	13.4±.6	0.58±.02	/3/
Spin-lattice relaxation	14.1±.1	0.611±.004	/5/
	13.4	0.58	/4/
Diffusion ^{18}O	15.7±3.0	0.68±.13	/8/
^{3}H	14.5±1.0	0.63±.04	/6,9/
	15.7±2.0	0.67±.08	/10/

Since an ice crystal is a three-dimensional network of hydrogen bonds, admittedly disordered as first suggested by Bernal and Fowler /11,12/, the study of these processes is rather quickly seen to be a study of lattice imperfections. It is not, however, so readily decided in every case just what sort of imperfection is involved. Comparisons of the various processes can be helpful in limiting the possibilities.

2. Classical Bjerrum defects

Historically the first type of lattice imperfection in ice to gain widespread acceptance was the orientational mismatch suggested by Bjerrum /13/ to account for dielectric relaxation (fig.1). Even if the geometry of the defect is not exactly as proposed by Bjerrum, the essential feature of the imperfection is the chain of molecular reorientations accompanying its migration through the crystal: a molecule visited may rotate its dipole

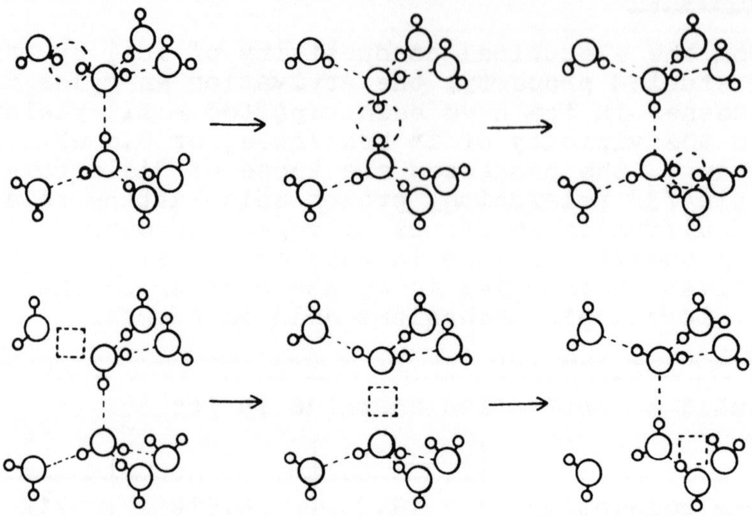

Fig.1 Orientational defects, or Bjerrum faults /13/. The mismatch at the top is known as a D fault, the one at the bottom as an L fault. The migration of either produces rotations of molecules in its path

moment to any of its six possibilities <u>except</u> the reverse of its original orientation.

This observation alone allows us to effect one correlation of the processes in table 1. Dielectric and elastic relaxation of assemblies of polar molecules are very closely related phenomena: the former /15/ measures the autocorrelation function of $P_1(\cos\theta)$, while the latter /14/ measures the autocorrelation function of $P_2(\cos\theta)$. (This statement is strictly true only for systems with cylindrical symmetry; more generally, spherical harmonics of order one and two, respectively, are involved). In a viscous, isotropic medium, rotational diffusion governs molecular reorientations and the autocorrelation functions of $P_j(\cos\theta)$ are exponentials whose time constants are eigenvalues of the surface Laplacian /16/. The more rapidly varying function P_2 must decay more quickly; in fact the ratio of time constants τ_{diel}/τ_{elas} would be 3 for such a system /14/.

In a crystal environment, where rotations are between discrete orientations, the time constants are eigenvalues of a transition matrix. In the cubic crystal thoria, the occasional occurrence of a CaO impurity in the ThO_2 lattice provides an oxygen vacancy which is polar and mobile,

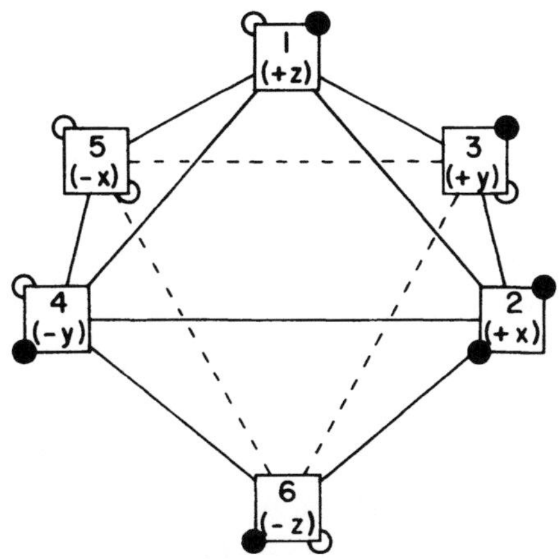

Fig.2 Bjerrum defect migrations produce transitions between orientation states joined by an edge of the octahedral diagram

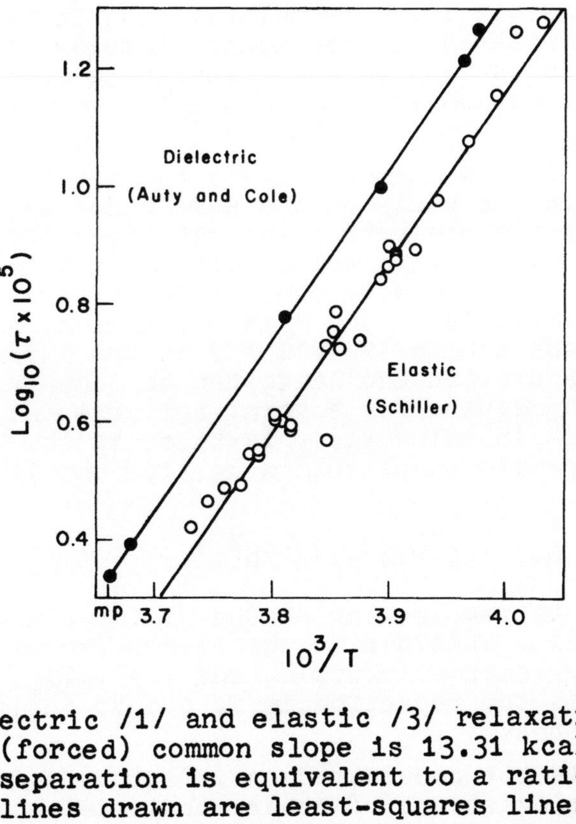

Fig.3 Dielectric /1/ and elastic /3/ relaxation times. The (forced) common slope is 13.31 kcal/mole and the separation is equivalent to a ratio of 1.45. The lines drawn are least-squares lines

hopping between eight possible positions /17/. The resulting ratio of rate constants τ_{diel}/τ_{elas} is 2 in this case, as has been borne out by experiment.

For the reorientation scheme produced by the Bjerrum defects (see fig.2) the corresponding ratio of eigenvalues is found /14/ to be 3/2. In fig.3 are shown the relaxation times of Auty and Cole /20/ and of Schiller /9/: The separation between the two lines is equivalent to a ratio of $\tau_{diel}/\tau_{elas} = 1.45$. This rather good agreement with the theoretical value of 1.5 is encouraging, but it should be realized that it reflects only the symmetry of the reorientation scheme produced by the Bjerrum faults and gives no detailed picture of their geometry.

It is clear that this mechanism as pictured so far provides no possibility for mass transport /18/ and so we pass over the diffusion process for the present. There has been considerable success with some polar molecules in correlating the dielectric relaxation time with the spin-lattice relaxation time T_1 of spins within the molecule /16/. This is the case when the same molecular motions are responsible for both relaxations. Basically spin-lattice relaxation in two spin molecules is an indirect measure of autocorrelation functions of second-order spherical harmonics - or rather their Fourier transforms.

In the case of ice, we can calculate the necessary spectral densities on the basis of the same model as previously employed for the dielectric and elastic relaxation phenomena. An added complication here is that the (magnetic) dipolar coupling between protons in different molecules also contributes to the spin relaxation, although the contributions typically fall off as the inverse sixth power of the separation and hence can be computed quite accurately. The results for several temperatures are shown in table 2 in which τ_{diel} is taken as the known independent parameter to calculate T_1 from the final formula /14/

$$(T_1)_{diel} = 0.7 \times (40/27) b^6 \omega^2 / \hbar^2 \gamma^4 \nu \;,$$

where b is the intramolecular proton-proton distance (1.62 Å), $\omega = 2\pi \times 10$Mc/sec is the Larmor frequency, γ is the proton gyromagnetic ratio, and $\nu = \tau_{diel}^{-1}$. The factor of 0.7 is the reduction in T_1 due to intermolecular interactions.

The agreement is not considered satisfactory and is taken to indicate the existence of other mobile defects besides the Bjerrum faults; additional evidence in the same direction will be provided later. To conclude this section, however, we offer a few speculative comments about the above comparison along lines which have yet to be studied thoroughly.

Table 2 Spin-lattice relaxation time in ice in sec., at 10 Mc/sec

	$-1.5^{\circ}C$	$-10.8^{\circ}C$	$-20.9^{\circ}C$	$-27.5^{\circ}C$
Calculated from τ_{diel}	3.2	7.8	21	43
Experimental /4,5/	0.5-0.8	1.2-1.9	3.4-5.1	7.2-10.2

There are two aspects of the Bjerrum fault motions which conceivably could contribute to the spin relaxation but be unnoticed by dielectric experiments. Study of fig.1 reveals that an L-fault by circumnavigating a molecule can permute the protons of that molecule. Clearly such a change would have no effect on the degree of polarization of a crystal, nor on the intramolecular spin Hamiltonian, but it would change the intermolecular Hamiltonian and make a slight contribution to the spin relaxation. Since this permutation cannot be effected by the D faults, it represents in principle an attack on the separation of the mobilities of the two types. In practice, however, it is probably too small to detect.

The spin relaxation effects offer another possibility of an interesting separation of factors. The dielectric relaxation time involves both the concentration and mobility of defects; the relationship, in fact, is /14/

$$\nu = \tau_{diel}^{-1} = \left((N_D/N)\nu_D + (N_L/N)\nu_L\right)/4 \quad ,$$

where N_D/N is the number of D faults per molecule and ν_D is the jump frequency of the defect; similarly for the L faults. The factor of one-fourth is understood most simply as the net fraction of defect jumps in the forward direction (one-half forward, one-fourth back and one-fourth "sideways" - see fig.1). This dielectric relaxation frequency ν is of the order of 10^5 sec^{-1} just below the melting point, and is the average rate of rotation of any given molecule. Of course, while in the process of rotating - that is, during the visit of a Bjerrum defect - a molecule undergoes motions of a much higher frequency, of the order of ν_D or ν_L. If we guess the unknown defect concentrations to be of the order of 10^{-6}, then the high frequency displacements are of the order of 10^{11} sec^{-1} below the melting point.

Now motions effective in producing spin-lattice relaxation are those with a significant component at the Larmor frequency $\omega_L \sim 10^8$ sec^{-1}. The spectral density of the Bjerrum fault-produced motions has a main peak near $\omega = 2\pi\nu$ which usually dominates the relaxation, but also secondary peaks near $2\pi\nu_D$ and $2\pi\nu_L$. These latter ones would eventually become the important frequencies at low enough temperatures, provided some different relaxation process did not take over first. If such a region were reached, some information would thereby be provided about the concentration of defects; moreover, in this region the activation energy shown would be that of ν_D (or ν_L) and hence characteristic of only the motion of the defect and not its creation. All low temperature measurements /4,5,19/ of T_1 do in fact show a decrease in activation energy, but it is not known whether or not this is due to the effect discussed or to the onset of some new relaxation mechanism.

3. Augmented Bjerrum defects

The diffusion process, which we skipped over in the preceding section, has been the object of considerable study recently /6-10/. Within experimental accuracy it appears that oxygen and hydrogen diffuse at the same rate in ice crystals, essentially forcing us to identify the diffusing species as an intact molecule and not an ion. Summarized in fig.4 are the several measurements reported for the diffusion coefficient D_{H_2O}.

It was pointed out by Gränicher /18/ that just as Bjerrum fault migration produces no mass transport, so the common diffusion mechanisms would in ice produce no displacement current. A new type of defect would thus be necessary if it were to provide a common mechanism for diffusion and dielectric (and elastic) relaxation. Suggestions which have been offered /20,21/ include (a) interstitial molecules bound to Bjerrum faults and (b) Bjerrum faults diluted by vacancies (see fig.5).

It may be seen that the molecular rotations produced by the displacement of the augmented defects follow exactly the same scheme as produced by the "classical" defects; hence the ratio of τ_{diel}/τ_{elas} would remain unaltered. The new feature, however, is that since the transfer frequency can be measured by τ_{diel}, the diffusion

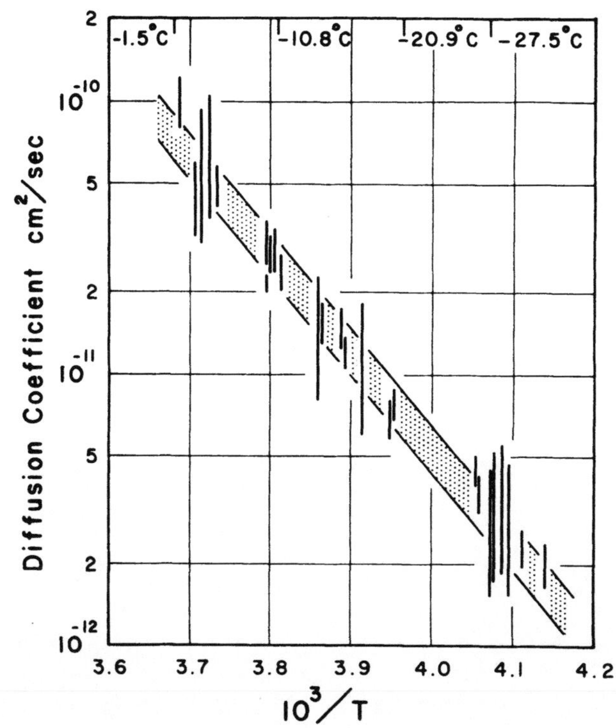

Fig.4 Experimental diffusion coefficients /6-10/ for both ^3H and ^{18}O

coefficient can be computed.

There are complications, however. The observable diffusion constant would be given by

$$D_{H_2O} = K\left((N_D/N)D_D + (N_L/N)D_L\right)$$

in terms of the defect diffusion constants D_D and D_L (related to ν_D and ν_L) and a correlation coefficient K. The correlation coefficient is of order one but somewhat dependent on the details of the propagation steps /30/. Of all the composite defects we have considered, the one at the top of fig.4 appears to be the most efficient in mass transport per molecular rotation. The relationship is /14/

$$D_{H_2O} = d^2/3\tau_{diel}$$

where d is the oxygen-oxygen distance, 2.76Å. The values at various temperatures are given in table 3.

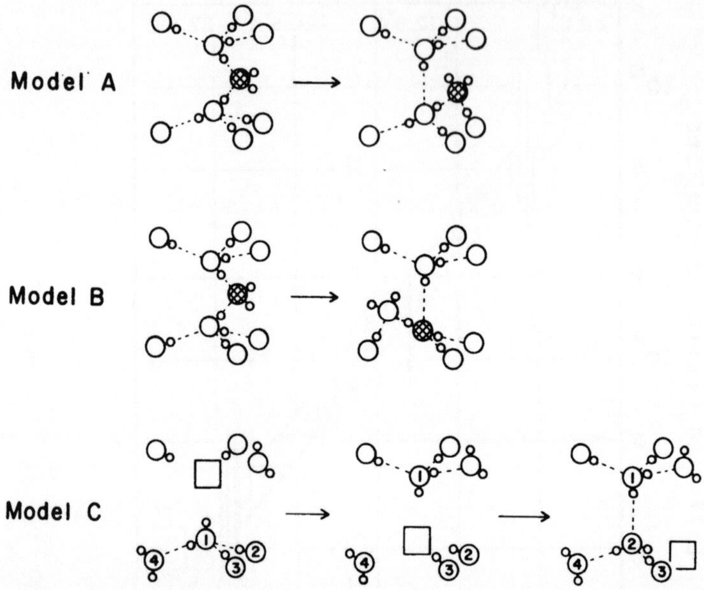

Fig.5 Composite defects. Types A and B, suggested by Haas /20/, differ in their manner of propagation. Type C is a Bjerrum fault diluted by a vacancy /21,14/

Table 3 Self-diffusion coefficient in ice (cm^2/sec) x 10^{12}

	$-1.5^{\circ}C$	$-10.8^{\circ}C$	$-20.9^{\circ}C$	$-27.4^{\circ}C$
Calculated from τ_{diel}	10	4.2	1.5	0.77
Experimental /6-10/	60-90	20-30	6-9	2.5-3.5

It is seen that there is a discrepancy of a factor of about 5 or more; that the agreement is better at lower temperatures reflects the likelihood, already evident from table 1, that the two processes probably have different activation energies.

It must of course be admitted that the inability to conceive of a viable composite defect does not preclude the existence of one. It does, however, appear unlikely that one sufficiently more efficient will be discovered, especially in view of the activation energies for the two processes. One also feels that the ratio of frequency factors /1/ for dielectric relaxation of H_2O and D_2O (1.45) is too large for the more massive aggregates.

4. Vacancies and interstitials

We have been forced to the conclusion, originally stated by Gränicher /18/, that diffusion and dielectric relaxation are governed by separate microscopic processes, the diffusive one being the faster of the two.

The need for a process faster than dielectric relaxation already has been learned from the spin-lattice relaxation measurements. It is now of interest to see if the diffusive motions can be successfully correlated with the spin-lattice relaxation time T_1. To compute the relevant autocorrelation functions, we assume that at an average frequency $\nu = 6D_{H_2O}/d^2$ a molecule jumps to a neighboring site, assuming an orientation consistent with its original one (hydrogen bond matching rule). The intermolecular contribution to T_1^{-1} must again be estimated, this time arising from relative translation /22/ as well as rotation of nearby molecules. The results at several temperatures are given in table 4, as calculated from /14/

$$(T_1)_{diff} = 0.6 \times (4/3)b^6\omega^2/\hbar^2\gamma^4\nu \quad .$$

Again the leading factor is the intermolecular reduction in T_1.

Table 4 Spin-lattice relaxation time in ice in sec., at 10 Mc/sec

	-1.5°C	-10.8°C	-20.9°C	-27.5°C
Calculated from D_{H_2O}	0.14-0.21	0.42-0.64	1.4-2.1	3.6-5.1
Experimental /4,5/	0.5-0.8	1.2-1.9	3.4-5.1	7.2-10.2

The agreement may be somewhat better than in table 2, but off in the opposite direction; for the diffusive motions seem too rapid as judged by the spin-lattice relaxation. There is, however, a way of rationalizing this discrepancy. If the mean squared displacement per transit were larger than d^2, such as could pertain to an interstitial mechanism, the jump frequency ν would be reduced accordingly, thereby increasing the calculated T_1.

Further evidence along these lines is provided by line-shape studies of magnetic resonance absorptions in ice, including proton /23/, deuteron /24,25/ and ^{17}O resonance /25,27/ studies. The subject of interest to us is the motional effects on the absorption, the general rule being /28/ that a splitting or broadening of order $\Delta\omega$ will be averaged away by large amplitude motions with a correlation time $\tau \ll (\Delta\omega)^{-1}$. From the closely related free induction decays, the correlation times for H_2O ice in table 5 have been obtained /23/. For comparision we show also the dielectric relaxation time and the "diffusion time" $d^2/6D_{H_2O}$.

Table 5 Correlation times in ice
µsec

	from linewidth	Dielectric relaxation time /1/ H_2O	Dielectric relaxation time /1/ D_2O	Diffusion time /6-10/ $d^2/6D$
-5°C	1-5	30	44	2.5
-25°C	30-50	220	320	30

Here we can see again the controlling influence of the diffusive motions, rather than the in-place rotations, on the nuclear spins.

Considerable quadrupole structure is shown by the deuteron resonance in D_2O single crystals /24/, reflecting the various molecular orientations. The splittings are sufficiently large (2π x 100 kc/sec) that they have not been motion averaged at the highest temperature studied (-10°C). In fact, the resolved lines show little if any of the broadening which might be expected /28/ from jumpings at the average rate $\nu = 6D/d^2$; if the jump rate were reduced considerably (by about a factor of ten) by an increase in the mean squared displacement, as mentioned above, the observed spectra could be reconciled with the other data we have on molecular motions in ice.

In a broad sense the rationalization of the ^{17}O spectrum /26,27/ (in D_2O ice) would follow the same lines as above, but the signal is much weaker and as yet the spectrum has not been analyzed completely.

In view of the present proposal of an interstitial mechanism for diffusion, the observation of an anisotropy in

D_{H_2O} with a larger value perpendicular to the c-axis is disquieting. Riehl and coworkers /6,8,9/ have felt this anisotropy (of the order of 10%) favors a vacancy mechanism. In light of the contradictory evidence above, it must be admitted that the diffusion mechanism is still in question.

The answer could be obtained by a modification of the Kirkendall (moving marker) experiment /29/. A speck of some substance (AgI) would be embedded in a cylindrical crystal of ice and subsequently observed. By keeping a temperature gradient across the sample, a higher concentration of the diffusing defect would be maintained at the hot end. The resulting (defect) concentration gradient would cause a net flux of defects into the cold end. The marker would appear to move toward the hot end if an interstitial mechanism were operative but toward the cold end with a vacancy mechanism. Unfortunately the experiment would likely be of several years' duration before the marker would have moved noticeably.

References

/1/ Auty, R.P., R.H.Cole, J.Chem.Phys. 20, 1309 (1952)

/2/ Humbel, F., F.Jona, P.Scherrer, Helv.Phys.Acta 26, 17 (53

/3/ For a review, see P.G.Owston, Adv. in Phys. 7, 171 (1958)

/4/ Krüger, G.J., Thesis, T.H. Stuttgart, 1961

/5/ Barnaal, D.E., I.J.Lowe, J.Chem.Phys. 48, 4614 (1968)

/6/ Blicks, H., O.Dengel, N.Riehl, Phys. kondens. Materie 4, 375 (1966)

/7/ Kuhn, W., M.Thürkauf, Helv.Chim.Acta 41, 938 (1958)

/8/ Delibaltas, P., O.Dengel, D.Helmreich, N.Riehl, H.Simon, Phys. kondens. Materie 5, 166 (1966)

/9/ Dengel, O., E.Jacobs, N.Riehl, Phys. kondens. Materie 5, 58 (1966)

/10/ Itagaki, K., J.Phys.Soc. Japan 19, 1081 (1964)

/11/ Bernal, J.D., R.H.Fowler, J.Chem.Phys. 1, 515 (1933)

/12/ Pauling, L., J.Am.Chem.Soc. 57, 2680 (1935)

/13/ Bjerrum, N., K. Danske Vidensk.Selsk. 27, 1 (1951)

/14/ Onsager, L., L.K.Runnels, to be published

/15/ Kubo., R., Lectures in Theoretical Physics, Vol.I, Interscience Publishers, Inc.,New York, 1959, p. 120

/16/ Bloembergen, N., Nuclear Magnetic Relaxation, W.A. Benjamin, Inc., New York, 1962; A.Abragam, The Principles of Nuclear Magnetism, Oxford University Press, London, 1961, Chap.8

/17/ Wachtman, J.B., Phys.Rev. 131, 517 (1963)

/18/ Gränicher, H., Z.Kristallogr. 110, 432 (1958); Proc.Roy. Soc. A247, 453 (1958)

/19/ Kume, K., J.Phys.Soc. Japan 15, 1493 (1960)

/20/ Haas, C., Phys.Letters 3, 126 (1962)

/21/ Kopp, M., private communications

/22/ Eisenstadt, M., A.G.Redfield, Phys.Rev. 132, 635 (1963)

/23/ Barnaal , D.E., I.J.Lowe, J.Chem.Phys. 46, 4800 (1967)

/24/ Waldstein, P., S.W.Rabideau, J.A.Jackson, J.Chem. Phys. 41, 3407 (1964)

/25/ Jackson, J.A., S.W.Rabideau, J.Chem.Phys. 41, 4008 (1964)

/26/ Rabideau, S.W., P.Waldstein. J.Chem.Phys. 44, 1304 (1966)

/27/ Waldstein, P., S.W.Rabideau, J.Chem.Phys. 47, 5338 (1967)

/28/ Abragam, A., The Principles of Nuclear Magnetism, Oxford University Press, London, 1961, Chap.10

/29/ Van Bueren, H.G., Imperfections in Crystals, North-Holland Publishing Company, Amsterdam, 1961, p.417

/30/ Friauf, R.T., J.Appl.Phys. 33, 494 (1962)

EVALUATION OF DIELECTRIC DISPERSION DATA

H. Gränicher

Eidg. Technische Hochschule

Zürich / Switzerland

This paper deals with the problem how true dielectric dispersion data of the bulk material can be derived from the apparent complex impedance of the sample as it is obtained from bridge measurements. Nearly all equations to be shown are already published in the literature. The need to present them in concentrated form comes from the fact that often experimentalists have overlooked the necessity to apply conductivity corrections and theorists have used experimental data which for the same reason were not correct.

The problem is due to the fact that ice is not an ideal insulator but is an ionic conductor. The effect of the dispersion behavior and the ionic conductivity is treated in the first section as inherent bulk properties. The ionic conductivity produces secondary effects because no truly ohmic contacts can be made. Therefore, the charge carriers in the ice are not or only partially discharged at the electrodes and thus give rise to space-charge polarization. In the measurements this results in an additional dispersion region which is at considerably lower frequencies for sufficiently pure ice than the Debye dispersion. The data of this space charge dispersion are not specific; they depend on the ice purity, the electrode material and separation and the maximum voltage applied to the ice capacitor. This more complex situation is considered in section 2.

1. Ice capacitor without space-charge effects

The Debye dispersion including a dc (ionic) conductivity

σ_o' is described by the following equations:

$$\varepsilon'(\omega) - \varepsilon_\infty' = \frac{\varepsilon'_s - \varepsilon'_\infty}{1+\omega^2\tau^2} ;$$

$$\varepsilon''(\omega) = \frac{\varepsilon'_s - \varepsilon'_\infty}{1+\omega^2\tau^2}\,\omega\tau + \frac{\sigma_o'}{\varepsilon_o\omega} .$$

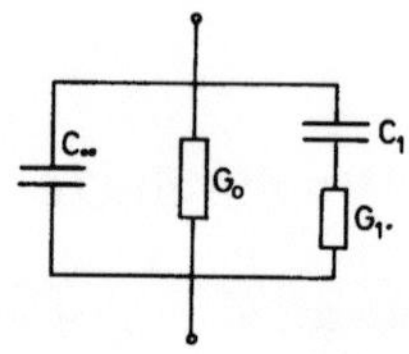

These equations correspond to the equivalent circuit shown with the following values of the components:

C_o = geometrical capacity = ε_o·area/electrode separation;

$C_\infty = \varepsilon_\infty' \cdot C_o$

$G_o = \sigma_o' \cdot C_o/\varepsilon_o$

$C_1 = (\varepsilon_s' - \varepsilon_\infty')C_o$

$G_1 = \varepsilon_o(\varepsilon_s' - \varepsilon_\infty')/\tau$.

From the impedance or admittance values obtained by bridge measurements by multiplication with the appropriate geometrical factor (and if necessary edge corrections) apparent, effective ε_m' and ε_m'' -values are obtained. In order to get the true bulk values several plots can be used.

1.1 Log $(\varepsilon'-\varepsilon_\infty')$ versus log ω plot /1/ (see fig. 1)

The extrapolation of the flat portion between the Debye region and the additional increase due to space charge effects intersects the ordinate at the true value:

$$(\varepsilon_m' - \varepsilon_\infty')_{\substack{\text{extrapolated}\\ \omega \to o}} = (\varepsilon_s' - \varepsilon_\infty')_{\text{true}} .$$

In the Debye region there is a straight line portion theoretically of slope -2. The relaxation time $\tau = 1/\omega_d$ is obtained from the ω_d value which corresponds to the ordinate log $[(\varepsilon_s' - \varepsilon_\infty')/2]$. It is advantageous to use a template for getting an optimum fit with experimental points /1/.

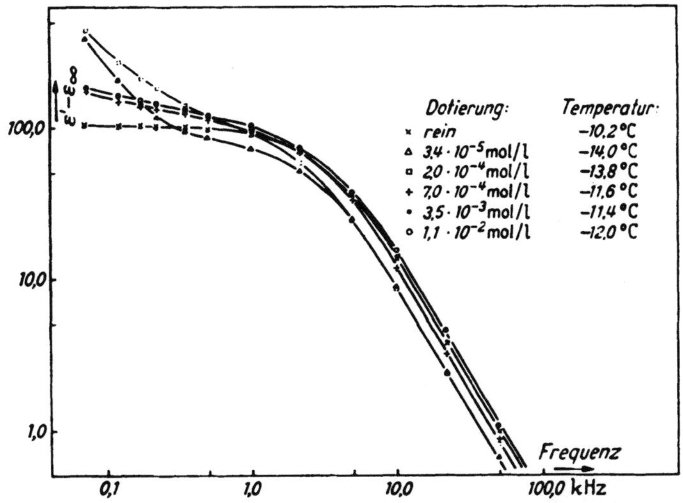

Fig.1 log($\varepsilon' - \varepsilon'_\infty$) versus log ν plot for NH_4F-doped ice (O. Dengel et al. /2/)

1.2 <u>Linear plot ε'_m versus ($\varepsilon''_m\ \omega$)</u> (see fig. 2)

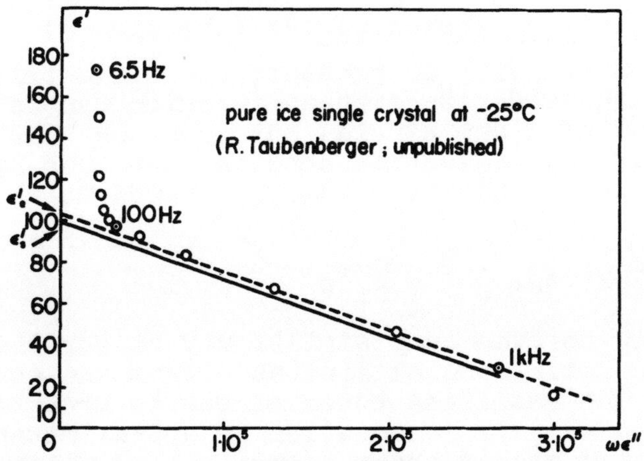

Fig.2 Linear plot ε'_m versus ($\varepsilon''_m\omega$) for a pure ice crystal at -25 °C

In the ideal case the following equation holds /3/

$$\varepsilon' = \varepsilon'_s - \tau\ (\varepsilon''\ \omega)\ ,$$

but if a frequency independent dc conductivity is present, the straight line is shifted parallel to the ideal line

$$\varepsilon'_m = \varepsilon'_s + (\tau\ \sigma'_o/\varepsilon_o) - \tau\ (\varepsilon''_m\ \omega)\ .$$

The extrapolation of the experimental line intersects at the too high value $\varepsilon'_{extrap.} = \varepsilon'_{s_{true}} + \tau\ \sigma'_o/\varepsilon_o$.

The true static dielectric constant can be calculated, if σ'_o has been determined. This plot is very favorable as the slope $-\tau$ is unaffected by σ'_o and because ε'_s and τ result from a linear fit of all experimental points (R. Taubenberger and H. Vetsch, forthcoming paper).

In the case of an ideal Debye dispersion an <u>alternative linear plot</u> is possible and sometimes particularly useful:

$$\varepsilon' = \varepsilon'_\infty + \frac{1}{\tau}\left(\frac{\varepsilon''}{\omega}\right).$$

However, if there is a $\sigma'_o \neq 0$, this plot gives no straight line any more due to the $1/\omega^2$-dependence of the conductivity term in the equation

$$\varepsilon'_m = \varepsilon'_\infty + \frac{1}{\tau}\left(\frac{\varepsilon''_m}{\omega}\right) - \frac{1}{\tau}\,\frac{\sigma'_o}{\varepsilon_o\omega^2}\ .$$

This representation cannot be used.

1.3 <u>Cole-Cole plot: ε''_m versus ε'_m</u> /4/ (see fig. 3)

Even if a dc conductivity is present, the dielectric data of the high and moderate frequency region can be fitted to a segment of a semicircle. But then the value of the real part of the dielectric constant obtained by the intersection of the semicircle $\varepsilon'_{extrap.}$ needs to be corrected:

$$\varepsilon'_{s_{true}} = \varepsilon'_{extrap.} - 2\tau\ \sigma'_o/\varepsilon_o\ .$$

This equation is derived in a similar way as for the ordinary Cole plot, i.e. one eliminates ω from the equations for ε' and ε''. The resulting equation can be brought to the form of a semicircle equation by approximations, namely by dropping higher order terms in $1/\omega^n$ which are large for low frequencies. In view of the approximate character of this correction and the fact that values of σ'_o and τ need to be known, this plot is not very useful for conductive dielectrics. The correction has not been used in many papers e.g. /2/ and /5/. Caution is required when interpretations of the data are attempted.

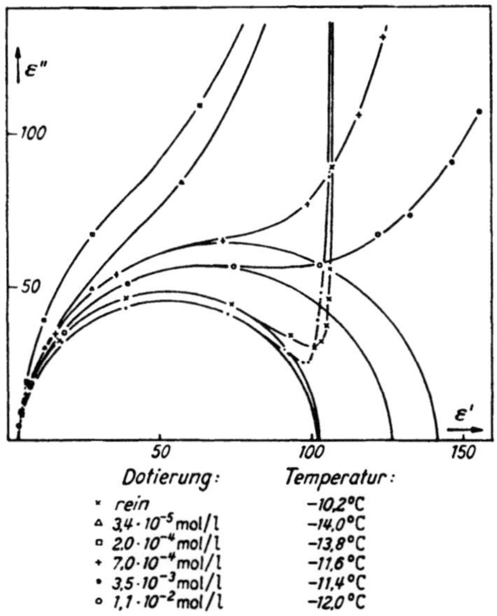

Fig.3 Cole-Cole plots of the same specimens as on figure 1 /2/

1.4 Conductivity plot: $(\varepsilon_m''\omega)$ versus ω^2 /6/

In the limit $\omega^2\tau^2 \ll 1$ we can get a linear plot with the following equation

$$(\varepsilon_m'' \omega) = (\varepsilon_s' - \varepsilon_\infty')\tau\, \omega^2 + \sigma_o'/\varepsilon_o \, .$$

The extrapolation to zero frequency intersects the ordinate at σ_o/ε_o .

The value obtained can be inserted in the corrections to ε_s' of § 1.2 and 1.3. The slope $(\varepsilon_s' - \varepsilon_\infty')\tau$ might be used as a check for consistency of the ε_s' and τ -values found from the linear plot of § 1.2.

2. Ice capacitor with space-charge effects (electrode polarization) /1/

The space-charge effects lead to an additional dispersion region at low frequencies. For not too strongly doped ice this region is sufficiently separated from the Debye dispersion region for temperatures above about -40 °C. The log-log plot of § 1.1 has been widely used in such cases /1/. But this is not satisfactory, if a high precision is required. The linear plot of § 1.2 would be more suitable, but the question arises, what value has to be inserted for σ_0' in the correction term.

We can approximate the space-charge dispersion by a Debye type dispersion. This means that we have to supplement our equivalent circuit by an additional C_2R_2 branch. Sufficient separation of the dispersion region means that there is a frequency range such that

$$G_1 > \omega C_1 \quad \text{and} \quad G_2 < \omega C_2 .$$

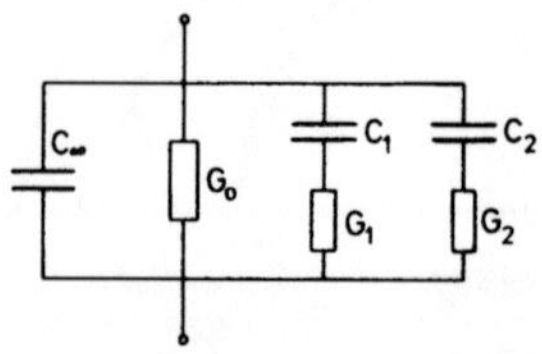

In this frequency range we get for the real part of the admittance

$$G'_{plateau} = G_0' + G_2 + \omega^2 C_1{}^2/G_1 .$$

This means that we get a flat plateau of $\sigma'(\omega)$, if the third term, the tail of Debye dispersion, $\varepsilon_0\tau(\varepsilon_s'-\varepsilon_\infty')\omega^2$ has become sufficiently small.

3. Conclusions

a) This shows that the conductivity plot (§ 1.4) in the presence of space charge effects leads to an extrapolated conductivity value $\sigma'_{plateau} = \sigma_0' + \sigma'_{space\ charge}$, where $\sigma'_{space\ charge}$ is the high frequency conductivity of the space charge dispersion $\sigma'_{sc} = \varepsilon_0 \cdot \Delta\varepsilon_{sc}/\tau_{sc}$.

b) Further, it can easily be proved that the value of σ_0' which has to be inserted into the correction terms of § 1.2 and 1.3, if space charge effects are present, is the value $\sigma'_{plateau}$ and not the true dc value, which can only be measured at frequencies below the space charge dispersion, i.e. below 10^{-2} Hz.

It may be noted that in contrast to the true dc conductivity value, $\sigma'_{plateau}$ merits no particular interest, if bulk properties of ice are studied. Chan et al. /6/ have mistaken the value obtained from a conductivity plot (§ 1.4) for the dc conductivity and have analyzed it for its pressure dependence in their chapter 3. They may not have been aware of the necessary conductivity correction to their ε_s'-values obtained by Cole plots. Fortunately the conductivity of their sample was so small that the correction was within the limits of accuracy.

An instructive example is shown in figure 3 /2/ where static dielectric constant values extrapolated on Cole plots are in a range from about 100 to 140, whereas the curves of figure 1 rather indicate the same true ε_s' value for all specimens. Even for their two purest crystals the correction to $\varepsilon'_{extrap.}$ is 3.

References

/1/ Steinemann, A., H. Gränicher, Helv.Phys.Acta 30, 553-580 (1957)

/2/ Dengel, O., N. Riehl, A. Schleippmann, Physik kondens. Materie 5, 83-8 (1966)

/3/ Cole, R.H., J.Chem.Phys. 23, 493 (1955)

/4/ Cole, K.S., R.H. Cole, J.Chem.Phys. 9, 341 (1941)

/5/ Humbel, F., F. Jona, P. Scherrer, Helv.Phys.Acta 26, 17-32 (1953)

/6/ Chan, R.K., D.W. Davidson, E. Whalley, J.Chem.Phys. 43, 2376-83 (1965)

ON THE INTERPRETATION OF THE PRESSURE DEPENDENCE OF PROPERTIES CONTROLLED BY LATTICE DEFECTS

H. Gränicher

Eidg. Technische Hochschule

Zürich / Switzerland

In the review paper given in this conference /1/ the following two problems were mentioned in the list of open questions:

a) What mechanisms are responsible for the selfdiffusion in ice and the diffusion of impurity molecules such as HF?

b) From electrical studies certain typical data of the defects, e.g. energies of formation and of activation of the mobility have been derived /2/ and for the theoretical description only basic properties of the defects need to be postulated /3/. However the details of the mechanism of formation and the structure of the defects are still not known.

This paper describes the possibility to gain information on these problems by the study of the relevant macroscopic properties as a function of hydrostatic pressure.

1. Let us first consider the case of diffusion and, in order to be specific, we assume for the moment that Schottky vacancies provide the predominant diffusion rate.

The diffusion coefficient D is expressed by the following relation /4/:

$$D = D_o \cdot \exp - \frac{\Delta G}{RT} = D_o \cdot \exp - \frac{\Delta G_v + \Delta G_m}{RT} . \qquad (1)$$

The constant D_o is considered as temperature and pressure independent. The activation energy ΔG can be split up into the energy of formation of the vacancies ΔG_v and the energy of activation of diffusion of the vacancy ΔG_m. Since ΔG is the Gibbs free energy (free enthalpy) of activation, the pressure derivative of ΔG is equal to the total volume change associated with the process. Therefore we get the following expression:

$$\left(\frac{\partial \ln(D/D_o)}{\partial p}\right)_T = -\frac{\Delta V_a}{RT} = -\frac{1}{RT}(\Delta V_v + \Delta V_m). \qquad (2)$$

ΔV_a is called the activation volume; it consists of ΔV_v, the partial molar volume of the vacancies and ΔV_m, the partial molar volume of the activated complex at the saddle point. This latter quantity is the volume change, when the moving molecule (or atom) squeezes itself through between neighboring molecules. This ΔV_m is probably small and positive for all types of diffusion mechanisms.

1.1 The quantity ΔV_v is more characteristic. For vacancies in a rigid lattice it is equal to the volume of one mole of the substance, but in a real lattice, there is always some elastic relaxation around the vacancy. Hence for Schottky vacancies we expect $\Delta V_v \leq V_M$, where V_M denotes the molar volume, and the activation volume is certainly positive. We can understand the effect of the hydrostatic pressure easily by invoking the principle of Le Chatelier. If vacancies are present in the crystal, its density is smaller than the one of the ideal perfect crystal. The pressure tending to increase the density, will produce a decrease of the concentration of vacancies and, for $\Delta V_m > 0$, a decrease of the number of jumps. Therefore, with increasing pressure the diffusion coefficient will decrease.

1.2 In the case of the Frenkel mechanism, vacancies and interstitial molecules will be present in equal number. ΔV_v will be close to zero and hence the defect concentration will have almost no effect on the density. The diffusion coefficient will be nearly pressure independent.

1.3 If interstitial molecules alone were present, the density would be higher than in the ideal crystal and would increase with pressure. Hence, in this case the diffusion coefficient increases with rising pressure, if the quantity ΔV_m does not contribute appreciably. ΔV_m depends on whether interstitial molecules move by the interstitial mechanism or by the interstitialcy mechanism.

This shows that the study of the pressure dependence of diffusion gives the possibility to prove, which type of defect mechanism is present. The argument will be most significant, if $\Delta V_v = \Delta V_a - \Delta V_m$ can be calculated based on a direct determination of ΔV_m.

1.4 In principle one can get ΔV_m by a pressure experiment on a sample quenched from a temperature close to the melting point down to a temperature where a relatively high defect concentration will stay frozen-in. One then has to measure the annealing rate, i.e. the rate with which the equilibrium defect concentration is reestablished. It is to be feared that such a determination of ΔV_m will be nearly impossible for ice. One will have to rely on theoretical estimates of this quantity. This limited knowledge of ΔV_m might be perhaps not a too serious problem; if the activation volume ΔV_a turns out to be strongly positive or negative. In addition to the self-diffusion experiments at various pressures suggested so far, Kirkendall type studies might be worthwhile with diffusion couples consisting of pure and doped single crystalline ice (e.g. with HF or NH_4F).

2. Now we turn to the case of the pressure dependence of the electrical properties. Dielectric investigations with poly-crystalline samples of hexagonal ice have been made by Chan, Davidson and Whalley /5/ and interpreted by a reaction rate theory approach similar to the one of section 1. From the pressure coefficient of the relaxation time they derived an activation volume of (2.9 ± 0.3) cm^3/mole and concluded that this value is good support for the Bjerrum mechanism as responsible for the majority mechanism. It can be seen from their equations that the comparison of pressure data taken on pure and HF- or NH_3-doped crystals enables the separate determination of the activation volume values for formation and defect diffusion.

2.1 In the following a more elaborate discussion is presented which is based on the formalism given in the theory of Jaccard /6/. First it should be noted that the high frequency conductivity $\sigma_\infty = \varepsilon_0(\varepsilon_s^\perp - \varepsilon_\infty^\perp)/\tau$, to be measured at frequencies sufficiently above the Debye dispersion range, is the most appropriate quantity to analyze, as it is simply the sum of the specific conductivities of the various defects: D,L, positive and negative ion-states.

$$\sigma_\infty = \sum_{i=1}^{4} \sigma_i = \sigma_D + \sigma_L + \sigma_+ + \sigma_- . \tag{3}$$

As an example, the value σ_L can be expressed as /6/:

$$\sigma_L = 4e_{DL}^2\, b^2\, n_L\, p_L \,/\, 9\, kT. \tag{4}$$

In this equation e_{DL} is the effective charge of Bjerrum defects, b the hydrogen-oxygen distance, n_L the concentration of L-defects and p_L the transfer probability for diffusion of such a defect.

As the elastic anisotropy in ice is small, we may assume that the oxygen-oxygen distances r vary isotropically if hydrostatic pressure is applied.

Hence
$$\kappa = -\frac{1}{r} \cdot \frac{\partial r}{\partial p} \quad , \tag{5}$$

where κ denotes the linear compressibility coefficient which is $4 \cdot 10^{-6}$ bar^{-1} at 0 °C /7/. The O-H-distance b will of course vary too. It is well known that hydrogen bonds tend to become symmetrical as the O-O distance decreases. Therefore, we have to introduce a factor $-\alpha$ and we define

$$\alpha\, \kappa = \frac{1}{b} \cdot \frac{\partial b}{\partial p} \,. \tag{6}$$

We can derive a value of α e.g. from semi-theoretical data /8/:

$$\alpha = -\frac{r}{b} \cdot \frac{\partial b}{\partial r} = +\,0.55\,. \tag{7}$$

2.2 Let us first treat the case of pure ice where $\sigma_D + \sigma_L >> \sigma_+ + \sigma_-$ and $n_D = n_L$ (intrinsic ice). From the mass action law we get

$$\frac{\partial \ln n_{DL}}{\partial p} = \frac{1}{n_{DL}} \cdot \frac{\partial n_{DL}}{\partial p} = 3\kappa - \frac{\Delta V_B}{2kT} \,. \tag{8}$$

The first term takes into consideration the increase of the crystal density by the applied pressure. The second term contains the molar volume of formation of a mole of Bjerrum defect pairs ΔV_B.

It is appropriate to introduce transfer numbers $u_D = \sigma_D/(\sigma_D+\sigma_L)$ and $u_L = \sigma_L/(\sigma_D+\sigma_L)$, which we can derive from conductivity measurements of doped ice. The relative pressure coefficient of the hf conductivity $S_{intrinsic}$ becomes

$$S_{intrinsic} \equiv \frac{1}{\sigma_\infty} \left(\frac{\partial \sigma_\infty}{\partial p}\right)_T = \tag{9}$$

$$= 2\,\frac{\partial \ln e_{DL}}{\partial p} + 2\alpha\,\kappa + 3\kappa - \frac{\Delta V_B}{2kT} + u_D \frac{\partial \ln p_D}{\partial p} + u_L \frac{\partial \ln p_L}{\partial p} \,.$$

2.3 Similarly we can derive $S_{extrinsic}$ for
a) HF-doped ice, where n_L is pressure and temperatur independent and where $\sigma_\infty \cong \sigma_L$, and
b) NH_3-doped ice, in which D-defects play the analogous role as the L's in HF-doped ice.

$$S_{ext_L} \equiv \left(\frac{\partial \ln \sigma_{\infty L}}{\partial p}\right)_T = 2\,\frac{\partial \ln e_{DL}}{\partial p} + 2\alpha\,\kappa + \frac{\partial \ln p_L}{\partial p}, \quad (10)$$

$$S_{ext_D} \equiv \left(\frac{\partial \ln \sigma_{\infty D}}{\partial p}\right)_T = 2\,\frac{\partial \ln e_{DL}}{\partial p} + 2\alpha\,\kappa + \frac{\partial \ln p_D}{\partial p}. \quad (11)$$

If we subtract the two equations, we get

$$\Delta S_{LD} \equiv S_{ext_L} - S_{ext_D} = \left(\frac{\partial \ln p_L}{\partial p} - \frac{\partial \ln p_D}{\partial p}\right) \quad (12)$$

since $p_L = \nu_{oL} \cdot \exp - E_L/kT$,

$$\frac{\partial \ln p_L}{\partial p} = \frac{\partial \ln \nu_{oL}}{\partial p} - \frac{\Delta V_L}{kT} \quad (13)$$

where ΔV_L is the partial molar volume of the saddle point configuration of the L diffusion. An analogous expression holds for D-defects. In section 1 the pressure dependence of the preexpontential factors is neglected. If we make the same approximation we may write

$$\Delta S_{LD} \cong - (\Delta V_L - \Delta V_D)/kT. \quad (14)$$

Similarly we can consider the pressure dependence of the effective charge e_{DL} as small and use the approximate equation to (10,11) for an evaluation of ΔV_L or ΔV_D.

$$S_{ext_L} \cong 2\alpha\,\kappa - \frac{\Delta V_L}{kT}. \quad (15)$$

$$S_{ext_D} \cong 2\alpha\,\kappa - \frac{\Delta V_D}{kT}. \quad (16)$$

2.4 Now we combine intrinsic with extrinsic data and get

$$\Delta S_{iL} \equiv S_{int} - S_{ext_L} = 3\kappa - \frac{\Delta V_B}{2kT} - u_D \left(\frac{\partial \ln p_L}{\partial p} - \frac{\partial \ln p_D}{\partial p}\right), \quad (17)$$

$$\Delta S_{iD} \equiv S_{int} - S_{ext_D} = 3\kappa - \frac{\Delta V_B}{2kT} + u_L \left(\frac{\partial \ln p_L}{\partial p} - \frac{\partial \ln p_D}{\partial p}\right). \quad (18)$$

By adding the two last equations we obtain, using (12),

$$\Delta S_{iL} + \Delta S_{iD} = 6\kappa - \frac{\Delta V_B}{kT} + (u_L - u_D) \cdot \Delta S_{LD} \, . \qquad (19)$$

As $(u_L-u_D) \leq 1$, the error in ΔS_{LD} may be expected to contribute little to the right hand side of the equation. It has been obtained without approximations and should enable one to get a reasonably good value for ΔV_B.

As pointed out by Jaccard /6/ for symmetry reasons $-e_L = e_D = e_{DL}$. This is the reason why in equations (12) and (17) to (19) the potential pressure dependence of the effective charge occurs no longer. In his paper of 1964, Jaccard /6/ discusses the possibility that the model quantities (e.g. e_i , σ_i) might have tensor character with respect to the crystal axes. The anisotropy of the pressure dependent electrical properties would be of particular interest. However, with the exception of the static dielectric constant no sufficiently reproducible electric measurements could be made even at normal pressure till now.

2.5 The influence of hydrostatic pressure on the ion-state mechanism has not been presented. The derivation follows exactly the same path. An essential difference shows up in the transfer probabilities p_+ and p_- . According to existing experimental and theoretical evidence, they are almost temperature independent and are due to proton tunneling. Therefore, these probabilities are expected to be strongly pressure sensitive.

2.6 It has been shown that the study of the pressure dependence of electrical quantities leads to molecular data which are strongly dependent of the actual structure of defects and the diffusion transfer mechanisms. Therefore, an experimental investigation with single crystals seems to be very rewarding. Such a study is in progress in our laboratory and preliminary measurements of R. Taubenberger and H. Vetsch are shown in figure 1 in comparison to published polycrystalline data /5/. One single crystal had evaporated gold electrodes and the other frozen-on Pt foils. The static dielectric constant agrees to within 2 % with literature values calculated for the particular orientation. The discrepancy in the absolute values of the relaxation time τ is probably due to impurities. If the difference in the activation volume can be considered as significant, the anisotropy of the activation volume seems to be quite considerable.

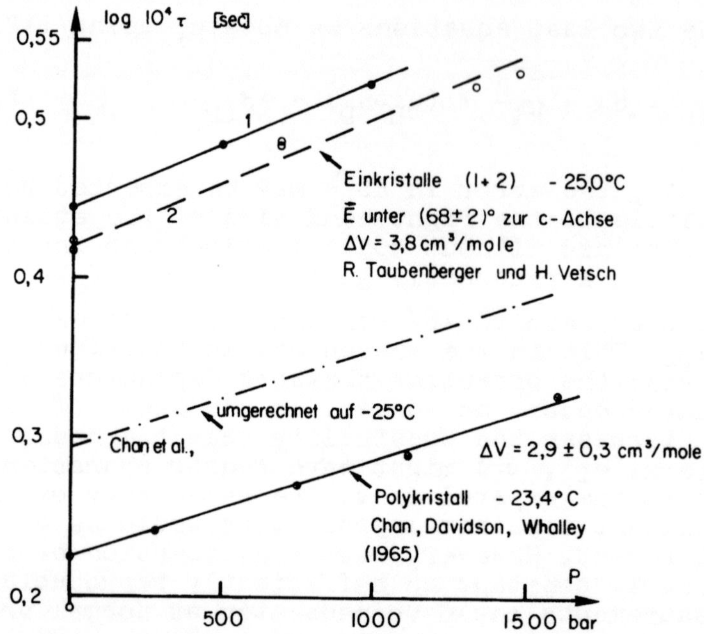

Fig.1 Pressure dependence of the dielectric relaxation time τ of hexagonal ice

References

/1/ Gränicher, H., Physics of Ice, Proc.Int.Symp., Munich, Plenum Press, New York (1969)

/2/ e.g. H. Gränicher, Physik kondens.Materie 1, 1-12 (1963)

/3/ Gränicher, H., Proc.Roy.Soc. A 247, 453-461 (1958) and Z.Kristallogr. 110, 432-71 (1958)

/4/ Shewmon, P.G., "Diffusion in Solids", McGraw-Hill Book Comp. 1963

/5/ Chan, R.K., D.W. Davidson, E. Whalley, J.Chem.Phys. 43, 2376-83 (1965)

/6/ Jaccard, C., Helv.Phys.Acta 32, 89-128 (1959) and Physik kondens.Materie 3, 99-118 (1964)

/7/ Dantl, G., Physik kondens.Materie 7, 390 (1968)

/8/ Lippincott, E.R., R. Schroeder, J.Chem.Phys. 23, 1099-1106 (1955)

ELECTRIC RESONANCE : APPLICATION TO THE HYDROGEN BOND

J. Amran Sussmann

Israel Atomic Energy Commission

Soreq Nuclear Research Center, Yavne, Israel

Abstract

The development of electric resonance is shortly reviewed. It is shown that protons in symmetrical hydrogen bonds should exhibit electric resonance absorption. The peculiarity of this system leading to a decrease in line intensity with increasing static electric field is described.

Conditions for the experimental observation are discussed.

1. Introduction

The resonant absorption of radiation, leading to transitions between quantum mechanical states of atomic or molecular systems is usually studied under two different headings: The first is optical (and infrared) spectroscopy and refers in general to transitions having electric dipole character. They involve photons in the infrared and visible range of frequencies. The other heading is magnetic resonance, and comprises transitions mostly of magnetic dipole character involving photons in the microwave range. Transitions of electric dipole character involving microwave frequency photons have been discussed and observed only recently. We shall put all such phenomena under a new heading called electric resonance. In analogy to magnetic resonance, in the electric resonance phenomenon, the frequency of the absorbed radiation can be changed, but now it is done by applying a static electric field.

Volger /1/ was the first to mention the possibility of electric resonance and Sussmann /2/ described a model of electrons trapped at vacancies, which exhibits such a behaviour. Känzig /3,4/ observed experimentally a phenomenon which is related to this process: the reorientation of O_2^- substitutional impurities in alkali halides by phonon absorption /5,6,7,8/. Kuhn and Luty /9/ predicted the reorientation of OH^- substitutional impurities in alkali halides by phonon absorption. This was then observed by Bron and Dreyfus /10/ and Feher, Shepherd and Shore /11/. Electric resonance absorption by trapped electrons has subsequently been observed in smoky quartz by Kerssen and Volger /12/.

The physical systems which do present electric resonance belong to a wide range of setups. There are ions which change their quantum state; there are transitions between states with different electric dipoles and between states without dipoles. A classification based on these differences turns out to be unsatisfactory. No system can be considered a pure case.

Atomic displacements are always involved, even when an electronic transition is considered. Thus the electron cannot be taken "naked" but must be "dressed" by the deformation of the lattice in its neighbourhood, deformation which depends on the state of the electron. Furthermore, even when the states involved in the transition do not have an electric dipole in the absence of an electric field, one or both of them will present an electric dipole when the static electric field is applied, or else no electric dipole transition would be possible.

Due to this mixing of features, each physical system presents its own characteristics and must be analysed on its own. Some may present peculiar behaviour as it will be shown in the case of the symmetrical hydrogen bond. In this case the system is linear and no state presents an electric dipole in zero electric field.

An applied static electric field changes the states of the system by mixing them. The overlap between the new states will depend on the amount of mixing and thus lead to a dependence of the photon absorption probability on the intensity of the applied static electric field.

2. The hydrogen bond

The proton in the hydrogen bond can be described as a particle in a double well potential. This potential may be a symmetrical or an asymmetrical one, relatively to the center of the bond, depending on the atoms on each side of the bond. We shall consider only the case of symmetrical double wells.

In a perfectly symmetrical double well potential, the wave functions of the proton are either symmetrical or antisymmetrical, leading to a total delocalization. Wave functions of different symmetry can be connected by a perturbation matrix element due to an electric field parallel to the bond. Thus a static electric field produces a mixing of the states, leading to localization in different wells of the wave functions resulting from rediagonalization.

Let us restrict ourselves to the lowest symmetric and antisymmetric wave functions Ψ_s and Ψ_a with an energy difference of δ. The static electric field E_o produces an asymmetry E_{as} between the two wells and leads to the new wave functions

$$\Psi_1 = \Psi_s \cos \tfrac{1}{2} \Phi + \Psi_a \sin \tfrac{1}{2} \Phi \qquad (1)$$

$$\Psi_2 = - \Psi_s \sin \tfrac{1}{2}\Phi + \Psi_a \cos \tfrac{1}{2} \Phi$$

where

$$\tan \Phi = E_{as}/\delta = 2 \mu\varepsilon_o/\delta \qquad (2)$$

and

$$\mu = e \int \Psi_s \, x \, \Psi_a \, dx. \qquad (3)$$

The energy splitting becoming

$$E_2 - E_1 = \sqrt{(\delta^2 + 4\mu^2\varepsilon_o^2)} = \sqrt{(\delta^2 + E_{as}^2)}. \qquad (4)$$

The degree of localization depends on the strength of the applied electric field.

The parameter $\lambda = \tan^2\Phi = \left(\frac{E_{as}}{\delta} \right)^2$ (5)

is very convenient to describe the degree of localizations, $\Phi = \pi/2$ standing for a perfect localization in one or the other potential well, and $\Phi = o$ for complete delocalization.

An alternating electric field ε_1 with a small amplitude compared to that of the static one will not change the above diagonalization and will lead to transitions between the new states (1), the transition matrix element being

$$\mu_{1,2}\varepsilon_1 = \mu\varepsilon_1\delta / (E_2 - E_1) = \mu\varepsilon_1 / \sqrt{(1 + 4\mu^2\varepsilon_o^2 \delta^{-2})}$$

$$= \mu\varepsilon_1 / \sqrt{(1 + \lambda)}. \qquad (6)$$

Through the above expression the transition probability is seen to decrease with increasing localization. Furthermore it turns out not to depend on the barrier height directly but only indirectly through λ.

It should be remembered that when the barrier increases in height and thickness, δ decreases. This means that a high barrier between the minimum does not hinder electric resonance in a perfectly symmetric potential, between non localized states; this unexpected result is the consequence of the extreme idealization of a perfectly symmetric potential well. As soon as $E_{as} \neq 0$, the barrier height does effect the transition probability. For a given asymmetry E_{as}, the higher the barrier, the stronger the localization λ and the smaller the transition probability: The effect is easily understood if we remember that with increasing localization the overlap betwen the initial and final states decreases.

Not only external electric field but also random defects in the crystal do lead to asymmetry in the potential. If the average asymmetry is small compared to δ an inhomogeneous broadening of the resonant absorption line results. If it is large compared to δ strong localization results and then the line is not only broadened but its intensity decreased.

Searching for the optimum conditions for the observation of electric resonance, we have to look for cases with low barrier between the wells. If the barrier is too high and consequently δ small, local random strains will produce strong localization and thus reduce the absorption line intensity, already in zero external static electric field. On the other extreme, a too low barrier, by increasing δ, may push the resonance frequency into such a high frequency region which is experimentally unconfortable. For the same reason we should look for systems with small random internal strains. In the high field limit, where static field intensity and resonance frequency are proportional, the absorption line intensity is inversely proportional to the square of static field intensity. Optimum conditions may require to restrict experiments to the weak field region where the resonance frequency varies with the square of the static field intensity.

From the isotope effect on the dc electric conductivity in ice, information can be obtained on the barrier in one particular case of symmetrical double well potential /5/. The charge carriers responsible for the dc conductivity in ice are protons, jumping in H_2O ...H.....OH_2

bond, and the analogous HO ... H ..... OH bond. The mobility of the proton in the former bond has been measured /13/, as well as the effect of deuteration. An isotope factor close to 7 results. This isotope factor can be interpreted using the theory of phonon induced tunneling /5/. It leads thus to a barrier height V = 0.18 eV and to $\frac{\delta}{k} = 15^{o}K$ if no effective mass correction for the proton is assumed. If the effective mass is larger than 1, V_o and δ will decrease.

The present day, techniques would require smaller values of δ, as the above one would require very short wavelength ($\sim$ 1 mm) and very high static electric field ($\sim 10^5$V/cm). Nevertheless the above considerations do show that the search for hydrogen bonded systems presenting electric resonance is by no means hopeless.

References

/1/ Volger, J., Inaugural Address, Utrecht 1959

/2/ Sussmann, J.A., Proc.Phys.Soc. 79, 758 (1962)

/3/ Känzig, W., Phys.Rev.Lett. 7, 304 (1961)

/4/ Känzig, W., Phys.Chem.Solids 23, 479 (1962)

/5/ Sussmann, J.A., Phys.kondens.Materie 2, 146 (1964)

/6/ Pirc, R., B.Zeks, P.Gosar, J.Phys.Chem.Solids 27, 1219 (1966)

/7/ Sussmann, J.A., J.Phys.Chem.Solids 28, 1643 (1967)

/8/ Gosar, P., R.Pirc, "Magnetic Resonance and Relaxation", Proc. of the XIVth Colloque Ampere, ed. R.Blinc, North Holland (1967), p.636

/9/ Kuhn, U., F.Luty, Solid State Commun. 3, 31 (1965)

/10/ Bron, W.E., R.W.Dreyfus, Phys.Rev.Lett. 16, 165 (1966)

/11/ Feher, G., L.W.Shepherd, H.B.Shore, Phys.Rev. Lett. 16, 500 (1966)

/12/ Kerssen, J., J.Volger, Phys.Lett. 24A, 647 (1967)

/13/ Eigen, M., L.De Maeyer, H.Ch.Spatz, Colloquium on Physics of Ice Crystals, Erlenbach, Switzerland (1962)

DIELECTRIC PROPERTIES OF ICE I

R.H. Cole and O. Wörz

Chemistry Department

Brown University, Providence, Rhode Island / USA

Abstract

A reinvestigation of the dielectric properties of H_2O-ice was performed covering a much wider range of temperatures than previous investigations. The frequency range 0.05 Hz < f < 200 kHz was covered by three- and four-terminal bridge measurements. This wide range of frequencies allowed unambiguous detection of any influence of electrode polarization on the results even at low temperatures. Highly purified samples were obtained by zone refining. We find the dc conductance to be much lower than generally assumed and to be strongly dependent on the purity of the sample. Experimental values of σ and ε and their f(T) will be presented. The improvements in accuracy compared to earlier investigations will be outlined and the significance of these more accurate values for existing theories will be discussed.

I. Introduction

It was found in the past that the dielectric properties of ice can well be described by the Debye equations using a single relaxation time. In an accurate determination of the parameters describing such a dispersion two major complications have to be overcome: (a) the dc conductivity causes a frequency independent contribution to all conductance readings and has to be taken into account, (b) electrode polarization causes frequency dependent deviations of the results from the true intrinsic values of the sample. The main aim of the present investigation was an accurate determination of the static dielectric constant ε_o of pure ice in a wide temperature range. In particular the determination of ε_o is complicated by the facts mentioned, since the relaxation frequencies at low temperatures are much lower than the frequency at which electrode polarisation is of serious influence. In some of the earlier measurements the separation of the different effects was further complicated by the fact, that only at high temperatures could the dispersion region be investigated by bridge measurements, whereas only transient measurements could be applied at low temperatures. The inconsistency of literature data is at least to some extent due to incomplete interpretation of measurements.

II. Four terminal measurements

For the determination of dc conductivity four terminal measurements are employed to eliminate the influence of electrode polarisation. The same principle can be employed for ac measurements. Fig.1 shows a block diagram of a bridge in which two electrometer amplifiers

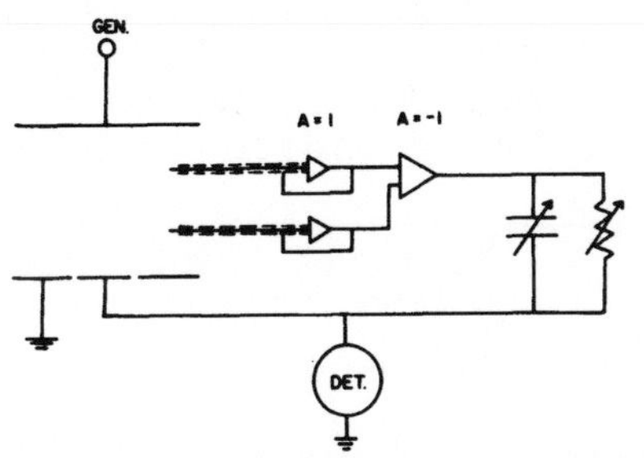

Fig. 1 Block diagram of four terminal ac bridge

with amplification factor one are used to measure the potential at two points in the sample /1/. The negative potential difference is fed into the adjustable side of the bridge. The resulting current to ground is compared with the sample current in the detector circuit. The probe currents are smaller than 10^{-14} amps and are assumed to be negligible compared to the sample current, so that no distortion of the electric field is produced. To eliminate capacitive shunting effects and errors from surface conduction the probe shields are driven from the output of the electrometer amplifiers. Since electrode polarisation is confined to a very thin layer near the surface of the electrodes, one expects to measure the true properties of the sample with such a device. Circuits of this type were first discussed by Schwan /2/, and promising experiments on biological samples were reported.

The bridge used in our investigation covers the frequency range 0.05 Hz up to 1 kHz, is direct reading in terms of conductance and capacitance of a parallel equivalent circuit and has a resolution of three parts in 10000 at all frequencies. A series of three- and four-terminal impedance measurements was made on ice samples. A three terminal transformer bridge was used to extend the measurements up to 200 kHz. **Although** the four terminal measurements were found not to yield the expected success in eliminating the influence of electrode polarisation, they still are of some interest and a short qualitative discussion will be given.

Fig.2 gives the capacitance readings versus frequency measured on three samples with different dc conductivity at -10°C. The purification of the ice was achieved by a zone refining process, and the curves shown represent **measurements after different zone refining runs. The Debye dispersion occurs at higher frequencies and is not** shown in the diagram. The lower part gives the three terminal results. The frequency at which electrode polarisation occurs is lower **the** lower the dc conductivity, the corresponding values being indicated on the diagram.

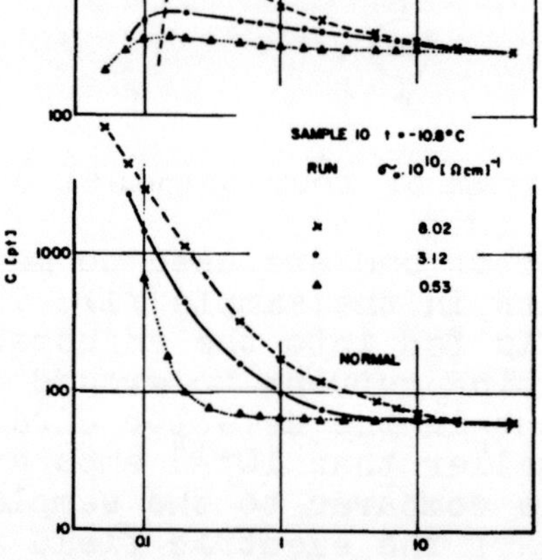

Fig.2 Capacitance versus frequency; lower part: three terminal result; upper part: four terminal result

Using the probe technique we find a small increase in capacitance reading with decreasing frequency which is followed by a strong decrease, and negative values are obtained for frequencies lower than a certain value. In trying to understand the observed behaviour another fact is helpful: In the case of a low conductance sample, which means small electrode polarisation effects, or at high enough frequencies, the bridge can be balanced very exactly and the Lissajous pattern used for null indication reduces to a straight line. When electrode polarisation is present, however, a strong second harmonic is observed. Since the bridge can be balanced for one frequency only, no complete zero is obtained. This observation suggests that the linear relation between the potential at the probes and the sample current, which is used in the derivation of the bridge balance condition, is not correct, and that the chemical potential due to a concentration gradient has to be taken into account. Regarding only one type of mobile charge carriers, the equation of balance including the effects of diffusion and of motion under the influence of the electric field can be written as /3/

$$\frac{\partial c}{\partial t} = D \frac{\partial^2 c}{\partial x^2} - u \frac{\partial (c \cdot E)}{\partial x}$$

where c is the concentration of the charge carriers, D the coefficient of diffusion and u the mobility. In addition we have Poissons equation

$$\frac{\partial E}{\partial x} = \frac{4\pi e}{\varepsilon} \cdot c$$

and the equation of continuity, which is in the case of blocking electrodes:

$$u\, c\, E - D \frac{\partial c}{\partial x} = 0 \quad \text{at } x = 0 \text{ and } x = L.$$

These equations are nonlinear, and since they govern the motion of charge carriers in the sample, the current flowing through the medium will contain all harmonics of the applied voltage. In theories of electrode polarisation it is in general assumed that the driving voltage is small enough to permit the neglect of all harmonic generation and approximate solutions containing the basic frequency only are worked out.

The maximum field strength in our measurements was about 1 V/cm. From the experimental results we conclude that this value is already too high to justify a linearisation of the problem in the case of ice, even if only small amounts of impurities are present. Harmonics of higher order would have to be included in the solution of the above equations to explain the experimental data as obtained by the probe measurements. For the study of the dielectric properties of pure ice, however, four terminal measurements are of little help at least with the probe arrangement which was used, and the results reported in the following are taken from three terminal measurements only.

III. Production of samples and handling of data

To be able to determine ε_0 in the whole temperature range with sufficient accuracy very pure samples had to be produced. This was achieved by zone refining. The electrode system was completely frozen into the ice. Zone refining was found to be very effective if the ice was first produced from very pure water and if the electrodes were just thin metallic plates. The use of a guard ring, however, made relatively thick supporting Teflon parts necessary, which made zone refining much less effective. Only a few samples with sufficiently low dc conductance could be produced, and the best sample (13C) obtained is used for discussion. At higher temperatures, where also data from less perfect samples could be analysed, we find good consistency of the results for ε_0.

Even with the purest samples electrode polarisation has to be taken into account. The following way of handling the data gave satisfactorily consistent and accurate values of the dc conductivity and the static dielectric constant.

If the dielectric properties can be described by one single relaxation time, a plot of conductance versus capacitance yields a straight line with slope $-\tau$ (fig.3, curve A). If there is a dc conductance we obtain curve B, but in general we then will also have electrode polarisation and get a plot of type C. In this case G_0 can not be read from the plot directly, but can be calculated from any two measured conductance values, using the Debye equations. It is easy to show that calculations of G_0 from different combinations of frequencies $f_1 f_2$, $f_1 f_3$ ... will yield a systematic increase of the result with increasing $(f_1 - f_1)$, if $G(f_1)$, the measured value at the lowest frequency used, is already smaller than the true conductance due to electrode polarisation. If a single value of G_0 is obtained from all combinations within experimental error one may assume that there is no serious error involved. Knowing G_0 the static dielectric constant is obtained from the intercept b in the diagram.

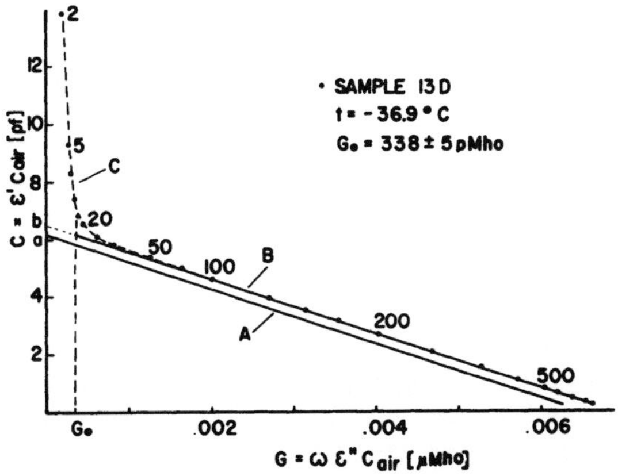

Fig.3 Schematic plots of capacitance versus conductance. A - no dc conductance, B - with dc conductance, C - dc conductance and electrode polarisation

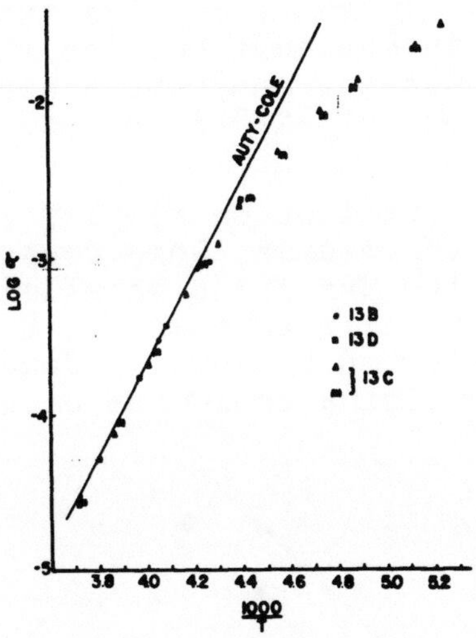

Fig.4 Relaxation time τ versus 1000/T

IV. Results

In fig.4 measured relaxation times are compared with the result of Auty and Cole /4/. Their result can well be described by a rate equation. We get very good agreement at high temperatures, but clear deviations for temperatures below -50°C. We know from experience that it is tempting to read relaxation times from plots of conductance versus capacitance, which fit the rate equation given by Auty and Cole in the whole temperature range studied, but which are obviously affected by electrode polarisation. We believe that our values are proper intrinsic relaxation times, since they were derived only from data which showed no evidence of significant polarisation errors.

The temperature dependence of ε_0 (fig.5) is well described by the equation

$$\varepsilon_0 - \varepsilon_\infty = 20715/(T-38),$$

thus yielding a Curie temperature of 38°K. A deviation from the above equation is seen for temperatures below -65°C; which we ascribe to a crack in the sample for several reasons. First, an easy calculation shows that an airgap of width $6 \cdot 10^{-4}$cm explains the observed deviation, and this value is easily attributable to the difference in expansion coefficients between Teflon and ice. The data obtained on raising the temperature of sample 13C from -80°C to 0°C indicate that the airgap was closed to some extent at temperatures higher than -40°C, and the differences in σ_0, σ_∞ and τ for the two runs are all consistent with such an explanation. The values of Auty and Cole, from which Onsager and Dupuis had inferred a formal Curie temperature of -146°K, are shown for comparison.

The results shown are obtained on samples with the electric field normal to the c-axis. Some measurements made in a different cell with the field parallel to the c-axis indicate a much smaller anisotropie of the dielectric constant than reported by Humbel, Jona, Scherrer /5/. But since zone refining could not be used for

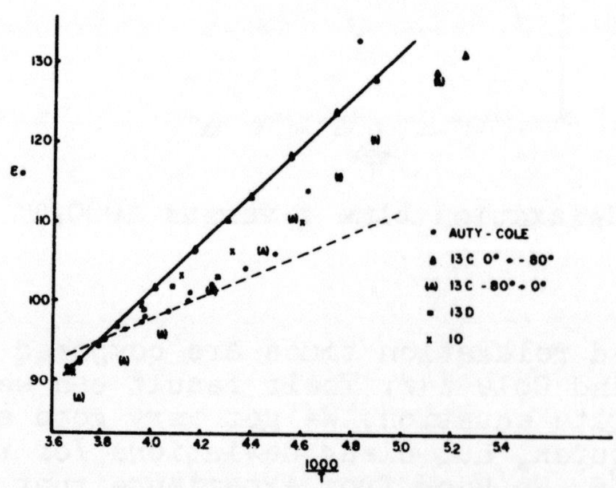

Fig.5 Static dielectric constant ε_0 versus 1000/T

purification with that cell our experimental error is too big for a quantitative decision.

Conductivity data obtained on three different samples are compared with Bradley's result in fig.6. For sample 13C the activation energy is found to be E_σ = 12.4 kcal/mole. In deriving this number low temperature data subject to polarisation errors were disregarded. Similar values were found on other samples with low conductivity, but the agreement is rather poor. A very different value of about 2 kcal/mole is found on samples with a conductivity of the order of 10^{-9} $(\Omega\ cm)^{-1}$ at $-10^{\circ}C$, whereas Bradley found a much higher value also on such high conducting samples. No explanation for the disagreement can be offered.

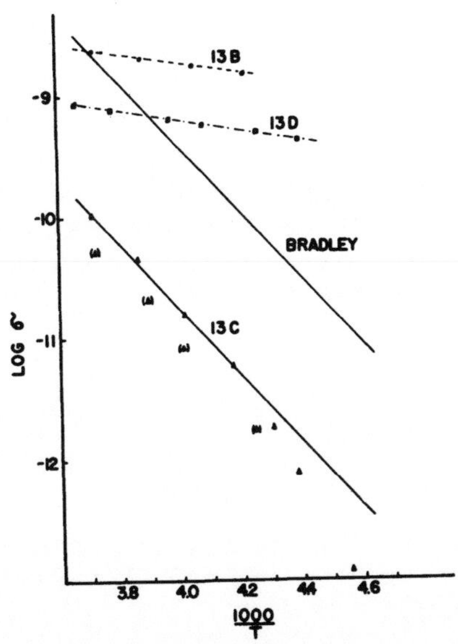

Fig.6 dc conductance σ_0 versus 1000/T

V. Discussion

From the field dependence of the saturation currents Eigen and De Maeyer /6/ could estimate the effective dielectric constant for the static interaction of the ions and found the limits $6.5 < \varepsilon_{eff.} < 11.5$. The value of $\varepsilon_{eff.}$ as calculated by Onsager and Dupuis /7/ depends on the assumed value for the dipole moment of the water molecule in the solid, which may be inferred from the Curie temperature. If we use $T_c = 38^{\circ}K$, we get

$\varepsilon_{eff.}$ = 7.34, when two types of mobile charge carriers are present, and $\varepsilon_{eff.}$ = 15.9 for the case of only one type of mobile charge carriers. Although the latter number is much smaller than the value 31 as used by Eigen and De Maeyer, it is still well beyond the experimental limits and the assumption of two types of mobile charge carriers in ice seems justified.

Estimates of the mobility and the concentration of the ions are obtained by combining the values of saturation current and conductivity. A lower conductivity yields a higher estimate for the concentration of the ions and a lower value for the mobility. A small compensation for the change in σ_0 results from the lower estimate for the dipole moment, but this is of minor importance. Since it seems a reasonable assumption that the saturation current is much less subject to errors from contamination than the conductivity we may combine Eigen and De Maeyer's value for the saturation current with our value for the conductivity $6 \cdot 10^{-11}$ $(\Omega\ \mathrm{cm})^{-1}$, which we regard as an upper limit. This yields a calculated value of $3.8 \cdot 10^{-4}$ cm^2/Vsec for the sum of the mobilities, and about $1 \cdot 10^{12}$ cm^{-3} for the concentration of the ions. However, the use of the value for the saturation current obtained on samples with a conductivity of the order of 10^{-9} $(\Omega\ \mathrm{cm})^{-1}$ in deriving these numbers should be kept in mind.

References

/1/ for details see: J.G. Berberian and R.H.Cole, to be submitted to Rev. Sci. Inst..

/2/ Schwan H.P., Determ. of Biol. Impedances in "Physical Techniques in Biological Research", vol. 6, Acad. Press (1963)

/3/ Jaffé G., Phys. Rev. 85, 354 (1952)

/4/ Auty R.P. and R.H. Cole, J.Chem.Phys. 20, 1309 (1952)

/5/ Humbel, F., F.Jona and P.Scherrer, Helv.Phys.Acta 26, 17 (1953)

/6/ Eigen M., L. De Maeyer and H.Ch. Spatz, Ber. Bunsenges. 68, 19 (1964)

/7/ Onsager L., M.Dupuis, Rendiconti S.I.F. X, 294-315 (1960)

DIELECTRIC RELAXATION, BULK AND SURFACE CONDUCTIVITY OF ICE SINGLE CRYSTALS

R. Ruepp and M. Käß

I. Physikalisches Institut

Universität Stuttgart / Germany

Abstract

Measurements were performed with ac bridges (frequency range: 0.06 cps to 300 kc; range of temperatures: -73 °C to -3.5 °C). A guard ring device allowed classification of conductivity into bulk and surface effects. The dielectric properties of ice at low frequencies exhibited an overlapping of the Debye dipole relaxation and of a process produced by space charge motion. By means of a special method of evaluation it has been possible to separate these two processes. These procedures provide information about dipole relaxation, bulk and surface conductivity. The activation energy of the relaxation is ≈14 kcal/mole(-3.5 °C>T>-55 °C) and 1o kcal/mole(-55 °C>T>-73 °C). The relaxation strength is inversely proportional to the temperature and thus has no ferroelectric behaviour. The value of dc-bulk conductivity is 10^{-10} to 10^{-11} Ω^{-1} cm^{-1} with an activation energy of 2 kcal/mole(-70 °C<T<-30 °C). The dc-surface conductivity becomes important at temperatures >-35 °C; its activation energy is 28 kcal/mole.

1. Introduction

This work represents an attempt to investigate dipole relaxation, bulk and surface conductivity simultaneously. The starting point for the following considerations was the presence of a double dispersion of the dielectric properties of ice I^h. This double dispersion was found more recently by Steinemann and Gränicher /1/ in hydrofluoric acid doped ice, and it was interpreted as an overlapping of the dipole relaxation and a dispersion produced by space charge motion. This interpretation has also been taken as a basis for the present work.

2. Experimental

The ice single crystals were grown by a modified Bridgemann method with a growth velocity of 0.2 μm/sec. The water, ion exchanged, thrice destilled and nitrogen purified had a conductivity of 3 to $7 \cdot 10^{-7}$ Ω^{-1} cm^{-1} at 25 °C. Samples of the remelted ice had the same conductivities.

A cylindric guard ring condenser was prepared by freezing on silver foil electrodes of 5 μm thickness. The vacuum capacity of this device has been calculated. Two and three terminal ac bridge measurements were performed to obtain the permittivity $\varepsilon^* = \varepsilon - \frac{\sigma}{\omega \varepsilon_0} \cdot i$.

The guard ring permits the separation of bulk and surface properties. The frequency was varied between 0.06 cps and 300 kc. The temperature range extends from -1.3 °C to - 73 °C. Typical data for permittivity ε and conductivity σ at -73 °C are shown in fig.1.

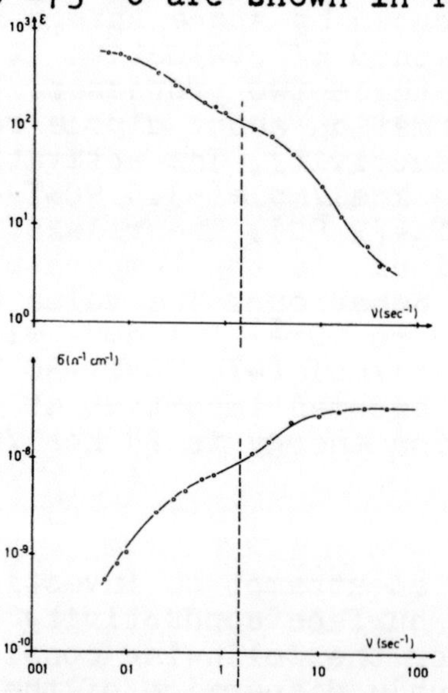

Fig.1 Frequency dependence of ε and σ. The dashed line marks the boundary between the dipole relaxation at high frequencies and the space charge dispersion at low frequencies

3. Discussion

It has been tried to compose the measured curves of two single relaxation processes. An explanation of this procedure is given by fig.2, remembering that in a double logarithmic plot the dispersion curves of single relaxation processes have the same shape for arbitrary values of the relaxation strength and the relaxation time. Therefore a simple graphical evaluation method is provided and makes it possible to separate the two processes.

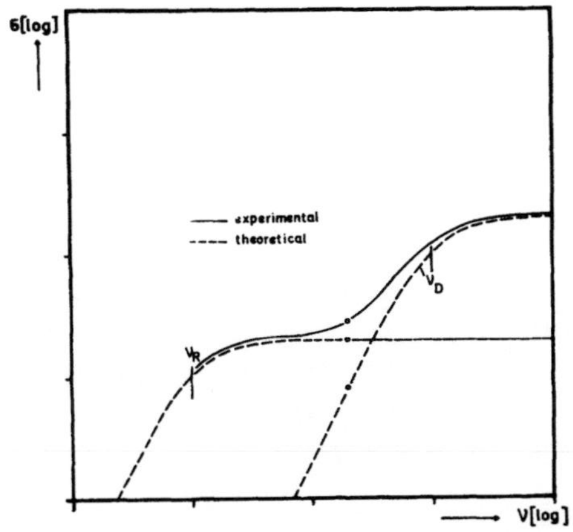

Fig.2 Example for the composition of a measured curve from two single relaxation processes

A. Low frequency dispersion

The theoretical discussion of the space charge motion is very difficult because it must be described by a system of nonlinear differential equations. A linearised approximation has been given by Macdonald /2/, making the supposition of the following boundary conditions:

1) blocking electrodes
2) total neutrality of charge
3) sinusoidal forcing voltage
4) homogeneous and frequency independent permittivity of the dielectric.

These four conditions will hold approximately if:
there is no, or little charge exchange on the electrodes,
at the beginning of the experiment the ice is neutral,
the applied voltage is sufficiently small, so that higher harmonics may be neglected,
and if the frequency range producing the space charge dispersion is sufficiently far from the range of the

dipole relaxation.

A considerable simplification of the analytical expression for the additional permittivity produced by space charge motion can be made taking into account, by order of magnitude, the values measured by Eigen, De Maeyer and Spatz /3/ for the rate constants of dissociation, recombination and the concentration of charge carriers at equilibrium. The result of all these considerations is, that the space charge dispersion must have a behaviour similar to that of a single relaxation process. This is in agreement with the experiment. The characteristic data of the process are:
The relaxation strength $\Delta\varepsilon_R = \varepsilon_{DS} \cdot M$,

where $M = \left(\frac{C_o e^2 L^2}{2 \varepsilon_o \varepsilon kT}\right)^{1/2}$.

ε_{DS} is the static permittivity of the dipole relaxation, C_o the concentration of the charge carriers at equilibrium, e the elementary charge, L the length of the sample, ε_o the vacuum permittivity, k the Boltzmann constant, and T the temperature in degrees Kelvin.

The relaxation time $\tau_R = \frac{\varepsilon_o \varepsilon_{DS}}{\sigma} \cdot M$,

σ is the dc conductivity.
The high frequency limiting conductivity

$$\sigma_{R\infty} = e\ (\mu+\mu')\ C_o ,$$

μ, μ' are the mobilities.

The high frequency limiting conductivity is equal to the dc conductivity σ because at high frequencies there is no space charge influencing the motion of the charge carriers.
This method of determination of the dc conductivity has two advantages:
1) The system is approximately in equilibrium. The elongations are small. Therefore, a defined concentration of charge carriers exists.
2) The applied forcing voltage can be made sufficiently small producing a conductivity independent of the applied voltage.

This result represents a possibility to determine the concentration of charge carriers and the sum of the mobilities, separately, if τ_R and $\sigma_{R\infty}$ can be determined.

However, most of the present measurements were not performed to sufficiently low frequencies - because of experimental difficulties - , so as to determine τ_R, and therefore only σ could be determined.

Fig.3 shows the dc conductivity σ as a function of the temperature of two different samples, but of the same ice crystal. The temperature dependence of the two samples is nearly the same, but there is a difference between the absolute value of the dc conductivity.

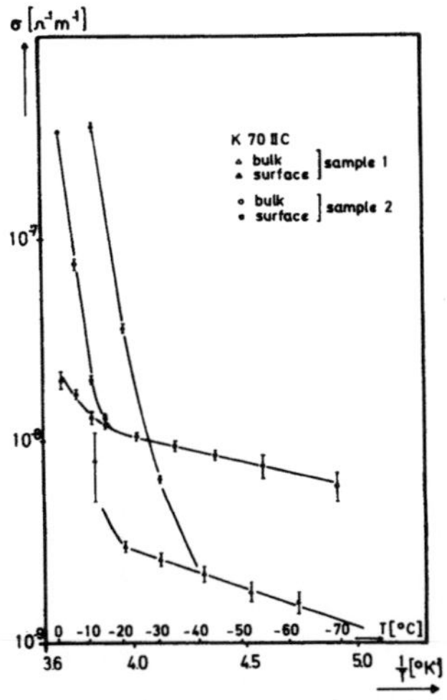

Fig.3 Temperature dependence of the dc conductivity

This is not yet understood, it may be that the different age of the two samples (sample 1: 9 months, sample 2: 18 months) is of some importance.

The apparent activation energy of the guard ring conductivity is about 30 kcal/mole. The surface conductivity will be dominant at $T > -38$ °C.

The activation energy of the bulk conductivity is about 7 kcal/mole in the temperature range of $-1.3 > T > -10$ °C and in the lower range of $-20 > T > -73$ °C about 2 kcal/mole.

A possible explanation for the low apparent activation energy is, that the concentration of charge carriers is given by impurities, and so, nearly independent of the temperature.
A definite interpretation can be given only, if measurements were performed to sufficiently low frequencies so that carrier concentration and mobility can be determined independently.

B. Debye dipole relaxation

Fig.4 shows the dependence of the reciprocal relaxation strength $\frac{1}{\varepsilon_S-\varepsilon_\infty}$ on the temperature.

ε_S is the static permittivity of the dipole relaxation and ε_∞ is the high frequency limiting permittivity.

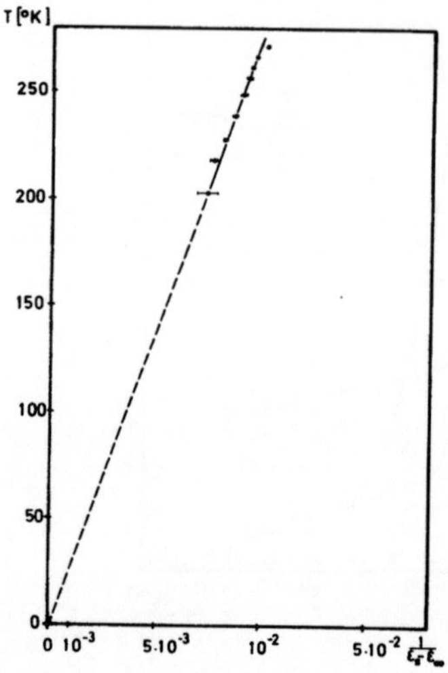

Fig.4 Temperature dependence of the reciprocal relaxation strength of the dipole relaxation

The values provide a straight line which can be extrapolated to absolute zero. Therefore it can be concluded that no ferroelectric behaviour, connected with this relaxation mechanism, exists.

Fig.5 is a plot of the dipole relaxation time τ_D versus the reciprocal temperature. The activation energy for the region of higher temperatures is 14.2 ± 0.2 kcal/mole (≙ 0.62 eV) and for the region of lower temperatures 10.2 ± 0.7 kcal/mole (≙ 0.45 eV) for all three samples. The lower activation energy at lower temperatures may be the result of impurities.

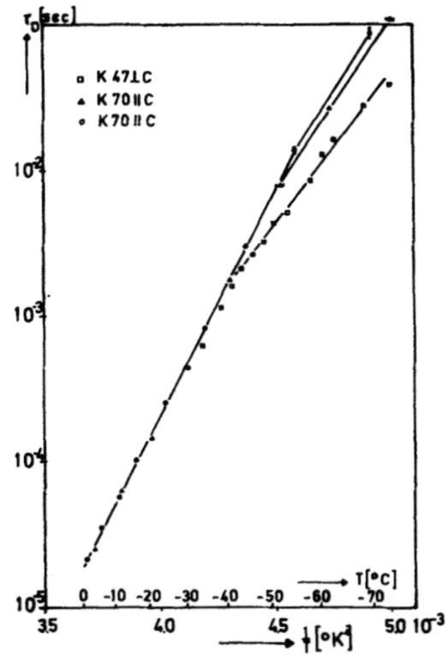

Fig.5 Temperature dependence of the relaxation time of the dipole relaxation

References

/1/ Steinemann, A., H. Gränicher, Helv.Phys.Acta 30, 553 (1957)

/2/ Macdonald, J.R., Phys. Rev. 92, 4 (1953)

/3/ Eigen, M., L. De Maeyer, H.Ch. Spatz, Ber. Bunsenges. 68, 19 (1964)

A CONTRIBUTION TO THE STUDY OF CONDUCTIVITY AND DIPOLAR RELAXATION IN DOPED ICE CRYSTALS

S. Mounier - P. Sixou

Laboratoire de Physique Electronique

Faculté des Sciences d'Orsay. 91 - France

Abstract

Having emphasized why the results of studies about certain properties, such as conductivity and Debye relaxation are inaccurate in the case of ice crystals and especially in doped ice crystals, it is shown how one of the principal causes of error - electrode polarization phenomena - can be eliminated. Particulary in the case of ice we employ a method based on the Maxwell-Wagner effect which allows the conductivity and the Debye relaxation to be obtained simultaneously and with good precision. By comparing them with former results, we are able to study the characteristics of these phenomena (relaxation time, static permittivity,...) in samples of pure and doped ice (doped with hydrofluoric acid) and with different parameters (temperature, concentration of impurities) which have been done in a large variety.

1. Introduction

Certain theoretical models /1, 2/ based upon the existence of different types of defects "Bjerrum or ionic" in pure and doped ice sometimes permit one to arrive at the characteristics of these defects (concentration, mobility, activation energy of diffusion ...) from experimental data in connection with the electrical properties of crystals (conduction and dielectric relaxation).

As we can see, the quantities which appear to be the most important are:

- static permittivity
- activation energies E and W in connection with the dielectric relaxation time and with conductivity, by relation:

$$\tau = \tau_0 \exp (E/kT)$$

$$\sigma = \sigma_0 \exp (-W/kT)$$

Unfortunately, measurements of these quantities are not very accurate (for instance, term W ranges from 8,5 to 24,2 kcal/mole according to the authors /3, 4/. Among several causes of error (difficulty in obtaining controlled pure crystals) it is agreed that the most important one is in connection with the preponderance of certain surface phenomena on volume phenomena which notably result from the partial discharge of the carrier at the electrodes. These phenomena are still unknown and the resulting effects in direct current (value of the apparent conductivity depending on the applied voltage /5/..) and in alternative current (apparent permittivity greater than the permittivity resulting from dielectric relaxation ...) often causes the experimental data to be quantitatively unworkable. In order to remedy this, Gränicher and his colleagues, in the study of ice crystals doped with hydrofluoric acid attempted to obtain "ohmic" electrodes by covering the metallic electrodes with a layer of hydrofluoric acid - an idea inspired by the electronic semi-conductor technology. This process permitted the attenuation of the blocking types of electrodes without liquidating them entirely; furthermore, there is a possibility of fluorine diffusing into the crystal. The process which we are using is quite different from that of Gränicher and his colleagues. Instead of trying to supress the partially blocking characteristic of the electrodes, we use electrodes which are totally blocking, with the help of practically perfectly insulating layers (mica...). With different examples, we have proven /6/ that:

- it is thus possible to eliminate the phenomena of polarization of the electrodes,

- it is possible to obtain the remaining phenomena (dielectric relaxation and conduction).

Now, we apply this process to study depolarization thermocurrent /7/ and complex permittivity in the case of ice and doped ice.

2. Method employed

We study the variations of the complex permittivity of a heterogeneous material composed of an ice layer (medium 2) between two layers of mica (medium 1) which is considered a nearly perfect insulator.

Due to this, we can write the permittivity of the two media:

$$\varepsilon_1^* = \varepsilon_1$$

$$\varepsilon_2^* = \varepsilon_{\infty 2} + \frac{\varepsilon_{s2} - \varepsilon_{\infty 2}}{1 + j\omega\tau_2} - j\,\frac{\sigma_2}{\omega\varepsilon_o} \; .$$

If $Q = d_2/d_1$ is the relation between the ice and mica thicknesses, it is quite easy to demonstrate that the permittivity of a homogeneous material, equivalent to all the layers together can be written as

$$\varepsilon^* = \varepsilon_{\infty\beta} + \frac{\varepsilon_{s\beta} - \varepsilon_{\infty\beta}}{1 + j\omega\tau_\beta} + \frac{\varepsilon_{s\alpha} - \varepsilon_{\infty\alpha}}{1 + j\omega\tau_\alpha}$$

with:

$$\varepsilon_{s\beta} = \varepsilon_{\infty\beta} \; .$$

By this fact the two relaxation ranges would be: Range of low frequency (α) linked to a Maxwell-Wagner effect; and we can show that τ_α is directly connected to the conductivity of the ice. The characteristics of high frequency range ($\varepsilon_{s\beta}$, ε_β, τ_β) are, on the other hand, in connection with those of the ice relaxation range and with Q.

On the other hand, we can demonstrate that by changing the thicknesses of the conducting and insulating layers (that is by changing Q) relative amplitudes of the two absorption ranges can be changed to facilitate the observation of one of the two phenomena, conduction and dielectric relaxation. If a relation between the thicknesses is chosen so that one has two ranges, with dispersion amplitudes $\varepsilon_s - \varepsilon_\infty$ nearly equivalent, one will have the possibility to observe the two phenomena simultaneously.

Experiments have been done with pure ice monocrystals and with polycrystalline samples of doped ice. We have to notice that even under best circumstances (ice highly doped with HF, polycrystalline samples) where the reproducibility is bad when working between metallic electrodes, it is possible to obtain very well reproducible results by using the offered method.

The figure 1 represents the graph of Cole and Cole with reduced coordinates for such a sheet in case of a high value of the thickness relation, doped ice/mica.

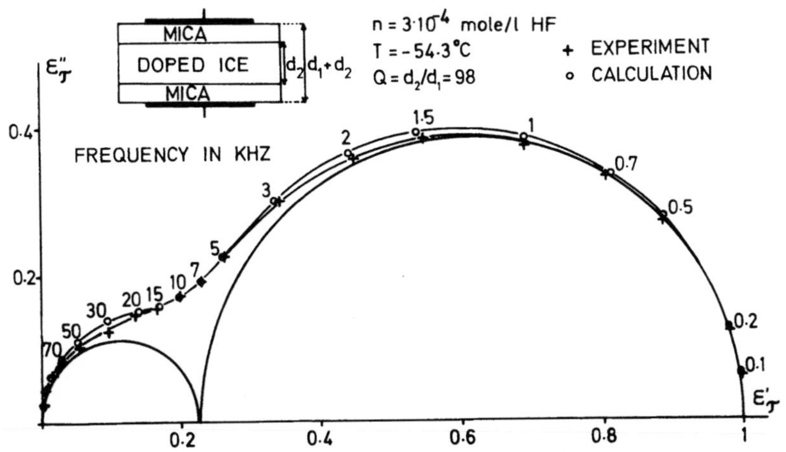

Figure 1: Cole- Cole plot (reduced coordinates) for a heterogeneous sample

In order to obtain with precision σ_2, τ_2... with the help of Univac 1108 computer we must try to separate the bulk relaxation spectrum into two elementary ranges with a unique relaxation time. After having found the most likely characteristics (ε_s, $\varepsilon_{\infty\beta}$, τ_α, $\varepsilon_{s\beta}$, $\varepsilon_{\infty\beta}$, τ_β) we are able to calculate the complex permittivity resulting from the composition of the two dispersion ranges. This calculation can be made at every frequency.

A relatively good conformity between experimental and calculated points can be seen. It is to be noticed that losses observed by experiment are smaller than those which have been calculated. This fact can often be observed in our experiments, particularly in the low frequency range. This may be in relation with the polycrystalline characteristics of the doped ice which we use.

Figure 2 shows by an example (of same concentration of impurities and almost the same temperature) the possibility of modifying the relative importance of the two ranges. In (a) with relatively thin insulating layers,

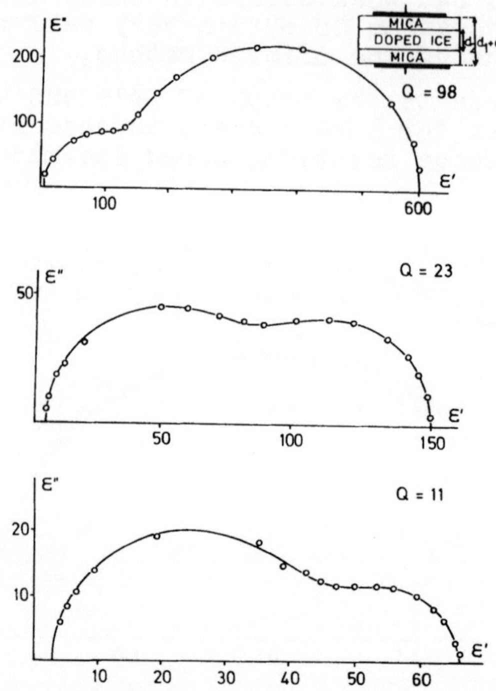

Figure 2: Cole - Cole plot for a heterogeneous sample

the low frequency range has been facilitated: we can then notice that the static permittivity is high (we can also show that the value of this static permittivity must be written: $\varepsilon_{s\alpha} = (1 + Q)\varepsilon_1$). In (c), on the contrary, the high frequency range has been facilitated by making the thickness relation Q smaller. In (b) the value of Q is ajusted in order to give both ranges the same amplitude of dispersion. We must say that these results are not dependent of the total thickness of the sample (as is the case for ice between metallic electrodes); only the relation between the thicknesses of the conducting and the insulating layers is important.

3. Influence of temperature

The above results are relative to a certain concentration but, similar results (exept numerical values) have been obtained with all used concentrations. The temperatures we have been experimenting range from 0° to $-100°$C.

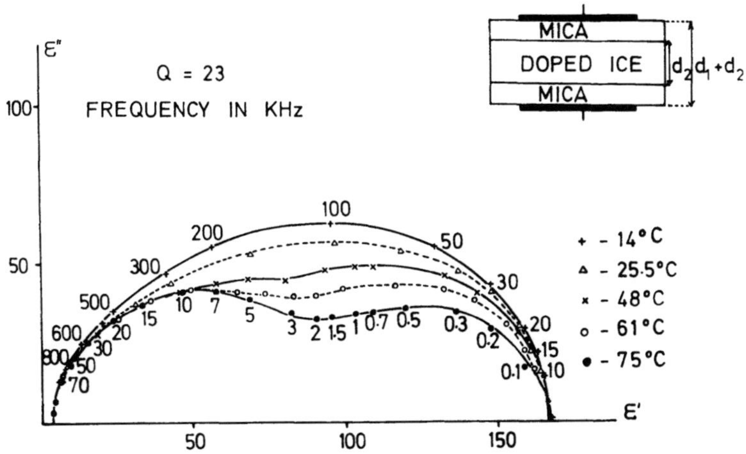

Figure 3: Cole - Cole plot for a heterogeneous sample; effect of temperature

Figure 3 shows the influence of temperature for a value of Q giving approximately equivalent relaxation ranges. We can see that the two ranges, overlapping at high temperatures, separate more and more as the temperature is lowered. This indicates that the activation energies E and W are different, and we can, without doubt, certify that W is higher than E. The separation of the experimental spectra with a computer enables us to get the values of E and W (figure 4). On that graph, we can notice two series of points corresponding to different values of Q: the conformity between these two experiments is, however, satisfying.

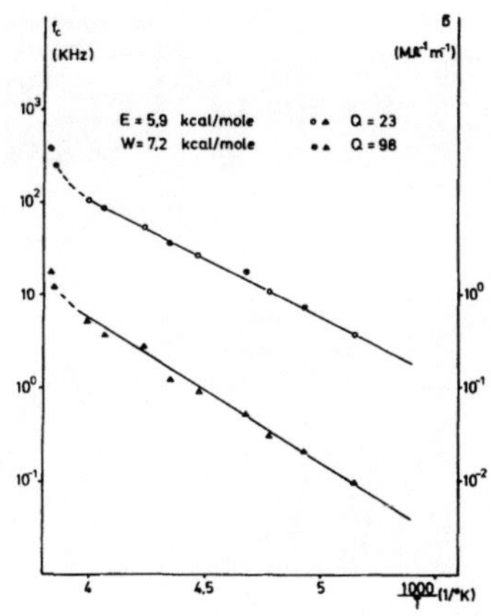

Figure 4: Curves $f_c = \phi(1/T)$ and $\sigma = \phi(1/T)$

On the other hand, the terminal $\varepsilon_{s\beta}$ gives ε_{s2}; we can thus obtain variations of the static permittivity with temperature in the case of doped ice samples.

4. Influence of concentration

a) Pure Ice

We have made experiments with pure ice monocrystals and a large thickness relation; the two ranges of dispersion are very well separated; however, the range linked to conductivity is then at very low frequencies. In that frequency band, certain measurements are being made in order to obtain the term W with precision. The activation energy of the dielectric range is a little higher than the value usually found in literature /8/.

b) Doped Ice

We have studied seven concentrations between $4.8\cdot10^{-5}$ mole /l HF and $7.5\cdot10^{-3}$ mole/l HF, in order to have the same variation of concentration as Gränicher and his colleagues.

When we change the concentration we observe that:

-- In the series of concentrations used, for a given temperature, the static permittivity of the dielectric range increases linearly with the concentration. Figure 5 represents the Cole and Cole graphs, separations have been obtained for similar temperatures and the same value of Q. In that figure we can see the increasing of $\varepsilon_{s\beta}$ and we can deduce ε_{s2}. For low temperatures and strong concentrations, ε_{s2} can be very high.

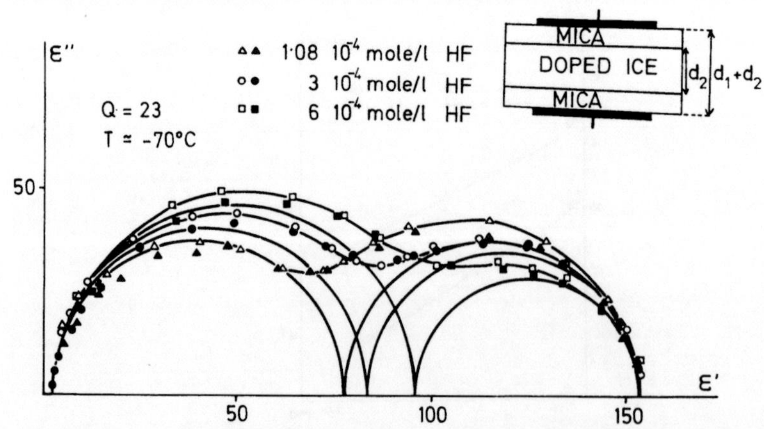

Figure 5: Cole - Cole plot for a heterogeneous sample; effect of concentration

-- The quotient of the critical frequencies of the two relaxation ranges resp. increases as the concentration decreases (figure 6).

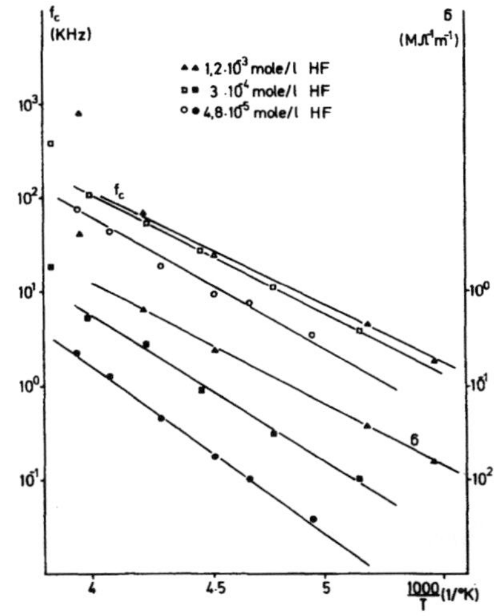

Figure 6: Effect of concentration on the curves $f_c = \phi(1/T)$ and $\sigma = \phi(1/T)$

-- Critical frequency of each of these two dispersion ranges increases with increasing concentration without allowing the deduction of an explicit variation law.

However, we notice that the critical frequency of the high frequency range differs little; the variation with concentration of the low frequency range, that is to say the variation of the conductivity, is more perceptible.

-- Activation energies do not differ much with concentration and have a tendency to become smaller for the highest concentrations.

The comparison between these results, and those of Gränicher, Steinemann, Jaccard,... /1, 2/ shows:

-- Similarities in the values of E and W for example. We see that E is smaller than its value in pure ice, and that E is lower than W in doped ice.

-- Differences (at least in appearance): the variation law of ε_{s2} does not agree with Gränicher's.The variation of σ in relation with the concentration is not expressed by $\sigma = A\sqrt{n_{HF}}$ = concentration in hydrofluoric acid.Whilst the increase of σ is smaller than $\sqrt{n_{HF}}$ for weaker concentrations, it is larger for stronger concentrations. We may explain these differences by the difference of the experimental conditions (lower temperatures, and, above all, the polycrystalline characteristics of the samples). In order to make a better comparison, we carry on our experiments with monocrystals by extending the TBF measurements and by using other impurities (NH_4F). Regarding the fact that this method allows the elimination of all the phenomena of electrodes (injections, chemical reactions), we hope to find, with precision, the variation law of ε_{s2} in relation to the concentration and to, thus, check Gränicher's theory which is the main experimental basis.

References

/1/ Gränicher, H., C.Jaccard, P.Scherrer, A.Steinemann, Discuss. Faraday Soc., 23, 50 (1957)

/2/ Jaccard, C., Helv.Phys.Acta, 32, 89 (1959)

/3/ Bullemer, B., N.Riehl, Sol. State Comm., 4, 447 (1966)

/4/ Heinmets, F., R.Blum, Trans. Far. Soc., 59, 1141-6 (1963)

/5/ Eigen, M., L. De Maeyer, Proc.Roy.Soc. 247, 505-533 (1958)

/6/ Dansas, P., P.Sixou,R.G.E. 76, 4 (1967)

/7/ Dansas, P., S.Mounier, P.Sixou, C.R. Acad.Sci.Fr., in print

/8/ Auty, R.P., R.M. Cole, J.Chem.Phys., 20, 1309-14 (1952)

INTERPRETATION OF THE PROTON SPIN-LATTICE RELAXATION IN HEXAGONAL ICE

G. Siegle+ and M. Weithase

I.Physikalisches Institut

Universität Stuttgart / Germany

Abstract

Experimentally, the spin-lattice relaxation time T_1 in ice is found to depend strongly on the temperature and on the external Zeeman field H_o. The results can be described by

$$T_1 = CH_o^2\tau_c \text{ with } \tau_c \sim \exp(\Delta E/kT)$$

(τ_c = correlation time, ΔE = activation energy). In pure ice (H_o || c-axis) ΔE is 0.62 eV, impurity controlled relaxation gives $\Delta E \approx$ 0.2 eV. These results are interpreted by the motion of defects in the crystal. The analysis of our nmr-measurements and additional data on the dielectric relaxation as well as the self-diffusion coefficient yield the result that T_1 in pure ice is determined mainly by Schottky defects as proposed earlier for the explanation of the self-diffusion coefficient, whereas Bjerrum faults give an additional but small contribution. Using this model taking into account intra- and intermolecular interaction - the constant C can be calculated. We found $C = 1.76 \cdot 10^{-2}$ Oe^{-2}, from this $\tau_c = 13.5 \pm 1$ µsec at -10°C which agrees approximately with the value of τ_c from measurements of the nmr-line width and with the τ_c estimated from self-diffusion and dielectric relaxation data. In doped ice the concentration of Bjerrum (and ionic) defects is assumed to be independent of temperature. Therefore, the activation energy ΔE = 0.2 eV

+Present address: Robert Bosch GmbH., Stuttgart
Breitscheidstrasse

is the energy of defect migration only without any contribution of the energy of formation.

The structure and the motion of the protons in hexagonal ice have been investigated repeatedly by the method of the nuclear magnetic resonance /1-8/. We refer to measurements of the reorientation of the nuclear magnetisation when the Boltzmann distribution of the proton spins on the two energy levels has been disturbed. If this reorientation process leads exponentially to thermal equilibrium we call the time constant T_1. The crystal lattice must absorb the abundant energy as the spins flip from the upper to the lower Zeeman level; T_1 is denoted the spin-lattice relaxation time.

This relaxation was measured by several authors /1-6/. The principle features of the results obtained in pure single crystals of ice can be seen in fig.1, which also shows some of our own results.

The relaxation time increases rapidly as the temperature of the sample is lowered. Below -50°C the activation energy decreases to a value of about 0.25 eV. We found the activation energy of the process α dominating at high temperatures by substracting the low temperature relaxation rate $T_{1ß}^{-1}$ from the whole rate:

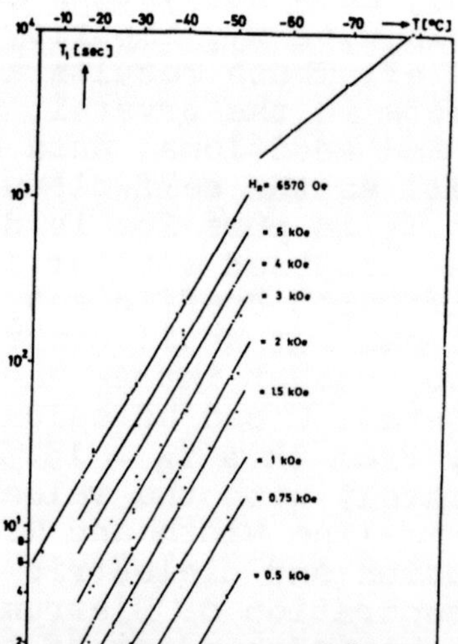

Fig.1 Spin-lattice relaxation times of pure ice vs. the reciprocal temperature at different values of the external magnetic field H_R

$$T_{1\alpha}^{-1} = T_1^{-1} - T_{1\beta}^{-1}.$$

Its value of (0.62 ± 0.02)eV agrees very well with that found by Barnaal /6/ in zone refined ice showing no second relaxation process down to -60°C. Doping the ice, e.g. with HF, reduces the relaxation time at which the second process will dominate.

The dependence of the relaxation time on the external magnetic field at constant temperature is proportional to H_R^2, so the experimental data can be described at temperatures above -50°C by

$$T_1 = C.H_R.\exp(\Delta E/kT). \tag{1}$$

To explain this result, and to calculate the constant C it is necessary to discuss the motions of the proton in the ice.

In an ideal crystal no motions of the protons between neighbouring lattice sites are possible. We thus have to discuss the lattice defects which have been proposed to exist in ice, and their migration through the lattice. Table 1 summarizes the most important ones.

All defects are capable of influencing the nuclear magnetic relaxation by their motion; because T_1 is determined by the intensity of the Fourier component of these motions at the transition frequencies of the nuclear spins. To decide which of the lattice defects cause the

Table 1 List of defects in pure ice

Type of defect	Proposed to explain
Ionic defects /15/ OH^-, H_3O^+	dc conductivity
Bjerrum defects /15/ (orientational defects)	dielectric relaxation
Schottky defects /14/	self-diffusion
combined defects /16/ (Frenkel + Bjerrum)	dielectric relaxation and self-diffusion
combined defects (Schottky + Bjerrum)	mech., diel. and nuclear magn. relaxation; self-diffusion /4/
free interstitial /17/	self-diffusion and nuclear magn. relaxation

nuclear magnetic relaxation it is necessary to take into account results of the measurement of self-diffusion and dielectric relaxation. By doing this we can hope to get a uniform interpretation of these processes. First of all the number of defects known from literature will be reduced to those which dominate T_1. After presenting the result of the calculation of T_1 it will be compared with the self-diffusion constant and the dielectric relaxation time.

1. Reduction of the number of defects

a) Ionic defects determine the dc conductivity, but their influence on the mechanical and dielectric relaxation is smaller than that of the Bjerrum defects, due to the smaller product of defect concentration x mobility /9/. As Barnaal /5/ pointed out, the influence of a single jump of the different defects on T_1 is equivalent, so the influence of the Bjerrum defects will outweigh that of the other defects.

b) The motion of interstitials and Schottky defects may cause self-diffusion. The latter should give the main contribution: On one side their concentration is about two orders of magnitude higher than that of the interstitials due to their lower formation energy /14/. On the other side the self-diffusion constant measured by the Munich group/10/ e.g. was found to be larger in the direction perpendicular to the c-axis, whereas interstitials should give a maximum value parallel to it, due to the channels in this direction. Jumps of the two defects yield a reorientation of water molecules with similar efficiency on T_1; the Schottky defects will predominate.

The existence of Haas defects seems unlikely due to two additional arguments: The assumption that the dielectric relaxation time equals the interval between two jumps which cause self-diffusion fails. As will be shown at the end of this report, these two times differ by more than an order of magnitude. Moreover, no dependence of the self-diffusion coefficient on the concentration of doping substances as proposed, exists. As the dielectric relaxation depends strongly on doping, different mechanisms govern dielectric relaxation and self-diffusion.

c) The combined Schottky-Bjerrum defect, which may contribute to the dielectric relaxation, is dropped because the measured relaxation time is about one order of magnitude longer than the calculated one (starting from this defect and the measured self-diffusion constant).

By this elimination Bjerrum and Schottky defects remain, and can explain the experimental data relatively well as is shown in the following section.

2. Calculation of the spin-lattice relaxation time

The calculation of T_1 caused by atomic jumps was done by Eisenstadt and Redfield /11/. Barnaal /5/ used it to determine the influence of molecular reorientations by Bjerrum defects. Because in ice the time τ_c between two jumps of a given molecule is long compared to the Larmor precession period, the general formula can be simplified,

$$T_1^{-1} = \frac{9}{4}\frac{\gamma^4\hbar^2}{\omega_R^2\tau_c}\Big\{\Big[\langle F^{(1)}F^{(1)*}\rangle_{n=o} - \langle F^{(1)}F^{(1)*}\rangle_{n=1}\Big] + \frac{1}{4}\Big[\langle F^{(2)}F^{(2)*}\rangle_{n=o} - \langle F^{(2)}F^{(2)*}\rangle_{n=1}\Big]\Big\} \qquad (2)$$

(n= number of jumps, F= coordinate functions of the dipolar interaction, $\omega_R = \gamma H_R$, γ = gyromagnetic ratio).

We calculated the influence of Schottky defects by eq.(2). According to a suggestion of Kopp /4/ only those molecules can jump into a vacancy which have already one proton in it (molecules 3 and 4 in fig.2). We accept this as a hypothesis. It should be energetically more favourable for the hydrogen bond to break near the oxygen than near the proton. The result of the calculation is nearly the same as that of Barnaal /5/, though the values for the

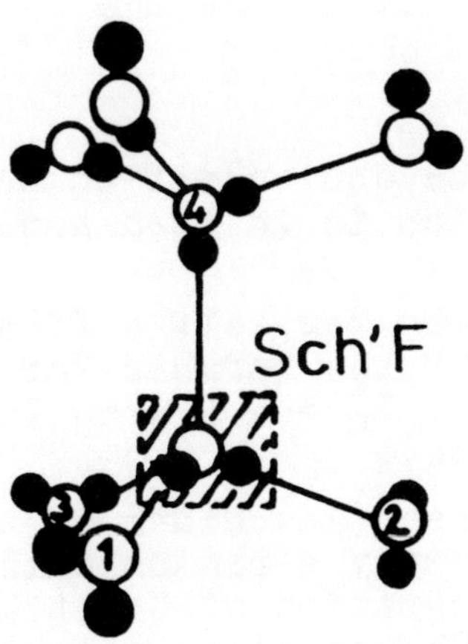

Fig.2 Schottky defect

different terms of eq. 2 differ. We thus get the relaxation time of isolated molecules which are reoriented by Schottky and Bjerrum defects in intervals τ_c

$$T_1^{\text{intra}} = 1.52 \frac{r^6 \omega_R^2}{\gamma^4 \hbar^2} \tau_c \qquad (3)$$

(r = proton-proton distance).

In order to take into account the intermolecular contribution too, we assume

$$\frac{T_1^{-1\,\text{intra}}}{T_1^{-1\,\text{inter}}} = \frac{M_2^{\text{intra}}}{M_2^{\text{inter}}}$$

$$(T_1^{-1} = T_1^{-1\,\text{intra}} + T_1^{-1\,\text{inter}};\; M_2 = M_2^{\text{intra}} + M_2^{\text{inter}})$$

with M_2 = second moment of the nmr absorption line. Using values of Barnaal /5,7/ (M_2=39.38 Oe^2, M_2^{intra}=22.24 Oe^2) we get

$$T_1 = 0.87 \frac{r^6 \omega_R^2}{\gamma^4 \hbar^2} \tau_c \;. \qquad (4)$$

From this we find at -10°C τ_c = 11.5 μs using the T_1 measured by us.

3. Comparison with dielectric and self-diffusion measurements

Because of the small concentration of defects, the correlation time τ_c between two jumps is composed of the mean times τ_S and τ_B between the visits of a Schottky or Bjerrum defect at a given molecule

$$\tau_c^{-1} = \tau_S^{-1} + \tau_B^{-1} \;. \qquad (5)$$

τ_B is easily shown to be equivalent to the dielectric relaxation time T_d, which is in turn about 50 μsec at -10°C /9,12/.

τ_S can be calculated from the self diffusion constant D: It is related to D by the formula for the mean path length $\sqrt{R_n^2}$ of a molecule $R_n^2 = 6\, D\, n\, \tau_S$ (n = number of jumps, isotropic lattice).

R_n^2 cannot be taken as na^2 (a = width of one jump = O-O-distance), this would imply a random walk of the spins. In the model for the migration of Schottky defects only a fraction of the neighbours of a vacancy will jump, so the way $\sqrt{R_n^2}$ changes. This is described by $R_n^2 = \Gamma n a^2$, the correlation factor

$$\Gamma = \overline{\left(1 + \frac{2}{\;} \sum_{j=1}^{n-1} \sum_{i=1}^{n-j} \cos\theta_{i,i+j} \right)} \qquad /13/$$

($\theta_{j,i+j}$ = angle between the directions of the jumps i and i+j) was calculated to be 1.94.

We thus get $\tau_s = \Gamma a^2/6D$, and with the value of $D=3\cdot10^{-11}$ $cm^2s^{-1} \pm 50\%$, $\tau_s = 5.5 - 16.5\ \mu s$ at $-10^{\circ}C$.

Using formula (5) τ_c is found to be $\tau_c = 5.4 - 12.4\ \mu s$ at $-10^{\circ}C$. This is in good agreement with the value calculated from T_1.

The temperature dependence of τ_c is given by comparing eq.(1) and (4) : $\tau_c \approx \exp(\Delta E/kT)$. This exponential behaviour can be measured, because the activation energies of self diffusion and dielectric relaxation show nearly no difference within the limits of error (0.68 ± 0.13 eV /10/ and 0.61±0.04 eV /12/.

At low temperatures the relaxation times show only a weak temperature dependence. We assume that impurities give a nearly temperature independent concentration of Bjerrum- (and perhaps other) defects. The activation energy measured, should be the migration energy of these defects only. We calculate it for Bjerrum-defects from the activation energy of the dielectric relaxation (0.61 eV) and the energy of formation of a pair of L- and D-defects (0.68 eV /9/) using a formula given by Gränicher /14/

$$\Delta E_B^{travel} = \Delta E_B^{diel.relax.} - \frac{1}{2}\,\Delta E_B^{formation} = 0.27\ eV.$$

This value agrees very well with that found by the T_1-measurements and confirms our model.

We wish to thank Prof.Dr.H.O. Kneser for the possibility of performing this work in his institute. The work was sponsored, in part, by the Deutsche Forschungsgemeinschaft.

References

/1/ Bloembergen, N., et al., Phys.Rev. 73, 679 (1948)

/2/ Kume, K., J.Phys.Soc.Japan 15, 1493 (1960)

/3/ Krüger, G.J., Thesis, Stuttgart 1961

/4/ Kopp, M., University of Pittsburgh, unpublished progress reports 1964/1967

/5/ Barnaal D.E., Ph.D. Thesis, University of Minnesota 1965

/6/ Weithase M., Master Thesis, Stuttgart 1968

/7/ Barnaal D.E., I.J. Lowe, J.Chem.Phys. 46 4800 (1967)

/8/ Ipsen P., Thesis, Stuttgart 1967

/9/ Gränicher H., Phys.kondens.Materie 1, 8 (1963)

/10/ Delibaltas P. et al. Phys.kondens.Materie 5, 166 (1966)

/11/ Eisenstadt M., A.G. Redfield, Phys.Rev. 132 635 (1963)

/12/ Ruepp R., Master Thesis, Stuttgart 1967

/13/ Shewmon P.G., Diffusion in Solids, McGraw Hill 1963

/14/ Gränicher H., Z.Krist. 110, 443 (1958)

/15/ Bjerrum N., Science 115, 385 (1952)

/16/ Haas C., Phys.Letters 3, 126 (1962)

/17/ Runnels L.K., Ph.D. Thesis, Yale University 1963; L. Onsager, L.K. Runnels, Proc.Nat.Acad.Sci. 50, 208 (1963)

DIFFUSION OF HYDROGEN FLUORIDE IN ICE

H. Haltenorth, J. Klinger +)

Physik-Department

Technische Hochschule München / Germany

Abstract

Large and nearly perfect single crystals of pure ice were doped with hydrogen fluoride by diffusion. Diffusion profiles were determined by successive ablation of 0.5 mm layers. The diffusion constant for HF in ice was determined with high accuracy to $(1.08 \pm 0.01) 10^{-7}$ cm^2/sec at $-10^{o}C$. The corresponding activation energy is 4.61 ± 0.04 kcal/mole in the range from -5 to $-90^{o}C$. Precautions taken exclude any surface diffusion. The HF diffusion constant of crystals containing small angle boundaries is larger by a factor 1.25.

Selfdiffusion of ice, measured with several tracers, tritium and oxygen 18 /1/, is independent of the amount of impurities /2,3/. But the diffusion coefficient of HF in ice was found to be many orders of magnitude larger than the diffusion coefficients of hydrogen and oxygen bound in a watermolecule /4/. On the other hand, we know that the concentration of incorporated impurities is much larger in disturbed parts of a single crystal /5/. So it could be expected that the large diffusion coefficient of hydrogen fluoride in ice is due to incorporation and diffusion in disturbed structures.

+) Present address: Faculté des Sciences de Grenoble, F-38 St.Martin-d'Hères / France

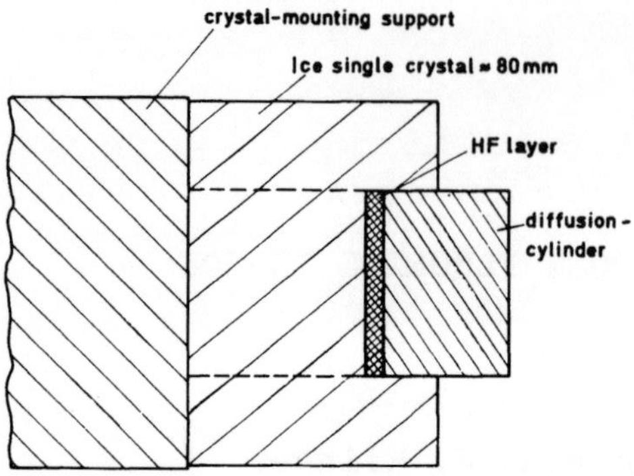

Fig.1 HF-diffusion sample

In order to prove this assumption we began in 1966 to dope ice single crystals by diffusion with hydrogen fluoride. For the diffusion measurements, ice single crystals were used. In order to test the quality of the ice samples to be used in these experiments we cut off one slice from either side of the samples. If both test slices showed few Tyndall flowers and no small angle boundaries, detectable by infrared irradiation, in a central part of about 50 mm in diameter, the sample was used for diffusion measurements. These ice samples containing no detectable small angle boundaries and with a very small amount of impurities were shown in a short movie by Müller-Krumbhaar /6/. The arrangements of the diffusion experiment is shown in figure 1. The borehole, into which the plexiglass diffusion cylinder had to be inserted, was cut with the aid of a lathe. The diffusion cylinder had a diameter of 40 mm, the borehole was larger by 0.1 mm. Highly pure, distilled hydrofluoric acid with a concentration of 0.35 mole/ℓ was frozen onto the diffusion cylinder. The smoothed surface of the frozen acid was connected to the bottom of the borehole by pressing the diffusion cylinder against the bottom of the borehole for several seconds while the ice single crystal was rotating on the lathe. Thereafter the samples were placed into polyethylene bags which were closed and welded. These samples were stored at different temperatures for times between 2 and 20 days depending on the storage temperature.

If one assumes, that the diffusion obeys the exponential law:

$$c = c_o \cdot \exp \; (-x^2/4\,Dt)$$

one can evaluate the diffusion coefficient by plotting logarithm c vs. x^2.

Only the core of the sample (the part within the dashed lines in fig.1) was used for diffusion measurements in order to prevent the measurement of surface diffusion. The thickness of the discarded ice layer was greater than the depth of penetration of HF. From the remaining core, layers with a thickness of 0.5 mm were removed with the lathe. The cuttings were collected in small containers. The conductivity of the melting water gives the concentration of hydrofluoric acid in the respective layer.

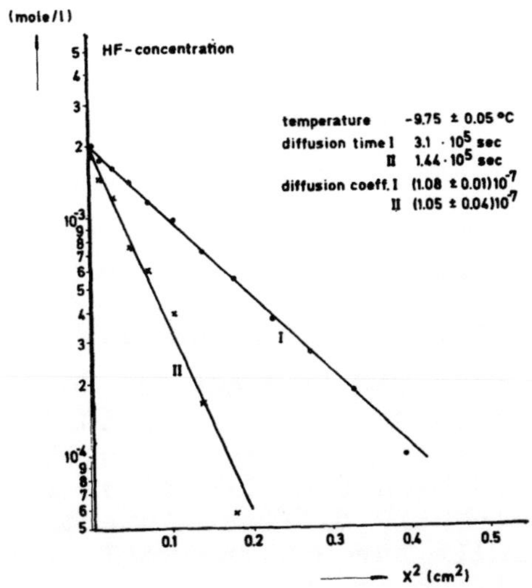

Fig.2 Typical plot of HF-concentration versus x^2

Fig.2 shows two typical plots of concentration vs. x^2. The two samples were stored at the same temperature but for different lengths of time.

Plotting logarithm D vs. 1/T yields the activation energy (fig.3):

$$E = (4.61 \pm 0.04) \text{ kcal/mole.}$$

The accuracy of our measurements was better than that of previous experiments . Independent results on HF-diffusion by the present authors agree within the limits of error. The values measured by Kopp et al. /4/ are about 10 times as large. Comparing, e.g., the diffusion coefficients parallel to the c-axis at -10°C:

Ref. / 4/: $D = 0.8 \cdot 10^{-6}$ cm^2/sec ± 100%

Our value: $D = 1.08 \cdot 10^{-7}$ cm^2/sec ± 1%

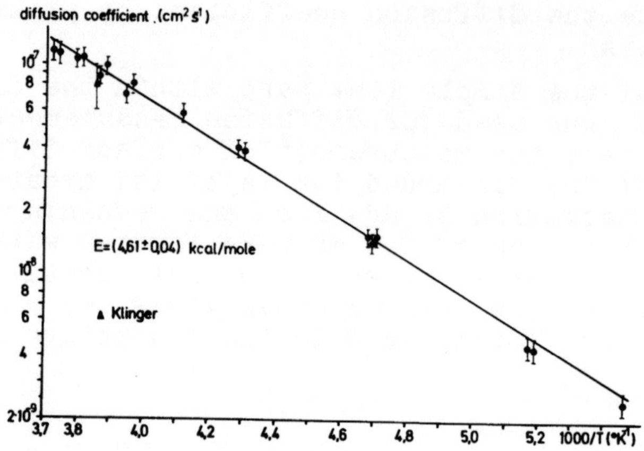

Fig.3 Plot of HF-diffusion coefficients versus 1/T

and the activation energies:

Ref. /4/: E = 13.4 kcal/mole

Our value: E = (4.61 ± 0.04) kcal/mole.

The large value of D measured by Kopp can easily be explained by surface diffusion. If one considers his small ice samples of 8 mm diameter and 60 mm length, one can easily imagine that hydrogen fluoride first spread over the surface and subsequently diffused into the interior. HF could diffuse radially through the whole sample during his relatively long diffusion times.

A subsequent experiment was performed to study diffusion parallel to the c-axis at -20°C in single crystals which contained an appreciable concentration of detectable small angle boundaries. The diffusion coefficient in this experiment was 25% larger than that measured in the undisturbed single crystals mentioned above.

A measurement of diffusion perpendicular to the c-axis at -20°C in undisturbed crystals resulted in a diffusion coefficient which was 20% larger than that for diffusion parallel to the c-axis at the same temperature.

The diffusion experiments on crystals containing detectable small angle boundaries as well as the experiments of Blicks et al./2/ and Dengel et al./3/ who grew ice single crystals from hydrogenfluoride and ammoniumfluoride solutions respectively showed the same result: The amount incorporated and the diffusion coefficient increase when ice crystals contain imperfections. Fig.4

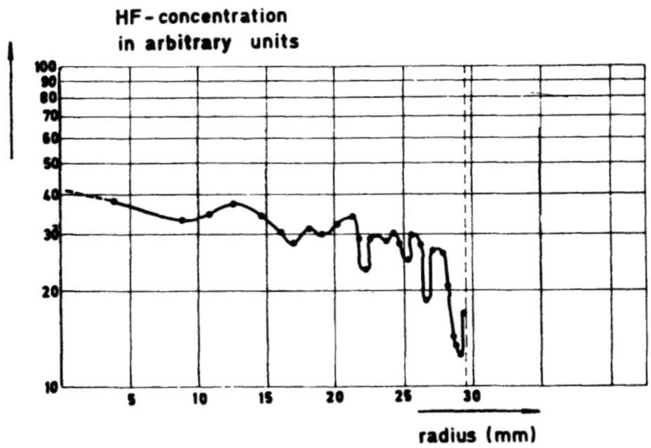

Fig.4 Fluctuation of radial HF-concentration in a disturbed ice single crystal

shows how small angle boundaries influence the incorporation of HF. In this experiment an ice single crystal was grown from diluted hydrofluoric acid and the radial distribution of HF measured. The HF concentration fluctuates in the outer part of the crystal which is less perfect than the inner part. We may assume that all ice single crystals which we were able to grow without detectable small angle boundaries and which incorporated a very small amount of impurities had a microstructure. Truby /7/ detected this mosaic structure (fig.5).

Our so called "ice single crystals" are composed of an immense number of hexagonal prisms, shown a little simplified in fig.5. According to Truby, all these hexagonal prisms are not exactly parallel with respect to their crystallographic axes. They deviate by few minutes of arc. These mosaic boundaries may have a great influence on HF-diffusion. The mosaic structure for example could explain the anisotropy between the diffusion coefficients resulting from diffusion parallel to the c-axis and that

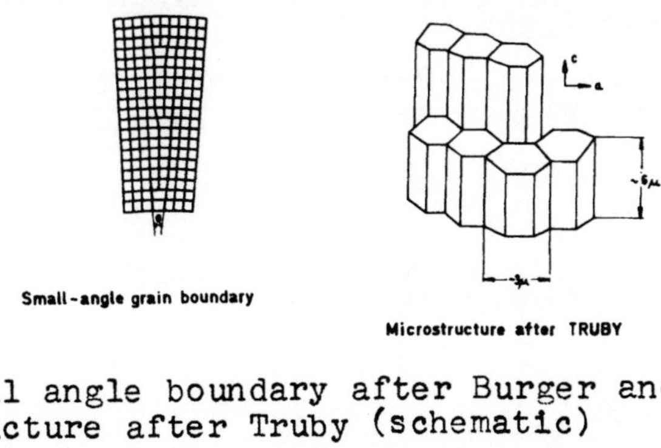

Fig.5 Small angle boundary after Burger and mosaic structure after Truby (schematic)

perpendicular to the c-axis.

A simple geometric consideration of the mosaic boundaries indicates that diffusion perpendicular to the c-axis should be larger than that parallel to the c-axis. Another result which makes HF-diffusion difficult to explain by a vacancy mechanism are the large values of D. The values for selfdiffusion and HF-diffusion, resp. at -10°C are about:

$$D_s = 1.5 \cdot 10^{-11}\ cm^2/sec$$

$$D_{HF} = 1.08 \cdot 10^{-7}\ cm^2/sec.$$

These values differ by about four orders of magnitude. The values of the respective activation energies also differ:

$$E_s = 14\ kcal/mole$$

$$E_{HF} = 4.6\ kcal/mole$$

Considering these results we think that the mosaic boundaries have great influence on HF-diffusion in ice single crystals.

Reference

/1/ Delibaltas, P., O. Dengel, D. Helmreich, N.Riehl, H. Simon, Physik kondens.Materie 5, 166 (1966)

/2/ Blicks, H., O. Dengel, N.Riehl, Physik kondens. Materie 4, 375 (1966)

/3/ Dengel, O., E. Jacobs, N. Riehl, Physik kondens. Materie 5, 58 (1966)

/4/ Kopp, M., D.E. Barnaal, I.J. Lowe, J.Chem.Phys. 43, 2965 (1965)

/5/ Blicks, H., H. Egger, N. Riehl, Physik kondens. Materie 2, 419 (1964)

/6/ Müller-Krumbhaar, H., Physics of Ice, Proc.Intern. Symp., Munich, Plenum Press, New York (1969)

/7/ Truby, F.K., Science 121, 104 (1955)

IMPLICATIONS OF ICE PHYSICS FOR PROBLEMS OF FIELD GLACIOLOGY

J.W. Glen

Department of Physics

University of Birmingham / England

Abstract

This paper reviews those problems of field glaciology which depend for their solution on the physical properties of ice. The classical case is the problem of glacier flow; experiments on the behaviour of ice under stress in the laboratory have provided flow laws which have been used in glacier theory, but many problems still remain including the treatment of transient creep, preferred orientations, slip on the bed and associated problems of glacier erosion and deposition, and the nature of ice sheets under extreme conditions such as those which may obtain on Venus. Apart from the plastic properties of ice, there are many other problems such as the formation of crevasses, determined by the fracture properties of ice, and the radiation properties of ice and snow, determined by the electromagnetic properties of ice. There are also physical techniques useful for field measurements. One of these, the electrical resistivity method, suffers at present from an unexplained discrepancy between the laboratory and field values for the resistivity of ice. Electrical properties of ice are also of vital importance in theories of thunderstorm electricity. The nucleation and growth of ice crystals are also fundamental to many problems in the study of snow, and of lake, river and sea ice.

1. Introduction

It is most fitting that a symposium in Munich on the physics of ice should devote some time to the implications of ice physics for problems of field glaciology. This is not only because here, so near to the majestic glaciers of the Alps, our minds turn naturally towards the geophysical applications of our physical studies, but also because Munich is associated with the early work on glacier flow and its relation with physical theory and the physical properties of ice by the pioneering work of Sebastian Finsterwalder, A.Blümcke,and H.Hess, a tradition continued in the Technische Hochschule München by Richard Finsterwalder until his death five years ago.

I must start by briefly explaining what I mean by field glaciology, for glaciology is a word that suffers from having a meaning that is not quite what over-simplified etymology would suggest; it is not simply the study of glaciers, but the study of ice in all its forms, being derived from the Latin word for ice (the Greek word, Krustallos, having been taken over by the crystallographers for their own purposes!). By field glaciology, I therefore mean all aspects of the study of ice in nature (I would include the subject of this symposium itself within the term glaciology unqualified by "field").

2. Mechanical properties of ice

As I have suggested in the introduction, the classical application of ice physics is to glacier flow theory. The problems of how glaciers flow, and the response of glaciers to changes in snowfall or melting (accumulation and ablation) are clearly ones where the physical properties of ice subjected to stress are fundamental. This was my own introduction to glaciological problems, when I attempted to make physical tests on randomly oriented, polycrystalline ice in order to provide a physically based flow law for use in glacier theory (Glen /1/). This attempt was broadly successful,in the sense that it was possible to give an approximate relation between the applied stress in a compressive test and the steady-state creep rate that resulted after transients had died out. This relation has been widely used, but there are many problems in its application that depend on physical factors, some of which have been properly studied, and some which have not. For example the use of a law derived from tests in uniaxial compression to situations of general triaxial stress requires a knowledge of the dependence of the various invariants of the strain-rate tensor on the corresponding invariants of the stress tensor, and may not in general be possible at all. Luckily experiments show (Rigsby /2/) that hydrostatic pressure has very little effect on the flow law, although Vyalov /3/ has more recently presented results which suggest a definite though small effect. The influence of the third invariant has not been investigated as fully as is desirable, and it must be said that most theories of glacier flow at present are based on the hope that the octahedral shear strain-rate is a function of octahedral shear stress.

Quite apart from questions of the applicability of uniaxial results to triaxial conditions, there are further problems which arise in using these laboratory results for real ice masses. The ice masses probably have a preferred crystal orientation, which may well affect their mechanical properties. For glaciers this preferred orientation may itself be the result of recrystallization in the stress field, and maytherefore be itself determined by stress, but because the stress is not uniform as a particle of ice moves down a glacier, we cannot necessarily resolve this problem by regarding the preferred orientation as itself part of the flow law, in other words a piece of ice may have a preferred orientation related to the stress it had at some earlier stage of its travel down-glacier, and may now find itself in a stress field oriented in some quite different direction. When we move to other ice masses, such as those formed by freezing water, there is a very strong preferred orientation from the start, and as it is quite common to find this taking the form of parallel c-axis in all the crystals, the mechanical properties of such an ice mass may be very different indeed. Since ice deforms by slip on (0001) planes, a polycrystal with all the (0001) planes parallel might be expected to behave more like a single crystal than like a randomly-oriented polycrystal. Laboratory experiments on single crystals (see for example Glen and Jones /4/) show that they have very different creep behaviour from polycrystals, not only are they much softer, but there is a transient acceleration, instead of a deceleration. It is not yet known to what extent crystals with strong preferred orientation approximate to single crystals; experiments to investigate this are in progress.

The whole question of transient creep raises a further problem for field applications, in which stresses may vary and so introduce complicated transients throughout the flow history of a sample. A further complication comes from the fact that temperate glaciers are virtually at the melting point throughout their depth, and may contain an appreciable fraction in the liquid phase. The question of how this melt water is disposed has been discussed on physical grounds by Steinemann /5/, but has not been investigated very fully experimentally. A somewhat connected question concerns the boundary conditions at the bottom of a glacier. The current view (Weertman /6/; Lliboutry /7/) is that there is a velocity discontinuity at the bottom of glaciers, and that the ice slips over the bedrock by a combination of plastic deformation round obstacles and heat flow through them. This second process leads to what is sometimes known as regelation, the melting of ice at one side of an obstacle and refreezing at the other side; the physics of this process has recently been studied in detail by Nye /8/ and Frank /9/, and curious discrepancies revealed. The actual laboratory study of this joint flow and regelation is very difficult, due to temperature control problems; it has been attempted at Grenoble (Brepson /10/). Until more is known about it, it will be very difficult to give a proper physical basis to theories of glacier erosion and deposition, which are at present based on rather crude physical pictures.

One recent paper on glacier deposition might be mentioned, since it depends on a physical concept of some interest. Smalley and Unwin /11/ have discussed the origin of drumlins, the oval-shaped hills of glacial deposits that are sometimes left behind by glaciers, and have concluded that they may be due to an instability of the flow of the hill deposit which has a dilatant creep curve, i.e. it requires a much larger stress to initiate flow than to sustain it. It is of interest that the same is true of ice single crystals, and maybe also of strongly oriented polycrystalline ice. The theory of the flow of three-dimensional masses of dilatant materials clearly deserves deeper investigation.

I might also mention a novel application of glacier flow theory. It has recently been suggested that there may be very large polar ice sheets on Venus, possibly as much as 10 km thick. Weertman /12/ has pointed out that such ice sheets would not be capable of existing; they would rapidly spread towards the equator under their own weight. Libby /13/has replied that the amount of CO_2 on Venus may be sufficient to make the bottom of these ice caps into the high-pressure phase CO_2. $5.75H_2O$, which could be more resistant to flow. Clearly there is here scope for a further application of ice flow physics!

Besides the creep behaviour, other mechanical properties of ice have important applications in field glaciology. The elastic behaviour of ice determines the seismic velocities used for depth sounding, and the fracture properties are of importance in the theory of crevasses, but I will not go into details of these now; the applications are rather obvious.

3. Electromagnetic properties of ice

Many of the more important thermal properties of glaciers, including the question of the heat balance of ice and snow, are closely connected with the high-frequency electromagnetic properties of ice, for although ice is a poor absorber of visible light (and therefore in dispersed form as snow it is very white), in the infra-red it is an extremely good absorber, and therefore also a very good emitter. Thus the infra-red absorption of ice, related as it is to the hydrogen disorder (Bertie and Whalley /14/) is responsible for the way in which glaciers can remain cold despite relatively high air temperatures, rejecting a large amount of the solar radiation by day, but cooling very efficiently by long-wave radiation by night.

The extreme transparency of ice to electromagnetic waves in the radar range of frequencies is also of importance, since it enables radio echo soundings to be made through cold ice. This method is now enabling very rapid depth soundings to be made, even from airborne radar equipment flown over the Antarctic Ice Sheet.

The very low frequency dielectric dispersion of ice has also been used to study various properties of ice and snow (Evans /15/), and Philberth has suggested /16/ the use of a permittivity probe to determine ice temperature at depth. Another electrical property that has been suggested for use in sounding is the dc electrical resistivity, and this was the subject of a series of papers in a recent issue of the Journal of Glaciology (Röthlisberger /17/, Röthlisberger and Vögtli /18/, Hochstein /19/, Vögtli /20/). The potential between two probes when current is passed between two other probes is converted to an "apparent resistivity" which is plotted against the separation of the probes. If the resistivity of the ice is very different from that of the underlying rock, it should be possible to determine glacier depth in this way. The resistivity which has to be used for temperate glaciers varies from glacier to glacier, and is always much larger than the value which had been obtained from extrapolated laboratory determinations. The field figures vary from 10 to about 170 MΩm, whereas the laboratory values were about 5 MΩm (see Röthlisberger and Vögtli /18/, p.619). The latest experiments of Bullemer et al. /21/ indicate however that the true bulk conductivity is of this magnitude when care is taken to exclude surface currents, but even their value is well below the highest field resistivities.

The electrical properties of ice are of course very complicated, as this conference bears witness, and the various possible applications in field glaciology of the surface effects of ice can only be seen in full when the physical picture is clearer. However the important subject of thunderstorm electricity is one major field application of some of the more unusual electrical properties of ice.

4. Growth properties of ice

One of the most interesting problems posed by nature to ice physicists is the shape of snow crystals. From the early theories of Kepler, recently made available in English (Kepler /22/), as to why the six-fold symmetry was so general, to the studies of Nakaya /22/ on the different types of snow crystal and the conditions under which they were formed, snow crystals have provided abundant problems for study. The answer as to why under certain conditions of temperature and supersaturation you get needles, under other cups, plates and dendrites, and why some of these forms are extended along the *c*-axis while others are planar structures lying in the plane perpendicular to the *c*-axis must be found in very detailed and subtle changes in the properties of ice. Hobbs and Scott /24/ suggest that the property of importance is the velocity of growth of a step formed on an otherwise flat ice surface, while Mason and others /25/ have postulated that it is the distance a molecule migrates after reaching the surface before it is incorporated into a step. In either case the quantity in question has to differ as between basal and prismatic planes, and the sign of the difference has to change twice as temperature varies from the melting point down to about -25°C.

Ice growing from the liquid poses its problems as well. Natural ice bodies frozen from water sometimes occur with a strong preferred orientation with *c*-axis vertical, but also sometimes occur with a strong preference for *c*-axis horizontal. There are good physical reasons why the *c*-axis vertical should occur; the first small ice crystals will be plates perpendicular to the *c*-axis and these will float on the surface serving as nuclei for columnar grains all with their c-axis vertical. It is also clear that since the direction of most rapid growth is in the (0001) plane, as the ice thickens these c-axis vertical grains will tend to be wedged out by others more steeply inclined. The theory of this process has recently been discussed by Ketcham and Hobbs /26/ and Ramseier /27/. It is not so clear why c-axis horizontal ice sometimes forms at the top of lake ice.

The extremely small quantities of impurity that can be dissolved in ice in practically every case (the exceptions being HF and NH_3), give rise to the nature of sea ice. Here the impurity content is not dissolved in the ice lattice but trapped in between crystals of relatively pure ice, giving in most cases a layered structure with varying salinity as well as gross trapped bubbles of salt water. The structure has been the subject of intensive study, both because it is responsible for the mechanical and electrical properties of sea ice, and because an understanding of the reasons for its appearance is of great importance if reliable methods of obtaining drinking water from sea-water by freezing are to be developed.

Another aspect of the physics of ice crystal growth that has received wide interest for field applications is the question of nucleation of ice crystals from supersaturated water vapour. Here the discovery of effective ice nuclei has been exploited in the techniques of cloud seeding for the artificial stimulation of rain. Early experiments made use of such nuclei as AgI, and had considerable success, but the search for more efficient nuclei, i.e. ones acting nearer to $0^{\circ}C$, has continued, and now a whole range of materials is known to act including some organic materials that work well at remarkably high temperatures.

5. Conclusion

In the necessarily short paper I have tried to give some idea of some of the areas in which work on the pure physics of ice has found applications in the geophysical field. Ice is not only a material with very many remarkable properties that continue to provide stimulating problems for pure physics, it is also one of the most important and widely dispersed materials on the Earth. The collaboration between pure physics, investigations and field applications has been most fruitful in the past , and shows every sign of remaining so in the future. Although the list of references quoted in this paper is necessarily limited, a full bibliography of papers on the physics of ice and of applications to field glaciology can be found in the Glaciological Literature section at the end of each issue of the Journal of Glaciology.

References

/1/ Glen, J.W., Proc.Roy.Soc. A 228, 519-538 (1955)

/2/ Rigsby, G.P., J.Glac. 3, 271-278 (1958)

/3/ Vyalov, S.S. (In Ōura, H.ed.,Physics of Snow and Ice: International Conference on Low Temperature Science, Sapporo, Japan (1966), vol.1, Pt. 1, pp. 349

/4/ Glen, J.W., S.J.Jones. In Ōura, H., ed.,Physics of Snow and Ice: International Conference on Low Temperature Science, Sapporo, Japan (1966) vol.1, Pt. 1, pp.267

/5/ Steinemann, S., Union Géodésique et Géophys.Internat. Association Internationale d'Hydrologie Scientifique. Symposium de Chamonix, 16-24 sept.1958, p.254-265

/6/ Weertman, J., J.Glac. 5, 287-303 (1964)

/7/ Lliboutry, L., J.Glac. 7, 21-58 (1968)

/8/ Nye, J.F., Phil.Mag.8, 16, 1249-1266 (1967)

/9/ Frank, F.C., Phil.Mag.8, 16, 1267-1274 (1967)

/10/ Brepson, R., Comptes Rendus Hebdomadaires des Séances de l'Académie des Sciences. Ser.B, Tom.263, 15, 876-879 (1966)

/11/ Smalley, I.J., D.J.Unwin, J.Glac.7, 000-00 (1968)

/12/ Weertman, J., Science 160, 1473-1474 (1968)

/13/ Libby, W.F., Science 160, 1474 (1968)

/14/ Bertie, J.E., E.Whalley, J.Chem.Phys. 46, 1271-1284 (1967)

/15/ Evans, S., J.Glac. 5, 773-792 (1965)

/16/ Philberth, B., J.Glac. 6, 765-766 (1967)

/17/ Röthlisberger, H., J.Glac. 6, 599-606 (1967)

/18/ Röthlisberger, H., K.Vögtli, J.Glac. 6, 607-621 (1967)

/19/ Hochstein, M., J.Glac. 6, 623-633 (1967)

/20/ Vögtli, K., J.Glac. 6, 635-642 (1967)

/21/ Bullemer, B., I.Eisele, H,Engelhardt, N.Riehl, P.Seige, Solid State Communications 6, 663-664 (1968)

/22/ Kepler , J., The six-cornered snowflake. Oxford, Clarendon Press (1966) XVI, 75 p.

/23/ Nakaya, U., Snow Crystals, natural and artifical. Cambridge, Mass., Harvard University Press (1954) XII, 510 p.

/24/ Hobbs, P.V., W.D.Scott, J.Geophys.Res. 70, 5025-5034 (1965)

/25/ Mason, B.J., G.B.Bryant, A.P.Van den Heuval, Phil.Mag.8, 8, 505-526 (1963)

/26/ Ketcham, W.M., P.V.Hobbs, J.Crystal Growth 1, 263-270 (1967)

/27/ Ramseier, R.O., In Peiser, H.S., ed.,Crystal Growth. Proceedings of an international conference... Birmingham, 15-19 July 1968. Oxford, etc., Pergamon Press; in press

ATMOSPHERIC ELECTRICAL EFFECTS RESULTING FROM THE COLLISION OF SUPERCOOLED WATER DROPS AND HAIL

E.J. Workman

Cloud Physics Observatory

University of Hawaii, Hilo, Hawaii / USA

Abstract

The separation of electric charge that occurs under a great variety of conditions when water freezes must be important in atmospheric processes. The autor has described a model for thunderstorm electrification in which electric charges are separated when cold water drops impinge on cold atmospheric ice. The resulting rapid freezing separates charge at the interface and the splash-off of small water drops completes the separation process. Quantities of electric charge separated are relatively large (3 to 15 esu per gram of impacting water) and of sign consistent with observed thunderstorm polarity. The behavior of water contaminated with ammonia illustrates features of the basic process of charge separation. Moreover , the introduction of ammonia into cloud water gives promise of modifying the electrical development of storms - decreasing the activity for moderate concentrations of ammonia and reversing the sign of the charge for concentrations of 10^{-4} M or more. Since the charge generating process involves the separation and subsequent neutralization of ions, the chemical state of the atmosphere must be influenced. Considering primary charge generation in thunderstorms alone, the world effect might be the electrochemical equivalent of 20,000 coulombs per second.

It is desirable to mention briefly some background observations before presenting the subject data of this paper. We are concerned with electrical charge separation

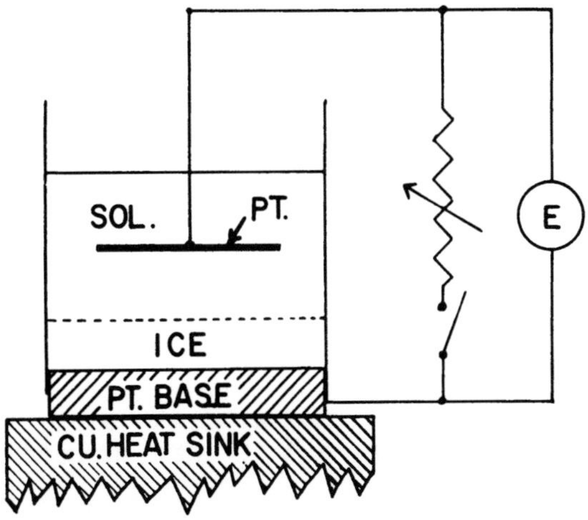

Fig.1 Apparatus used for demonstrating the separation of electric charge by freezing

by the freezing of dilute aqueous solutions. This effect can be demonstrated in a variety of ways, but it may be useful to refer to one of our early laboratory devices. Consider freezing a 2 x 10^{-5} M solution of CsF in a cup as shown in fig.1. The base of the cup is of heavy inert metal and the side walls of a thin plastic, like Teflon (polytetrafluorethylene). The solution is prepared by using a good grade of conductivity water (0.2 to 0.5 μmho) and reagent grade CsF, handled in such a way as to minimize possible contamination by CO_2 and ammonium compounds.

If the freezing rate is 0.5 to 1 mm per minute a potential difference of about 30 volts will be generated, as indicated by a high impedance electrometer (E in fig.1). The observed potential difference resides across the water-ice interface in such a shallow layer as to give a very high potential gradient. The potential difference becomes fully developed very soon after freezing begins and is maintained until freezing stops or until the ion population of the residual solution is significantly altered.

The new ice layers selectively incorporate F^- ions into their structure and these ions are continuously neutralized by a proton current. The protons, or proton states, move upward from the layers of ice near the base or they enter the ice across the interface, depending principally upon the constants of the external circuit. If a low external resistance is maintained and a somewhat more concentrated solution (3 x 10^{-4} M) is used a maximum ion current can be realized, equivalent to approximately

5×10^{16} F^- ions per cm^3 of ice frozen. If the pH of the solution is increased to about 7.8 the decreased hydrogen ion population may cause the open-circuit potential difference to be as much as 150 volts /1/.

Ice that contains 3×10^{16} HF impurity centers in each cm^3, one in 10^6 water molecules, is different from "pure" water ice. It is less adhesive, much more friable, it exhibits less diffuse Laue spots, and it is a relatively good electrical conductor. The microstructure of ice appears as hexagonal prisms 0.5 to 20 microns in diameter. In the case of pure water ice the etched patterns show depressed centers. As the HF content of the ice increases the depth of the center depression decreases, disappearing entirely for ice frozen from a 10^{-3} CsF solution /2/.

In general, growing ice layers appear to have a strong affinity for negative ions. Negative ions are incorporated if they are structurally acceptable as substitutes for oxygen in the lattice. Halogen ions are most suitable.

An important variation is produced when weak ammonium solutions are frozen. In this case the ice incorporates the NH_4^+ ion as a substitute for O^{--}. Ice containing NH_4^+ or NH_3 impurity centers is highly resistive. It is difficult to neutralize the positive charge centers as the interface moves and, as a consequence, high potential differences, often reaching 150 volts, are generated when ammonium solutions are frozen. Unfortunately, the electrical properties of ammonium doped ice are not well known.

The procedures for demonstrating the effects just mentioned are exacting and frequently fail to produce the expected results. It is interesting to observe however, that ice, regardless of how it is frozen, is active as an ion separator. This is shown by the fact that ice usually is found to contain more hydrogen ions than the solution from which it was frozen /1/. Ice that grows in lakes and ponds with its c-axis in the vertical or near vertical direction will show, upon analysis of the melt, a greater concentration of salts than ice grown from the same water at the same time, but with a largely horizontal c-axis /3/.

These observations raise the question of what takes place during the formation of the first layers of ice growth. Experiments bearing on this question cannot be performed using the cup procedure described earlier because freezing usually starts in a confusion of dendrites on the metal base under conditions of supercooling.

A test was arranged /4/ in which charge transfer was measured when a cold ($-10^{\circ}C$ to $-25^{\circ}C$) ice surface was dipped repeatedly into a cup of cold solution and quickly removed. It was found that almost any water sample that one might select would give negative ice and positive residual water. A sustained potential difference resulted for those solutions likely to work in the earlier cup experiments. In nearly all cases, however, a negative potential developed on the ice with a rise time of from 10 to 30 milliseconds and subsequently fell to zero or occasionally reversed in sign.

The property of growing ice that causes it to incorporate negative ions from the liquid when new ice layers are formed on a cold substratum gave promise of being effective in generating thunderstorm electricity. Workman and Reynolds /5/ had proposed that the growth of wet hail might be responsible for charge generation if the washing away of positive drops could produce the necessary separation.

Although this proposal was attractive to many investigators it had two principal disadvantages: (1) It imposed strict limits on the composition of atmospheric water and (2) the possibility of reversed polarity due to ammonium contamination left unresolved questions.

In view of well founded notions about the sequence of events in thunderstorms it was possible to specify, in general terms, the design features of an electrical charge mechanism. These arguments have been presented elsewhere /6/ and they will not be duplicated here. The requirements could be met if the thunderstorm environment could provide for the growth of new ice on cold dry ice particles through collisions with cold water drops; water drops that would provide splash-off after some freezing.

Laboratory tests of the proposed mechanism were devised as follows (fig.2). A copper tube standing vertically and extending through the top of a side opening deep freeze box serves as a drop tower. The tube is supported over a hole in a cylindrical copper chamber situated at the bottom of the deep freeze box. The vertical deep freeze is capable of maintaining an inside temperature of $-26^{\circ}C$ with the copper tube in place. The lower copper chamber is provided with heating coils and nickel-resistance temperature sensors. Two additional temperature sensors are fixed to the copper tube as shown. The upper end of the vertical copper tube has a plastic cover on which is installed a water dropping device.

The essential features of the water dropping device are

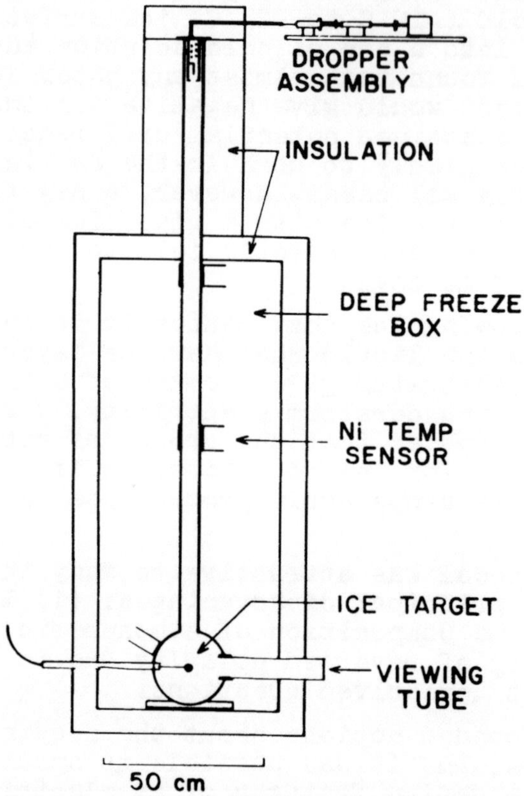

Fig.2 Laboratory drop tower

as follows. A synchronous gear train motor, through a screw arrangement, moves the plunger of a 2.5 ml hypodermic syringe. A small Tygon (Furane resin) tube leads from the needle of the hypodermic syringe through the upper plastic top. The Tygon tube then winds spirally on a thin plastic tube and terminates on a stainless steel tube on the end of which drops are formed as the water is exuded. The end of this small tube is cut square and the walls honed down to provide drops averaging 1.67 mm in diameter. The Tygon tube assembly is tightly fitted into a heavy copper tube to provide good thermal contact in an isothermal environment. This enclosing copper tube is provided with nickel-resistance temperature sensors and an electrical heating coil. The small inside diameter of the water conduit permits operation at about -5°C without freezing, when 1.67 mm diameter drops are forming at the rate of 5 drops per minute. Subsequent cooling of the drops as they fall results in an additional temperature depression of 3°C to 4°C before they enter the lower copper chamber.

A brass ball (0.79 cm diameter) attached to a thin metal shaft and coated with ice, serves as a simulated hailstone. The shaft leads directly to the gate of a 2N3631 field effect transistor, the output of which is fed to a high-speed chart recorder or to an oscilloscope or to both.

The calculated striking velocity of the water drops on the simulated hailstone is 365 cm/sec, compared with an expected terminal velocity at sea level pressure of about 540 cm/sec. The amount of new ice formed by the impact of each drop will vary somewhat with the temperature of the target ice but, in the working range of -15°C to -20°C, eighty percent of the water drop freezes and twenty percent splashes away in small drops on impact. These values are determined quite accurately by counting the number of drops and determining the accumulated ice by weighing. Drop size is determined by counting the number of drops for a specified piston displacement of the hypodermic syringe.

The solutes used in the drop forming water have been varied in kind and concentration to illustrate the electrical process, and to cover the range of impurities in thunderstorm water as they may become known. In all cases the substratum of the target ice was formed from the relevant drop forming liquid.

Representative natural waters and a few solutions having definite composition were tested. Distilled water, having electrical conductivity from 0.3 to 0.5 μmho, and analyzed chemical reagents were used in preparing specific solutions.The amount of charge generated, expressed in electrostatic units per gram of water dropped (the aggregate mass of water drops impacting the hail particle), for various solutions is shown in table 1.

Of very special interest are the results for ammonia solutions since ammonium compounds have been known to give reversed polarity (positive ice and negative water) in the laboratory tests previously described. In an earlier paper /6/ I described a model for the production of thunderstorm electricity and postulated that, since the electrical conductivity of ammoniated ice is very low, the process of reverse charge generation would not be effective because the rate of separation of charge was thought to be too slow for the impacting drop mechanism. It now appears that the action of ammonia is much more complicated than previously thought.

Table 1

Sample Number	Drop forming solution	Molarity	Conductivity (μmho)	Charge (esu/gram)
1.		10^{-7}	0.77	- 4.95
2.		10^{-6}	0.92	- 4.46
3.		10^{-5}	2.04	- 6.28
4.	NaCl	10^{-4}	32.2	- 6.70
5.		10^{-3}	91.0	- 5.34
6.		10^{-2}	910.0	0.0
7.		10^{-1}	1.720.0	+ 0.03
8.		10^{-8}	0.89	- 1.05
9.		10^{-7}	0.71	- 1.68
10.	$(NH_4)_2CO_3$	10^{-6}	1.31	- 1.14
11.		10^{-5}	2.50	- 0.73
12.		10^{-4}	17.8	+ 9.30
13.	Sea water collected 6.5 km off coast		74,600.0	+ 0.04
14.	Sea water diluted 1 to 100,000		1.72	- 1.54
15.	Rain water collected at 80 m MSL, fresh		3.85	- 3.34
16.	Same as sample (15), exposed 12 hours		2.50	- 1.98
17.	Rain water collected at 1310 m MSL, fresh		2.56	- 2.81
18.	Water from melted hail collected at 3400 m MSL		2.40	- 6.79
19.	Tap water -- Hilo		33.0	-13.2
20.	Tap water -- Hilo, 1 month after sample (19)		55.0	- 4.00
21.	Water from a fish pond, after sedimentation		111:0	-14.0
22.	Distilled water		0.34	- 1.12
23.	Distilled water		0.40	- 1.17
24.	Distilled water		0.50	- 1.43
25.	$(NH_4)_2CO_3(10^{-7}M)$ mixed 1 to 1 with NaCl $(10^{-7}M)$		1.14	- 0.19
26.	Same as sample (25), exposed 12 hrs at cool temp.		1.72	- 0.13
27.	$(NH_4)_2CO_3(10^{-4}M)$ mixed 1 to 1 with NaCl $(10^{-5}M)$		11.0	+ 8.00
28.	$(NH_4)_2CO_3(10^{-5}M)$ mixed 1 to 1 with NaCl $(10^{-5}M)$		1.85	- 1.24
29.	$NaNO_3$, molarities from 10^{-7} to 10^{-4}			<- 0.10

The behavior of ammonia in the freezing effect, as it might occur in thunderstorms, is very important because all atmospheric water is expected to contain traces of this substance. The charge transfer process for concentrations of ammonia equal to or less than 5×10^{-5} molar are normal (negative ice, positive water) under the conditions of these tests for all temperatures warmer than about $-23^{o}C$. It would appear, therefore, that ammonium compounds, regardless of charge transfer rates, if effective at all, will make a thunderstorm of normal polarity. It is important to note that ammonium ions give their characteristic effect when they are present in solutions of other salts -- salts that normally give large negative charges to the ice. The results for ammonium sulfate do not differ significantly from those of the carbonate.

While the behavior of ammonia invites much further study, speculation, based on these preliminary results, suggests the possibility of reversing, or conceivably preventing, thunderstorm charge separation by overdosing the thunderstorm water with ammonium compounds to produce concentrations of 10^{-4} molar or stronger. This possibility, based upon a much more limited knowledge of the behavior of ammonia, was suggested by Workman and Reynolds in reports written about twenty years ago.

If we assume that 10^{11} grams of water would be involved in a particular thunderstorm cell, something like 170 kg of NH_3, would be required to give a concentration of 10^{-4} molar. It would, of course, be necessary to disperse the NH_3 at the right place and at the right time.

It appears that the mechanism here described is capable of producing the observed electrical behavior of thunderstorms when the meteorological conditions essential to other cloud processes are satisfied. Adequate electric charge can be generated and transported under conditions that can be specified.

It should be noted that the electrical effects associated with freezing are likely to occur in any situation where ice grows in contact with aqueous solutions. It is interesting to note that the electrical process here described, as well as many naturally occuring freezing processes, involves the electrolysis of water.

Consider, for example, the production of free H_2 by thunderstorms. Estimating the efficiency (ratio of total air-earth current to actual charge generation) to be twenty percent, there are 6×10^{-4} coulomb-m^{-2} produced each year. It would thus require 80×10^{6} years to make the 5000 cm^3 of atmospheric H_2 observed above each square meter of the earth's surface. This would seem to be an adequate rate for the supply of free H_2 to the atmosphere.

References

/1/ Cobb, A.W., Unpublished report, Research and Development Division, New Mexico Institute of Mining and Technology, Socorro, New Mexico (1963)

/2/ Truby, F.K., J.Appl.Phys. 26, 1416 (1955)

/3/ Cobb, A.W., Science 141, 733 (1963)

/4/ Workman, E.J., Proceedings of the Third International Conference on Atmospheric and Space Electricity, American Elsevier Publishing Company, pp. 296-303 (1965)

/5/ Workman, E.J., S.E. Reynolds, Phys.Rev. 78, 254-259 (1950)

/6/ Workman, E.J., J.Franklin Institute 283, 540-557 (1967)

THE SEPARATION OF CHARGE DUE TO THE FRACTURE OF FREEZING WATER DROPS

D.A. Johnson

Meteorological Office, Bracknell

Berkshire, England

Abstract

The charge separated when freezing water drops shatter in large fragments has been studied by several workers. Two alternative mechanisms have been suggested to account for the charging - the thermoelectric effect in ice and charge separation in dilute aqueous solutions (the Workman-Reynolds effect). During an experimental study of the fracture of freezing drops it was found that small charged particles were sometimes ejected, even when the drops did not break into large fragments. The charge densities observed were too large to be explained by the thermoelectric effect, and were estimated to approach the limit imposed by the dielectric strength of air. Neither the charges due to the ejection of small particles nor those due to shattering in large fragments were significantly affected by the presence of salts (NH_4OH, NaCl) in solution showing that the Workman-Reynolds effect was not operative. It appears that some other mechanism is required to explain the charges observed.

1. Introduction

The electric charge separated when freezing water drops of about 1 mm diameter shatter into large fragments has been studied by Mason and Maybank /1/, Kachurin and Bekryaev /2/, Evans and Hutchinson /3/ and by Stott and Hutchinson /4/. No charging was ever observed if the drops did not shatter. It was shown by Johnson and Hallet /5/ that the conditions under which previous workers observed shattering were not representative of conditions in natural clouds, where drops are uniformly supercooled t the temperature of the environment prior to nucleation and are ventilated as they fall. Under these conditions drops do not shatter unless they rotate as they fall, leading to a uniform heat transfer distribution. It is, however, possible that under certain conditions drops freezing on accretion may fracture and eject ice splinters (Brownscombe and Hallett, /6/) and if so it is desirable to understand the mechanism responsible for separating charge when drops fracture on freezing.

Mason and Maybank /1/ interpreted their results in terms of the thermoelectric effect in ice (Latham and Mason, /7/) due to a radial temperature gradient in the ice shell of the freezing drop, although they pointed out that some of the charges measured were larger than the maximum predicted by the theory. Stott and Hutchinson /4/ obtained broadly similar results, but they considered that the charges could be explained only by the charge separation in freezing dilute solutions as observed by Workman and Reynolds /8/.

2. Experimental

The present work was undertaken in order to determine the mechanism responsible for charge separation in freezing drops, by freezing both drops of pure water (in which the Workman-Reynolds effect would be unlikely to be operative) and drops of ionic solutions chosen to show a large Workman-Reynolds effect. The apparatus (fig.1) was the refrigerated cell described by Johnson and Hallett /5/. An induction electrode in the form of a 5 mm diameter ring of fine copper wire situated just below the freezing drop was connected to a vibrating reed electrometer of 1.3 pF input capacity. The calibrated sensitivity was 1.1×10^{-5} esu mV^{-1} and the noise level of the most sensitive range was about 0.5 mV. The charge on the drop was measured before and after freezing by lowering it into the wire ring and then noting the change of potential as it was suddenly raised out of the electrode. Changes in drop charge during freezing were detected by the resulting induced charge on the electrode. The output of the electrometer was fed to a galvanometer recorder with an event marker which enabled events during freezing to be correlated with the record. The initial charge on the drop was reduced to near or below the sensitivity of the electrometer by exposing it to ionising radiation. Drops were frozen at environment temperatures between -5 and -20°C.

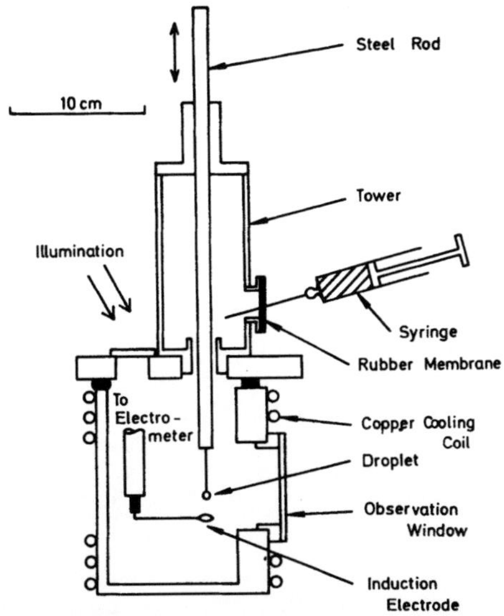

Fig.1 Apparatus for observing the freezing and charging of water drops

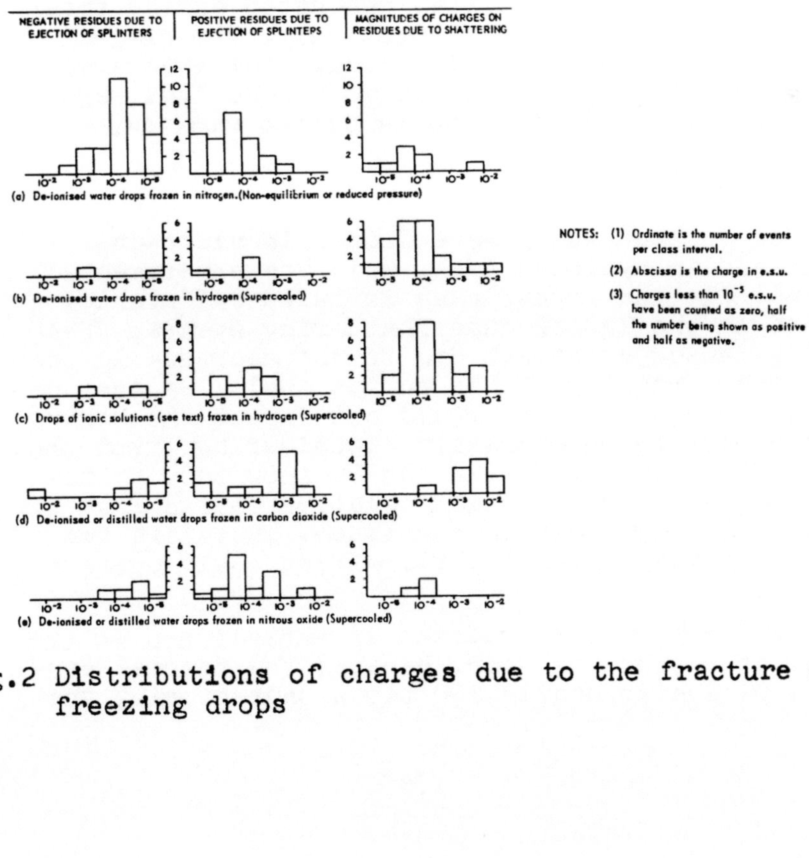

Fig.2 Distributions of charges due to the fracture of freezing drops

It was observed that charges up to 10^{-2} esu were separated when drops shattered into large fragments. Only the magnitudes of the charges separated in this case are shown in fig.2, since the sign depends merely on which part of the drop happened to be left on the supporting fibre. It was also observed that the ejection of small particles during freezing (splintering) was sometimes accompanied by significant charge separation even if the drop did not shatter. These particles were often seen as they were ejected and the event usually resulted in a slight vibration of the suspended drop.

The experimental conditions chosen were those in which shattering or splintering occurred with reasonable frequency. The distribution of charges on the drops after freezing are shown in fig.2. The "pure" water used was distilled once and passed through a mixed bed ion exchange column. The water was transferred to the cold cell in a syringe surrounded by a plastic bag full of nitrogen. The ionic solutions were 7×10^{-5} M ammonium chloride and 10^{-4} M sodium chloride. These solutions were frozen in bulk under conditions which permitted measurement of the potential difference between the liquid and the ice during freezing, after Workman and Reynolds /8/. The maximum potential differences (of the liquid with respect to the ice) generated during freezing were -100 V and +1.2 V respectively. The de-ionised water frozen under the same conditions generated a maximum potential of -0.3 V, but no precautions were taken to exclude contamination by carbon dioxide.

3. Discussion of results

Drops of de-ionised water were frozen in nitrogen (fig. 2a) and in hydrogen (fig. 2b). The experimental conditions in the former case favoured splintering events and in the latter case shattering events, fracture being caused by stress due to the decrease in density of water freezing within an ice shell. Because de-ionised water was used one would not expect the Workman-Reynolds effect to be operative. Considering first the ejection of splinters, fig.2a and 2b show that splinters removed both positive and negative charge from the drops up to 3×10^{-3} esu in magnitude. There are two ways in which the thermoelectric effect could operate in this case.

a) A radial temperature gradient is established in the ice shell of a freezing drop. Because the thermal conductivity of ice is nearly 100 times that of air, the gradient is small until near the completion of freezing, when it rises rapidly to its maximum value $(T_o-T_e)/a$ (where T_o is the temperature of the mainly liquid interior (assumed 0^oC), T_e is that of the remote environment and a is the drop radius). Putting $T_e = 20^oC$ and a = 0.05 cm, typical values, the maximum possible gradient is $400^oC\ cm^{-1}$. Splinters were usually ejected before the gradient could have reached its maximum value. The theory of Latham and Mason /7/ predicts that the outer surface of the drop would acquire a charge density of $+ 2 \times 10^{-2}$ esu cm^{-2}. The ejection of splinters would therefore leave the residue negatively charged.

b) If a splinter slides along the drop surface after ejection, it may become warmer than the surface due to asymmetric rubbing. According to the thermoelectric effect negative charge will be transferred to the splinter. If the temperature difference between the splinter and the drop surface were as high as 10^oC, the theoretical charge density would be 3×10^{-2} esu cm^{-2}. In this case the residue would become positively charged.

In both situations (a) and (b) the maximum charge density due to the thermoelectric effect is of the order 10^{-2} esu cm^{-2}. Even if the area of the splinter involved were as large as 100μ x 100μ, the maximum estimated charge per splinter is only 10^{-6} esu, much smaller than the largest charges measured.

In order to explain the largest charges observed when drops shattered into large fragments, more than the total charge on the outer surface of the drop (maximum possible value 10^{-3} esu) would have to be removed.

The distribution of charges in fig.2c (drops of ionic solutions frozen in hydrogen) is not significantly different from that in fig.2b (drops of de-ionised water frozen under the same conditions). This suggests that the charges on the fragments were not appreciably influenced by separation of charge at the liquid-ice interface due to the Workman-Reynolds effect. In any case it is difficult to see how the charges on splinters ejected from the drop surface could be due to this mechanism.

Figures 2d and 2e show the distribution of charges when drops fractured in carbon dioxide and nitrous oxide gases respectively. In both these gases drops fractured because of the release of large volumes of dissolved gas during freezing. The charges on drops frozen in carbon dioxide are generally higher than those on drops frozen in nitrous oxide.It should also be noted that only 1 of the 56 charges due to shattering shown in fig. 2a, 2b and 2c (no carbon dioxide present) was greater than 10^{-2} esu, whereas in the experiments of Evans and Hutchinson /3/ and Stott and Hutchinson/4/ , 7 of the 32 drops which shattered in large fragments had charges exceeding 10^{-2} esu. There was probably a large amount of carbon dioxide dissolved in their drops, since they used dry ice to produce the ice crystals which nucleated their drops.

The charges separated when de-ionised water drops shattered or splintered on freezing were generally too large to be explained by the thermoelectric effect in ice. Since dissolved salts did not change the level of charging appreciably it seems that the Workman-Reynolds effect does not contribute to the charging. To explain these results we require some mechanism capable of generating charge densities of the order 10 esu cm^{-2}.(Note that this corresponds to a surface potential gradient of the same order as the dielectric strength of air). There is some evidence that the presence of carbon dioxide increases the charging. Similar charge densities were observed by Takahashi /9/, who fractured ice rods by tensile stress and measured densities from 5×10^{-3} esu cm^{-2} (for single crystal specimen) to 1 esu cm^{-2} (for polycrystalline specimens). He suggested that the charging was due to friction between the separating fragments.

Latham /10/ postulated that the charges on fractured drops observed in earlier work were the result of two simultaneous charging processes - positive charge on the outer fragments due to the thermoelectric effect (mechanism (a) discussed earlier) plus a much more intense, unspecified, charging process imparting a random sign to the fragments. He showed that the overall average charge on smaller fragments was positive and not inconsistent with the thermoelectric effect (although, with Evans and Hutchinson /3/, he overestimated the maximum possible temperature gradient in the ice shell of a freezing drop by a factor of 10). The present results, with the new evidence of charging due to splintering, support the idea of a charging process much more intense than the thermoelectric effect. However, before these results can be applied to charge generation in clouds it will be necessary to obtain a full understanding of the nature of the charging mechanism - whether it is a random process or whether it is influenced, especially in sign, by factors such as the crystal structure and purity of the ice and by the precise mode of fracture. None of these factors was controlled in these experiments.

4. Conclusion

The charges separated when water drops fractured on freezing were found to be too large by 2 or 3 orders of magnitude to be explained by the thermoelectric effect in ice. The charges were insensitive to the presence of salts chosen to show a large Workman-Reynolds effect. Some other unspecified charging mechanism dominates both these charging processes. If drops of water fracture on freezing in clouds, whether individually or on accretion on an ice substrate, there seems to be no difficulty in principle in separating charges up to the maximum that can be carried by the fragments without breakdown of the air. However it has yet to be demonstrated conclusively that drop fracture can occur under atmospheric conditions (Brownscombe and Hallett, /6/) and that the charges separated have a predominate sign, as is required by the gravitational theory of thunderstorm electrification (Mason, /11/).

Acknowledgements

This work was commenced at Imperial College, London. The author would like to thank Dr. B.J. Mason for advice and discussions during the course of this work. This paper is published by permission of the Director General , Meteorological Office.

References

/1/ Mason, B.J., J. Maybank, Quart. J.R. Met. Soc. 86, 176-186 (1960)

/2/ Kachurin, L.G., V.I. Bekryaev, Dokl. Akad. Nauk. SSSR 130, 57-60 (1960)

/3/ Evans, D.G., W.C.A. Hutchinson, Quart. J.R. Met. Soc. 89, 370-375 (1963)

/4/ Stott, D., W.C.A. Hutchinson, Quart. J.R. Met. Soc. 91, 80-86 (1965)

/5/ Johnson, D.A., J. Hallett, Quart. J.R. Met. Soc. (in press)

/6/ Brownscombe, J.L. and J. Hallett, Quart. J.R. Met. Soc. 93, 455-473 (1967)

/7/ Latham, J., B.J. Mason, Proc. Roy. Soc. 260A, 523-536 (1961)

/8/ Workman, E.J., S.E. Reynolds, Phys. Rev. 78, 254-259 (1950)

/9/ Takahashi, T., J. Met. Soc. Japan (2) 40, 277-286 (1962)

/10/ Latham, J., Quart. J.R. Met. Soc. 90, 209-211 (1964)

/11/ Mason, B.J., Tellus 5, 446-460 (1953)

CHARGE SEPARATION IN ICE NEEDLES CONTAINING TRACES OF NO_3^- IONS

R.Reiter, W.Carnuth

Physikalisch-Bioklimatische Forschungsstelle Garmisch-Partenkirchen der Fraunhofer-Gesellschaft zur Förderung der angewandten Forschung e.V. München / Germany

Abstract

If needle-shaped ice crystals are grown in an atmosphere containing traces of nitrous gases, or if point discharges are allowed to occur at the points of the crystals during growth the amount of electrical charge generated when sharp crystal points break away is between 10 and 50 times as high as in the case of crystals grown in the absence of nitrous gases. Laboratory experiments showed that the amount of charge generated in this manner depends on the NO_3^- ion concentration gradient along the crystal needles. An explanation of this effect based on different mobilities of NO_3^- ions and protons predicts the correct sign but not the observed amount of the charge. - Since nitrous gases are produced by electrical discharges in the atmosphere, this effect may be important for charge generation in thunderstorms.

Problems of cloud physics, which will be discussed briefly later on, suggested us to investigate the charge separation occuring upon fragmentation of ice crystals and how it is influenced by small amounts of incorporated nitrate ions.

In a closed box, which we call cloud chamber, we created

needle-shaped ice crystals on an electrically grounded copper plate which can be cooled by a peltier cell down to $-30^{\circ}C$. Below this breeder plate a cup of water is mounted in order to supply the water vapor which is necessary for the ice crystals to grow. Traces of nitrate ions were incorporated into the crystals either by producing nitrous gases within the cloud chamber by a spark gap or by applying an electric field, ac or dc, between the grounded breeder plate and the water surface below it. Weak silent discharges at the growing crystal points created small amounts of nitrous gases and thereby NO_3^- ions were incorporated into the crystals during their growing phase. The temperatures of the plate and the surrounding air are recorded continuously by means of two thermocouples.

After crystal growth is terminated, the water cup below the breeder plate is replaced by a gold-plated bowl which is insulated from the grounded chamber by teflon and connected with a vibrating reed electrometer. The electrometer sensitivity is such that 10^{-14} Coulombs can be detected with certainty. Crystal fragments breaking away spontaneously or by intentionally applied mechanical shocks will fall into the gold-plated bowl and their charge can be measured with the electrometer. After disconnecting the voltage applied between the water cup and the breeder plate and before fragmentation of the crystal points, sufficient time was allowed to elapse to be certain that any crystal charge which may be inducted by the electric field or by the well-known Costa Ribeiro effect /1/ had been equalized. The broken off and collected crystal fragments were weighed quickly after the charge measurement and removed by rinsing with some milliliters of extra pure distilled water. The nitrate ion content was determined by a special micro-chemical method.

In fig.1 the measured charges in Cb/g are plotted vs. the concentration of nitrate ions. It is found a marked increase of the amount of liberated charge with increasing NO_3^- ion concentration, the points, however, being found to lie on two branches A and B of the graph, which represent negative resp. positive charges. Points on branch A were mainly obtained with thin, needle shaped crystals grown in a saturated atmosphere containing no suspended water droplets. Points belong to branch B if crystals were coarse and dense.

Before we try to explain the results of the experiments it is to be pointed out that, if the nitrate concentration is homogeneously distributed along the ice needle, charge separation upon fragmentation should be expected

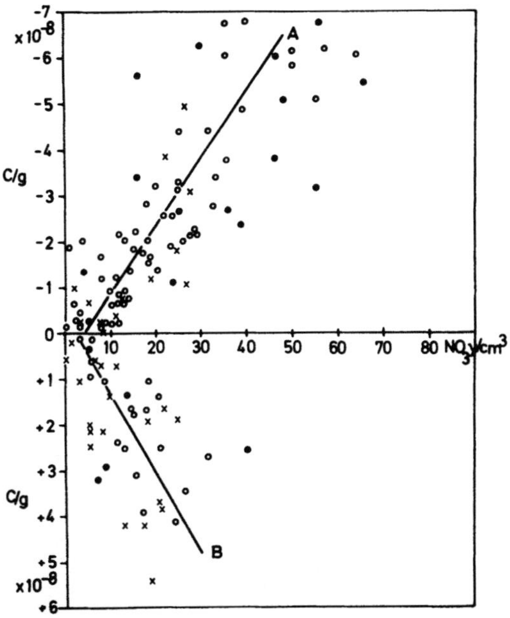

Fig.1 Relation between NO_3^- ion concentration of separated ice fragments and specific electric charge generated upon fragmentation.

- ● Externally produced nitrous gases supplied to chamber.
- o dc voltage applied between crystal points and counterelectrode (water surface charged to +5 - +6 kV).
- x ac voltage applied between crystal points and counterelectrode (water surface connected to 3 kV, 50 c/s)

to occur either not at all or in very small amounts only. It must be rather assumed that there is a concentration gradient along the crystal. In order to confirm this we tried to fractionate the fragmentation process, that is to get some information about the existing average nitrate concentration gradient by successively removing crystal fragments. Although this process proved to be very difficult, in a sufficient number of cases the attempts were successful. In fig.2 the quotient

$$Q_K' = \frac{NO_3^- \text{ ion concentration in the crystal points}}{NO_3^- \text{ ion concentration in the truncated crystals}}$$

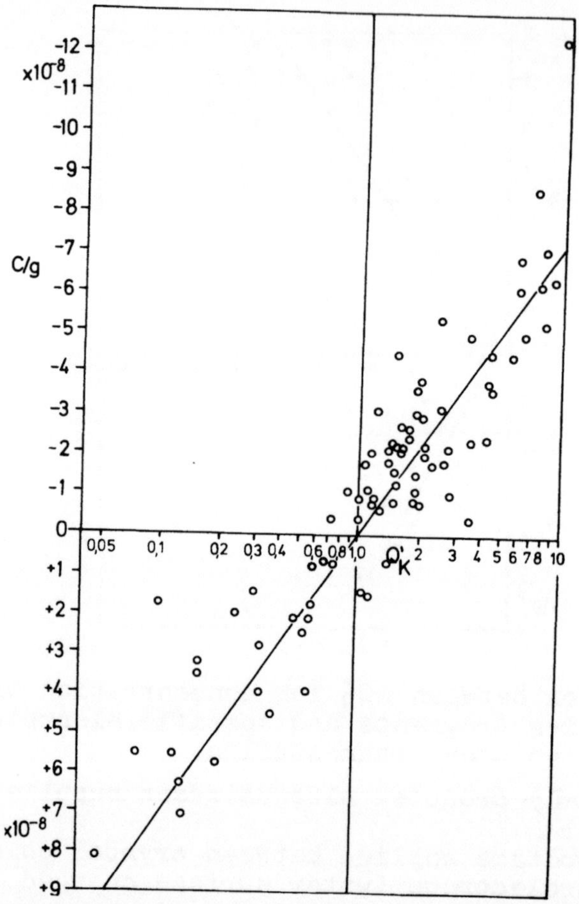

Fig.2 Relation between electric charge liberated upon fragmentation of ice crystals and NO_3^- ion concentration quotient Q'_K (see text)

is plotted in logarithmic scale against the amount of charge, Cb/g, which is stored in the first fraction of ice splinters removed. It is to be seen that the relation between the two quantities is practically linear in this scale.

Further we see that different signs of the charges are caused by different directions of the concentration gradient. A very important fact is that a surplus of nitrate ions is always connected with a negative charge.

These findings suggest an explanation of the observed charge separation by diffusion of protons, assuming the negatively charged NO_3^- ions being practically immobile in comparison to the positive protons. The higher proton density existing in the regions of higher nitrate ion concentration tends to equalize by diffusion until the space charge accumulated thereby and the ensuing electric counter field will lead to an equilibrium. In that case the diffusion current $\underline{j}_D = -eD\,\mathrm{grad}\,n$, with e denoting elementary charge, n number of protons and D = diffusion constant, is equal to the electric current $\underline{j}_E = \sigma\underline{E}$:

$$\sigma\underline{E} - eD\,\mathrm{grad}\,n = 0.$$

Upon introducing the Poisson integral we get

$$\frac{\sigma e}{4\pi\varepsilon\varepsilon_0}\,\mathrm{grad}\int\frac{n(\underline{r}')-H(\underline{r}')}{|\underline{r}-\underline{r}'|}\,d\tau' + eD\,\mathrm{grad}\,n = 0$$

where H = NO_3^- ion concentration, $\underline{r}$ and $\underline{r}'$ = radius vectors. A general solution of this integro-differential equation can only be found by approximations and at the cost of considerable amount of work. For a rough estimate, however, the following simplified consideration should be sufficient.

We consider a needle-shaped ice crystal of length L and radius $R \ll L$. Let the nitrate ion concentration be distributed in such a way that it increases linearly from $H = H_0 - \Delta H$ at the one end of the crystal to $H = H_0 + \Delta H$ at the other (fig.3). Under these conditions a diffusion current

$$j_D = \frac{2eD\Delta H}{L}$$

will flow. To estimate the order of magnitude of the electric current j_E we assume for this time the electric charges to be located at the ends of the crystal needle. Then the electric field strength will be

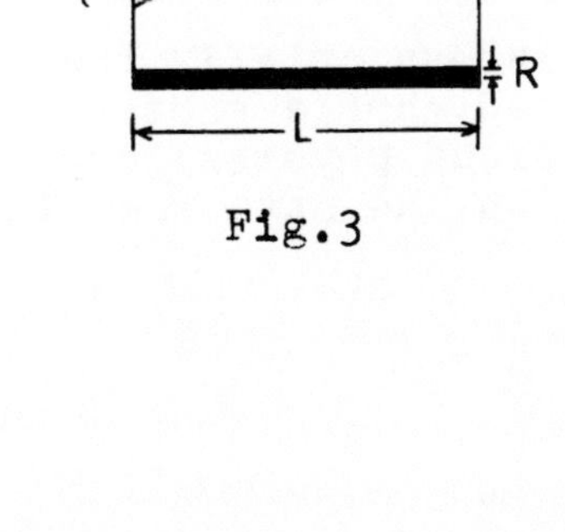

Fig.3

$$E \geqq \frac{2\,q}{\pi \varepsilon \varepsilon_0 L^2}\,.$$

E and grad n may be considered as scalar quantities in the present case. The above expression for E yields a charge

$$q = \pm\, \pi \varepsilon \varepsilon_0 e D \Delta H / L\,.$$

If the crystal is then broken apart in the middle, each of the two halves will carry the space charge (specific charge):

$$\rho = \pm \frac{2\,q}{R^2 \pi L} = \pm \frac{2 \varepsilon \varepsilon_0 e D \Delta H}{\sigma\, R^2}$$

the part containing more NO_3^- ions being charged negatively. For a numerical estimate we assume the concentration gradient to have the highest value compatible with our model, that is $\Delta H = H_0$. A typical measurement of our experiments yielded a specific charge of 10^{-7} Cb/cm^3. The measured nitrate ion content corresponds to a value of H_0 of 5.10^{-4} mole/ℓ or 3.10^{17} NO_3^- ions/cm^3. We can take further

$D = 2.10^{-11}$ cm/sec (after Dengel and Riehl, /2/)
$\varepsilon = 100$ and
$\sigma = 10^{-9}$ ohm^{-1} cm^{-1}.

With these values and $R = 10^{-2}$ cm we get

$$\rho \leqq 2.10^{-10}\,\text{Cb/cm}^3.$$

This value is almost three orders of magnitude smaller than the measured specific charge of 10^{-7} Cb/cm^3.

As in this approximation only the most favorable conditions and values of constants were assumed, it is obvious that the diffusion process alone is insufficient to explain the observed amounts of charge separated. Any effects involving contact potentials between the solid and liquid phase may safely be excluded. It is, therefore necessary to look for another more effective mechanism of charge separation or charge storage. It should be recalled here that, according to a paper of Gross from 1954 /3/, the diffusion potential has also proved to be insufficient to explain the potential difference between a growing solid and the corresponding liquid phase of water, namely the Costa-Ribeiro effect.

Concluding we shall discuss briefly the consequences which the just described charge separation process might have with respect to cloud physics, especially to thunderstorm electrification. We have seen that nitrous gases influence the charge generation upon fragmentation of ice particles. This process may occur in the upper parts of thunderstorm clouds which are accompanied with heavy turbulent motion of the air. If ice crystals of different sizes carry charges of opposite sign, space charges can be generated by the action of gravity in updraft air motion. These charges in turn may give rise to silent discharges between the cloud and precipitation particles, producing again more nitrous gases. Sartor was able to show such silent discharges between water droplets experimentally /4/.

The increase of nitrate content in the air and in the ice particles may again raise the charge separation upon breaking of ice particles. This feedback process /5-8/ would lead to an exponential-like increase of space charges which indeed is often observed in thunderstorm clouds. Some further evidence for the existence of this charge generation process emerges from figures 4 to 7. We measured the nitrate ion content of precipitation water at our three measuring stations Garmisch, Wank and Zugspitze.

In fig.4 and 5 the values are plotted versus the vertical

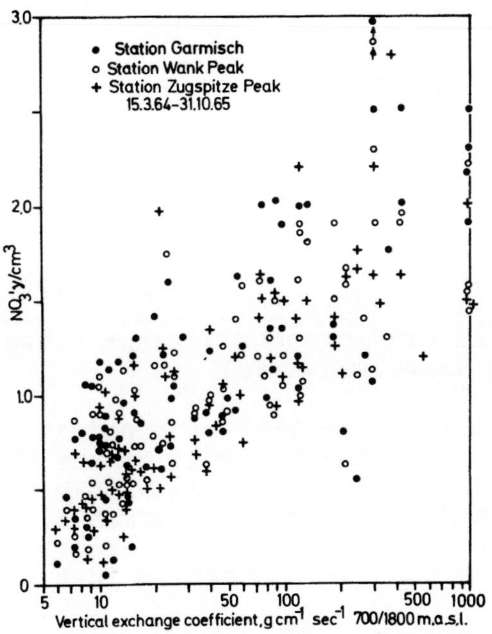

Fig.4 NO_3^- ion concentration of precipitation collected at stations Garmisch, Wank peak and Zugspitze peak as a function of the vertical exchange coefficient between 730 and 1780 m above sea-level

exchange coefficient, which is a measure of the intensity of turbulent air motion. The NO_3^- ion content increases with increasing turbulence, but doesn't show any decrease with height, as it would be expected if the nitrous gases were produced by combustion or other processes near the ground level and not within the cloud itself.

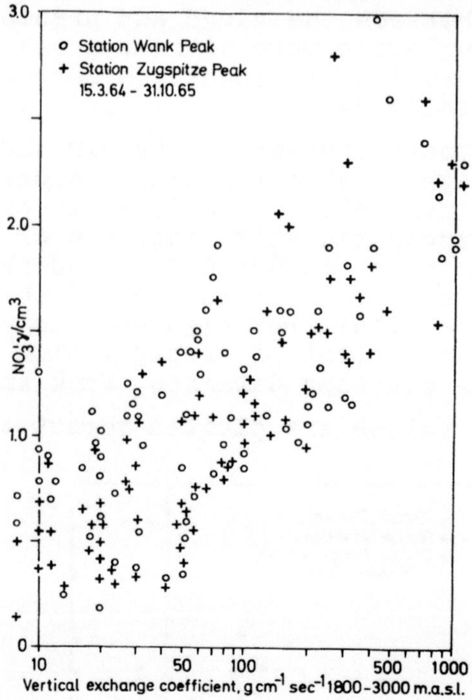

Fig.5 NO_3^- ion concentration of precipitation collected at Wank peak and Zugspitze peak as a function of the vertical exchange coefficient between 1780 and 2964 m above sea-level

In fig.6 the nitrate ion content of precipitation water is plotted against the time integral of the electric field strength as measured by a field mill within the thunderstorm cell. The increase is very remarkable. Fig.7 shows a typical field mill recording of the electric field strength from which the time integrals just mentioned are taken.

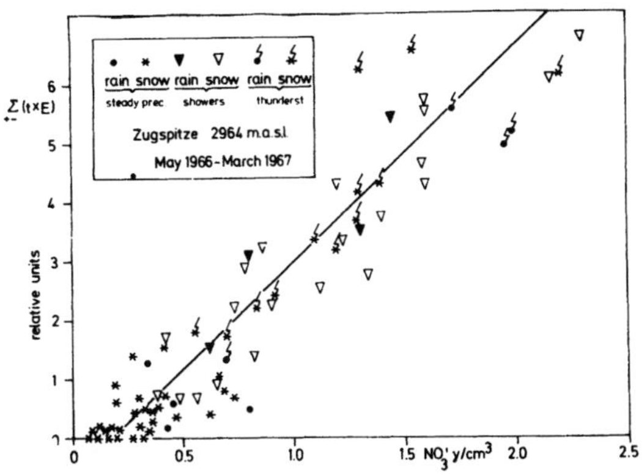

Fig.6 Relation between time field strength product area recorded at Zugspitze peak station during precipitation and NO_3^- ion concentration of simultaneous precipitation sample

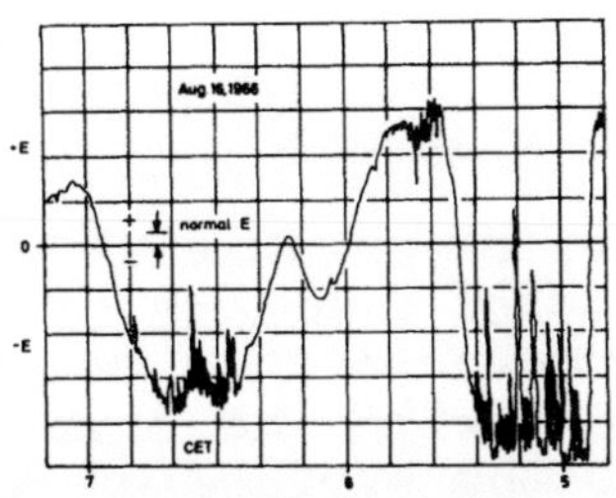

Fig.7 Typical field strength curve recorded at Zugspitze peak station during shower precipitation

References

/1/ Costa Ribeiro, J., Ann.BrazilAcad.Ci. 17, 2 (1945) 22, 321 (1950)

/2/ Dengel, O., N.Riehl, Phys.kondens.Mat. 1,191 (1963)

/3/ Gross, B., Phys.Rev. 94, 1545 (1954)

/4/ Sartor, J.D. J.Atmosph.Sci. 24, 601 (1967)

/5/ Reiter, R., M.Reiter, Recent Advances in Atmospheric Electricity, Pergamon Press, London (1959)

/6/ Reiter, R., Pure and Applied Geophysics (PAGEOPH) 57, 206 (1964)

/7/ Reiter, R., W.Carnuth; J.Atmosph. Terrestrial Phys. 27, 673 (1965)

/8/ Reiter, R., J.Atmosph. Terrestrial Phys. 28 (1966)

ORIENTATION OF ICE CRYSTALS GROWN BY ACCRETION OF SUPERCOOLED DROPLETS

L. Levi - A.N. Aufdermaur

National Research Council of Argentine, Buenos Aires / Argentine; Eidgen. Institut für Schnee- und Lawinenforschung, Weissfluhjoch, Davos / Switzerland

Abstract

Studies of the structure of artificial hailstones indicate that, when supercooled droplets of 20 μ to 100 μ diamter impinge on a substrate at a temperature $<0^oC$ (dry growth), they freeze with the crystallographic c-axis at large angles to the surface of the substrate. This angle decreases from 90° to 45° with increased supercooling. These results are interpreted considering the ice crystal habit observed in dendrites grown in supercooled water (Macklin , Pruppacher) and the features of the heat exchange of the freezing droplets with the surroundings. The change of the preferred orientation of the crystals with supercooling supports the hypothesis that the inclined dendrites observed by the cited authors are not basal but form an angle $\phi = 1/2\ \alpha$ with the basal plane, where α is the angle between dendrites starting at the same point. The largest deviations ($\phi = 45^o$) exceed the largest value $1/2\ \alpha \approx 22^o$ observed by Pruppacher. They possibly indicate a change of growth habit corresponding to extreme supercooling where direct measurements were not possible.

I - Introduction

Phenomena on freezing of supercooled droplets have been especially studied by Hallett /1/, Higuchi and Yosida /2/, Pruppacher /3/ and Aufdermaur and Mayes /4/. An important purpose of these studies was to obtain information relative to the process of hailstone formation. In fact, the relation between the structure of hailstones and the conditions of their growth is still not well understood. A special problem concerns the crystal orientation with c-axes in the direction of growth which has been observed several times in hailstones and is in apparent discrepancy with the general tendency of ice to grow preferentially in the basal plane.

Hallett /1/ studied the structure and orientation of ice formed in drops frozen after impinging on an ice substrate; Pruppacher /3/ nucleated ice in the drops by touching them with the tip of a previously formed dendrite. Both authors agree in the observation that large drops of about 1 mm and 1 cm diameter respectively freeze as single crystals with the same orientation as the substrate or as the nucleating dendrite, down to a supercooling of approximately 5°C. With higher supercooling, a polycrystalline structure was formed which was related by Pruppacher to the multiple splitting of dendrites taking place at this temperature. According to Hallett, when new crystals were formed, these were oriented with the c-axis parallel to the surface of the substrate. However, when the substrate had this latter orientation the droplets froze as single crystals with the same orientation down to -20°C.

Hallett's paper also shows some pictures of freezing droplets. In these examples the c-axis of the substrate was vertical (= normal to the surface). The formation of new orientations in the droplets is represented by the growth of dendrites which emerge from the substrate with their c-axis approximately horizontal. On the other hand, bulk ice which forms during the second stage of freezing when the droplet is at 0°C starts growing from the substrate with the orientation as the latter. The dendrites do not develop further. They may be observed between crossed polaroids being already imbedded in the bulk ice and contrasting with it because of their different orientation.

Higuchi and Yosida also studied the orientation of crystals formed in droplets and freezing on an ice substrate. Although they do not indicate the droplet diameter it seems from the pictures that it was smaller than in previous cases, probably some 100 μ. The air temperature was -8°C. Different ice substrates were used prepared with the c-axis parallel and normal to the surface. Many droplets were observed to freeze with the orientation of the substrate, especially when the c-axis was parallel to the surface. Droplets frozen in different orientations were also found; especially noted is the case of crystals formed with the c-axis vertical on a substrate with the c-axis horizontal and vice versa. In all cases the authors indicate a simple matching between the crystal lattices at the base of the droplet.

Finally, Aufdermaur and Mayes studied the size of crystals in accretions obtained in a wind tunnel using droplets of diameter Ø = 80 μ. They observed crystals much larger than single droplets down to temperatures of about -15°C. The orientation of these crystals was not studied.

As a conclusion we see that previous results do not indicate an explanation of the special orientation of crystals observed in several hailstones with the c-axis approximately in the direction of growth. On the other hand, the comparison of the results obtained by Aufdermaur and Mayes with the results of Hallett and Pruppacher indicates that the freezing features of large droplets may not be directly extrapolated to small droplets and to wind tunnel experiments.

In the present work accretions have again been obtained in a wind tunnel and their structures have been analyzed with special attention to the crystal orientation.

We will consider here only the experiments where the surface temperature of the accretion was < 0°C (dry growth). In fact, the orientation of crystals in wet growth (surface temperature 0°C and accretion formed by a mixture of water and ice) has been studied previously (Aufdermaur et al., /5/) and has been explained e.g. by Knight /6/. Typically, the c-axis of crystals is approximately parallel to the accreting surface. This was confirmed in the present experiments as will be shown in a more general paper on the structure of artificial hailstones.

II - Methods and results

Accretions were obtained on rotating cylinders consisting of plastic material at their extremities a brass tube in their central part. A thermistor fixed at the inner surface of the tube allowed the measurement of the accretion temperature. Droplet spectra were used of median volume diameters Ø = 20, 40 and 100 μ. The wind speed was 10, 20 and 40 m/sec. The air (and droplet) temperature was varied between -2° and -22°C. Due to the heat of crystallization, the temperature of the accretion during growth was several degrees higher than that of the air.

After finishing the process of accretion, the ice cylinders were taken to a cold laboratory (at a temperature of -10°C). Thin sections of the cylinders were prepared and observed between crossed polaroids. A replica method (Aufdermaur et al., /5/) was used to determine the angle ϕ between the c-axes of the crystals and the direction of growth. Statistical analysis of these orientations was then performed to obtain the number N (ϕ) of crystals corresponding to different values of the angle ϕ.

More details of the experimental condition will be given elsewhere.

Table I shows some results obtained by the replica analysis of several cylinders grown under different conditions of temperature. In column 3 of the table, ϕ_m represents that 10° interval in which the maximum of N (ϕ) was found. In column 4 we tabulate the percentage of crystals observed with the c-axis oriented more than 50° from the radial direction. In the same table, the two first columns indicate the air and surface temperatures corresponding to each experiment and the last columns give the droplet spectrum and wind speed resp. The last two parameters are not very important in the present discussion although they have some influence on the crystal size.

Table I

t_a [°C]	t_s [°C]	ϕ_m	N(ϕ>50°) [%]	Ø [μ]	v [m/sec]
- 7	-2 to -4	0-10	0	20	20
- 9.5	-6.5	10-20	8	100	15
-14	-1 to -5	10-20	0	20	40
-18	-7	20-30	10	100	15
-22	-2	40-50	30	100	15

The table shows that the angle Φ_m is in most cases <20°. It was found in the interval 0°-10° for air temperatures relatively near 0°C, and in the interval 10°-20° for air temperatures between -9° and -15°C. ϕ_m was shifted further for lower temperatures, reaching the interval 40-50° for temperatures near -20°C.

In spite of this progressive shifting of ϕ_m with decreasing temperature, it may be seen that the percentage of crystals forming an angle larger than 50° is always rather small. In some cases nearly all crystals were found in the interval 0-20° or 0-30°. For instance in the first example of the table, 80% of the crystals were found with their c-axes in the interval 0°-20°.

This high degree of order made it possible to recognize the prevailing orientation in many structures by a simple observation of the thin sections between crossed polaroids and by quickly checking the orientation of some crystals on the replica. If a thin section of a cylindrical accretion formed by crystals all oriented with radial c-axes is observed between crossed polaroids, a dark cross must appear with its arms parallel to the polarization planes of the polaroids. This cross was seen in most of the cylinders obtained in dry growth*, for air temperatures $t_a > -15°$ (fig.1) and we could

Fig.1 Thin section between crossed polaroids showing the accretion on a rotating cylindrical single crystal of ice of 18 mm diameter. Air temperature -6°C, accretion temperature -2°C, median volume diameter of accreted droplets 20μ, wind speed 20 m/s

* We do not consider here loose structure obtained with small droplets and low surface temperatures, where the crystals were small and probably disordered.

prove by comparison with the results of some replica analyses that in these cases the orientation of the c-axes was approximately radial with $\phi_m < 20^\circ$. In general, a detailed statistical analysis was not needed when the cross was observed and the orientation of only a few crystals in the replica had to be checked to make sure that the prevailing orientation actually was with radial and not with tangential c-axes••

For $t_a \leq -15^\circ C$ the maximum of $N(\phi)$ was shifted to angles $\phi_m > 20^\circ$, as shown by the last lines in table I. In these cases the cross disappeared; in some cases of temperatures near $-20^\circ C$ it appeared again, but shifted by 45° with respect to the normal cross. Here, the replica analysis gave values of ϕ_m in the interval $40\text{-}50^\circ$ (last line in the table). An example of the shifted cross is given in fig.2. In the present experiments the shifted

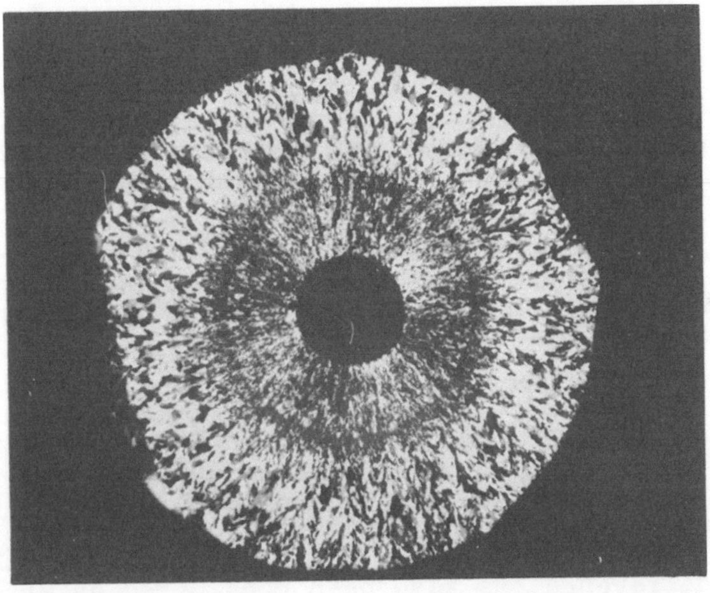

Fig.2 Accretion showing in the inner part a dark cross shifted by 45° with respect to the planes of polarization. Air temperature $-22^\circ C$, accretion temperatures -10 (inner) resp. $-4^\circ C$ (outer part), median volume diameter of accreted droplets 100μ, wind speed 10 m/s

•• The tangential orientations of wet growth did with a few exceptions not produce a dark cross.

cross has been observed at $t_a = -18^oC$, for low surface temperature, and at $t_a = -22^oC$.

Finally, we have performed some experiments in dry growth at air temperatures between -3° and -6^oC using ice cylinders of different orientations as a substrate. It was observed that the dry accretion was characterized in polarized light by immediate formation of the cross (fig.1). The typical orientation was formed very quickly indicating that the crystals of the growing bulk follow the orientation of the substrate only when the latter is oriented correctly. This selectivity of the process was strong also at temperatures close to 0^oC.

III - Discussion

According to the present results, the structure of layers within natural hailstones formed by crystals with their c-axes oriented in the direction of growth is well reproduced by artificial accretions obtained in dry growth. Since some of the hailstones analysed (Aufdermaur et al. /5/, Knight and Knight /7/) contained large crystals with the maximum of $N(\phi)$ in the interval 0°-10°, it may be speculated that the air temperature was relatively high, probably -10^oC, during the accretion of these layers.

It is now of interest to discuss the observed features taking into account also previous results in order to obtain a possible interpretation of the corresponding growth mechanism.

We have already seen in the introduction that in wind tunnel experiments crystals large in comparison to the accreted droplets have been obtained down to temperatures of the air (and of the arriving droplets) of -15^oC (Aufdermaur and Mayes /4/) while large droplets in laboratory experiments begin to split into more than one crystal for temperatures $< -5^oC$ (Hallett, /1/, Pruppacher, /3/). This difference has also been considered by Brownscombe and Hallett /8/. They suggested that the limit between the temperatures where single crystals are formed and the temperatures where the freezing droplets produce several crystals is shifted toward lower temperatures when the droplet diameter is reduced. This limit was found at -5^oC for droplets of about 1 mm in diameter but it must be reduced to -15^oC for droplets of the order of 100μ in order to fit the present results. They indicate that the discrepancy does not consist only in a shifting of the limit temperatures. In fact, also the selectivity of the orientation is much stronger in our wind tunnel experiments and the preferred orientation which appeared to be with the c-axis parallel to the surface in Hallett's experiments is now rotated by 90°, i.e. with the c-axis normal to this surface. The orientation in large droplets as observed by Hallet seems easier to be understood at first sight. A relatively simple explanation, however may be proposed for the present results which is at the same time consistent with the habit of crystallization of ice.

It is known that during the first stage of freezing of a supercooled droplet a fast dendritic growth takes place at the expense of the supercooling. During the second stage of freezing the droplet is at 0°C and bulk ice is formed the growth being controlled by heat exchange with the surroundings. It has been shown by Macklin and Payne /9/ and by Brownscombe and Hallett /8/ that droplets freezing on a substrate of a mean temperature of <0°C liberate the heat of crystallization first into the heat reservoir of the underlying colder ice. If dendrites are occasionally nucleated parallel to the accreting surface, they cover a relatively large part of the area from which the formation of bulk ice starts. Those dendrites which are nearest to the heat sink will be in the best conditions to grow in thickness and to form the bulk ice. For small supercooling, the dendrites of ice are formed in the basal plane or at a small angle with respect to it. If the bulk ice grows preferably on dendrites parallel to the accreting surface, we expect the c-axis of the bulk about normal to the surface. This selective growth mechanism explains the present results.

Other dendrites, formed at larger angles to the surface, would not be in favourable conditions to grow further. This is supported also by the pictures of freezing droplets presented by Hallett /1/ where dendrites emerging from the substrate are visible imbedded in the bulk ice. In our case, we can not decide whether such dendrites are not formed or not preserved or too small to be observed in the bulk of the accretion.

On the other hand, the results of studies of dendritic growth in water as a function of supercooling offer an interpretation of the shifting of ϕ_m observed with the decrease of the air temperature. Chalmers /10/, Lindenmayer et al. /11/, Macklin and Ryan /12/ and Pruppacher /3/ showed that nucleation of supercooled water produces dendrites split into symmetrical branches. The angle α between these branches increases with supercooling. Correspondingly, the dendrites are considered to grow at an angle $\alpha/2$ with respect to the basal plane. Since in the present case the important dendrites are supposed to develop parallel to the accreting surface, the crystals they generate would have their c-axes at the angle $\phi_m = \alpha/2$.

It could be objected that the shifting of ϕ_m with decreasing air temperature is not in exact accordance with the measurements of Macklin and Ryan and of Pruppacher. In fact, these authors found $\alpha > 20^\circ$ for supercooling $\Delta T > 4^\circ C$, so that one might expect $\phi_m > 10^\circ$, for air temperatures $t_a < -4^\circ C$. However, we found $\phi_m < 10^\circ$ down to air temperatures of about $-8^\circ C$ (first line in the table I). On the other hand, according to Pruppacher the angle $\alpha = 45^\circ$ looks like a limit value reached at a supercooling $\Delta T = 9^\circ C$ and remaining approximately constant up to $\Delta T = 14^\circ C$, the maximum supercooling of Pruppacher's measurements. Our results suggest that α may further increase reaching 90° for air temperatures near $-20^\circ C$.

A quantitative comparison between α and ϕ_m is complicated by the fact that the actual supercooling of the droplets is diminished by heat exchange with the warmer substrate immediately after the impact. Therefore, the actual supercooling encountered by the growing dendrites lies somewhere between the temperature of the air and the substrate. The loss of supercooling to the substrate is less important at low air temperatures, because the growth velocity of the dendrites increases approximately proportional to the square of the supercooling. This explains qualitatively why ϕ_m was smaller than $\alpha/2$ at air temperatures down to $-10^\circ C$.

If we adopt the interpretation of α in terms of the growth velocities in the basal plane and normal to it (Macklin and Ryan /12 ,13/), we may conclude from our maximum angle $\phi_m = 45^\circ$ observed at a supercooling of about $20^\circ C$ that the growth velocity in the c-axis has reached that in the basal plane. This result would confirm previous observations by Hallett /1/ and Macklin and Ryan /13/.

Considering the rather asymptotical value of $\alpha = 45^\circ$ obtained by Pruppacher at a supercooling in the interval between 9 and $14^\circ C$, we may also look at the dendritic growth velocities measured by Pruppacher /14/. As we find a discontinuity at a supercooling of 10 to $13^\circ C$, where the growth velocity remains about constant, we may speculate that α could increase again at higher supercooling according to our results.

IV - Conclusions

The c-axis distribution normal to the surface of a growing hailstone has been reproduced in wind tunnel experiments. It is attributed to dry growth, i.e. to the case when the mean temperature of the accreting surface is below 0°C. It was found that the maximum of the distribution deviates from the normal to the surface with increasing supercooling reaching 45° at air temperatures around -20°C. This shift is qualitatively consistent with the angle α measured between dendritic sheets growing in supercooled water. The observed c-axis distributions are explained on the hypothesis that the bulk ice forms on dendrites nucleated parallel to the surface of the growing substrate. This hypothesis is derived from considerations of the heat transfer of the individual freezing droplets.

V - Acknowledgements

The reported work has been carried out at the Federal Institute for Snow and Avalanche Research. It was part of the program of the Federal Commission for the Study of Hail Formation and Prevention and was supported by the Swiss Hail Insurance Company. The authors wish to thank Dr.M. de Quervain and Mr. O.Buser for helpful discussions and assistance.

Literature

/1/ Hallett, J., J.Atm.Sci. 21, 671-682 (1964)

/2/ Higuchi, K., T.Yosida, Physics of Snow and Ice, Proc.Int.Conf.Low Temp.Sci.Sapporo, Japan, 79-93 (1966)

/3/ Pruppacher, H.R., J.Glaciol. 6, 651-662 (1967)

/4/ Aufdermaur, A.N., W.C.Mayes, Proc.Int.Conf. Cloud Physics, Tokyo and Sapporo, 281-285 (1965)

/5/ Aufdermaur, A.N., R.List, W.C.Mayes, M.R. de Quervain, Z.Ang.Math.Phys. 14, 474-589 (1963)

/6/ Knight, C.A., J.Atm.Sci. 25, 440-444 (1968)

/7/ Knight, C.A., N.C.Knight, J.Atm.Sci. 25, 445-452 (1968)

/8/ Brownscombe, J.L., J.Hallett. Quart.J.R.Met.Soc. 93, 455-473 (1967)

/9/ Macklin, W.C., G.S.Payne, Quart.J.R.Met.Soc. 93, 195-214 (1967)

/10/ Chalmers, B., A.S.T.M. reprint of Edgar Marbury Lecture, 9 pp. (1961)

/11/ Lindenmeyer, C.S., G.T.Orrok, K.A.Jackson. B.Chalers, J.Chem.Phys. 27, 822 (1957)

/12/ Macklin, W.C., B.F.Ryan, J.Atm.Sci. 22, 452-459 (1965)

/13/ Macklin, W.C., B.F.Ryan, Phil.Mag.14, 847-860 (1966)

/14/ Pruppacher, H.R., J.Chem.Phys. 47, 1807-1813 (1967)

THE INFLUENCE OF AN ELECTRIC FIELD ON THE FREEZING OF WATER

Maud Roulleau

State University of New York at Albany A.S.R.C.

Scotia, New York / USA

Up until now the influence that an electric field could have on the freezing of water has not interested many researchers.

In 1951, Rau /1/ found that very high electric fields could cause the nucleation of big supercooled drops. He explained his results by a dielectric polarisation of water.

More recently, Pruppacher /2/ studied the freezing of water samples contained in glass or plastic tubing, placed between two electrodes submerged in paraffin oil. He concluded that the electric field causes ice nucleation only indirectly namely by deforming the drops in contact with the solid surface. So this phenomenon could not have any effect on physics of clouds.

We began to study this subject a few years ago and, now, we have made several series of measurements with different experimental set-ups.

1. Experiments with uniform fields

I should first briefly remember some of the oldest results.

Experiments with big drops

The first experiments /3/were done on big distilled supercooled water drops of about half millimeter diameter.

x) On leave from: Laboratoire de Météorologie et Physique de l'Atmosphère, Paris /France

They were set on a glass plate lying at the bottom of a metallic basin containing some silicone oil, a dc voltage was applied between the basin and a grid made of platinum wires placed above the drops in the silicone oil. The electric fields used vary from 1 kV/cm to 9 kV/cm. The experiments were done either at -10°C or -15°C.
It had been found that:
Under the action of an electric field, the drops freeze at a temperature nearer 0°C than without the field.
The drops freeze more quickly as the intensity of the electric field is increased.
The drops freeze more quickly as the supercooling increases (fig.1, 2).

Experiments on artificial fogs

Some other measurements were made on artificial fogs in an ice nucleus counter, using the method of supercooled sugar solutions (P_{oc} /4/).

The apparatus is mainly composed of a cylindrical 10 liter tank, uniformly cooled by a refrigerant. The top of the tank is closed with a plexiglass cover in the center of which is a hole which permits the damping device to be brought in. The electrodes used to set up the dc electric field are made of two parallel grids one centimeter apart. They are made of nickel wires of 0.3 mm diameter (fig.3 and 4).

The sugar solution is poured on a black painted brass tray which is put down at the bottom of the tank. At the end of the experiment, the tray is taken out of the apparatus and a picture is made at once. Crystals are counted on the pictures obtained.

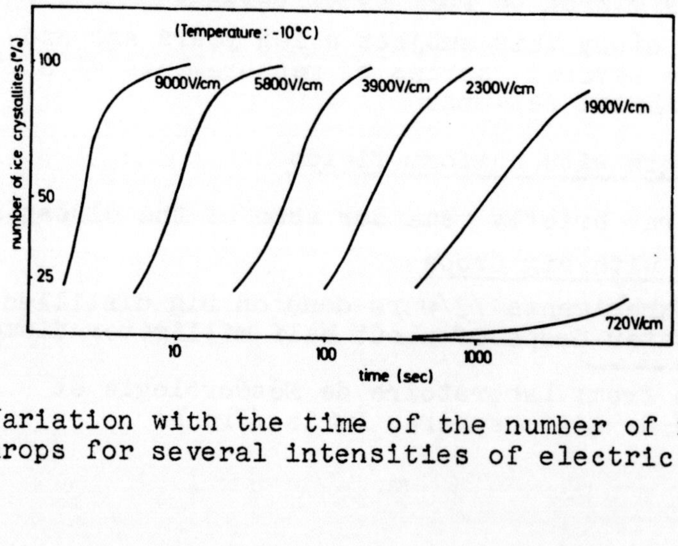

Fig.1 Variation with the time of the number of frozen drops for several intensities of electric field

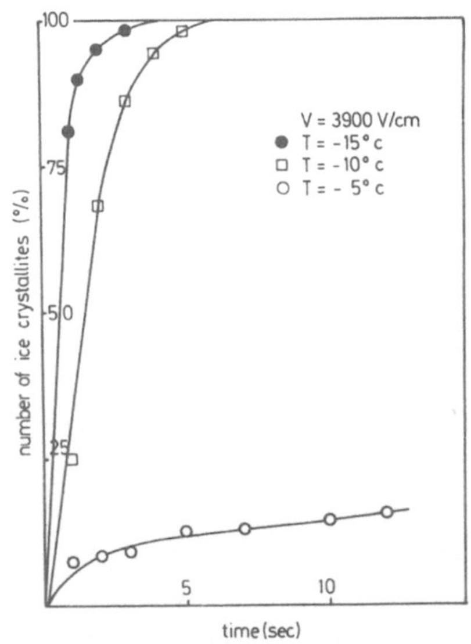

Fig.2 Variation with the time of the number of frozen drops for different temperatures

Fig.3 Ice nucleus counter

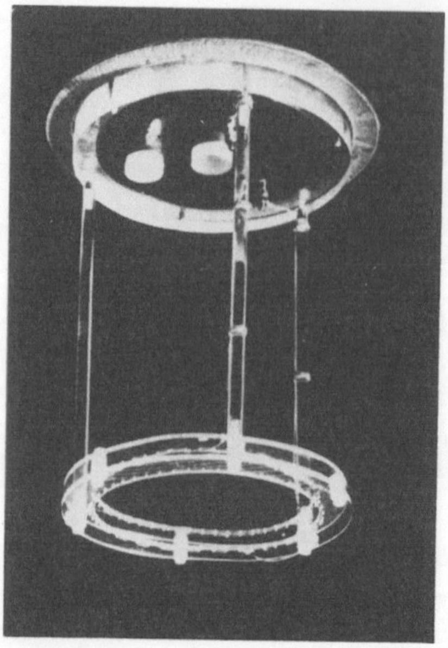

Fig.4 Electrodes used to set up the dc electric field

The experiments are made with electric fields varying from 1 kV/cm to 5 kV/cm. The measurements are done at -5°C, -10°C, -15°C, -20°C.

- for a given value of the intensity of the electric field, the lower the temperature, the greater is the number of ice crystals.
- the results seem also to depend on the properties of the fog. To increase the quantity of water, the humidification time is increased. The thicker the fog, the greater is the number of ice crystals (fig. 5,6,7,8).

2. Experiments with non-uniform fields

All the previous experiments are made with uniform fields.

I shall report now some more recent measurements made with cylindrical electrodes connected to a dc high voltage source. There are two kinds; some are made with Schaefer's /6/ cold chamber and the others with a NCAR ice nucleus counter (Langer, Rosinski, Edwards /5/).

Experiments in the cold chamber

These experiments are made with the Schaefer technique in a box cooled between -20°C and -25°C, lined with black velvet. The fog is formed simply by breathing and a light inclined at 45° permits the detection of ice crystals.

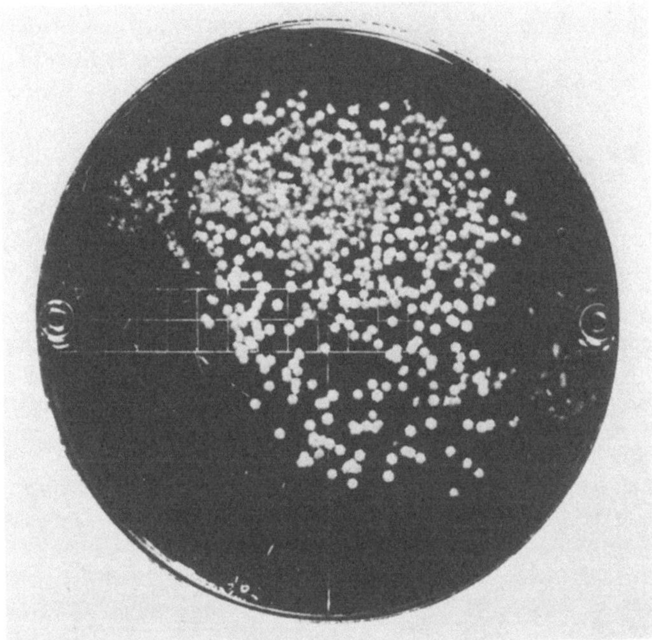

Fig.5 Experiment at -10°C with a field of 3000 V/cm

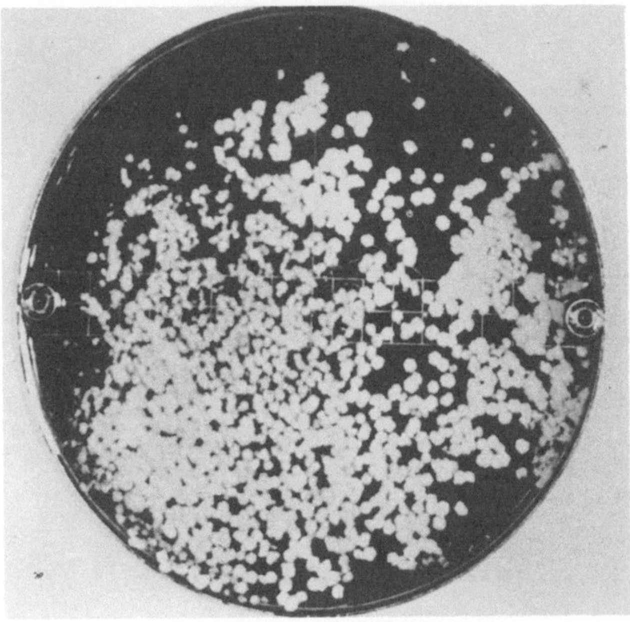

Fig.6 Experiment at -10°C with a field of 5000 V/cm

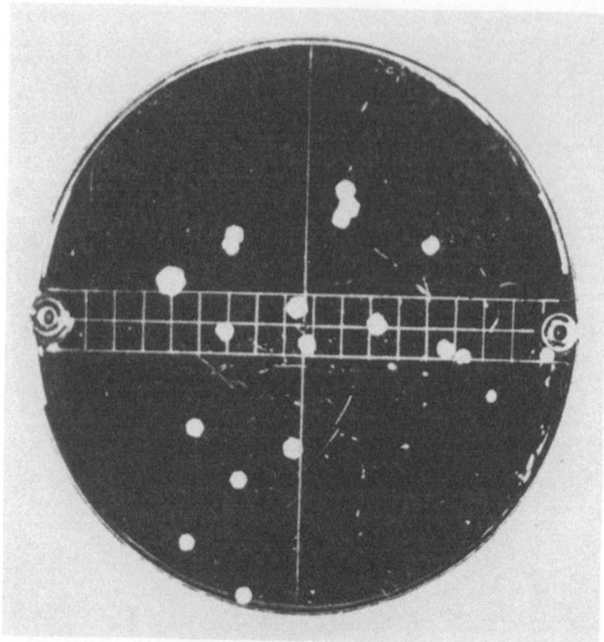

Fig.7 Experiment at -20°C without field

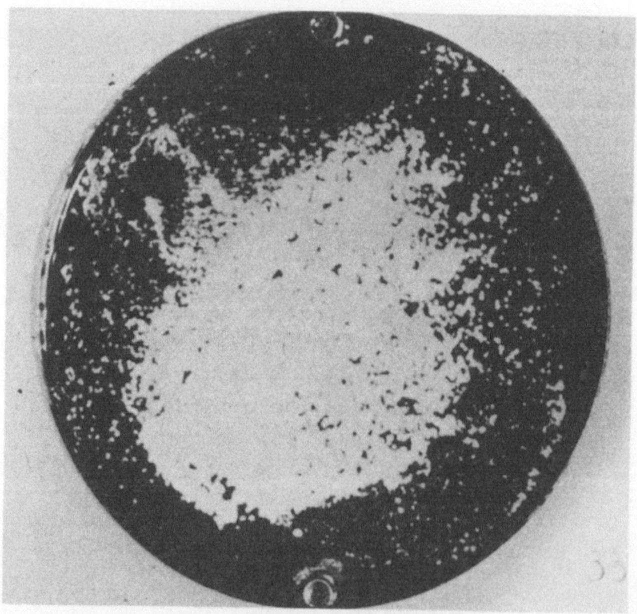

Fig.8 Experiment at -20°C with a field of 5000 V/cm

If a cylindrical electrode connected to a dc high voltage is put in the supercooled fog, in certain conditions, the formation of a great number of ice crystals is obtained.

If the diameter of the wire is small enough, with the proper value of the high voltage, a spray-like formation of ice crystals can be seen at a small distance of the wire (Schaefer /6/).

- The ice crystal formation does not appear to depend on the nature of the electrode. The same phenomenon is observed with three different kinds of metals: copper, platinum, tungsten.
- The phenomenon does not appear to depend on the polarity of the high voltage. The same results are obtained either with a positive or negative electrode resp.
- The ice crystals formed do not carry a very high electrical charge. They do not appear to be attracted or repelled by a grounded electrode.
- The voltage required to obtain the formation of ice crystals increases with the diameter of the electrode. Since quantitative measurements are difficult in the cold box, the experiments are repeated afterwards in the ice nucleus counter.
- When the fog is warmer than the wire, ice crystals grow on the wire surface which becomes rapidly coated with ice. This coating appears to prevent ice crystal formation in the fog near the electrode.

A possible interpretation of these experiments is that ice crystals can be simply formed when water droplets come in contact with the electrode. They would be repelled by the electrode when they acquire a sufficiently large electric charge. If this is the case, they would be highly charged and, under the influence of the field, they should move with a high speed towards the grounded walls of the chamber. It has been seen that they are not highly charged and that they only float in the fog until they are big enough to fall down to the bottom of the cold box.

It is also possible to get the ice crystal formation with electrodes warmer than $0^{\circ}C$. A very small diameter (25 μ) tungsten or platinum wire is stretched between two plexiglass supports and is heated by means of a small ac current which can be adjusted. Its temperature can be obtained by measuring the resistance. It has been found possible to warm the wire electrode as high as $50^{\circ}C$ and to obtain the formation of ice crystals with somewhat higher voltage than when the electrode is cold.

Experiments in the ice nucleus counter

To try to obtain quantitative measurements, an NCAR ice nucleus counter is used. The experiments are made at -20°C with filtered air. Under normal conditions, without electric field, the number of natural ice crystals is always very small, less than one per liter at -20°C. A metallic wire is inserted in the lower part of the apparatus with the wire end carefully insulated to avoid point ionization. If the wire is connected to a high voltage dc source, with the chamber walls grounded, a non-uniform field is produced.

The test results show that the following conclusions can be drawn:
- the phenomenon does not appear to depend on the polarity of the high voltage,
- the voltage required to get the ice crystal formation increases with the diameter of the electrode. The minimum value of the necessary voltage for obtaining ice crystals are determined for different diameters of wires and are as following:

diameter (μ)	high voltage (kV)
25	1.3
100	1.5
300	2.0
1000	3.0
2000	4.0

- the number per liter of ice crystals formed seems also to depend on the properties of the fog. The results are variable. The following table gives examples of the results of the number of ice crystals per liter obtained for different diameters of wires:

H V (kV)	25μ	0.1mm	0.3mm	1 mm	2mm
1.5	20	3			
2	44	10			
2.5		30	15		
3.			20	6	
3.5			35	11	
4				17	6
5					13
6					60

Ionization effect

In the experiments reported, one might think that there is perhaps ionization and that the ionization current is responsible for the phenomenon.
An electrometer placed in series with the electrode never shows the existence of a current greater than 10^{-8}A. If the same electrodes are placed in the cold box at about 2.5 cm above a plate connected to an electrometer, which is grounded on the other side, it is possible to measure the minimum value to obtain an ionization current with the different electrodes. The results obtained are shown in the following table:

diameter (μ)	HV (kV)
25	1.8
100	3
300	6
1000	8

It is readily evident that these voltages are much higher than the values necessary to get the ice crystal formation.

Role of electric charges

It appears that the mere presence of concentrated electric charges on water drops is not sufficient to cause them to freeze. With a glass capillary tubing filled with distilled water and connected with a wire to a dc positive high voltage source(Vonnegut and Neubauer /7/) it is possible to obtain a spray of very highly charged droplets. Even at -20°C, in the cold box, we never observe any of these highly charged droplets to freeze.

3. Conclusion

These experiments appear to be consistent with the results obtained with uniform fields. However, the mechanism by which the electric field causes the nucleation of water remains unknown.

References

/1/ Rau, W.,Zeits.Naturforsch. 6a,649-654 (1951)

/2/ Pruppacher, H.R., J.Geophys.Res.68, 4463-74 (1963)

/3/ Roulleau, M., Ann.Geophys. 20, 319-324 (1964)

/4/ Poc, M., J.R.A. 4, 127-137 (1967)

/5/ Langer, G., J.Rosinski, C.P.Edwards, J.Appl.Met. 6, 114-125 (1967)
/6/ Schaefer, V., J.Appl.Met. 7, 452-455 (1968)
/7/ Vonnegut, B., R.L.Neubauer, J.Coll.Sci. 7, 616-622 (1952)

CONCLUDING REMARKS

L. Onsager

Department of Chemistry, Yale University

New Haven, Connecticut

Herr Professor Riehl, it is not fitting that we should leave without thanking you . Some participants have compared the situation this week with that in Erlenbach six years ago. It has been like a couple of milestones and I don't think there has been anything comparable in between. It was a bit different then, quite a bit, because at that time, some fundamental ideas were new and possibly became much more widely disseminated and assimilated than it would have been the case without that meeting. It is not so much that seems to be new at this meeting. It would appear, however, that most of the ideas that were new at the time of the Erlenbach meeting have to some extent stood the test of time and somehow inspired or helped to catalyze an expansion of the experimental work. Because what we knew then was that we needed a great many numbers of which we hoped we knew the order of magnitude. We have learned a good deal more about them and still there have been conflicts: Different folks still get different results. And in addition, the research has branched out into many new problems and applications. Some of the most imaginatory work has been done right here in Munich in Professor Riehl's laboratory; also, some of the prettiest techniques have been invented here. So it is very fitting we should meet here. And there was a great need to come together and swap experience, and to have listened to lively exchange of experience in and out of the lecture room. There has been plenty of it. I think all participants have drawn quite rich benefits from this meeting. The schedule has been heavy, and I am quite happy you did not make it any heavier. On the other hand, there is very little I wanted to miss. We have been fortunate that everything has been running so nice and smoothly. I know that it has taken an effort to prepare this meeting. And the better you make it the less it shows. I should congratulate you on the success of it and thank you for making the effort.

MIX
Papier aus verantwortungsvollen Quellen
Paper from responsible sources
FSC® C105338

If you have any concerns about our products,
you can contact us on
ProductSafety@springernature.com

In case Publisher is established outside the EU,
the EU authorized representative is:
Springer Nature Customer Service Center GmbH
Europaplatz 3, 69115 Heidelberg, Germany

Printed by Libri Plureos GmbH
in Hamburg, Germany